T0211416

Thermodynamics and Statistical Mechanics

Learn classical thermodynamics alongside statistical mechanics with this fresh approach to the subjects. Molecular and macroscopic principles are explained in an integrated, side-by-side manner to give students a deep, intuitive understanding of thermodynamics and equip them to tackle future research topics that focus on the nanoscale. Entropy is introduced from the get-go, providing a clear explanation of how the classical thermodynamic laws connect to molecular principles, and closing the gap between the atomic world and the macroscale. Notation is streamlined throughout, with a focus on general concepts and simple models, for building basic physical intuition and gaining confidence in problem analysis and model development.

Well over 400 guided end-of-chapter problems are included, addressing conceptual, fundamental, and applied skill sets. Numerous worked examples are also provided, together with handy shaded boxes to emphasize key concepts, making this the complete teaching package for students in chemical engineering and the chemical sciences.

M. Scott Shell is an Associate Professor in the Chemical Engineering Department at the University of California, Santa Barbara. He earned his PhD in Chemical Engineering from Princeton in 2005 and is well known for his ability to communicate complex ideas and teach in an engaging manner. He is the recipient of a Dreyfus Foundation New Faculty Award, an NSF CAREER Award, a Hellman Family Faculty Fellowship, a Northrop-Grumman Teaching Award, a Sloan Research Fellowship, and a UCSB Distinguished Teaching Award.

"This textbook presents an accessible (but still rigorous) treatment of the material at a beginning-graduate level, including many worked examples. By making the concept of entropy central to the book, Prof. Shell provides an organizing principle that makes it easier for the students to achieve mastery of this important area."

Athanassios Z. Panagiotopoulos
Princeton University

"Other integrated treatments of thermodynamics and statistical mechanics exist, but this one stands out as remarkably thoughtful and clear in its selection and illumination of key concepts needed for understanding and modeling materials and processes."

Thomas Truskett
University of Texas, Austin

"This text provides a long-awaited and modern approach that integrates statistical mechanics with classical thermodynamics, rather than the traditional sequential approach, in which teaching of the molecular origins of thermodynamic laws and models only follows later, after classical thermodynamics. The author clearly shows how classical thermodynamic concepts result from the underlying behavior of the molecules themselves."

Keith E. Gubbins
North Carolina State University

Cambridge Series in Chemical Engineering

SERIES EDITOR

Arvind Varma, *Purdue University*

EDITORIAL BOARD

Christopher Bowman, *University of Colorado*
Edward Cussler, *University of Minnesota*
Chaitan Khosla, *Stanford University*
Athanassios Z. Panagiotopoulos, *Princeton University*
Gregory Stephanopoulos, *Massachusetts Institute of Technology*
Jackie Ying, *Institute of Bioengineering and Nanotechnology, Singapore*

BOOKS IN SERIES

Baldea and Daoutidis, *Dynamics and Nonlinear Control of Integrated Process Systems*
Chau, *Process Control: A First Course with MATLAB*
Cussler, *Diffusion: Mass Transfer in Fluid Systems, Third Edition*
Cussler and Moggridge, *Chemical Product Design, Second Edition*
De Pablo and Schieber, *Molecular Engineering Thermodynamics*
Denn, *Chemical Engineering: An Introduction*
Denn, *Polymer Melt Processing: Foundations in Fluid Mechanics and Heat Transfer*
Duncan and Reimer, *Chemical Engineering Design and Analysis: An Introduction*
Fan and Zhu, *Principles of Gas–Solid Flows*
Fox, *Computational Models for Turbulent Reacting Flows*
Franses, *Thermodynamics with Chemical Engineering Applications*
Leal, *Advanced Transport Phenomena: Fluid Mechanics and Convective Transport Processes*
Lim and Shin, *Fed-Batch Cultures: Principles and Applications of Semi-Batch Bioreactors*
Marchisio and Fox, *Computational Models for Polydisperse Particulate and Multiphase Systems*
Mewis and Wagner, *Colloidal Suspension Rheology*
Morbidelli, Gavriilidis, and Varma, *Catalyst Design: Optimal Distribution of Catalyst in Pellets, Reactors, and Membranes*
Noble and Terry, *Principles of Chemical Separations with Environmental Applications*
Orbey and Sandler, *Modeling Vapor–Liquid Equilibria: Cubic Equations of State and their Mixing Rules*
Petyluk, *Distillation Theory and its Applications to Optimal Design of Separation Units*
Rao and Nott, *An Introduction to Granular Flow*
Russell, Robinson, and Wagner, *Mass and Heat Transfer: Analysis of Mass Contactors and Heat Exchangers*
Schobert, *Chemistry of Fossil Fuels and Biofuels*
Sirkar, *Separation of Molecules, Macromolecules and Particles: Principles, Phenomena and Processes*
Slattery, *Advanced Transport Phenomena*
Varma, Morbidelli, and Wu, *Parametric Sensitivity in Chemical Systems*

To Janet, Mike, Rox, and the entire Southern Circus

Thermodynamics and Statistical Mechanics

An Integrated Approach

M. SCOTT SHELL

University of California, Santa Barbara

CAMBRIDGE
UNIVERSITY PRESS

CAMBRIDGE
UNIVERSITY PRESS

University Printing House, Cambridge CB2 8BS, United Kingdom

One Liberty Plaza, 20th Floor, New York, NY 10006, USA

477 Williamstown Road, Port Melbourne, VIC 3207, Australia

4843/24, 2nd Floor, Ansari Road, Daryaganj, Delhi - 110002, India

79 Anson Road, #06-04/06, Singapore 079906

Cambridge University Press is part of the University of Cambridge.

It furthers the University's mission by disseminating knowledge in the pursuit of education, learning and research at the highest international levels of excellence.

www.cambridge.org
Information on this title: www.cambridge.org/9781107656789

First published 2015

A catalogue record for this publication is available from the British Library

Library of Congress Cataloging in Publication data
Shell, M. Scott (Michael Scott), 1978–
Thermodynamics and statistical mechanics : an integrated approach /
M. Scott Shell.
    pages cm - (Cambridge series in chemical engineering)
ISBN 978-1-107-01453-4 (Hardback) – ISBN 978-1-107-65678-9 (Paperback)
1. Thermodynamics.   2. Statistical mechanics.   I. Title.
QC311.S5136 2014
536'.7–dc23   2014010872

ISBN 978-1-107-01453-4 Hardback
ISBN 978-1-107-65678-9 Paperback

Additional resources for this publication at www.engr.ucsb.edu/~shell/book

Contents

Preface

Like so many texts, this book grew out of lecture notes and problems that I developed through teaching, specifically, graduate thermodynamics over the past seven years. These notes were originally motivated by my difficulty in finding a satisfactory introductory text to both classical thermodynamics and statistical mechanics that could be used for a quarter-long course for first-year chemical engineering graduate students. However, as the years pressed forward, it became apparent that there was a greater opportunity to construct a new presentation of these classic subjects that addressed the needs of the modern student. Namely, few existing books seem to provide an integrated view of both classical and molecular perspectives on thermodynamics, at a sufficient level of rigor to address graduate-level problems.

It has become clear to me that first-year graduate students respond best to a molecular-level "explanation" of the classic laws, at least upon initial discussion. For them this imparts a more intuitive understanding of thermodynamic potentials and, in particular, entropy and the second law. Moreover, students' most frequent hurdles are conceptual in nature, not mathematical, and I sense that many older presentations are inaccessible to them because concepts are buried deep under patinas of unnecessarily complex notation and equations.

With this book, therefore, I aim for a different kind of storytelling than the conventional *classical first, statistical second* approach. Namely, I have endeavored to organize the material in a way that presents classical thermodynamics and statistical mechanics side-by-side throughout. In a manner of speaking, I have thus eschewed the venerable *postulatory approach* that is so central to the development of the classical theory, instead providing a bottom-up, molecular rationale for the three laws. This is not to say that I reject the former and its impressive elegance, or that I view it as an unnecessary component of a graduate-level education in thermodynamics. It is merely a pedagogical choice, as I strongly believe one can only truly appreciate the postulatory perspective once one has a "gut feel" and a solid foundation for thermodynamics, and this is best served by a molecular introduction. Moreover, the topics of modern graduate research are increasingly focused on the nanoscale, and therefore it is essential that all students understand exactly how macroscopic and microscopic thermodynamic ideas interweave.

At the same time, this book seeks to provide a contemporary exposure to these topics that is complementary to classic and more detailed texts in the chemical thermodynamics canon. Here, I place heavy emphasis on *concepts* rather than formalisms, mathematics, or applications. My experience has been that complex notation and long analyses of intricate models at the outset get in the way of students' understanding of the basic conceptual foundations and physical behaviors. Therefore, I have tried to streamline notation and focus on simple qualitative models (e.g., lattice models) for building basic physical intuition and student confidence in model development and refinement. By the same token, this narrative does not try to be comprehensive in covering many applied thermodynamic property models, which I feel are best left in existing and specialist texts. I also deliberately use a straightforward, casual voice for clarity.

I have included a number of problems at the end of each chapter, most of which are entirely original. Many of these are guided and multi-step problems that walk students through the analysis of different kinds of systems, including modern problems in biophysics and materials, for example. These are divided into three categories: *conceptual and thought problems* that address the basic origins, behaviors, and trends in various thermodynamic quantities; *fundamentals problems* that develop classic and general thermodynamic relations and equations; and, finally, *applied problems* that develop and analyze simple models of specific systems.

I owe tremendous thanks to the many students over the years in my group and course who have provided great amounts of feedback on my notes. Perhaps unbeknownst to them, it has been their questions, discussions, and epiphanies that have shaped this text more than anything else – inspiring a seemingly unending but happy circumstance of repeated revisions and improvements. I am also deeply indebted to my mentors Pablo, Thanos, Frank, and Ken, who not only chaperoned my own appreciation for thermodynamics, but also provided immaculate examples of clear and concise communication. Finally, I am profoundly fortunate to have the love and support of my family, and it is returned to them many times over.

As with any first edition, I am under no illusion that this book will be entirely free of errors, typographical or otherwise, despite the repeated edits it has received from many different eyes. I am grateful to future readers for pointing these out to me, and I welcome any form of feedback, positive or negative.

M.S.S.
Santa Barbara, CA.

Reference tables

Table A Counting and combinatorics formulae

Description	Example	Formula
Number of ways to pick k ordered objects from n without replacement	How many ways are there to put k distinctly colored marbles in n separate buckets, with at most one marble per bucket?	$_nP_k = \dfrac{n!}{(n-k)!}$
Number of ways to pick k unordered objects from n without replacement	How many ways are there to put k identical blue marbles in n separate buckets, with at most one marble per bucket?	$_nC_k = \dfrac{n!}{k!(n-k)!}$
Number of ways to pick k ordered objects from n with replacement	How many ways are there to put k distinctly-colored marbles in n separate buckets, with any number of marbles per bucket?	n^k
Number of ways to pick k unordered objects from n with replacement	How many ways are there to put k identical orange marbles in n separate buckets, with any number of marbles per bucket?	$\dfrac{(k+n-1)!}{k!(n-1)!}$
Number of ways to pick k_1 objects of type 1, k_2 of type 2, etc., out of $n = k_1 + k_2 + \cdots$ in an unordered manner and without replacement	How many ways are there to put k_1 blue, k_2 orange, and k_3 red marbles in $k_1 + k_2 + k_3$ buckets, with at most one marble per bucket?	$\dfrac{n!}{\prod_i k_i!} = \dfrac{\left(\sum_i k_i\right)!}{\prod_i k_i!}$

Table B Useful integrals, expansions, and approximations

$\ln n! \approx n \ln n - n$

$n! \approx (n/e)^n$

$e^x = \displaystyle\sum_{n=0}^{\infty} \frac{x^n}{n!}$

$(1+x)^n = \displaystyle\sum_{k=0}^{n} \frac{n!}{k!(n-k)!} x^k$

$\displaystyle\int_0^\infty x^n e^{-x}\, dx = n! = \Gamma(n+1)$

$\ln(1+x) \approx x$ for small x

$(1+x)^{-1} \approx 1 - x$ for small x

$\displaystyle\int_0^\infty e^{-cx^2}\, dx = \frac{\pi^{1/2}}{2c^{1/2}}$

$\displaystyle\int_0^\infty x e^{-cx^2}\, dx = \frac{1}{2c}$

$\displaystyle\int_0^\infty x^2 e^{-cx^2}\, dx = \frac{\pi^{1/2}}{4c^{3/2}}$

$\displaystyle\int_0^\infty x^3 e^{-cx^2}\, dx = \frac{1}{2c^2}$

$\displaystyle\int_0^\infty x^4 e^{-cx^2}\, dx = \frac{3\pi^{1/2}}{8c^{5/2}}$

$\displaystyle\int_0^\infty x^n e^{-cx^2}\, dx = \frac{\pi^{1/2}(n-1)!!}{2^{n/2+1}c^{(n+1)/2}}$ (n even)

Table C Extensive thermodynamic potentials

Name	Independent variables	Differential form	Integrated form
Entropy	$S(E, V, \{N\})$	$dS = \dfrac{1}{T}\,dE + \dfrac{P}{T}\,dV - \sum_i \dfrac{\mu_i}{T}\,dN_i$	$S = \dfrac{E}{T} + \dfrac{PV}{T} - \sum_i \dfrac{\mu_i N_i}{T}$
Energy	$E(S, V, \{N\})$	$dE = T\,dS - P\,dV + \sum_i \mu_i\,dN_i$	$E = TS - PV + \sum_i \mu_i N_i$
Enthalpy	$H(S, P, \{N\})$	$dH = T\,dS + V\,dP + \sum_i \mu_i\,dN_i$	$H = E + PV = TS + \sum_i \mu_i N_i$
Helmholtz free energy	$A(T, V, \{N\})$	$dA = -S\,dT - P\,dV + \sum_i \mu_i\,dN_i$	$A = E - TS = -PV + \sum_i \mu_i N_i$
Gibbs free energy	$G(T, P, \{N\})$	$dG = -S\,dT + V\,dP + \sum_i \mu_i\,dN_i$	$G = E + PV - TS = A + PV$ $= H - TS = \sum_i \mu_i N_i$

Table D Intensive per-particle thermodynamic potentials for single-component systems

Name	Independent variables	Differential form	Integrated relations
Entropy per particle	$s(e, v)$	$ds = \dfrac{1}{T}\,de + \dfrac{P}{T}\,dv$	$\dfrac{\mu}{T} = -s + \dfrac{e}{T} + \dfrac{Pv}{T}$
Energy per particle	$e(s, v)$	$de = T\,ds - P\,dv$	$\mu = e - Ts + Pv$
Enthalpy per particle	$h(s, P)$	$dh = T\,ds + v\,dP$	$h = e + Pv$ $\mu = h - Ts$
Helmholtz free energy per particle	$a(T, v)$	$da = -s\,dT - P\,dv$	$a = e - Ts$ $\mu = a + Pv$
Gibbs free energy per particle	$g(T, P)$	$dg = -s\,dT + v\,dP$	$g = e + Pv - Ts = a + Pv$ $= h - Ts$ $\mu = g$

Table E Thermodynamic calculus manipulations

Name	Applies to	Functional form	Example
Inversion	Anything	$\left(\dfrac{\partial X}{\partial Y}\right)_Z = 1 \Big/ \left(\dfrac{\partial Y}{\partial X}\right)_Z$	$\left(\dfrac{\partial P}{\partial S}\right)_T = 1 \Big/ \left(\dfrac{\partial S}{\partial P}\right)_T$
Triple product rule	Anything	$\left(\dfrac{\partial X}{\partial Y}\right)_Z \left(\dfrac{\partial Z}{\partial X}\right)_Y \left(\dfrac{\partial Y}{\partial Z}\right)_X = -1$	$\left(\dfrac{\partial P}{\partial T}\right)_S = -\left(\dfrac{\partial S}{\partial T}\right)_P \Big/ \left(\dfrac{\partial S}{\partial P}\right)_T$
Addition of variable	Anything	$\left(\dfrac{\partial X}{\partial Y}\right)_Z = \left(\dfrac{\partial X}{\partial W}\right)_Z \Big/ \left(\dfrac{\partial Y}{\partial W}\right)_Z$	$\left(\dfrac{\partial H}{\partial V}\right)_P = \left(\dfrac{\partial H}{\partial T}\right)_P \Big/ \left(\dfrac{\partial V}{\partial T}\right)_P$
Non-natural derivative	Anything	$Z(X,Y) \rightarrow \left(\dfrac{\partial Z}{\partial Y}\right)_W = \left(\dfrac{\partial Z}{\partial X}\right)_Y \left(\dfrac{\partial X}{\partial Y}\right)_W + \left(\dfrac{\partial Z}{\partial Y}\right)_X$	$\left(\dfrac{\partial E}{\partial V}\right)_P = \left(\dfrac{\partial E}{\partial S}\right)_V \left(\dfrac{\partial S}{\partial V}\right)_P + \left(\dfrac{\partial E}{\partial V}\right)_S = T\left(\dfrac{\partial S}{\partial V}\right)_P - P$
Potential transformation	Potentials	$\dfrac{\partial}{\partial X}\left(\dfrac{F_1}{X}\right)_Y = -\dfrac{F_2}{X^2}$	$\dfrac{\partial}{\partial T}\left(\dfrac{A}{T}\right)_V = -\dfrac{E}{T^2}$
Maxwell relations	Potential second derivatives	$\left(\dfrac{\partial^2 F}{\partial X \, \partial Y}\right) = \left(\dfrac{\partial^2 F}{\partial Y \, \partial X}\right) \rightarrow \left(\dfrac{\partial A}{\partial X}\right)_Y = \left(\dfrac{\partial B}{\partial Y}\right)_X$	$\left(\dfrac{\partial S}{\partial P}\right)_T = -\left(\dfrac{\partial V}{\partial T}\right)_P$

The term "anything" indicates any complete state function.

Table F Measurable quantities

Name	Notation and definition
Pressure	P
Temperature	T
Volume	V
Total mass of species i	m_i
Total moles of species i	n_i
Molecular weight of species i	$\mathcal{M}_i$
Molecules of species i	$N_i = m_i/\mathcal{M}_i$
Mole fraction of species i	$x_i,\ y_i,$ or z_i
Enthalpy or latent heat of phase change	ΔH_{latent}
per particle or per mole	Δh_{latent}
Constant-volume heat capacity	$C_V \equiv \left(\dfrac{\partial E}{\partial T}\right)_{V,N} = T\left(\dfrac{\partial S}{\partial T}\right)_{V,N}$
per particle or per mole	$c_V \equiv \left(\dfrac{\partial e}{\partial T}\right)_{v} = T\left(\dfrac{\partial s}{\partial T}\right)_{v}$
Constant-pressure heat capacity	$C_P \equiv \left(\dfrac{\partial H}{\partial T}\right)_{P,N} = T\left(\dfrac{\partial S}{\partial T}\right)_{P,N}$
per particle or per mole	$c_P \equiv \left(\dfrac{\partial h}{\partial T}\right)_{P} = T\left(\dfrac{\partial s}{\partial T}\right)_{P}$
Isothermal compressibility	$\kappa_T \equiv -\dfrac{1}{V}\left(\dfrac{\partial V}{\partial P}\right)_{T,N} = -\left(\dfrac{\partial \ln V}{\partial P}\right)_{T,N} = -\left(\dfrac{\partial \ln v}{\partial P}\right)_{T}$
Thermal expansivity or thermal expansion coefficient	$\alpha_P \equiv \dfrac{1}{V}\left(\dfrac{\partial V}{\partial T}\right)_{P,N} = \left(\dfrac{\partial \ln V}{\partial T}\right)_{P,N} = \left(\dfrac{\partial \ln v}{\partial T}\right)_{P}$

Table G Common single-component statistical-mechanical ensembles

Property	Microcanonical	Canonical	Grand canonical	Isothermal–isobaric
Constant conditions	E, V, N	T, V, N	T, V, μ	T, P, N
Fluctuations	None	E	E, N	E, V
Microstate probabilities	$\wp_m = \dfrac{\delta_{E_m,E}}{\Omega(E,V,N)}$	$\wp_m = \dfrac{e^{-\beta E_m}}{Q(T,V,N)}$	$\wp_m = \dfrac{e^{-\beta E_m + \beta \mu N_m}}{\Xi(T,V,\mu)}$	$\wp_m = \dfrac{e^{-\beta E_m - \beta P V_m}}{\Delta(T,P,N)}$
Partition function	$\Omega(E,V,N) = \sum_n \delta_{E_n,E}$	$Q(T,V,N) = \sum_n e^{-\beta E_n}$	$\Xi(T,V,\mu) = \sum_N \sum_n e^{-\beta E_n + \beta \mu N}$	$\Delta(T,P,N) = \sum_V \sum_n e^{-\beta E_n - \beta P V}$
Relations to other partition functions	None	$Q = \sum_E e^{-\beta E}\,\Omega(E,V,N)$	$\Xi = \sum_N \lambda^N Q(T,V,N)$ $= \sum_N \sum_E \lambda^N e^{-\beta E}\,\Omega(E,V,N)$ where $\lambda \equiv \exp(\beta\mu)$	$\Delta = \sum_V e^{-\beta P V} Q(T,V,N)$ $= \sum_V \sum_E e^{-\beta E - \beta P V}\,\Omega(E,V,N)$
Potential	$S = k_{\mathrm B} \ln \Omega(E,V,N)$	$A = -k_{\mathrm B} T \ln Q(T,V,N)$	$PV = k_{\mathrm B} T \ln \Xi(T,V,\mu)$	$G = -k_{\mathrm B} T \ln \Delta(T,P,N)$
Classical partition function	$\Omega = \dfrac{1}{h^{3N} N!} \int \delta[H(\mathbf{p}^N, \mathbf{r}^N) - E]\, d\mathbf{p}^N\, d\mathbf{r}^N$	$Q = \dfrac{Z(T,V,N)}{\Lambda^{3N} N!}$ $Z \equiv \int e^{-\beta U(\mathbf{r}^N)}\, d\mathbf{r}^N$ $\Lambda \equiv [h^2/(2\pi m k_{\mathrm B} T)]^{1/2}$	$\Xi = \displaystyle\sum_{N=0}^{\infty} \dfrac{\lambda^N Z(T,V,N)}{\Lambda^{3N} N!}$ where $\lambda \equiv \exp(\beta\mu)$	$\Delta = \dfrac{1}{\Lambda^{3N} N!} \int_0^{\infty} e^{-\beta P V} Z(T,V,N)\, dV$

Sums over n are sums over all microstates at a given V and N.

Sums over N are from 0 to ∞, sums over V are from 0 to ∞, and sums over E are from $-\infty$ to ∞.

Classical partition functions are given for a monatomic system of indistinguishable, structureless particles.

Table H Fundamental physical constants

Name	Notation and definition
Boltzmann constant	$k_B = 1.38065 \times 10^{-23}$ J/K
Gas constant	$R = 8.31446$ J/mol $\cdot$ K
Avogadro constant	$\mathcal{N}_A = 6.02214 \times 10^{23}$ mol^{-1}
Elementary unit of charge	$e = 1.60218 \times 10^{-19}$ C
Planck constant	$h = 6.62607 \times 10^{-34}$ J $\cdot$ s
Reduced Planck constant	$\hbar = h/(2\pi) = 1.05457 \times 10^{-34}$ J $\cdot$ s
Standard gravitational acceleration	$g = 9.80665$ m/s^2
Vacuum permittivity	$\epsilon_0 = 8.8542 \times 10^{-12}$ C^2/J $\cdot$ m

1 Introduction and guide for this text

Thermodynamics is a remarkable subject, both in its pervasiveness throughout the pure and engineering sciences, and in the striking simplicity and elegance of its principles. Indeed, it is hard to underestimate the significance of thermodynamics to virtually any physical problem of interest, even if its role appears only indirectly through derivative theories or models. As a testament to its importance, Einstein made the rather potent statement that thermodynamics is "the only physical theory of universal content concerning which I am convinced that within the framework of the applicability of its basic concepts, it will never be overthrown."

At the same time, thermodynamics can be surprisingly difficult to grasp at a fundamental level, even for the experienced student. Unlike many other advanced scientific subjects, its main challenges are not mathematical in nature; a working knowledge of multivariate calculus is usually quite sufficient. Instead, the most difficult aspects of thermodynamics are its conceptual underpinnings. Students often struggle with the seemingly simple task of how to begin thinking about a problem, finding it difficult to answer questions such as the following. *What constitutes the system? What is constant or constrained? What thermodynamic variables are equal across a boundary? What assumptions and models are reasonable?* All of these questions precede the analytical analysis and are concerned with how to transform the physical problem into a mathematical one. When this is done, the solutions often present themselves in a rather straightforward manner, at least for an introductory treatment.

It is exactly these conceptual ideations on which this book is focused. This text presents an advanced undergraduate or early graduate-level overview of thermodynamics aimed at students in chemical science and engineering. It is designed to provide a fundamental understanding of thermodynamic principles that emphasizes general concepts and approaches, rather than notations, mathematical frameworks, or solution strategies for specific kinds of applications. It adopts the philosophy that the most important step a student can take in this area is to gain basic physical intuition and confidence in problem analysis and model-development. To some extent, this book is designed to "fill in the gaps" from earlier, introductory exposure to the subject and to help students see and become comfortable with the "big picture." That being said, it is assumed that the reader is equipped with some prior exposure and training in the following areas.

- *Multivariate differential and integral calculus*: familiarity with total differentials, partial derivatives, single- and multiple-variable integrations, and solutions to very simple differential equations.
- *Basic statistics and probability*: familiarity with the concepts of probability, probability distributions (including multivariate), and combinatorics.
- *Introductory exposure to thermodynamic concepts and terminology*: familiarity with the concepts of a system, surroundings, boundary, absolute temperature, pressure, heat, work, and processes.

It is very likely that any reader with at least three or four years of undergraduate coursework in physics, chemistry, or engineering will be sufficiently prepared for this book. Regardless, most of the requisite background material is reviewed or explained with examples in an as-you-go manner.

The most distinguishing feature of this text is that it integrates macroscopic principles (classical thermodynamics) and molecular aspects of thermodynamics (statistical mechanics) throughout. This constitutes a different perspective than many traditional treatments of the material that begin purely at the macroscopic level with the so-called *postulatory approach*. The latter gives a beautiful formulation of classical thermodynamics that makes no reference to the molecular world and hence is independent of the particular nature of microscopic interactions. Instead, the postulatory approach proposes several general principles, reinforced many times over by empirical observation, with which any problem can be analyzed. In other words, that approach begins *a priori* with the laws of thermodynamics. Although they may be phrased in different ways, the following list gives some usual possibilities for the laws.

1. No process can operate in such a way as to create or destroy energy.
   The total energy of an isolated system is constant. If the internal energy of a closed system changes, the difference must exactly equal the sum of the heat added and work done on it: $dE = \delta Q + \delta W$.
2. No process can operate so as to completely convert the heat absorbed by a system into usable work.
   Systems have a quantity called the entropy that is a function of state and that can be measured using reversible heat transfers: $dS = \delta Q_{rev}/T$. In an isolated system, the entropy of spontaneous processes can only increase with time.
3. No process can operate so as to bring a system to absolute zero in a finite number of steps and in finite time.
   The entropy of all perfect, pure, monatomic crystalline substances at absolute zero is zero.

The brilliance of these statements is that, despite their simplicity, they have profound implications that can be studied in great mathematical detail for every physical process. Furthermore, they require no understanding of molecular interactions or the fundamental theory thereof.

Unfortunately, while the postulatory formulation is often greatly esteemed by experienced scientists, it can be a challenging starting point for the early learner. The main problem is that it requires students to merely accept these statements, without explanation as to how they connect to molecular principles that they surely have seen in other courses. *What really is the entropy and why does it exist?* The gap between the atomic world and thermodynamics often leaves students feeling unsatisfied, confused, lacking intuition, and missing the big picture.

This book therefore takes a different, *integrated approach* to teaching thermodynamics that blends molecular and statistical-mechanical concepts with the exposition of the classical laws. It attempts to be bottom-up rather than top-down by first presenting and then rationalizing ideas on the basis of atomic-scale interactions. In this sense, it aims to give the reader some feeling for the *why* of thermodynamics. Of course, this approach itself is not devoid of postulates. To begin, one must accept some level of atomic theory, whether quantum or classical or other. Moreover, ultimately the second law requires the *equal a priori* assumption that is the foundation of statistical mechanics, as discussed in Chapter 4. In this sense the approach taken by this text is motivated on pedagogical grounds, not scientific ones. After understanding the material, therefore, the reader is highly encouraged to revisit the postulatory presentation to appreciate the generality of thermodynamics as an empirical natural science that is independent of the microscopic world.

The reader should be cautioned that a deep understanding of thermodynamics does not simply evolve from the page, but rather requires a concerted effort to explore the material outside of the main narrative. Some recommendations for working through this text are the following:

- **Pay particular attention to general, broad concepts.** The most important ones are highlighted for you in gray call-out boxes. Ask yourself the following questions. *Does this make sense intuitively? Does this make sense mathematically?* Challenge yourself to understand and apply the ideas, initially in very simple examples. Make sure that you feel comfortable with the concepts before proceeding too far ahead, since it is easy to become lost. Check yourself with questions in the *Conceptual and thought problems* section at the end of each chapter.
- **Work through the end-of-chapter problems.** One simply cannot appreciate thermodynamics without tackling actual problems. If you are reading this text outside of class, a suggested course of study is the end-of-chapter problems that are indicated with boxed numbers. You will likely struggle through them, and it is important that you do so! It is through this process of struggle that subtleties bubble to the surface. This book has been written so that many important results and implications are explicitly left for you to discover in this engaged, problem-driven manner. Note that some chapters have many more problems than others, and this is because they are natural synthesis points for incorporating earlier material.
- **Think in terms of functions and variables, not values.** Students often become complacent with plugging numbers into off-the-shelf equations. This is a *value-focused* way to solve problems, and it is entirely the wrong way to understand thermodynamics. Instead, much of the beauty of this subject is that properties are interrelated through a systematic calculus of functions. Indeed, any equilibrium

property is a function of state and thus also a *mathematical* function. This means that such properties have multiple independent variables and partial derivatives. At first it may seem unsettling to consider temperature as a function, for example $T(E, V, N)$, but keep in mind that its behavior is no different than that of the generic $f(x, y, z)$.

- **Always solve problems from general principles, not specialized equations.** One of the attractive features of thermodynamics is that there are just a few fundamental equations from which essentially all other results can be derived. Therefore, you should not try to memorize every equation, but instead strive to be able to quickly pinpoint the underlying assumptions so that you would be able to re-derive the key ones in isolation. Throughout this book, the most important equations are numbered in bold.

Finally, the reader is encouraged to explore other texts as a means to broaden understanding and clarify confusing points. Some especially useful texts are referenced at the end of this chapter. In particular, many parts of the present book follow closely and were inspired by the brilliant texts of Denbigh, Hill, and McQuarrie. While these seminal works have a less modern tone, they present the material with great care and in significantly greater depth. In addition, the text by Dill gives a terrific introduction to thermodynamics suited to a more general audience, and it addresses many background concepts that are not covered in detail by the present work.

FURTHER READING

Highly recommended supplementary reading for early graduate-level students in the chemical sciences and engineering

K. Denbigh, *The Principles of Chemical Equilibrium*, 4th edn. New York: Cambridge University Press (1981).

K. Dill and S. Bromberg, *Molecular Driving Forces: Statistical Thermodynamics in Biology, Chemistry, Physics, and Nanoscience*, 2nd edn. New York: Garland Science (2010).

T. L. Hill, *An Introduction to Statistical Thermodynamics*. Reading, MA: Addison-Wesley (1960); New York: Dover (1986).

D. A. McQuarrie, *Quantum Chemistry*. Mill Valley, CA: University Science Books (1983).

D. A. McQuarrie, *Statistical Mechanics*. Sausalito, CA: University Science Books (2000).

Also recommended

H. Callen, *Thermodynamics and an Introduction to Thermostatistics*, 3rd edn. New York: Wiley (1985).

D. Chandler, *Introduction to Modern Statistical Mechanics*. New York: Oxford University Press (1987).

J. R. Elliot and C. T. Lira, *Introductory Chemical Engineering Thermodynamics*, 2nd edn. Upper Saddle River, NJ: Prentice Hall (2012).

J. Israelachvili, *Intermolecular and Surface Forces*, 3rd edn. Burlington, MA: Academic Press (2011).

C. Kittel and H. Kroemer, *Thermal Physics*. New York: W. H. Freeman (1980).

A. Z. Panagiotopoulos, *Essential Thermodynamics*. Princeton, NJ: Drios Press (2011).

J. M. Smith, H. V. Ness, and M. Abbott, *Introduction to Chemical Engineering Thermodynamics*, 7th edn. New York: McGraw-Hill (2005).

J. W. Tester and M. Modell, *Thermodynamics and Its Applications*, 3rd edn. Upper Saddle River, NJ: Prentice Hall (1997).

For this chapter

A. Einstein, "Autobiographical notes" in *Albert Einstein: Philosopher-Scientist*, P. A. Schlipp, ed. Evanston, IL: Library of Living Philosophers (1949).

2   Equilibrium and entropy

2.1   What is equilibrium?

At its most basic level, the subject of thermodynamics is the study of the properties of systems and substances at *equilibrium*. What do we mean by equilibrium? A simple way of thinking about this concept is that it represents the state where time is an irrelevant variable.

> We can think of **thermodynamic equilibrium** as the condition where the following statements hold.
>
> (1) The properties of a system do not change with time.
> (2) The properties of a system do not depend on how it was prepared, but instead depend only on the current conditions of state, that is, a short list of parameters such as temperature, pressure, density, and composition that summarize the current equilibrium. A system brought to a specific equilibrium state always behaves identically, and such states are *history-independent*. The notion of history-independence is more restrictive than the statement that properties do not change with time. Indeed, history-independence is an important factor of *thermodynamic* equilibrium.
> (3) The properties of a large number of copies of the same system at the same state conditions are identical, irrespective of whether or not each copy had a distinct preparation and history.

On the other hand, one might question whether these statements are compatible with the molecular nature of reality. Do not the molecules in a glass of water rotate and move about? Are not their positions, orientations, and velocities constantly changing? How then can the glass of water ever be at equilibrium, given this ongoing evolution?

The resolution to this seeming conundrum is that thermodynamic equilibrium is concerned with certain *average* properties that become time-invariant. By average, we mean two things. First, these properties are measured at a bulk, *macroscopic* level, and are due to the interactions of many molecules. For example, the pressure that a gas exerts on the interior of a container is due to the average rate of collisions and momentum transfer of many molecules with a vessel wall. Such macroscopic properties are typically averaged over very many ($\sim 10^{23}$) molecular interactions.

Second, equilibrium properties are measured over some window of time that is much greater than the time scales of the molecular motion. If we could measure the instantaneous density of a gas at any single moment, we would find that some very small, microscopic regions of space would have fewer molecules and hence lower density than others, while some spaces would have more molecules and higher density, due to random atomic motions. However, measured over a time scale greater than the average collision time, the time-averaged density would appear uniform in space.

In fact, the mere concept of equilibrium requires there to be some set of choices that a system can make in response to environmental conditions or perturbations. These choices are the kinds of positions, orientations, and velocities experienced by the constituent molecules. Of course, a system does not make a literal, cognitive choice, but rather the behavior of the molecules is determined naturally through their energetic interactions with each other and the surroundings.

So far, we have hinted at a very important set of concepts that involve two distinct perspectives of any given system.

> **Macroscopic** properties are those that depend on the bulk features of a system of many molecules, such as the pressure or mean density. **Microscopic** properties are those that pertain to individual molecules, such as the position and velocity of a particular atom. The equilibrium properties of a system measured at a macroscopic level actually derive from the **average** behavior of many molecules (typically $\sim 10^{23}$), over periods of time.

The connection between macroscopic equilibrium properties and the molecular nature of reality is the theme of this book, and the basis of thermodynamics. In particular, we will learn exactly how to connect averages over molecular behavior to bulk properties, a task that forms the basis of statistical mechanics. Moreover, we will learn that, due to the particular ways in which molecules interact, the bulk properties that emerge when macroscopic amounts of them interact are subject to a number of simple laws, which form the principles of classical thermodynamics.

Note that we have not defined thermodynamics as the study of heat and energy specifically. In fact, equilibrium is more general than this. Thermodynamics deals with heat and energy because these are mechanisms by which systems and molecules can interact with one another to come to equilibrium. Other mechanisms include the exchange of mass (e.g., diffusion) and the exchange of volume (e.g., expansion or contraction).

2.2 Classical thermodynamics

Classical thermodynamics provides laws and a mathematical structure that govern the behavior of bulk, macroscopic systems. While its basic principles ultimately emerge from molecular interactions, classical thermodynamics makes no reference to the atomic scale and, in fact, its core was developed before the molecular nature of matter was generally accepted. That is to say, classical thermodynamics provides a set of laws and relationships exclusively among macroscopic properties, and can be developed entirely on the basis of just a few postulates without consideration of the molecular world.

In our discussion of equilibrium above, we did not say anything about the concepts of heat, temperature, and entropy. Why? These are all macroscopic variables that are a *consequence* of equilibrium, and do not quite exist at the level of individual molecules. For the most part, these quantities have real significance only in systems containing numerous molecules, or in systems in contact with "baths" that themselves are macroscopically large. In other words, when large numbers of molecules interact and come to equilibrium, it turns out that there are new relevant quantities that can be used to describe their behavior, just as the quantities of momentum and kinetic energy emerge as important ways to describe mechanical collisions.

The concept of *entropy*, in particular, is central to thermodynamics. Entropy tends to be confusing because it does not have an intuitive connection to mechanical quantities, such as velocity and position, and because it is not conserved, like energy. Entropy is also frequently described using qualitative metrics such as "disorder" that are imprecise and difficult to interpret in practice. Not only do such descriptions do a terrible disservice to the elegant mathematics of thermodynamics, but also the notion of entropy as "disorder" is sometimes outright wrong. Indeed, there are many counter-examples where entropy increases while subjective interpretations would consider order to increase as well. Self-assembly processes are particularly prominent cases, such as the tendency of surfactants to form micelles and vesicles, or the autonomous hybridization of complementary DNA strands into helical structures.

In reality, entropy is not terribly complicated. It is simply a mathematical *function* that emerges naturally for equilibrium in *isolated systems*, that is, systems that cannot exchange energy or particles with their surroundings and that are at fixed volume. For a single-component system, that function is

$$S = S(E, V, N) \qquad (2.1)$$

which states that the entropy is dependent on three macroscopic quantities: the total internal energy of the system E, the total volume of the system V, and the number of particles (molecules or atoms) N. The internal energy stems from all of the molecular interactions present in the system: the kinetic energies of all of the molecules plus the potential energies due to their interactions with each other and with the container walls. For multicomponent systems, one incurs N additional variables for each species,

$$S = S(E, V, N_1, N_2, N_3, \ldots) \qquad (2.2)$$

At this point, let us think of the entropy not as some mysterious physical quantity, but simply as a mathematical function that exists for all systems and substances at equilibrium. We do not necessarily know the analytical form of this function, but nonetheless such a function exists. That is, for any one system with specific values of E, V, and N, there is a unique value of the entropy.

The reason why the entropy depends on these particular variables relates, in part, to the fact that E, V, and N are all rigorously constant for an isolated system due to the absence of heat, volume, and mass transfer. What about non-isolated systems? In those cases, we have to consider the total entropy of the system of interest *plus* its surroundings, which together constitute a net isolated system. We will perform such analyses later on, but will continue to focus on isolated systems for the time being.

The specific form of this entropy function is different for every system, irrespective of whether it is a pure substance or mixture. However, *all* entropy functions have some shared properties, and these common features underlie the power of thermodynamics. These properties are mathematical in nature.

(1) **Partial derivatives**

$$\left(\frac{\partial S}{\partial E}\right)_{V,N} = \frac{1}{T} \text{ where } T \text{ is the absolute temperature}$$

$$\left(\frac{\partial S}{\partial V}\right)_{E,N} = \frac{P}{T} \text{ where } P \text{ is the absolute pressure}$$

$$\left(\frac{\partial S}{\partial N}\right)_{E,V} = -\frac{\mu}{T} \text{ where } \mu \text{ is the chemical potential}$$

(2) **Extensivity**

$S(\lambda E, \lambda V, \lambda N) = \lambda S(E, V, N)$ where λ is an arbitrary positive real number

(3) **Concavity**

$$\left(\frac{\partial^2 S}{\partial X^2}\right)_{Y,Z} \leq 0 \text{ where } X \text{ is } E, V, \text{ or } N, \text{ and } Y \text{ and } Z \text{ are the remaining two}$$

Properties (2) and (3) arise from the nature of molecular interactions in large systems, and two problems at the end of Chapter 3 consider a simple approach to their derivation. For now, the most important of these common features is the partial derivatives of property (1). Instead of listing these separately as above, we often write them using the total differential,

$$dS = \frac{1}{T} dE + \frac{P}{T} dV - \frac{\mu}{T} dN \tag{2.3}$$

The differential form simply summarizes the set of partials. Strictly speaking, the variables T, P, and μ are not independent in the entropy function (like E, V, and N), but rather are themselves functions that stem from the derivatives of S,

$$T(E, V, N), P(E, V, N), \mu(E, V, N) \tag{2.4}$$

This result, of course, is a basic mathematical fact. The differential form for the entropy can be rearranged into an expression for the energy,

$$dE = T\, dS - P\, dV + \mu\, dN \tag{2.5}$$

Both forms imply the same relationships. Equations (2.3) and (2.5) are simple transformations of each other, equivalent to inverting the function $S(E, V, N)$ to $E(S, V, N)$.

That is, (2.5) implies a function that returns the equilibrium internal energy corresponding to a specified entropy (and V and N). This form will become more useful and intuitive later on, but it contains exactly the same information. In fact, both of these differential forms are called the *fundamental equation*, being labeled the entropy and energy versions, respectively. For multicomponent systems, there is a distinct chemical potential corresponding to each particle number. In this case, the fundamental equations are

$$dS = \frac{1}{T}\, dE + \frac{P}{T}\, dV - \sum_i \frac{\mu_i}{T}\, dN_i \tag{2.6}$$

$$dE = T\, dS - P\, dV + \sum_i \mu_i\, dN_i \tag{2.7}$$

The entropy is the basic function describing equilibrium. It is natural to wonder what this means. What does one do with the entropy function? The relationships we have described thus far already tell us a few important facts. First, they indicate what is necessary to determine the equilibrium state of a system.

Only three pieces of information are necessary to specify the entropy for a single-component system, E, V, and N. By extension, only three pieces of information are necessary to completely specify the **equilibrium state** of the system. For a C-component system, we need $C + 2$ pieces of information. As we will learn later, intensive properties that do not scale with the size of the system require one less piece of information. That information is essentially the system size.

Moreover, the common properties of the entropy provide a way to interrelate thermodynamic properties.

We can calculate the pressure, temperature, and chemical potential from the entropy function if we know a system's E, V, and N, simply by evaluating the three partial derivatives at these values.

Alternatively, if we knew T, V, and N, we could find the energy of the system as the value that satisfied $(\partial S/\partial E)_{V,\,N} = 1/T$. Note here that E appears, along with V and N, as the point at which the derivative is evaluated. A similar construction can be made with the other partial derivatives.

A key idea is that we can relate the entropy to measurable properties like the temperature or heat capacity by exploiting connections to its derivatives. In fact, often we can use data measured in experiments to back-out the form of the entropy through integration. For example, if we integrate the energy partial derivative, we find

$$S = \int \frac{1}{T(E,V,N)}\, dE + f(V,N) \tag{2.8}$$

where f is an integration constant that depends on V and N because the partial derivative that was integrated held these constant. This expression requires us to know how the temperature varies with energy, which we could construct from calorimetric data. If now we take the volume derivative of this result, we obtain

$$\frac{P}{T} = \frac{\partial}{\partial V} \int \frac{1}{T(E, V, N)} \, dE + \frac{\partial f(V, N)}{\partial V} \tag{2.9}$$

Experimental data on the behavior of the pressure might then enable us to substitute for the left-hand side (LHS) of this equation, integrating again to determine the function f to within an N-dependent constant, and so on. Several problems at the end of this chapter demonstrate this approach. Though the procedure may seem complicated at present, we will later find in Chapters 7 and 8 that there are simpler ways to utilize experimental data by making connections to other thermodynamic properties.

Finally, the common properties of entropy functions suggest constraints on the manner in which thermodynamic properties can vary with respect to each other.

T, P, and μ are all related to derivatives of the *same* function. This has important mathematical consequences for relationships between these variables, as we will find later.

Extensivity and concavity also place limits on the kinds of values that T, P, and μ can adopt, and in particular, how they can depend on E, V, and N.

The main idea is that there exist quantitative, mathematical relationships among the macroscopic properties E, V, N, T, P, and μ at equilibrium. The study of these relationships constitutes classical thermodynamics.

Example 2.1 *An ideal gas has the equation of state $P = Nk_BT/V$. Find the volume-dependence of its entropy.*

Note that $P/T = Nk_B/V$. Conveniently, the right-hand side (RHS) of this expression is a function only of the independent variables in the entropy. This means that we can integrate Eqn. (2.3) directly at constant E, N conditions,

$$dS = \frac{P}{T} \, dV = \frac{Nk_B}{V} \, dV \text{ (constant } E, N)$$

Performing the integration gives

$$S = Nk_B \ln V + f(E, N)$$

2.3   Statistical mechanics

Of course, we have said nothing about the origins of the entropy. Why does it emerge as the important quantity at equilibrium? Why do all entropy functions have the set of common properties described above? How does one determine the entropy function for a specific substance or system?

These are all questions that are not part of classical thermodynamics because they require an understanding of the molecular nature of matter. In short, the entropy emerges because systems consist of an extraordinarily large number of molecules. The molecular origins of thermodynamics are rooted in *statistical mechanics*, which gives an exact recipe for the entropy for isolated systems called Boltzmann's equation,

$$S = k_B \ln \Omega \, (E, \, V, \, N) \tag{2.10}$$

Here, k_B is Boltzmann's constant, which has a value near 1.38×10^{-23} J/K such that the entropy has units of energy per temperature. It is related to the gas constant $R = 8.314$ J/mol K through Avogadro's number, $\mathcal{N}_A$,

$$k_B = R/\mathcal{N}_A \tag{2.11}$$

One can think of k_B as the per-molecule version of the per-mole constant R. For multi-component cases, the dependent variables in (2.10) are extended with all of the molecule numbers N_i of the different species. The function Ω is called the *density of states*, and it is specific to the particular system or substance of interest. The density of states gives the number of microscopic, molecular configurations of a system that are possible for particular values of E, V, and N. Essentially, Ω is a counting of microscopic states, or *microstates*, for given E, V, and N.

A microstate is one "snapshot" of the system at a given instant in time. To specify a particular microstate, one must list all the molecular details of a system. At a classical level, this means the specification of the position and velocity of every atom. A system proceeds through many different microstates as time evolves, even at equilibrium. The density of states counts the number of such snapshots, or configurations, that are possible for a given total internal energy E with a number of molecules N in a volume V. Example 2.2 illustrates this with a simple model.

Example 2.2 *Two molecules are adsorbed on a simple two-dimensional square crystalline surface, each at one particular lattice site. There are in total V lattice sites where the molecules can adsorb. If the two molecules are at adjacent sites, attractive interactions between them result in a negative potential energy $-\epsilon$; otherwise, the molecules do not interact. Consider the volume to be fixed, and neglect all other energies, such as kinetic energies and those of the molecules with the surface. What are $\Omega(E = 0)$ and $\Omega(E = -\epsilon)$?*

For this problem, we must think about what microstates are possible in the system. Some possibilities for a system with $V = 16$ are shown in Fig. 2.1. We notice that each lattice site has a total of four neighbors, with the exception of the edge sites. However, the number of edge sites will be much less than the total number of sites V in a macroscopic system ($V \sim 10^{23}$); therefore, we will neglect this subtle effect for now.

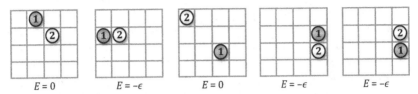

Figure 2.1. Some microstates for two molecules adsorbed onto a square lattice.

The number of possible $E = -\epsilon$ microstates is simply the total number of ways we can place the first molecule times the number of possible neighbors to it,

$$\Omega(E = -\epsilon) = V \times 4$$

On the other hand, all remaining possible configurations have an energy of zero. The total number of configurations is the number V of spots at which to put the first molecule, times the number $(V - 1)$ of spots at which to then place the second. Therefore,

$$\Omega(E = 0) = V(V - 1) - 4V$$

What if the two molecules are indistinguishable? That is, what if the fourth and fifth microstates above look exactly the same? In that case, there are fewer distinct micro-states in total and we have overcounted by the number of ways of swapping the places of different particles. For indistinguishable molecules, we would obtain

$$\Omega(E = -\epsilon) = \frac{V \times 4}{2} \qquad \Omega(E = 0) = \frac{V(V - 1) - 4V}{2}$$

We will find out later on that indistinguishability has an actual effect on the properties of a system, and is due to the rules of quantum mechanics.

How does the density of states give rise to distinct properties in different systems? As you may have anticipated from this simple example, the reason lies in the particular ways in which molecules interact energetically. Distinct molecular architectures modulate the potential energies that can be stored within bonded interactions and that can be exerted on other molecules through nonbonded forces. Consider, for example, two hypothetical containers, one filled with liquid water and one with liquid mercury. Let us presume that each is prepared with the same total E, V, and N. The value of the density of states will differ for these two systems because the numbers of ways of storing the energy E will be entirely distinct. Water molecules have two O—H bonds that can affect the energy by stretching and bending, whereas mercury is monatomic and has no such modes. Moreover, water can store energy in hydrogen-bonding interactions, which depend sensitively on the orientations of the molecules, whereas mercury atoms interact with each other in a more isotropic manner.

How does one calculate the density of states? From the example above, we could imagine the following recipe, in principle, for computing Ω given a set of values E, V, and N. First, place N molecules in a fixed volume V, a rigid container of some kind. Then systematically enumerate every possible microstate, e.g., every possible combination of positions, orientations, and velocities for *all* of the atoms. For each microstate, determine the total molecular energy due to that particular set of positions, orientations, and velocities. If that microstate's energy equals E, then count it. Otherwise, ignore it. At the end of the enumeration, the total counts will give the density of states. In principle, it's that simple.

In reality, this is not usually a feasible calculation because the number of microstates is overwhelmingly huge, scaling exponentially with the number of molecules, which itself may be of order 10^{23}. Even using sophisticated computer simulation techniques, direct counting of this kind is generally not possible when more than a few atoms are present. This is not a complete dead-end, however, because there are other ways to approach the thermodynamic properties of systems that skirt around the issue of counting microstates. We will discuss these mathematical approaches later in this book.

2.4   Comparison of classical thermodynamics and statistical mechanics

We have seen that classical thermodynamics describes equilibrium for macroscopic systems, and that statistical mechanics gives us a way to understand the origins of bulk equilibrium behavior in terms of molecular properties. There is a close connection between the two that it will be critical to keep in mind throughout this book. While classical thermodynamics deals with bulk properties such as temperature and pressure, statistical mechanics is concerned with microstates and molecular configurations, but clearly the two are connected through Boltzmann's equation. Figure 2.2 illustrates the relationships.

macroscopic point of view	microscopic point of view
"bulk" properties	many molecules
 E, V, N T, P, μ	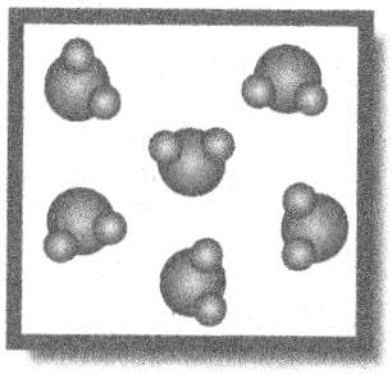
fundamental equation $dS = \frac{1}{T}\, dE + \frac{P}{T}\, dV - \frac{\mu}{T}\, dN$	density of states $S(E,V,N) = k_B \ln \Omega(E,V,N)$
S can be determined using measured relationships	S can be derived from molecular properties
ideal gas example: $\left(\frac{dS}{dV}\right)_{E,N} = \frac{P}{T} = \frac{k_B N}{V}$ $\rightarrow\; S = Nk_B \ln V + constant\ (E,N)$	*ideal gas example:* $S = Nk_B \ln\left[\left(\frac{4\pi mE}{3h^2}\right)^{3/2}\frac{V e^{5/2}}{N^{5/2}}\right]$
classical thermodynamics *history in 1800s*	statistical mechanics *history in 1800s and 1900s*

Figure 2.2. A comparison of macroscopic and microscopic perspectives on equilibrium.

What we have presented in this chapter already encapsulates the most important concepts in thermodynamics and statistical mechanics, such that a large fraction of the remainder of this book derives directly from these core ideas. We will discuss specific models of various physical behaviors, such as those for gases and solutions, but most of what you will learn will come from understanding the implications of the ideas and equations presented in this chapter. In particular, you will need to understand the mathematical consequences of the equations, which form the basis of *thermodynamic calculus*. This is essentially multivariate calculus combined with several kinds of transforms, and it requires understanding how to carefully translate physical manipulations and properties of systems into a coherent mathematical structure.

There is a vocabulary of thermodynamics that accompanies the discussion of equilibrium states and processes that change them. You will need to become familiar with this language as well. We have already touched on some of these, but let us summarize the most important terms:

system – a collection of things with a physical or conceptual boundary between it and the outside world; the particular focus of the thermodynamic analysis at hand
environment or surroundings – what is outside the system
universe/world – everything; the system plus its surroundings
isolated system – a system where absolutely nothing in the surroundings interacts with it and where nothing crosses the system boundaries
closed system – energy (heat) can cross the boundaries, but matter cannot
open system – matter and energy (heat) can cross the boundaries
change of state – a change in any of the macroscopic thermodynamic properties E, V, N, T, P, or μ; a change to a different set of equilibrium conditions
path – a progression in changes in the state of a system (e.g., changes in variables like E, V, N, T, P, and μ) that describes the process of moving from one equilibrium state to another
adiabatic process – a process in which there is no heat transfer between the system and its surroundings
isothermal process – a process that occurs at constant temperature T
isochoric process – a process that occurs at constant volume V
isobaric process – a process that occurs at constant pressure P
isentropic process – a process in which the entropy S is constant

2.5 Combinatorial approaches to counting

Equation (2.10) shows that the entropy is ultimately tied to a counting of microscopic configurations. Probability and statistics offer a number of formulae that help develop closed-form expressions for the density of states in simple models. The most useful counting expressions involve the selection of k objects or choices out of a total number of possibilities n. Depending on whether the order of selection matters and whether the same object can be picked more than once, the total number of possibilities can vary.

The number of ways to pick k objects from n is given in the following cases:

(1) without replacement and their order matters,

$$_nP_k = \frac{n!}{(n-k)!}$$

(2) without replacement and their order is irrelevant,

$$_nC_k = \binom{n}{k} = \frac{n!}{k!(n-k)!}$$

(3) with replacement (each object can be picked multiple times) and their order matters,

$$n^k$$

(4) with replacement and their order is irrelevant,

$$\frac{(k+n-1)!}{k!(n-1)!}$$

These combinatorial expressions involve factorials that can be difficult to manipulate mathematically. Instead, we frequently use Stirling's approximation for large n,

$$\ln n! \approx n \ln n - n \rightarrow n! \approx (n/e)^n \tag{2.12}$$

Stirling's approximation is a convenient way to treat factorials that is widely used in statistical mechanics. It is an excellent approximation for large values of n, and since often n will be of the order of the number of molecules, it is extremely reliable.

Two other useful mathematical identities involving factorials are the exponential expansion,

$$e^x = \sum_{n=0}^{\infty} \frac{x^n}{n!} \tag{2.13}$$

and the binomial theorem,

$$(1+x)^n = \sum_{k=0}^{n} \frac{n!}{k!(n-k)!} x^k \tag{2.14}$$

Often in the development of molecular models, x will be an exponential term involving the temperature T and a constant of units energy ϵ, such as $x = \exp(-\epsilon/k_BT)$. Equations (2.13) and (2.14) will become helpful in our discussion of statistical mechanical models in Chapters 16–19.

Example 2.3   *N particles are absorbed on a lattice consisting of V possible adsorption sites, $V \gg N$. Write an expression for the number of distinct molecular configurations Ω in this simple system in the following cases.*

(a) *Only one particle can sit at each site and the particles are all identical (e.g., all blue).*

In this case, we have V options in total. We can think of a process of selecting N of the V sites and then placing particles on them. Since the particles are identical, the order in which we put them on sites does not matter. In other words, if we swap two particles' positions, we do not incur a new configuration. Therefore,

$$\Omega = \frac{V!}{N!(V-N)!}$$

(b) *Only one particle can sit at each site and the particles each have a unique identity (e.g., different colors).*

We still have V options, but now the order in which we placed the particles on the N selected sites does matter because we can tell the particles apart from one another. As a result,

$$\Omega = \frac{V!}{(V-N)!}$$

(c) *More than one particle can sit at each site and the particles each have a unique identity.*

This scenario is equivalent to the one in which particles do not interact and thus do not "see" each other. Every particle has V possible locations, and the independence of their positions means that the total number of possibilities is multiplicative, giving

$$\Omega = V^N$$

(d) *More than one particle can sit at each site and the particles are all identical.*

This is the trickiest case of the four. It is helpful to consider a specific example to illustrate the complications. Let us take $N = 2$ and $V = 2$. It is easy to see that $\Omega = 3$ because there are only three distinct states: both particles in the first site, both in the second site, and one in each site. However, the expression V^N fails here, giving $2^2 = 4$. That expression incorrectly counts twice the state in which there is one particle in each site; it counts it once when the first particle is in the first site and once when it is in the second site. It does not, however, overcount the other two states. It turns out that the correct expression is in fact

$$\Omega = \frac{(N+V-1)!}{N!(V-1)!}$$

It is easy to see that this relation gives the right answer for the specific numbers considered here, $\Omega = 3!/(2!1!) = 3$. The derivation of this expression, however, is left to the reader as an end-of-chapter problem.

Problems

Conceptual and thought problems

2.1. The fundamental function $E(S, V, N)$ implies that one finds the value E such that $S = S(E, V, N)$, where S on the LHS and V and N on the RHS are all specified. In principle, if the entropy could be any mathematical function, there could be multiple solutions of E that could satisfy this relationship and hence the internal energy would not be a well-defined function of the entropy. Explain why in fact this is hardly ever the case.

2.2. Consider the particles-on-a-lattice system posed in Example 2.3, and the case in which the particles are all identical and more than one can occupy a lattice site. Provide a derivation showing that the correct expression for the number of configurations is indeed $\Omega = (N + V - 1)!/(N!(V - 1)!)$.

2.3. Systems have degrees of freedom, or variables that can be changed to modify the microstate or configuration, such as particle positions and velocities. When counting microstates we might say that two degrees of freedom are independent if we can choose values for one irrespective of the values of the other. For the cases in Example 2.2, indicate whether or not the particles' positions are independent degrees of freedom. What is a general expression for counting numbers of configurations of distinguishable particles when n degrees of freedom $g_1, g_2, \ldots, g_n$ are independent?

Fundamentals problems

2.4. In calorimetric measurements, simple forms are often used to correlate the energy of a substance with temperature. Given the following expressions for the energy E as a function of temperature T, find the underlying entropy function, $S(E)$, up to an integration constant. Assume constant volume.
(a) $E = C_V T + E_0$, where C_V and E_0 are constants. This expression invokes the so-called constant-heat-capacity approximation. It correctly describes the energies of monatomic ideal gases and perfect crystals, at a classical (non-quantum) level.
(b) $E = aT^{3/5} + E_0$, where a and E_0 are constants. This modern theoretical expression for liquids is due to Rosenfeld and Tarazona [*Molec. Phys.* **95**, 141 (1998)].

2.5. In one set of experiments, it is found that a pure substance closely adheres to the following heat-capacity and equation-of-state relations:

$$E/N = cT^2 + e_0$$
$$P = aT\rho^5$$

where $\rho = N/V$ and a, c, and e_0 are N-, E-, and V-independent constants. The first of these expressions corresponds to a heat capacity that is linear in temperature.

(a) Find the underlying entropy function $S(E, V, N)$, up to an E- and V-independent constant. Be sure to consider that S must have proper extensive behavior.

(b) What if the equation of state is instead $P = b\rho^5$, where b is a constant? Is it possible for a substance to simultaneously obey this and the energy relation above? Why, or why not?

2.6. In another set of experiments, it is found that a pure substance obeys the following heat-capacity and equation-of-state relations:

$$E/N = cT + e_0$$
$$P = aT\rho^3$$

where $\rho = N/V$ and a, c, and e_0 are N-, E-, and V-independent constants. The first of these expressions invokes the so-called constant-heat-capacity approximation. Find the underlying entropy function $S(E, V, N)$, up to an E-, and V-independent constant. Be sure to consider that S must have proper extensive behavior.

2.7. Provide a rough derivation of Stirling's approximation. There are at least two ways to approach this.

(a) Approximate the derivative of $\ln n!$ using the finite-difference expression $\ln n! - \ln(n-1)!$, and then integrate your result.

(b) Express $\ln n!$ as $\ln n + \ln(n-1) + \ln(n-2) + \cdots$, then approximate the sum using an integral.

2.8. Show that the combinations formula for $_nC_k$ can be expressed as

$$\ln{}_nC_k = -n[x \ln x + (1-x)\ln(1-x)]$$

where $x \equiv k/n$. You will need to employ Stirling's approximation.

2.9. A simple model of a ferromagnet supposes that it consists of a macroscopic number N of so-called "spins." Each spin can have one of two orientations, either up or down. Assume here that the spins do not interact with one another; that is, their energies are independent of the up or down state of their neighbors. Instead, for every up spin, there is an energy of $-h$ and for every down one, an energy of $+h$.

(a) Find an expression for the density of states of this system.

(b) Find the fraction of spins that are up, f, as a function of the temperature.

(c) What is the behavior of f as $T \to \infty$ and as $T \to 0$?

2.10. A substance has the following form for its entropy fundamental relation: $S = aE^b V^c N^d$, where a is a positive constant. What constraints exist for the constants b, c, and d?

FURTHER READING

H. Callen, *Thermodynamics and an Introduction to Thermostatistics*, 3rd edn. New York: Wiley (1985).

K. Denbigh, *The Principles of Chemical Equilibrium*, 4th edn. New York: Cambridge University Press (1981).

J. W. Tester and M. Modell, *Thermodynamics and Its Applications*, 3rd edn. Upper Saddle River, NJ: Prentice Hall, 1997.

3 Energy and how the microscopic world works

3.1   Quantum theory

To conceptualize the molecular origins of thermodynamic equilibrium, one must first understand the elemental ways by which molecules interact. How does the world really work? What are the most fundamental principles that form the basis of reality as we know it?

Currently our understanding of reality rests upon two principal concepts in physics: quantum theory and relativity. Both of these have been subjected to stringent experimental tests over the past century, and their combination has in part led to a deep understanding of elementary particles. There still remain some incompatibilities between the two, namely in understanding the nature of gravity, and there have been intense efforts to find new fundamental physical explanations. However, for the purposes of our discussion, we will focus solely on quantum theory since for nearly all of the models and systems that we will discuss one can safely avoid considerations of relativistic effects.

Quantum mechanics describes the complete time evolution of a system in a quantum sense, in a manner analogous to what Newtonian mechanics does for classical systems. It is most easily described in terms of a system of fundamental particles, such as electrons and protons. Particles have associated position variables, $\mathbf{r}_1, \mathbf{r}_2, \ldots$, where $\mathbf{r} = (x, y, z)$ is a vector. The complete state of the system at any point in time is then specified by a function $\Psi(\mathbf{r}_1, \mathbf{r}_2, \ldots, t)$ called the *wavefunction*. The wavefunction takes on complex values of the form $a + bi$. Its physical significance is that its norm, $\Psi^*(\mathbf{r}_1, \mathbf{r}_2, \ldots, t)$ $\Psi(\mathbf{r}_1, \mathbf{r}_2, \ldots, t)$, gives the *joint probability* that particle 1 is at $\mathbf{r}_1$, particle 2 is at $\mathbf{r}_2$, etc., at the time t. Here the superscript * denotes the complex conjugate, i.e., $a - bi$. The quantity $\Psi^*\Psi$ therefore is always real and positive, $a^2 + b^2$, as we would expect for a probabilistic interpretation.

The essence of quantum mechanics is that the wavefunction describes the evolution of probabilities. This is rather different from Newtonian mechanics, in which each particle has an exact position at time t and not a distribution of possible positions. Quantum mechanics asserts that this distribution is the most we can know about the system; we cannot predict the positions of the particles to more accuracy. There is some inherent randomness in nature, and the best that we can do is understand the probabilities of different possible outcomes. This may sound a bit strange, because we are not used to this kind of behavior at the macroscopic scale. Indeed, at the macroscale, these probability distributions appear narrowly peaked such that we can usually say, from our perspective, "exactly" where a large-scale object lies in space. For microscopic objects, like atoms and electrons, these distributions look much broader by comparison.

It is beyond the scope of our inquiry to enter a deep discussion about the origins of the wavefunction and quantum theory. Instead, we will briefly describe the main equations that govern it. The time evolution of the wavefunction – and hence the time evolution of the probability distribution – is determined by the *Schrödinger equation*, the quantum-mechanical analogue of Newton's equations of motion. For a single particle traveling in a time-independent potential energy field $U(\mathbf{r})$, Schrödinger's equation reads

$$-\frac{\hbar^2}{2m}\nabla^2\Psi(\mathbf{r}, t) + U(\mathbf{r})\Psi(\mathbf{r}, t) = i\hbar\frac{\partial\Psi(\mathbf{r}, t)}{\partial t} \tag{3.1}$$

Here, $\hbar = h/(2\pi)$, where h is Planck's constant, a fundamental constant that in a loose sense governs the scale of quantum randomness. m is the mass of the particle. The Laplacian operator ∇^2 is given in Cartesian coordinates by

$$\nabla^2 = \frac{\partial^2}{\partial x^2} + \frac{\partial^2}{\partial y^2} + \frac{\partial^2}{\partial z^2} \tag{3.2}$$

Finally, $U(\mathbf{r})$ is the potential energy function, or just the *potential*, which gives the potential energy of the particle as a function of its position within the field. For example, if the field were gravity, we would have $U(\mathbf{r}) = mgz$.

Note that the Schrödinger equation is a differential one that is first order in time and second order in position. To solve it, we use the separation-of-variables technique, writing the wavefunction as the product of spatially dependent and time-dependent functions,

$$\Psi(\mathbf{r}, t) = \psi(\mathbf{r})T(t) \tag{3.3}$$

With this substitution, we have

$$-\frac{\hbar^2}{2m}T(t)\nabla^2\psi(\mathbf{r}) + U(\mathbf{r})\psi(\mathbf{r})T(t) = i\hbar\psi(\mathbf{r})\frac{\partial T(t)}{\partial t} \tag{3.4}$$

By rearranging so that all spatial terms are on the left and all time-dependent terms on the right, we obtain

$$-\frac{\hbar^2}{2m}\frac{1}{\psi(\mathbf{r})}\nabla^2\psi(\mathbf{r}) + U(\mathbf{r}) = i\hbar\frac{1}{T(t)}\frac{\partial T(t)}{\partial t} \tag{3.5}$$

This equation can be satisfied only if both sides equal the same time- and position-independent constant. Let this constant be E. It turns out that the constant is indeed the energy of the particle, which is the reason for our choice of notation. Setting the LHS of this equation equal to the constant, and performing some rearrangement, yields

$$-\frac{\hbar^2}{2m}\nabla^2\psi(\mathbf{r}) + U(\mathbf{r})\psi(\mathbf{r}) = E\psi(\mathbf{r}) \tag{3.6}$$

This particular equation is called the *time-independent Schrödinger equation*. It gives the part of the Schrödinger equation that depends only on position. It is common to rewrite the LHS as an operator called the *Hamiltonian*, *H*, such that

$$H\psi(\mathbf{r}) = E\psi(\mathbf{r}) \qquad \text{where } H = -\frac{\hbar^2}{2m}\nabla^2 + U(\mathbf{r}) \qquad (3.7)$$

When applied to the wavefunction, the Hamiltonian operator returns the same wavefunction multiplied by the scalar energy. This is a kind of eigenvalue problem; an operator applied to a function yields a scalar value times that function. Such problems give rise to a discrete set of solutions, a set of eigenfunctions possible for ψ and eigenvalues possible for *E*, but not a continuous range of values. In other words, for a given potential *U* and hence a specific Hamiltonian *H*, the energy of a particle can adopt only certain discrete values.

This is an important physical aspect of quantum mechanics – energies are discretized. A particle cannot take on any arbitrary energy as is possible in Newtonian physics. Instead, its energy must derive from the spectrum of discrete eigenvalues. In fact, the observation of discrete energetics in the study of radiation and atomic spectra was one of the initial motivations that led to the discovery of quantum mechanics.

You might wonder how the Schrödinger equation was derived. In fact, as a truly fundamental equation, it cannot be derived, just as Newton's laws could not be derived upon their discovery. We must accept it on the basis of the copious predictions it makes that have remarkable agreement with experiments. The question of how it was discovered, however, is more tractable. In brief, the discovery grew from the early postulate by de Broglie that particles have wavelike properties akin to those of photons, and from Schrödinger's subsequent attempts to make analogies with the then-established equations of wave mechanics.

In this book, we will not solve many specific instances of the Schrödinger equation in detail. For this task, the reader is referred to any of a number of physical chemistry texts, several of which are recommended in the references. However, we present without derivation the results for two simple examples that illustrate the energy eigenvalues resulting from the application of particular potential fields.

A single particle inside a box with side lengths. Here, there are no interactions and field other than confinement, so $U = 0$ everywhere inside the box and $U = \infty$ outside. The box boundaries demand that $\psi = 0$ beyond them where the particle must have zero probability, which determines the boundary conditions for the second-order Schrödinger equation. One finds that the energy spectrum is

$$E = \frac{h^2}{8mL^2}\left(n_x^2 + n_y^2 + n_z^2\right) \qquad (3.8)$$

with $n_x = 1, 2, 3, \ldots$; $n_y = 1, 2, 3, \ldots$; $n_z = 1, 2, 3\ldots$, where *m* is the mass of the particle and the *n* values are positive integers.

A single electron orbiting a proton fixed at the origin. Here, the potential energy is due to Coulombic interactions between the electron and proton, $U(\mathbf{r}) = -e^2/4\pi\epsilon_0|\mathbf{r}|$. One finds that the energy eigenvalues are

$$E = -\frac{m_r e^4}{8\epsilon_0^2 h^2 n^2} \quad \text{with } m_r = \frac{m_p m_e}{m_p + m_e} \tag{3.9}$$

and $n = 1,2,3,\ldots$, where e is the elementary charge, ϵ_0 is the vacuum permittivity, m_p and m_e are the masses of the proton and electron, and m_r is the so-called reduced mass. Importantly, the eigenfunction solutions for ψ given by the Schrödinger equation give rise to the atomic orbitals (s, p, d, etc.). Here, part of what is discrete is the distinction between the orbitals themselves. In fact, this simple model reproduces many properties of the hydrogen atom and was one of the early successes of quantum mechanics. It turns out that it is impossible to solve the Schrödinger equation analytically for atoms with more than one electron. Instead, approximate or numerical methods must be used.

Our main point is that quantum mechanics produces a set of discrete energy levels for any given system, frequently termed the *energy spectrum*. As in the cases illustrated above, the eigenvalues involve a set of integer variables that are indexed, for example, by n. The set of these integers for a given system is called the set of *quantum numbers* and they determine the quantum state of a system.

These examples considered only a single particle, but what about many-particle systems typical of the ones we will want to study at equilibrium? The time-independent Schrödinger equation for a system of N fundamental particles can be written as

$$H\psi(\mathbf{r}_1, \mathbf{r}_2,\ldots,\mathbf{r}_N) = E\psi(\mathbf{r}_1,\mathbf{r}_2,\ldots,\mathbf{r}_N) \tag{3.10}$$

where

$$H = -\sum_i \frac{\hbar^2}{2m_i} \nabla_i^2 + U(\mathbf{r}_1, \mathbf{r}_2, \ldots, \mathbf{r}_N)$$

Notice that the complexity of the wavefunction increases substantially, incurring three additional independent variables for each new particle. The potential also depends on all of these variables. A system of molecules can be viewed, in part, as a collection of charged electrons and protons that interact electrostatically. For that case, the potential energy involves Coulomb's law for each pair of particles,

$$U(\mathbf{r}^N) = \frac{e^2}{4\pi\epsilon_0} \sum_{i<j}^{N} \frac{q_i q_j}{r_{ij}} \tag{3.11}$$

where q_i and q_j are the charges of particles i and j (-1 for electrons, $+1$ for protons), and r_{ij} is the distance between them. The sum is taken over all particle pairs denoted as $i < j$. We also abbreviated the collection of particle coordinates as $\mathbf{r}^N = (\mathbf{r}_1, \mathbf{r}_2,\ldots,\mathbf{r}_N) = (x_1, y_1,\ldots,z_N)$. We will use this notation throughout this book.

Clearly, solving the Schrödinger equation for molecular systems of even tens of atoms becomes a very challenging task. In such cases, the wavefunction will be of high dimensionality, with a large number of independent position variables, and the Schrödinger equation will involve many second derivatives. There are sophisticated numerical techniques to approach this problem, and such first-principles computational strategies are indeed widely pursued. We will not address such complications further, but now proceed to discuss approximate classical models that are easier to solve and in many cases still provide an adequate picture.

3.2   The classical picture

Do we need to solve the Schrödinger equation every time we want to understand the behavior of a system? Clearly the answer is "no," since we can accurately tackle many problems in physics, such as the trajectories of large projectiles, without quantum-mechanical ideas. We also do not need quantum mechanics in order to understand many molecular-level interactions. Why?

Quantum effects are dominant at very small scales, typically ångström units (Å) or less. By quantum effects, we mean that the probability distribution implied by the wavefunction is not sharply peaked, and uncertainties in the positions and velocities of the atoms are significant. Moreover, quantum behavior is often not as important at moderate to high temperatures because, above absolute zero, systems sit at energies that are significantly higher than their ground-state values. At these elevated energies, the energy eigenvalue spectrum becomes dense such that it appears effectively continuous rather than discrete. Figure 3.1 shows the spectrum for a particle in a box, where it is clear that the energy levels begin to look continuous at elevated energies.

In addition, for systems in which electrons are *localized*, to a good approximation we often do not need quantum theory to describe atom–atom interactions. A quantum treatment, however, is always necessary when electron clouds are nonlocalized, as in metals, or when there is bond formation or bond breaking. That is, classical models cannot capture chemical reactions at the molecular level.

When quantum effects are not important, we can model the world to a good approximation using a classical description. Here, the term "classical" in the sense of a system is distinct from that of "classical thermodynamics." Classical systems are composed of particles with definite positions and momenta, and their time evolution is governed by deterministic laws. Indeed, the classical picture forms the basis of

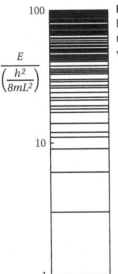

Figure 3.1. The energy eigenvalue spectrum for the classic particle-in-a-box problem. Each horizontal line gives an energy eigenvalue. Notice that the energy levels appear more closely spaced together when viewed at larger energy scales.

mechanics in physics. At the molecular level, a classical description has the following characteristics. The fundamental particle is the atom, and subatomic particles are not considered. Each atom has a definite position $\mathbf{r}$ and momentum $\mathbf{p}$. We could consider the velocity $\mathbf{v}$ instead of the momentum, since the two simply differ by the mass of the atom, but the momentum is actually a more natural choice in statistical mechanics.

Interactions between atoms are described by a *classical potential energy function* that depends on the positions but not the momenta of all of the atoms, $U(\mathbf{r}_1, \mathbf{r}_2, \ldots) = U(\mathbf{r}^N)$. Conceptually, the potential is an approximation to the lowest, ground-state quantum energy that would be found by solving the Schrödinger equation for the electron wavefunction given fixed positions of the nuclei, using Eqns. (3.10) and (3.11). Importantly, the gradient of the potential energy gives rise to forces on each atom i,

$$\mathbf{f}_i = -\nabla_i U\left(\mathbf{r}^N\right)$$

$$= -\frac{\partial U(\mathbf{r}^N)}{\partial \mathbf{r}_i}$$

$$= \left(-\frac{\partial U(\mathbf{r}^N)}{\partial x_i}, \; -\frac{\partial U(\mathbf{r}^N)}{\partial y_i}, \; -\frac{\partial U(\mathbf{r}^N)}{\partial z_i}\right) \tag{3.12}$$

where the gradient is expressed in several equivalent ways. For this reason, the classical potential is often also called the *force field*.

Given the position-dependent forces, the time evolution of the system is deterministic and described by Newton's second law,

$$m_i \frac{\partial^2 \mathbf{r}_i}{\partial t^2} = \mathbf{f}_i \tag{3.13}$$

for all atoms i. Alternatively, using Eqn. (3.12),

$$\frac{\partial \mathbf{p}_i}{\partial t} = -\frac{\partial U(\mathbf{r}^N)}{\partial \mathbf{r}_i} \tag{3.14}$$

When an atomic system evolves according to Newton's laws, its total energy E is given by a *classical Hamiltonian*, in analogy with the quantum Hamiltonian operator. This is simply a sum of the kinetic and potential energies of all atoms,

$$H\left(\mathbf{p}^N, \mathbf{r}^N\right) = U\left(\mathbf{r}^N\right) + K\left(\mathbf{p}^N\right)$$

$$= U\left(\mathbf{r}^N\right) + \sum_i \frac{\mathbf{p}_i^2}{2m_i} \tag{3.15}$$

Note that H depends on the entire set of positions and momenta of the atoms. The Hamiltonian adopts a constant value during the system's time evolution. This is a statement of the conservation of energy due to Newton's laws. We can prove that H is independent of time using the laws of motion,

$$\frac{dH}{dt} = \sum_i \frac{\partial U}{\partial \mathbf{r}_i}\frac{d\mathbf{r}_i}{dt} + \sum_i \frac{d}{dt}\left(\frac{\mathbf{p}_i^2}{2m_i}\right)$$

$$= \sum_i \left(\frac{\partial U}{\partial \mathbf{r}_i}\frac{d\mathbf{r}_i}{dt} + \frac{\mathbf{p}_i}{m_i}\frac{d\mathbf{p}_i}{dt}\right)$$

$$= \sum_i \left(\frac{dU}{d\mathbf{r}_i}\frac{d\mathbf{r}_i}{dt} + \frac{d\mathbf{r}_i}{dt}\frac{d\mathbf{p}_i}{dt}\right) \quad \text{using} \quad \mathbf{p}_i = m_i\frac{d\mathbf{r}_i}{dt}$$

$$= \sum_i \frac{d\mathbf{r}_i}{dt}\left(\frac{\partial U}{\partial \mathbf{r}_i} + \frac{d\mathbf{p}_i}{dt}\right)$$

$$= \sum_i \frac{d\mathbf{r}_i}{dt}\cdot 0 = 0 \quad \text{using} \quad \frac{d\mathbf{p}_i}{dt} = -\frac{\partial U}{\partial \mathbf{r}_i} \tag{3.16}$$

The most important aspect of the classical description is the potential energy U. This function takes the set of all atomic positions and returns an energy due to the interactions. The success of the classical approximation hinges on finding an accurate form for U. Often this is achieved by examining the various modes by which atoms can interact according to the Schrödinger equation, and patching simple, perhaps "first-order," theoretical expressions for energies of these modes together.

It is often convenient to classify components of U into two main categories: energies due to bonded and nonbonded interactions. This division itself is an approximation, since the Schrödinger equation makes no such distinction. It is, however, a convenient way to think about contributions to interaction energies. A schematic representation of the different contributions is given in Fig. 3.2. The bonded interactions typically describe the change in potential energies due to the following effects.

Bond stretching. A common functionality is $u = a(d - d_0)^2$, where d is the length of the bond and a and d_0 are constants.

Bond-angle bending. Deviations from a preferred hybridization geometry (e.g., sp^3) should incur energetic penalties. A common form is $u = b(\theta - \theta_0)^2$, where θ is the bond angle formed by three atoms and b and θ_0 are constants.

Bond torsions. These interactions occur among four atoms and account for the energies of rotations along a central bond. A common approximation is a cosine expansion, $u = \sum_n c_n \cos(\omega)^n$, where ω is the torsional angle, n is a summation index, and c_n are summation coefficients.

On the other hand, the nonbonded interactions apply to atoms that are not connected by bonds, either within the same molecule or in two different molecules. The most frequent kinds of nonbonded energies are the following.

Electrostatics. In the classical approximation atoms can have charges, which may be formal (integer) charges due to ionized groups or partial (fractional) ones due to polarization among atoms with different electronegativities. Atoms with charges interact through Coulomb's law, $u = q_i q_j/(4\pi\epsilon_0 r_{ij})$ for atoms i and j separated by distance r_{ij}. This interaction is *pairwise* between atoms.

Van der Waals attractions. Correlations between the instantaneous electron densities surrounding two atoms gives rise to an attractive energy. This is a general attractive

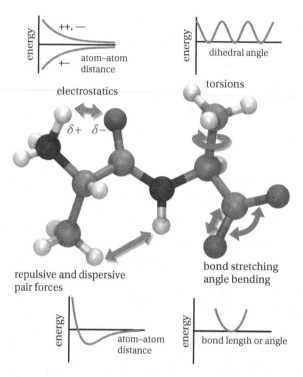

Figure 3.2. Typical components of a classical model of molecular interactions, including both bonded and nonbonded terms. The graphs give a schematic representation of the dependence of the component energies on distances and angles. Loosely adapted from Boas and Harbury, *Curr. Opinion Struct. Biol.* **17**, 199 (2007).

force between all atoms. Solving the Schrödinger equation for models of this effect shows that the attraction has the functional form $u = \text{constant} \times r_{ij}^{-6}$, where r_{ij} is the distance between the two atoms and the constant depends on their identities (the elements and chemical environments). The r^{-6} model is pairwise in nature. The van der Waals forces are also called *London (dispersion) forces*.

Short-range repulsions. When two atoms make a close approach, they experience a steep increase in energy and a correspondingly strong repulsion. This occurs because the electron clouds of the atoms begin to overlap, and the Pauli exclusion principle requires electrons to move to higher energy states. The functional form for this repulsion that comes from theory is not as simple as that for van der Waals forces. It is often successfully approximated as a simple power law, $u = \text{constant} \times r_{ij}^{-m}$, where m is greater than 6.

All of these nonbonded components assume that energies can be modeled as *pairwise*, that is, by directly considering pairs of atoms and their mutual distances. Indeed, this too is yet another approximation because nearby neighboring atoms can affect the direct interaction energy of a given atom pair. In many systems, however, the pairwise approximation is often reasonably successful. In particular, a longstanding way to model both van der Waals and repulsive forces is the combined *Lennard-Jones* interaction,

$$u = 4\epsilon \left[\left(\frac{r_{ij}}{\sigma} \right)^{-12} - \left(\frac{r_{ij}}{\sigma} \right)^{-6} \right] \tag{3.17}$$

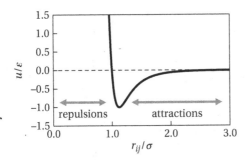

Figure 3.3. The Lennard-Jones interaction potential, as in Eqn. (3.17). The energy between an atom pair, u, is expressed in terms of the Lennard-Jones coefficient ϵ; similarly, the pair distance is in terms of the size parameter σ. The simple form includes both attractive van der Waals forces at large interatomic separation distances and repulsive ones at short distances.

where ϵ and σ are constants that depend on the particular types of atoms i and j. The prefactor of 4 ensures that the minimum value of the potential is $-\epsilon$. Here, the attractive force enters with the r_{ij}^{-6} term, while the repulsive energy is given by r_{ij}^{-12}. The rationale for this particular repulsive form is merely convenience and simplicity, since it is just the square of the van der Waals term; it has no deep theoretical backing. Figure 3.3 depicts the potential.

Putting all of these considerations together, we arrive at a classical picture of molecular systems described by a minimalist potential energy function, as depicted in Fig. 3.4. Notice that the potential energy depends on all of the atomic positions because the bond distances d_i, angles θ_j, torsions ω_k, and pairwise distances r_{ij} are functions of the coordinates $\mathbf{r}^N$.

How are all of the parameters in this classical model for U determined? In quantum mechanics, one needs to know only the atom types, giving the numbers and masses of electrons and nuclear particles. There are no parameters to fit in quantum theory. Since the classical potential is an approximation to the true quantum-mechanical behavior, in principle its parameters could be derived by solving the Schrödinger equation for a multi-atom system in many different atomic configurations, and then performing some kind of best-fit procedure to match energies or other properties. In practice, the classical parameters are often fitted both to detailed numerical solutions of the Schrödinger equation on small, computationally tractable molecules – which itself requires many approximations for multiple-electron systems – and to experimental spectroscopic and thermodynamic measurements.

3.3 Classical microstates illustrated with the ideal gas

Armed with a classical picture of matter, we now show that the ideal gas law can be derived from very basic considerations of the form of the density of states. An ideal monatomic gas consists of particles that are point masses and do not interact with each other; therefore, there are no intermolecular potential energies and the total energy

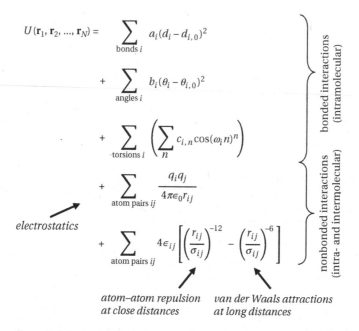

The following appears to the right of and around the equation:

bonded interactions (intramolecular)

nonbonded interactions (intra- and intermolecular)

$$U(\mathbf{r}_1, \mathbf{r}_2, ..., \mathbf{r}_N) = \sum_{\text{bonds } i} a_i(d_i - d_{i,0})^2$$

$$+ \sum_{\text{angles } i} b_i(\theta_i - \theta_{i,0})^2$$

$$+ \sum_{\text{torsions } i} \left(\sum_n c_{i,n} \cos(\omega_i n)^n \right)$$

$$+ \sum_{\text{atom pairs } ij} \frac{q_i q_j}{4\pi\epsilon_0 r_{ij}}$$

electrostatics

$$+ \sum_{\text{atom pairs } ij} 4\epsilon_{ij} \left[\left(\frac{r_{ij}}{\sigma_{ij}}\right)^{-12} - \left(\frac{r_{ij}}{\sigma_{ij}}\right)^{-6} \right]$$

atom–atom repulsion at close distances   van der Waals attractions at long distances

Figure 3.4. A simple classical potential energy function for describing molecular systems. The variables a_i, $d_{i,0}$, b_i, $\theta_{i,0}$, $c_{i,n}$, q_i, ϵ_{ij}, and σ_{ij} are all parameters fitted to reproduce experimental data or actual calculated quantum-mechanical energies on small molecular fragments. The variables d_i, θ_i, ω_i, and r_{ij} are the bond lengths, bond angles, torsion angles, and atom-pair distances, respectively. These all depend on the specific configuration of atoms at any moment in time, $\mathbf{r}^N$.

of the gas E simply equals the molecules' kinetic energy K. For a given set of values E, V, and N, we desire to compute $\Omega(E, V, N)$, the number of compatible microstates or microscopic configurations of the system. This requires us to count the number of ways there are to choose the positions and momenta of N ideal gas molecules in a volume V such that the total kinetic energy is the given value E.

The fact that the gas is ideal enables an important simplification: the number of ways of picking the momenta is independent of the number of ways of picking the positions. This occurs because the choice of positions does not affect E, since there are no intermolecular energies. If the molecules did interact, then the choice of positions would determine a potential energy U; the choice of momenta would then be constrained to satisfy a kinetic energy that gave $K = E - U$. Here, because the total energy is always equal to the kinetic energy regardless of positions, we can write

$$\Omega(E, V, N) = \Omega_{\text{momenta}}(E, N) \times \Omega_{\text{positions}}(V, N) \tag{3.18}$$

where $\Omega_{\text{momenta}}(E, N)$ counts the ways of assigning momenta to N molecules in volume V with $K = E$. Note that Ω_{momenta} does not depend on the volume of the container. While the available positions for each molecule depend on the location of the system boundaries, the momenta have no such constraint.

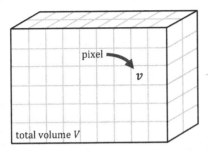

Figure 3.5. Discretization of the continuous space of a classical system. By subdividing the overall system volume V into many tiny pixels of volume v, we are able to count microstates. The continuum limit is recovered as we let $v \to 0$.

Let us focus on $\Omega_{\text{positions}}(V, N)$, the number of ways of placing N ideal gas molecules in a volume V. In order to count configurations, we must discretize space (Fig. 3.5). We slice up the volume V into tiny cubes, each with a minuscule volume v. You can think of these cubes as discrete pixels that make up the total volume. How many ways are there to arrange N molecules in the pixels? We first assume that only a single particle can occupy each pixel. This may seem inconsistent because ideal gases are point masses and have no volume. However, as we ultimately let $v \to 0$, the pixels will become small enough that effectively each particle occupies its own space, neglecting the infinitesimally small probability that two or more have exactly the same position. We will also assume that the number of particles is far less than the number of pixels, such that we can ignore the already-occupied pixels when counting the spots available to a molecule. As a result, there are essentially V/v pixels in which to place each particle. Since the particles can be placed independently, the total number of positional configurations becomes

$$\Omega_{\text{positions}}(V, N) = \left(\frac{V}{v}\right)^N \tag{3.19}$$

This expression inherently considers particles to be distinguishable since the counting includes all of the configurations that differ only by swapping the identities of particles (e.g., switching particles 1 and 2 between their lattice sites). The number of ways to permute the identities of particles that are all in separate pixels is $N!$, so we correct for our overcounting using this factor,

$$\Omega_{\text{positions}}(V, N) = \frac{1}{N!} \left(\frac{V}{v}\right)^N \tag{3.20}$$

To simplify the factorial term, we can use Stirling's approximation, $N! \approx (N/e)^N$, which will be accurate for large N. These considerations are enough for us to derive the equation of state of the gas. With Stirling's approximation and $S = k_B \ln \Omega$, we find that

$$S = k_B \ln \Omega_{\text{momenta}}(E, N) + N k_B \ln(V/N) + N k_B \ln(e/v) \tag{3.21}$$

Now, taking the partial derivative with respect to volume, we have

$$\frac{P}{T} = \left(\frac{\partial S}{\partial V}\right)_{E,N}$$
$$= \frac{Nk_B}{V} = \frac{nR}{V} \qquad (3.22)$$

which gives the ideal gas law. Here, $n = N/\mathcal{N}_A$ is the number of moles and R is the gas constant. Notice that the momentum component of the density of states drops out completely since it is independent of volume. In addition, the size of the pixel v becomes irrelevant to the equation of state after all. It vanishes when the derivative is taken. It turns out that, classically, there is no way to define a physically meaningful size of volume elements like this, which ultimately means that there is no way to define an absolute entropy. This might be evident in the last term in (3.21). However, the world is not classical but quantum in nature, and thus there is something of a natural "pixel" size that emerges from Heisenberg uncertainty and depends on the value of Planck's constant. In quantum-mechanical systems, therefore, absolute entropy exists. We will address this topic in more detail in Chapter 15.

3.4   Ranges of microscopic interactions and scaling with system size

Two features of nearly all molecular interactions result in important behaviors as the size of a system is scaled up. First, atoms have a *repulsive core*, which disfavors their overlapping. Second, interactions are frequently *short-ranged* in nature, meaning that two molecules separated by several molecular diameters typically interact very weakly and thus do not really "feel" each other's presence. These features figure naturally from the form of the interactions described above. Consider, for example, the Lennard-Jones potential in Fig. 3.3; this pairwise energy increases sharply below $r = \sigma$, but also decays to nearly zero beyond $r \approx 3\sigma$.

In a condensed phase such as a liquid, it is reasonable to think that a single "central" molecule interacts only with others that lie within a distance δ surrounding it, where δ is the range of the interaction (Fig. 3.6). The existence of this molecularly sized length scale means that the energy of a homogeneous system of molecules scales linearly with the size of the system. In other words, the energy is *extensive*. Each molecule interacts with a small number of neighbors within the range δ, which is insensitive to the size of the system. Thus, a molecule's contribution to the total potential energy does not depend on how many particles there are overall, but rather is of the order of a modest number of interactions. Upon scaling up the system size, increasing both V and N by the same factor, the total energy should thus also scale with N. Alternatively, if each molecule were to interact substantially with every other, the energy would scale like N^2, with the number of pairs. These are of course merely scaling arguments and not predictions of exact energies, which depend on the detailed molecular configurations.

A further effect of short-ranged interactions is that the energetics corresponding to surfaces and boundaries in macroscopic systems are often negligible relative to the total

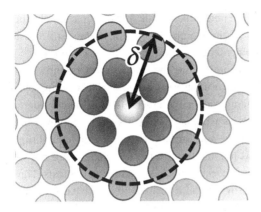

Figure 3.6. An illustration of the range of interaction for a central particle and neighboring ones in a bulk, liquid-like phase. The variable δ is the distance over which interactions are significant; beyond it, pair energies with other molecules are nearly negligible.

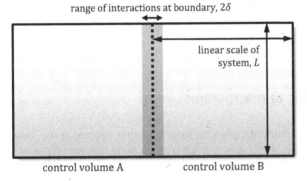

Figure 3.7. A schematic representation of two systems or control volumes that interact at a boundary. The energies within each system scale as L^3 but the boundary interactions scale as δL^2.

energy. Consider two systems in cubic boxes interacting along one boundary, as seen in Fig. 3.7. We assume their contents are homogeneous and at some fixed overall density. The number of molecules that interact across the interface is proportional to the boundary area L^2 times the range of interaction 2δ. Thus the interaction energy across the boundary scales as $E_{AB} \propto 2\delta L^2$. On the other hand, the total number of molecules in system A is proportional to the volume of A. Per the extensivity arguments put forth above, the interaction energy of all molecules *within* A scales as $E_A \propto L^3$. Similarly, $E_B \propto L^3$. The total energy of the supersystem is the sum of these terms, $E = E_A + E_B + E_{AB}$.

This scaling analysis shows that as the size of the systems grows (as L is increased while maintaining the density), the boundary energy becomes insignificant relative to the total energy. Namely, E_{AB} scales as L^2, whereas E_A and E_B scale as L^3. This is a further manifestation of extensivity: if we grow a system by doubling it, such that we effectively have two copies that then interact, we essentially also double the total internal energy.

Because boundary interactions are negligible in bulk macroscopic systems of molecules, it also turns out that their equilibrium properties are insensitive to the shape of the containers that they inhabit, being dependent only on the total volume.

Of course, if the properties of interest are the interfaces or boundaries themselves, if the systems are not macroscopic, or if these interactions play a strong role in driving a particular physical behavior, then interfacial interactions must be considered explicitly. Thermodynamics allows for this as well, and we will see in Chapter 5 that there are ways to incorporate surface areas and surface tensions into equilibrium analyses. There are also special situations in which molecular interactions are not short-ranged, but we leave these topics to more specialized statistical mechanics texts. The main point of these considerations is to provide some justification and basis for the existence of extensive scaling behavior in molecular systems.

> Molecular interactions are often effectively **short-ranged**, which means that boundary and interfacial interactions are negligible in macroscopic systems relative to the total internal energy, that the internal energy scales linearly with system size, and that the properties of a system are insensitive to its boundary shape.

3.5   From microscopic to macroscopic

We just described fundamental physical principles that operate at the molecular level. We now arrive at an important question: what about temperature? Where in this classical description of atomic processes does temperature come into play? Is it the velocities? Is temperature another parameter that we forgot to include, or is it somehow missing from the Schrödinger equation?

Here is the big conceptual leap. Temperature is not a fundamental variable in microscopic physics. That is, molecules do not individually have a temperature. Instead, temperature is a property that *emerges* when we put large numbers of molecules together. Similarly, entropy and heat are properties that emerge in large molecular systems. For simplicity, let us elaborate by considering a classical system of molecules, although it should be noted that these ideas readily extend to the quantum world.

Why do we acquire new variables like temperature when we assemble macroscopic amounts of molecules? The reason stems from the fact that we do not care about the microscopic details of most systems, i.e., the exact positions and velocities of each and every atom. We usually care only about a few bulk properties that we measure at the macroscopic level, like density or pressure. When we give up knowing all of the details, in order to make predictions about a system we must pay for this loss of information by instead using quantities like temperature, heat, and entropy.

For example, imagine a bottle of water whose walls are rigid and insulating. If we were to fully characterize this system, we would need to know information about the position and momentum of every atom. At any instant in time, there are $3N$ position variables $+ 3N$ momentum variables $= 6N$ pieces of information, where N is the number of atoms. These $6N$ variables completely specify the state of the system at the microscopic level. Namely, they form a complete set of information required to predict the time evolution according to Newton's equations. That is, we need $6N$ initial conditions for the $3N$ coupled, second-order differential equations of (3.14).

> At a classical level, the **microscopic state** of a system at any instant in time is characterized by a list of the $3N$ positions $\mathbf{r}^N$ and $3N$ momenta $\mathbf{p}^N$, for a total of $6N$ pieces of information. Collectively, the positions and momenta comprise the **microscopic degrees of freedom** of the system.

A particular set of values for the degrees of freedom – a list of specific numbers for each and every position and momentum – could be considered one particular configuration of the system, as if we suddenly stopped time and were able to zoom in and study every water molecule. We call such a configuration, or one possible snapshot in time, a *microstate*.

> A **microstate** is simply one "configuration" of the system, the specification of values for all microscopic degrees of freedom.

How much information is in a microstate? For a one-liter bottle of water, there are about 3×10^{25} molecules, which implies $\sim 5.9 \times 10^{26}$ degrees of freedom. Thus if we wanted to take a snapshot of a microstate and store all of these variable values to second-order precision, we would need nearly 10^{16} 100-GB hard drives. That is a *lot* of information.

Usually for large systems, we are not interested in such detail. We live at a macroscopic level and the kinds of measurements that we make, both as empiricists using lab equipment and as cognizant beings using our blunt senses, typically detect only *macroscopic* changes in the bottle of water. That is, rarely does it matter to us where each molecule is at every moment in time. (In principle, we cannot even measure this information according to Heisenberg uncertainty.) Instead, we are more interested in the average effects of multitudes of water molecules exerted over periods, not instants, of time.

Moreover, we frequently care about equilibrium properties that do not change with time. For one, we cannot make an accurate measurement if the system fluctuates too significantly. Some properties rigorously do not change with time, especially for isolated systems. For the hypothetical bottle of water, the value N and hence the number of degrees of freedom is constant since no water molecules enter or leave the bottle. Provided that no interactions persist through the bottle walls, Newtonian mechanics states that the total internal energy of the system E is rigorously constant. Moreover, given the rigidity of the container, the volume V is constant as well. This example shows that three variables in isolated systems are strictly constant: E, V, and N. These should look familiar since they are the independent variables of the entropy function. Indeed, their constancy is what makes them important to the entropy.

There are also macroscopic properties for which, although they are not rigorously constant, the variations in them are so small that they go unnoticed. These variations are called *fluctuations*, and they tend to be extremely small relative to macroscopic scales. One example might be the pressure, which in our example is due to the net force exerted by water molecules on atoms in the bottle wall. The force of course stems from the water–wall interactions in the potential energy function. Because water molecules are constantly bouncing about, coming near to and then away from the wall atoms, the forces they exert change with time. However, there are so many water molecules abutting the wall at any one moment that these force fluctuations are averaged out and minuscule relative to their net effect. Thus, the pressure appears constant from a macroscopic point of view.

Instead of characterizing a system using detailed information about molecular configurations, we tend to use constant or apparently constant bulk equilibrium properties. In other words, we tend to characterize the state of a system at a *macroscopic* level. When we label a system's conditions using macroscopic descriptors, we are specifying the macroscopic state or macrostate of the system. In isolated systems, it turns out that the macrostate is naturally specified by the variables that are rigorously constant.

A **macrostate** is a set of bulk descriptors characterizing the current equilibrium state of a system. Quantities like the number of molecules N, the internal energy E, the volume V, the temperature T, and the pressure P are all macroscopic variables that can define a macrostate.

For a single-component isolated system, the macrostate is completely specified by the variables E, V, and N, which remain rigorously constant throughout molecular motion in a classical sense. T and P are redundant here, in that it is impossible to separately choose their values once E, V, and N have been specified. By extension, multicomponent systems require the specification of all molecular numbers N_i.

Because a macrostate specifies only bulk properties, it says nothing about what is happening at the molecular level. This is the loss of information that occurs when we take a macroscopic perspective. Indeed, a macrostate does not correspond to a single microstate or single atomic configuration. Rather, a system can evolve through many different microstates while its macrostate (i.e., the values of N, E, V, T, and P or any other thermodynamic state property) remains constant.

What is the relationship between microstates and macrostates? Macrostates are collections of microstates, called *ensembles* in statistical mechanics. If we are able to characterize the current equilibrium state of a system using particular values of macroscopic properties like E, V, and N, then the specific microstates through which the system evolves must produce those average properties. If the system is isolated, only those microstates whose total energy equals E will be visited by the system since the energy is rigorously constant. Any other microstates will not be visited. Another way of thinking about this is the following: if we initialize an isolated system in one particular microstate with a particular total energy, we know that with time it will evolve only through other microstates of the same energy.

In a general sense, ensembles are essentially collections of microstates that are compatible with the macrostate. Later on we will find out that ensembles do not always filter microstates in a binary manner (visited versus not visited), but can be probabilistic in nature, meaning that microstates are assigned weights in the ensemble.

> **Ensembles** are collections of microstates that are compatible with a given macrostate. Ensembles can be viewed as collections of weights assigned to each possible microstate. The weights give the contributions of those microstates to equilibrium properties.

Let us summarize what we have discussed so far. We usually look at the world in terms of macrostates. We characterize the state of a system using a small number of variables like E, V, or N, rather than by an exhaustive list of atomic positions and momenta. For a particular set of macrostate variables, there are many microstates that are compatible with it. There is one important piece of information that is missing here, and it underlies all of macroscopic thermodynamics. That information is exactly how ensembles of microstates are related to macroscopic properties. What determines which microstates are part of an ensemble and their weights in it? How do thermodynamic properties such as pressure and temperature emerge in this picture? We will explore these issues in depth in Chapter 4.

3.6 Simple and lattice molecular models

Before proceeding further, we introduce the notion of simple or "toy" models of molecular physics. Just as we saw that a classical picture of matter is a simplification and approximation of quantum behavior, it is common to develop even coarser models than those of the classical description. In fact, many successful statistical-mechanical theories rely on highly simplified models that do not attempt to resolve the physics quantitatively at the atomic level.

Simple models often treat molecules as sitting on fixed lattice sites in space. So-called *lattice models* are particularly convenient for statistical-mechanical theory because they make it easy to count microstates. Figure 3.8 shows a few examples of lattice models of polymers, gases, and solutions that have been important to statistical mechanics and will be described in later chapters.

In any lattice model, one must specify a potential energy function, in the same spirit as those used for classical systems. In these cases the potential is often highly simplified and reflects an average over the kinds of interactions present in the true atomic physics. For example, a simple energy function that reproduces the effect of van der Waals attractions is one that assigns neighboring molecule pairs on the lattice a fixed, negative energy. A constraint that no two particles can occupy the same lattice site simulates repulsive interactions upon the close approach of atoms.

Note that it is difficult to define velocities using lattice models, since particles cannot move continuously throughout space but rather are constrained to sit on a fixed set of

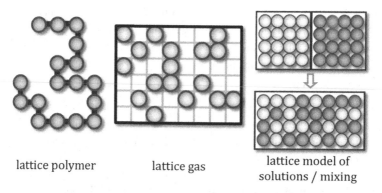

lattice polymer   lattice gas   lattice model of
                                              solutions / mixing

Figure 3.8. Several simple lattice models of different systems. In each, the positions of different coarse molecular groups are constrained to a square grid.

possible sites. As a result, velocities and kinetic energies are typically neglected in these approaches. It turns out that the contribution of kinetic degrees of freedom to the overall properties of a system can usually be solved analytically, so that the lattice models can be "patched" with this correction later on. We will discuss a natural approach to separating kinetic from configurational or positional contributions later in Chapter 18.

Why are such models successful if they neglect the detailed physics and are so coarse-grained? Should not the assumption of a lattice have a profound effect on the resulting properties of the system since it imposes certain structures and symmetries at the molecular level? Fortunately for us, many bulk physical phenomena are not sensitive to the fine details of molecular interactions but rather depend on very basic properties, such as the scales of repulsions and attractions, the range of interactions between atom pairs, or the nature of chain connectivity (e.g., in polymers). In particular, most of the basic trends of many phase transitions involving fluid states can be predicted from lattice-model arguments. In fact, at the liquid–vapor critical point, lattice models describe the properties of many systems *exactly* because they behave in a universal way near this part of the phase diagram. On the other hand, there are cases that may be poorly described using such simple approaches. These often involve the existence of phase transitions to *structured* phases, where precise molecular configurations dictate bulk properties. Examples include crystal phases and the self-organizing behavior of many biomolecular systems, such as proteins. Even so, there do exist simplified thermodynamic approaches and models for tackling such systems.

3.7   A simple and widely relevant example: the two-state system

We conclude with an example that illustrates the concepts of this chapter. The so-called two-state system is one of the simplest statistical mechanical models and describes a

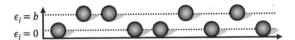

Figure 3.9. An illustration of the classic two-state system. Each particle i can exist in either a low-energy state with $\epsilon_i = 0$ or a high-energy one with $\epsilon_i = b$. Particles are independent of each other such that the total energy is given by $E = \sum_i \epsilon_i$.

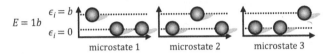

Figure 3.10. Several distinct microstates of the two-state model with the same total energy.

wide variety of physical phenomena at a basic level. It envisions a collection of atoms, particles, molecules, or groups thereof, in which any microscopic degree of freedom can exist in one of two configurations: a state of low energy 0 or one of high energy b. Let us consider a collection of structureless particles and denote the energy of particle i as ϵ_i. In this model particles are independent of each other; that is, the state of one particle does not affect the energy of the others. The total energy of the system is simply

$$E = \sum_i \epsilon_i \qquad (3.23)$$

A way to depict the two-state model visually is shown in Fig. 3.9. Here, the low- and high-energy states correspond to the particle being in a down or up position, respectively.

The two-state model captures systems with two physical ingredients: the existence of a set of degrees of freedom that can toggle between a low-energy and an excited high-energy state, and the independence of these degrees of freedom from each other. In reality, these features might emerge under conditions for which molecules can individually populate two different conformers, or can move between two different spatial regions, such as in different phases or at an interface versus in the bulk.

Consider the snapshots of a two-state system of three particles depicted in Fig. 3.10. It should be easy to see how terms defined in this chapter relate to the microscopic details of this system. A microstate is simply one of these configurations, such as microstate 1. On the other hand, a macrostate is the specification $E = 1b$, $N = 3$. The ensemble is the collection of microstates in a given macrostate. So, the ensemble for macrostate $E = 1b$ and $N = 3$ is the set of microstates 1, 2, and 3. Any other configurations have zero weight or probability in this ensemble since their energies or number of particles would not match the macrostate prescription. Finally, the density of states counts the number of compatible microstates, $\Omega(E = 1b, N = 3) = 3$.

Example 3.1 *Compute the density of states and the entropy S(E, N) for the two-state system, for any values of E and N. Simplify your result in the limit of large N.*

For a given value of E, we can find the number of particles in the up position, denoted by M,

$$M = E/b$$

We now must consider how many ways there are to pick which particles are up. This is a combinatorics problem of picking M objects from N, where the order in which we pick them does not matter. Thus, we use the well-known combinations formula,

$$\Omega = \frac{N!}{M!(N-M)!}$$

$$= \frac{N!}{(E/b)!(N-E/b)!}$$

For large values, it becomes possible to use Stirling's approximation to the factorial,

$$N! \approx N^N/e^N$$

or, alternatively,

$$\ln N! \approx N \ln N - N$$

This approximation is used universally in statistical mechanics, and it is extremely accurate due to the fact that N is typically on the order of 10^{23}. Very commonly, we need to apply Stirling's approximation to the combination formula as here. There is a simple way of writing the result,

$$\Omega = \frac{N!}{M!(N-M)!}$$

$$\approx \frac{e^M e^{N-M}}{e^N} \frac{N^N}{M^M (N-M)^{N-M}}$$

$$= \left[\left(\frac{M}{N}\right)^{M/N} \left(1 - \frac{M}{N}\right)^{1-M/N} \right]^{-N}$$

$$= \left[x^x (1-x)^{1-x} \right]^{-N}$$

where x is shorthand for M/N and thus varies between 0 and 1. Keep this formula in mind because it will reappear many times in a variety of statistical-mechanical models.

Applying Stirling's approximation to the two-state system and taking the logarithm, we can then compute the entropy,

$$S = k_B \ln \Omega$$

$$= N k_B \left[-\left(\frac{E}{Nb}\right) \ln\left(\frac{E}{Nb}\right) - \left(1 - \frac{E}{Nb}\right) \ln\left(1 - \frac{E}{Nb}\right) \right]$$

Notice that S is extensive; that is, $S(\lambda E, \lambda N) = \lambda S(E, N)$.

Problems

Conceptual and thought problems

3.1. Prove that Newton's laws for a classical atomic system also conserve net total momentum,

$$\mathbf{P} = \sum_i \mathbf{p}_i$$

3.2. Do atoms have a size? Is there such thing as a well-defined "diameter" of, say, a helium atom? Consider a classical model of helium in which two atoms interact solely through the Lennard-Jones potential. Explain how you might find an approximate, but perhaps not rigorous, definition of atomic diameter using this model.

3.3. Consider a system of liquid argon. At a classical level, we might model atomic interactions using a *pairwise* decomposition,

$$U(\mathbf{r}^N) \approx \sum_{i<j} u_{\mathrm{pair}}(r_{ij})$$

The Lennard-Jones potential, for example, could be used for u_{pair}. The pairwise approach is an approximation that considers only two-body interactions. When might this approach break down? Can you think of physical origins of three-body interactions that may be important? How do the numbers of pairwise and three-body interactions scale with the number of argon atoms?

3.4. Return to the ideal gas example presented in this chapter, in which we discretized space in order to count microstates. What happens if we explicitly account for the case in which multiple particles are able to occupy the same pixel? Keep in mind that the particles remain indistinguishable.

(a) Convince yourself that the number of unique microstates in this case is *not* given by $M^N/N!$, where $M = V/v$ is the number of pixels. Show that there are only ten microstates when $N = 2$ and $M = 4$, which does not equal $4^2/2!$. What makes this situation different?

(b) Show that, if multiple particles can occupy each site, the positional density of states is

$$\Omega_{\mathrm{positions}} = \frac{(N+M-1)!}{N!(M-1)!}$$

(c) Use Stirling's approximation to show that the limit of this expression as $v \to 0$ ($M \to \infty$) is in fact given by $M^N/N!$. You will need to use the limit that

$$\lim_{M\to\infty} (N/M + 1)^{N+M-1} = e^N$$

3.5. One property of the entropy is that it is extensive, $S(\lambda E, \lambda V, \lambda N) = \lambda S(E, V, N)$ for a single-component system. Why is this always the case? You can proceed to show that $\ln \Omega$ is extensive. A rough argument is to consider a large, macroscopic system of $\sigma(10^{23})$ molecules. For conceptual purposes, we will consider this system to be a

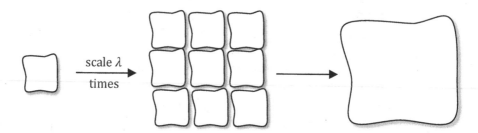

Figure 3.11. A schematic representation of scaling the size of a system for a cube of sugar.

cube of sugar. You then "scale" the system by copying it λ times over into a much larger cube, that is, by merging multiple cubes into one large one (Fig. 3.11). What can be said about the interfacial interactions between the different copies, relative to the total energy? Write down the final density of states in terms of the single-system one, and show that the resulting overall S is linear in λ. Do you expect the same, linear behavior for very, very small systems, with $o(1)$ molecules? Why, or why not?

3.6. Another property of the entropy is that it is a concave function of its arguments. Ultimately this property stems from the fact that an isolated macroscopic system with a *convex* entropy function will spontaneously split into two subsystems (each still macroscopic) so as to increase the number of possible microstates and in effect remove the convexity. Here, you will prove this in a general manner.

To keep things simple, ignore the volume contribution to the entropy. Consider the scenario in Fig. 3.12, where the dark lines in the bottom figures give an intensive, per-particle version of the entropy as a function of the per-particle energy, $s(e)$. According to this line, a system with total energy E_T and number of molecules N_T will have a total entropy $S_T = N_T s(E_T/N_T)$. Ostensibly the system has $\exp(S_T/k_B)$ microstates at those conditions.

Now consider the case in which the system splits into two subsystems, or phases, with E_1, N_1 and E_2, N_2 as shown on the RHS of Fig. 3.12. Note that $N_1 + N_2 = N_T$ and $E_1 + E_2 = E_T$ due to conservation principles. Since the subsystems are macroscopic and of the same substance, they are described by the same intensive entropy function, $s(e) = S(E, N)/N$. Make an argument that the system can actually reach a higher number of microstates and hence a higher entropy by splitting into the two subsystems. In other words, using its connection to the density of states, show that the total entropy of the two-subsystem scenario forms the thin tangent line in Fig. 3.12,

$$S/N = s_1 + \left(\frac{s_2 - s_1}{e_2 - e_1}\right)(e_T - e_1)$$

Thus, the entropy function must actually be augmented to follow the tangent line, removing its convex character, since the overall system always has states available to it where its constituent particles can self-segregate into two subsystems.

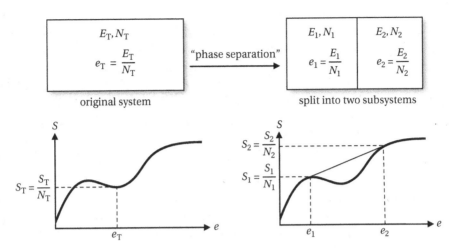

Figure 3.12. A system with a concave entropy function will spontaneously split or phase-separate into two separate subsystems in order to increase the overall number of microstates. The notations e and s indicate the per-particle internal energy and entropy, respectively.

Hint: first ask yourself what the total number of microstates is for the subsystem scenario, assuming that the interfacial interactions between the two subsystems are small.

3.7. An alternative and more general approach to the classical equations of motion is the Lagrangian formulation. The Lagrangian for a system is defined as the kinetic energy minus the potential energy, $L = K - U$. In Cartesian coordinates, it is expressed as a function of the atomic positions $\mathbf{r}^N$ and velocities $\mathbf{v}^N$, but more generally it can be written in terms of any arbitrary set of positions and velocities in any coordinate system. In such cases, the equation of motion for a generalized position coordinate q and its corresponding velocity $\dot{q}$ is given by the relation

$$\frac{d}{dt}\frac{\partial L}{\partial \dot{q}} = \frac{\partial L}{\partial q}$$

Show that this expression returns Newton's familiar second law for the Cartesian case.

3.8. We said that the entropy depends on E, V, and N because all three of these variables are rigorously constant for an isolated system. That is, given that they have initial values at some time $t = 0$, they can never change as the system approaches equilibrium so long as it remains isolated. Can you think of other variables that similarly, rigorously remain constant and thus, in principle, should contribute to the entropy function?

Fundamentals problems

3.9. Consider a particle that lives in an isolated one-dimensional world and is tethered to the origin by a harmonic spring. The total energy of the particle at any point in time is due to its kinetic plus potential energy,

$$E = K + U = \frac{p^2}{2m} + \frac{kx^2}{2}$$

where p is the momentum of the particle and x is its position, and k is the force constant of the spring. We define the *phase space* of a system as the multidimensional space of all of its microscopic degrees of freedom. Here the particle has two degrees of freedom, p and x, and the phase space is a two-dimensional space.

(a) For this isolated system, show that the particle is confined to an ellipse in phase space. Draw a graph illustrating the shape of the ellipse with principal radii indicated in terms of E, k, and m as appropriate.

(b) Consider all points in phase space for which the particle can have energy between E and $E + \delta E$, where δE is a small number. Show that this is equivalent to an area between two ellipses in phase space. Find an expression for this area in terms of E, δE, k, and m.

(c) What if the particle lives not in a one-dimensional world, but a d-dimensional one, still tethered to the origin by a harmonic spring? For example, $d = 2$ for a particle confined to a plane, or $d = 3$ for one in three-dimensional space. By analogy with part (b), give an expression for the $2d$-dimensional volume in phase space for energies between E and $E + \delta E$. Expand your answer and omit terms in δE with powers higher than one. Note that the volume of a $2d$-dimensional ellipse is given by the product of all radii with C_{2d}, a d-dependent constant.

(d) You now want to count the number of possible microscopic configurations (microstates) that lie inside this volume so that you can compute the entropy. In general, a microstate is defined as a unique set of positions and momenta. You will need to discretize or "pixelate" phase space in order to perform this counting. Discretize positions into blocks of size δx and momenta into blocks of size δp. Find an expression for the entropy.

(e) Your expression should contain the term $(\delta p \, \delta x)$, which is essentially a discrete area in phase space. To what physical constant might this be related?

(f) Using the entropy, predict the temperature dependence of the total energy, $E(T)$. Do the values of δp, δx, or δE matter?

3.10. In this problem, you will determine the analytical entropy function $S(E, V, N)$ for the classical monatomic ideal gas, that is, for non-interacting, volumeless point particles of mass m. In Eqn. (3.18) we found that the ideal gas density of states could be written as the product of a momentum part and a position part. To count configurations, we must discretize the momenta and position variables of each particle such that they can take on finite values. Assume that each component of a particle's momentum (p_x, p_y, p_z) can adopt only integer multiples of a small value δp, and that similarly each position component can only take on integer multiples of δx.

(a) Show that the number of ways of picking momenta for N particles with total energy less than or equal to E is approximately given by

$$\Xi \approx C_{3N} \times \left(\frac{2Em}{(\delta p)^2} \right)^{3N/2}$$

where C_n is such that $C_n R^n$ gives the volume of an n-dimensional sphere with radius R. Hint: the energy can be written as the squared length of a vector in the $3N$-dimensional space of momenta, $E = (2m)^{-1}|\mathbf{p}^N|^2$, where $\mathbf{p}^N = (p_{1,x}, p_{1,y} \ldots, p_{N,z})$. The number of momenta microstates is the number of vectors with length equal to or less than this.

(b) To a good approximation, we can count the number of ways of picking momenta within a range of energy δE centered around a value of E by taking the derivative of Ξ, such that $\Omega_{\text{momenta}} \approx \delta E \times (d\Xi/dE)$. With this approximation, show that

$$\frac{S(E,V,N)}{k_B} \approx N \ln \left[V \left(\frac{2Em}{(\delta p \, \delta x)^2} \right)^{3/2} \right] + \ln \left[\frac{C_{3N}}{N!} \right] + \ln \left[\frac{3N}{2} \left(\frac{\delta E}{E} \right) \right]$$

$$\approx N \ln \left[V \left(\frac{2Em}{(\delta p \, \delta x)^2} \right)^{3/2} \right] + \ln \left[\frac{C_{3N}}{N!} \right] + \sigma(\ln N)$$

In the second line, one can ignore the last term since it is much smaller than the other terms, which are of order N, and N is typically $\sigma(10^{23})$.

(c) Show that the constant-volume heat capacity, $C_V \equiv (dE/dT)_{V,N}$, is given by $3Nk_B/2$.

(d) To what physical constant might the term $(\delta p \, \delta x)$ be related? With this fact, Stirling's approximation, and $C_n \approx \pi^{n/2}/(n/2)!$, show that

$$\frac{S(E,V,N)}{k_B} = N \ln \left[\left(\frac{E}{N} \right)^{3/2} \left(\frac{V}{N} \right) \right] + \frac{3}{2} N \left(\frac{5}{3} + \ln \left[\frac{4\pi m}{3h^2} \right] \right)$$

This famous formula is called the *Sackur–Tetrode equation*. It gives the exact functionality for the ideal monatomic gas entropy, and all thermodynamic properties of ideal monatomic gases can be derived from it. Notice that the entropy is extensive, that is, it scales as λ if E, V, and N all scale as λ.

3.11. In this problem you will determine an expression for the binary entropy of mixing. Consider an isolated system composed of two types of particles A and B initially separated by a partition, as in Fig. 3.13. The particles are constrained to occupy sites on a regular lattice, and there are exactly as many lattice sites as there are initial particles. The use of a lattice is an approximation, but it will give a very general expression for the entropy of mixing. There are no potential or kinetic energies, and swapping two particles of the same color is not considered a new configuration, i.e., particles of the same type are indistinguishable from each other. At some point, the partition is removed, and particles freely mix. Using its relationship to the density of states, write expressions for the entropy before and after the wall is removed. Find the entropy difference, $\Delta S_{\text{mix}} = S_{\text{after}} - S_{\text{before}}$. Use Stirling's approximation to simplify your answer. Express your final answer in terms of the mole fractions, $x_A = N_A/N_T$ and $x_B = N_B/N_T$, and the total number of molecules $N_T = N_A + N_B$.

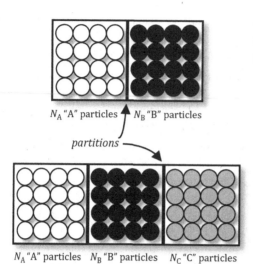

Figure 3.13. The initial configuration of lattice systems for entropy-of-mixing problems.

N_A "A" particles N_B "B" particles

partitions

N_A "A" particles N_B "B" particles N_C "C" particles

3.12. Using methods similar to those in the previous problem, determine an expression for the entropy of mixing of three types of molecules A, B, and C, initially separated by partitions as in Fig. 3.13. Find the entropy difference, $\Delta S_{mix} = S_{after} - S_{before}$, in terms of the three mole fractions and the total number of molecules $N_T = N_A + N_B + N_C$.

3.13. Two charge-neutral carbon atoms in vacuum interact through the Lennard-Jones potential of Eqn. (3.17), where $\epsilon = 0.15$ kcal/mol and $\sigma = 4.0$ Å.

(a) Sketch the interaction. At what atomic separation distance is this energy a minimum? What is the value of the minimum energy? What can be said about the force that the atoms exert on each other at this distance?

(b) Assume that, instead of being neutral, the atoms have equal partial charges q (in terms of the unit electron charge e) and thus additionally interact through Coulomb's law. For some magnitude of the charge, the repulsive like-charge interaction will overwhelm the attractive van der Waals interaction such that the energy minimum vanishes. Find this magnitude of charge q.

3.14. Consider the particle labeled A in Fig. 3.14 that approaches a uniform solid interface made of a lattice of atoms. The surface extends indefinitely in the lateral directions and is infinitely thick. In this problem, you will compute the effective interaction of the particle with the entire surface as a function of its distance from it, summing over individual pair interactions $u(r_{Ai})$ with *all* constituent surface atoms i. To do so, you will approximate the surface as being "smeared-out" in the sense that the atoms are distributed evenly and randomly throughout.

Let the number density of the atoms in the surface be ρ_S. For a differential volume element dV in the surface, the average number of atoms is $\rho_S\, dV$ and their interaction energy is given by $u(r)$, where r is defined as in Fig. 3.14. Let z be the distance of A from the interface.

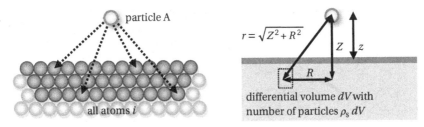

Figure 3.14. A particle interacts with a solid surface built from many atoms. The net interaction of the approaching particle with all surface atoms depends only on the distance z, and can be found by integrating over the entire volume of a smoothed version of the surface.

(a) Let the pair interaction between A and surface atoms be of a Lennard-Jones form, Eqn. (3.17), that captures van der Waals attractions and steric repulsions. Show that the net interaction of the particle with the surface is then

$$U(z) = \frac{4\pi\epsilon\rho_s\sigma^3}{3}\left[\frac{1}{15}\left(\frac{z}{\sigma}\right)^{-9} - \frac{1}{2}\left(\frac{z}{\sigma}\right)^{-3}\right]$$

(b) Find an expression for the z-directional force the wall exerts on the particle, as a function of z. Find the distance and energy at which the force is zero.

(c) Now instead, assume that the interface is coated in positive charge with number density *per area* given by γ. Let particle A have a negative charge such that it interacts favorably through Coulomb's law with surface atoms. Show that the average force that the wall exerts on the particle in the z direction is constant. This is a general result from electrostatics that states that the electric field due to a flat plate is constant. Hint: first show that the force is given by

$$f_z = -\iint \gamma \frac{du_{\text{Coulomb}}(r)}{dr}\frac{z}{r}\, dx\, dy$$

Applied problems

3.15. In the largely aqueous environment of the cell, complementary DNA strands spontaneously self-assemble to form the iconic B-helical structure. At first glance, this may seem counterintuitive because the phosphate–deoxyribose backbone of each DNA strand contains negative charges that would result in repulsive like-charge interactions.

(a) The human genome consists of roughly 3 billion base-paired nucleotides. In eukaryotes, DNA exists inside of the cell nucleus, and a typical mammalian nucleus is spherical with a diameter of about 6 microns. What total concentration of positively charged, monovalent species (e.g., Na$^+$ or K$^+$) must be present inside the nuclei of human cells in order to maintain an overall charge-neutral environment there? Express your answer in millimolar units. Compare it with the value of ~150 mM often found in the cytosol.

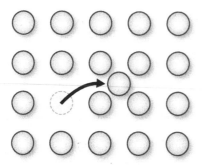

Figure 3.15. A hypothetical scenario involving the formation of a single-site defect in a crystalline structure, an interstitial–vacancy pair.

(b) In the B-helical structure, negative charges on strand backbones opposite of each other are separated by about 12 Å in distance. Compute the potential energy of this interaction in kcal/mol assuming that the charges interact in a vacuum through Coulomb's law, $u(r) = q_1 q_2/(4\pi\epsilon_r\epsilon_0 r)$, where $\epsilon_r = 1$ for a vacuum. Compare this with the average strength of a hydrogen bond in water, roughly 2 kcal/mol.

(c) Repeat part (b) for the case in which the charges interact in water, where $\epsilon_r \approx 80$. How does water help stabilize the double-helix structure?

(d) In dehydrated *in vitro* DNA samples, one often finds that DNA forms an alternate, wider A-helical structure. In this structure, negative charges on strand backbones opposite of each other are separated by about 19 Å. Is this consistent with your calculations in (b) and (c)?

3.16. Most crystalline substances have a number of defects in which the lattice structure is broken or disrupted. In some technologies, such as in the microelectronics industry, the formation of large defect-free crystal structures is vital to the performance of the final product. One kind of defect is an interstitial–vacancy pair in which a single atom becomes misplaced from its regular lattice site, as shown in Fig. 3.15. A very simple model of defects is the following: (1) the number of possible defects is given by the number of atoms in the crystal N; (2) each defect results in an energy penalty ϵ; and (3) defects are non-interacting, meaning that the presence of one defect does not affect defects adjacent to it. This latter assumption is valid at low defect density.

(a) Develop an expression for the entropy of this simple model. First, write an expression for the number of microstates as a function of the number of defects, n. Then, replace n with the total energy using the fact that $E = n\epsilon$. Simplify your answer using Stirling's approximation, and rewrite it in terms of the defect density $x \equiv n/N$ and total number of particles N. Hint: this is a problem of combinatorics of placing defects.

(b) Find the defect density as a function of temperature. Hint: first find the temperature from the entropy.

(c) What value of defect energy ϵ is required for 1% of possible defects to be present at near-ambient temperature, $T = 300$K? For this value of ϵ, at what temperature will there be a 50% defect density?

FURTHER READING

T. L. Hill, *An Introduction to Statistical Thermodynamics*. Reading, MA: Addison-Wesley (1960); New York: Dover (1986).

T. L. Hill, *Statistical Mechanics: Principles and Selected Applications*. New York: McGraw-Hill (1956); New York: Dover (1987).

J. Israelachvili, *Intermolecular and Surface Forces*, 3rd edn. Burlington, MA: Academic Press (2011).

E. A. Jackson, *Equilibrium Statistical Mechanics*. Mineola, NY: Dover (1968).

C. Kittel and H. Kroemer, *Thermal Physics*. New York: W. H. Freeman (1980).

D. A. McQuarrie, *Statistical Mechanics*. Sausalito, CA: University Science Books (2000).

D. A. McQuarrie, *Quantum Chemistry*. Mill Valley, CA: University Science Books (1983).

D. Ruelle, *Statistical Mechanics: Rigorous Results*. River Edge, NJ: World Scientific (1999); London: Imperial College Press (1999).

R. C. Tolman, *The Principles of Statistical Mechanics*. New York: Dover (1979).

4 Entropy and how the macroscopic world works

4.1 Microstate probabilities

In Chapter 3 we discussed the way the world works at a *microscopic* level: the interactions and laws governing the time evolution of atoms and molecules. We found that an important, unifying perspective for both quantum and classical descriptions is the concept of energy. Now we take a *macroscopic* point of view. What happens when many ($\sim 10^{23}$) molecules come together, when we cannot hope to measure individual atomic properties but can probe only bulk material ones? As the title of this chapter suggests, the relevant concept at the macro-resolution is entropy.

Remember that from a macroscopic perspective, we care about macrostates, that is, states of a system characterized by a few macroscopic variables, like E, V, N, T, or P. Empirical measurements generally establish values for properties that are the net result of many atomic interactions averaged over time, and we are thus able to describe a system only in terms of these large-scale, reduced-information metrics that smooth over the molecular world. The statement that a system is at one specific macrostate actually implies that it is evolving through a particular ensemble of many microscopic configurations.

We will focus on classical isolated systems because these offer the simplest introductory perspective. Let us imagine that a closed, insulated container of molecules evolves in time. The microstate of this system constantly changes, as the positions and velocities of each molecule vary under the influence of their mutual interactions and those with the container walls (Fig. 4.1). However, the macrostate of this system – specifically, the set of values of E, V, and N – remains constant since each of these parameters is fixed per the microscopic Newtonian evolution in the isolated case.

For the moment, let us suppose a hypothetical scenario in which we are able to measure the position and velocity of every atom. Assume also that we have available a list of all possible microstates for this system, each indexed by a unique identification number. Figuratively speaking, this could be a catalogue of some sort where each page corresponded to one microstate and presented either a picture or a table of values of all of the positions and momenta. Of course, such a book may be very long (or, as you might anticipate, even infinite), but let us ignore this complication for the moment.

If we were to probe the ensemble of configurations visited by this isolated system, we could conceptualize a test in which we froze time and interrogated the instantaneous microstate. We could repeat the test a huge number of instances, each time noting which microstate we saw. By tabulating the number of appearances of each microstate in

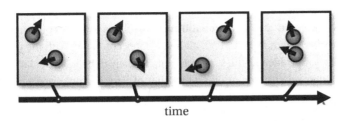

time

Figure 4.1. An illustration of some distinct microstates that might be found at instantaneous time points for two particles in a box. Each microstate involves a specific set of both positions and momenta for each of the particles. If the particles exist in three dimensions, there are twelve degrees of freedom, whose combined values specify the microstate.

the catalogue, using tick marks for example, we could evaluate their frequency. Specifically, we could construct probabilities from the tally numbers divided by the total number of test repetitions. We will call this all-important probability the *microstate probability*. It gives the likelihood that the system would pass through different microstates, for a particular macrostate. Our microstate probabilities would change if we altered the macrostate, for example, if the system started with a different total energy or volume.

The remarkable feature of these conceptualizations is that at equilibrium the microstate probabilities do not change with time. That is, it does not matter when we start making tick marks, so long as we are able to do so for long enough to gather accurate statistics. Regardless of the extensiveness of our own tallying, the true underlying probabilities that we approximate through counting remain constant at equilibrium. These statements may seem redundant since we have already defined the concept of equilibrium using the idea of time invariance. However, here the time independence emerges in a microscopic property, not a macroscopic one. Thus while the microscopic degrees of freedom continually vary – the atomic positions and velocities are in constant flux – the microstate probabilities do not, at equilibrium.

The **microstate probabilities** are a collection of probabilities assigned to all microstates that give the likelihood each will be visited by the system. At equilibrium, these probabilities do not change with time, although they are dependent on the current macrostate conditions.

4.2   The principle of equal a priori probabilities

The power of statistical thermodynamics lies in the fact that large collections of molecules manifest a behavior that greatly simplifies the microstate probabilities. The fundamental rule governing the behavior of macroscopic systems of molecules at equilibrium is called the principle of equal a priori probabilities.

The **principle of equal a priori probabilities** postulates that, at equilibrium in an isolated system (constant E, V, and N), all microstates consistent with the macrostate are equally likely. Specifically, the system is equally likely to be in any of its $\Omega(E, V, N)$ microstates. Other microstates, at different values of E for example, are not visited by the system.

In other words, the principle gives us an exact recipe for computing the microstate probabilities in isolated systems. For our exercise in making tally marks, we would expect to find the same number of tallies on average for each microstate in the catalogue that had the same energy as the initial energy, which is also the macroscopic energy E. For the other microstates, we expect zero tally marks. That is, eventually the system will explore all microstates that have the energy E.

What leads to this principle? We can rationalize some aspects of it at an intuitive level. The total energy E is constant for the classical microscopic equations of motion; therefore, the probability for microstates with energies other than the observed macroscopic one is rigorously zero. Moreover, large collections of molecules display a kind of chaotic behavior. Owing to this, small differences in the initial positions or velocities of the atoms can result in huge divergences in the ultimate molecular trajectories. That is, a nearly infinitesimal perturbation to the initial conditions can ultimately lead to a very distinct progression of microstates as time evolves. In very approximate terms, this chaotic behavior enforces a "democracy" among microstates such that, in the limit of long periods of time, the system has no particular preference for certain configurations over others. Finally, one can show that Newton's equations of motion preserve the equal a priori probability distribution, which we discuss at the end of this chapter.

In simple systems like ideal gases, one can prove the principle of equal a priori probabilities through an analysis of particle collisions, a body of work that extends from Ludwig Boltzmann's original ideas (called the H-theorem). Mathematical theory has made good cases for the principle for a variety of other models; however, in general it cannot be proven rigorously for all systems. Ultimately, we must accept the principle as a general feature of collections of molecules, given the large body of empirical evidence that shows that its consequences and predictions are correct, especially the existence of thermodynamic behavior and history-independent equilibrium states.

Even so, not all systems obey the principle of equal a priori probabilities. The main exceptions are those that do not reach equilibrium because they are slow-evolving systems. For example, glasses and amorphous materials are essentially ultra-high-viscosity liquids whose internal relaxation time scales (e.g., the characteristic times for diffusion and flow) are astronomically large compared with the time scales of our observations. These systems are often metastable with respect to some crystalline phase; that is, in an infinite amount of time they would crystallize. Thus, glasses are not at equilibrium and do not visit equally all of their microstates as would be prescribed by the principle due to their sluggish dynamics. In statistical mechanics, we use a specific terminology for systems that obey the principle.

Systems that are at equilibrium and obey the principle of equal a priori probabilities are called **ergodic**. Those that do not are **non-ergodic.**

The existence of non-ergodic systems should not be discouraging, since a huge range of systems can be considered ergodic, particularly those involving bulk fluid phases. For many systems, it is important only that they are *locally* at equilibrium. Indeed, this planet and the very universe in which we live are ever-changing, but we are still able to describe systems of more modest scale in the lab at an equilibrium level, even if we do not expect them to be the same or even exist eons from now. Moreover, the equilibrium thermo-dynamic framework developed for ergodic systems remains a major starting point for the analysis of non-ergodic ones.

Example 4.1 *Consider the three-molecule two-state system discussed in Chapter 3. Compute the microstate probabilities in the case for which $E = b$.*

Figure 4.2 shows a complete enumeration of all microstates for the system. We see that there are three configurations with $E = 1b$. Therefore, according to the principle, the probability of each of configurations 2, 3, and 4 equals 1/3. For the other microstates – configurations 1 and 5 through 8 – the probability is zero.

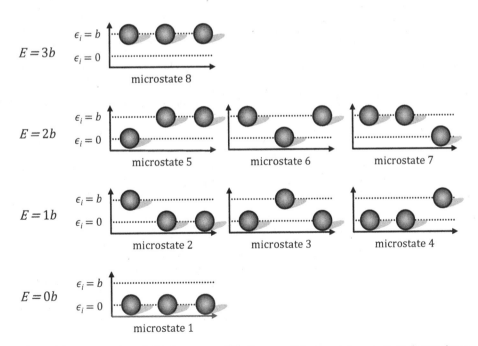

Figure 4.2. Enumeration of all microstates of the three-particle two-state system. Each row shows the collection of microstates for a single macrostate described by the total energy E.

4.3   Ensemble averages and time averages in isolated systems

The principle of equal a priori probabilities gives us an exact way to characterize ensembles of isolated systems by allowing us to compute the microstate probabilities from the density of states $\Omega(N, V, E)$. Let us introduce a formal mathematical approach for this task. Let the quantity $\wp_m$ give the probability that the system is in microstate m, where m is simply some index that runs over the total list of microstates (e.g., it ran from 1 to 8 in the previous two-state example). The principle tells us that we can write $\wp_m$ in the following manner:

$$\wp_m = \begin{cases} \dfrac{1}{\Omega(E, V, N)} & \text{if } E_m = E \\ 0 & \text{if } E_m \neq E \end{cases} \tag{4.1}$$

where E_m is the total energy of microstate m. In classical systems, E_m is the sum of the kinetic energy (dependent on the instantaneous atomic velocities) and the potential energy (dependent on the instantaneous positions) of microstate m. From where do E, V, and N originate? These values give the macrostate of the system for any point in time or as given by the initial conditions of Newtonian evolution. Recall that the principle tells us how to construct an ensemble corresponding to a state that is characterized from the *macroscopic* level so it is essential that we specify these variables.

We can write Eqn. (4.1) in a more compact form,

$$\wp_m = \frac{\delta_{E_m, E}}{\Omega(E, V, N)} \tag{4.2}$$

where the Kronecker delta function has the properties

$$\delta_{x, y} = \begin{cases} 1 & \text{if } x = y \\ 0 & \text{if } x \neq y \end{cases} \tag{4.3}$$

With this approach, we can also write a compact expression for the density of states,

$$\Omega(E, V, N) = \sum_{\text{all microstates } n \text{ at } V, N} \delta_{E_n, E} \tag{4.4}$$

Equation (4.4) has a conceptual interpretation. The summation proceeds through a list of all possible microstates of a system, indexed by n, with N molecules in a volume V. For each microstate, we examine whether or not its energy equals that of the macrostate energy. If it does, we count it by adding one to our running summation. If not, we ignore it. In the two-state example above, the sum would extend over eight configurations.

In a classical system, we compute the microstate energy using the Hamiltonian,

$$E_m = H\left(\mathbf{r}_m^N, \mathbf{p}_m^N\right) = K\left(\mathbf{p}_m^N\right) + U\left(\mathbf{r}_m^N\right) \tag{4.5}$$

In this case, the microstate index m denotes a specific set of all position and momenta values, $\left(\mathbf{r}_m^N, \mathbf{p}_m^N\right)$. At this point, you may be a bit perplexed about applying these ideas to classical systems. Is not the number of microstates infinite since we can continuously change the position and velocity of each atom? The answer is, of course, yes. In this case, we could approximate the sum above by finely discretizing the positions and momenta, allowing them to take on values only in a regularly spaced manner, in order to make the total number of microstates finite and countable. However, we would find that as we were to shrink the size of the discretization to zero, the density of states would scale in a predictable way. For comparison, you may want to consider the rough ideal gas arguments in Chapter 3.

It turns out that we can determine *relative* values of the density of states in classical systems, for example the value of $\Omega(E, V, N)/\Omega(E, 2V, N)$. Here the size of any discretization will appear identically in the numerator and denominator and thus cancel out. The infinite number of classical configurations means that we can know the density of states only to an arbitrary N-dependent multiplicative constant. As a result, we can never know the absolute value of the entropy since the logarithm of Ω turns the multiplicative constant into an additive one. Ultimately, however, the world is quantum in nature, and we can indeed compute absolute entropies. We will discuss this subtle point further in our consideration of crystals and the third law in Chapters 14 and 15.

Now that we have a recipe for evaluating microstate probabilities, we can actually develop a systematic way to determine system properties at equilibrium by taking something called an *ensemble average*. Suppose that there is a macroscopic observable that we are interested in measuring, such as the pressure exerted by the system on the surroundings. We must connect the observable to interactions at the molecular level. Namely, we need a relationship between what we measure macroscopically and the properties of microstates.

If we were able to freeze time, we might be able to make a specific measurement of the observable for that instantaneous microscopic configuration. For example, we could measure the net force that all of the molecules exert on the container walls, which is connected to the pressure. Or, we could measure the density in different spatial regions or near walls and interfaces. Indeed, there is a huge range of properties that we could observe and to which we could assign specific values for a given molecular configuration. Mathematically, let there be a property X that we can measure at any instant in time. If we were to follow the exact microscopic trajectory of the system, we could construct a history of this measurement with time, $X(t)$. If we then did this for a long enough observation interval τ, ultimately to infinity for equilibrium, we could construct a *time average* of the property,

$$\langle X \rangle_{\text{time}} = \lim_{\tau \to \infty} \frac{1}{\tau} \int_0^\tau X(t)\,dt \qquad (4.6)$$

Here, the angle brackets indicate averaging, a notation that is used throughout this book. An average gives the value that would be measured macroscopically for a property when fluctuations in it are too small to detect. It is an average that smoothes over microscopic variations in X as the system evolves through different microstates.

On the other hand, we could construct a different kind of average in terms of microstate probabilities and equilibrium behavior. Let the value of X for microstate m be denoted by X_m. As the system evolves in time at equilibrium, it explores different microstates according to $\wp_m$. Thus, another mean is the weighted *ensemble average*,

$$\langle X \rangle_{ens} = \sum_{\text{all microstates } m \text{ at } V, N} \wp_m X_m \tag{4.7}$$

In isolated systems the value $\wp_m$ will be zero for many of the configurations in the sum, namely those with $E_m \neq E$, and their values X_m will not contribute to the average. Along the same lines, $\langle X \rangle_{ens}$ clearly depends on the macroscopic state characterized by E, V, and N since the probabilities $\wp_m$ do as well. The values X_m, however, depend only on the characteristics (i.e., positions and momenta) of each microstate m.

The power of the principle of equal a priori probabilities is that it dictates that ensemble averages equal time averages at equilibrium,

$$\langle X \rangle_{ens} = \langle X \rangle_{time} \tag{4.8}$$

That ensemble averages equal time averages is a tremendous simplification. It says that we do not need details of the molecular trajectory in order to connect equilibrium properties to the microscopic world; we can do away with the precise time-history of the atomic positions, such as when certain atoms "collide." Instead, we only need to know the current macroscopic state of the system. This allows us to construct the exact probabilities with which each microstate will appear. At equilibrium, it is as if the system proceeds through microscopic configurations randomly, and we can use probabilities rather than the actual trajectory to compute macroscopic averages of properties.

> Because **time averages** are equal to **ensemble averages** at equilibrium, one does not need to know the complete time-history of an observable to compute its average, macroscopic value. Instead, one need only know the ensemble probabilities. When viewed at long times, the trajectory of a system explores microstate configurations in a probabilistic, random fashion.

Example 4.2   *Consider again the two-state system with $N = 3$. Let us say that whenever two neighboring molecules are in the up position, some kind of signal is emitted, such as light. Assume that the amount of signal (its intensity) is given by the number of molecule pairs in the up state. Denote the signal intensity by I. Compute the average value of I for the $E = 0b$, $E = 1b$, $E = 2b$, and $E = 3b$ macrostates.*

From the problem statement, I is the total number of adjacent molecule pairs in the up position. We can assign a value of I to every microstate in the system, as is shown graphically in Fig. 4.3.

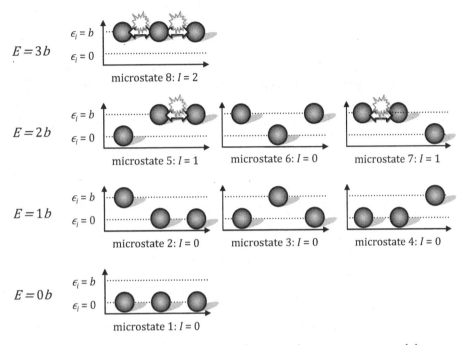

Figure 4.3. A depiction of the microstates of the three-particle two-state system and the hypothetical measurable I, which counts the number of neighboring pairs simultaneously in the up position.

To compute the average $\langle I \rangle$ at equilibrium, we can use the ensemble formula,

$$\langle I \rangle = \sum \wp_m I_m$$

Note that the microstate probabilities depend on the macrostate under consideration. Take the macrostate $E = 2b$. Here, only three configurations satisfy that energy, and thus we have

$$\wp_m = \begin{cases} 1/3 & \text{if } m = 5, 6, 7 \\ 0 & \text{otherwise} \end{cases}$$

Therefore, we can write

$$\langle I \rangle_{E=2b} = (1/3)(1) + (1/3)(0) + (1/3)(1)$$
$$= 2/3$$

Here we have omitted the other five terms in the sum for which the microstate probability is zero. Similarly, we can evaluate the average signal in each of the other macrostate cases,

$$\langle I \rangle_{E=0b} = (1)(0) = 0$$
$$\langle I \rangle_{E=1b} = (1/3)(0) + (1/3)(0) + (1/3)(0) = 0$$
$$\langle I \rangle_{E=3b} = (1)(2) = 2$$

Thus, for this system, we would expect that adding energy to it would sharply increase the signal emitted.

4.4   Thermal equilibrium upon energy exchange

We are already beginning to see how entropy figures into the world of equilibrium, via the principle of equal a priori probabilities. In isolated equilibrium systems, the microstate probability is given by Eqn. (4.2), which involves the density of states. Thus, not only does $\Omega(E, V, N)$ count the number of microstates compatible with a set of macroscopic descriptors E, V, and N, but also it serves as a normalizing factor for the microstate probabilities, ensuring that they sum to one. As such, the entropy also has a direct role in the microstate probabilities via Boltzmann's equation,

$$S(E, V, N) = k_B \ln \Omega(E, V, N) \tag{4.9}$$

Before proceeding, we address the question of why the entropy has this particular relationship to the density of states. It turns out that the logarithm emerges naturally when two systems come to mutual equilibrium, which we will show momentarily. The logarithm also makes the entropy extensive, scaling linearly with the size of the system. This occurs because the density of states grows exponentially with system size due to combinatorics. The argument is simple: if we double the size of a macroscopic system, to leading order the number of microstates increases as $\Omega \times \Omega = \Omega^2$. If we triple the size, we obtain $\Omega \times \Omega \times \Omega = \Omega^3$, and so on and so forth.

What about the universal constant k_B? There is no fundamental physical basis for the value of this constant; rather, it simply defines temperature units in terms of energy units. Historically, the concept of temperature emerged before it was understood to be a statistical property of molecular systems. If we were willing to give up temperature as a separate unit, we could easily measure "hotness" in units of energies, setting $k_B = 1$.

With these considerations in mind, we can now show that systems tend towards states that *maximize* the total entropy. In order to do this, we will need to consider the case in which two systems 1 and 2 come to mutual equilibrium, as shown in Fig. 4.4. When both systems are isolated, system 1 has energy E_1 and system 2 has E_2, both of which are constant and fixed. The number of microstates available to each is $\Omega_1(E_1, V_1, N_1)$ and $\Omega_2(E_2, V_2, N_2)$, respectively. Notice that the two densities of states have subscripts 1 and 2 themselves. This is because we have not said that the two systems are identical; they could contain completely distinct components.

Now we will take the perspective that systems 1 and 2 actually constitute one supersystem. How many combined microstates are there in this supersystem? A combined microstate consists of a microstate each in systems 1 and 2; it is a microstate as if we did not know that 1 and 2 were separate entities but had originally drawn our system boundary around both. We still maintain the systems as mutually isolated within the supersystem, but this new perspective now sees a large number of configurations that stem from all possible combinations of microstates in 1 and 2. Since the subsystems remain individually isolated and independent, we merely count the ways of pairing a microstate in 1 with another in 2 to obtain the total number of supersystem microstates Ω_T,

$$\Omega_T = \Omega_1(E_1, V_1, N_1)\Omega_2(E_2, V_2, N_2) \tag{4.10}$$

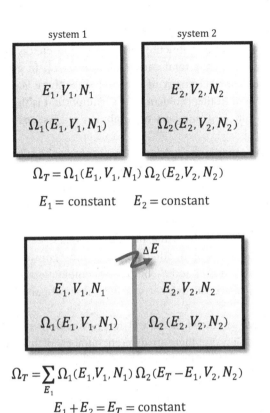

$$\Omega_T = \Omega_1(E_1, V_1, N_1)\, \Omega_2(E_2, V_2, N_2)$$

$$E_1 = \text{constant} \qquad E_2 = \text{constant}$$

$$\Omega_T = \sum_{E_1} \Omega_1(E_1, V_1, N_1)\, \Omega_2(E_T - E_1, V_2, N_2)$$

$$E_1 + E_2 = E_T = \text{constant}$$

Figure 4.4. Two initially isolated systems 1 and 2 (top) are brought together and allowed to exchange energy (bottom). The equations underneath each system describe the total number of microstates as well as the constraints on the energies.

By taking the logarithm of this equation and multiplying by k_B, we can rewrite this expression in terms of entropies,

$$S_\mathrm{T} = S_1(E_1,\, V_1,\, N_1) + S_2(E_2,\, V_2,\, N_2) \tag{4.11}$$

Let us now bring the two systems together and allow them to freely exchange energy through a common, heat-conducting wall. We can define a quantity ΔE that measures the amount of energy that transfers from system 1 to system 2. We do not know yet what ΔE will be; it can fluctuate with time and be positive or negative depending on which system gains energy. These fluctuations are governed by the atomic interactions at the boundary, and how each system distributes energy to them. However, both systems together still constitute an isolated supersystem. While the energies E_1 and E_2 can change due to the transfer of energy between the two systems, the total energy $E_T = E_1 + E_2$ must remain constant. Because the supersystem is isolated, the principle of equal a priori probabilities tells us *exactly* how ΔE will change and fluctuate.

What would happen if a precise amount of energy ΔE were transferred, such that the energies changed to $E_1' = E_1 - \Delta E$ and $E_2' = E_2 + \Delta E$? We compute the new number of combined microstates as before, $\Omega'_T = \Omega_1(E'_1, V_1, N_1)\Omega_2(E_T - E'_1, V_2, N_2)$. We write this expression as a function of E'_1 since the energy of system 2 is constrained by the constancy of the total energy, $E'_2 = E_T - E'_1$. In general it will be convenient for us to focus on the energy of system 1 since there is only a single macroscopic degree of freedom in this problem. What is important is that the super-system will have access to different numbers of combined microstates for different values of E'_1.

In a real thermal-contact scenario, the amount of heat transferred is not externally fixed because energy can move freely back and forth between the systems according to the boundary molecular dynamics. That is, ΔE fluctuates with time. Here is where the principle of equal a priori probabilities helps us. It says that each microstate in the combined isolated system – any of the supersystem microstates for *any* value of energy transferred ΔE – is assigned the same probability weight or is visited with the same frequency by the supersystem. In other words, we must consider all possible microstates, not for one set of values E_1 and E_2, but for all combinations due to energy transfer. Each microstate in this collective set has the same probability according to the principle.

The key implication is that variations in ΔE will not be uniform: each individual microstate has the same probability, but there are different numbers of them for different values of E_1. Specifically, the fraction of time spent at one value of E_1 is proportional to the number of microstates there. With these considerations we can now develop an expression for the frequency of the macroscopic variable E_1,

$$\wp(E_1) \propto \Omega_1(E_1, V_1, N_1)\Omega_2(E_T - E_1, V_2, N_2) \tag{4.12}$$

Here, $\wp(E_1)$ gives the probability with which we would expect to see a given value of E_1 at a random moment in time. Because time averages are equal to ensemble averages, $\wp(E_1)$ also gives the fraction of the time that the supersystem spends in states where system 1 has energy E_1. Keep in mind that these probabilities are also implicitly a function of the total energy; if we were to change E_T, we would change the distribution $\wp(E_1)$ by the way in which it figures into Eqn. (4.12).

Example 4.3 *Compute the distribution of E_1 upon free energy exchange if the systems are both two-state models. Let the initial states of the systems be given by $N_1 = 6$, $E_1 = 1b$, $N_2 = 12$, and $E_2 = 5b$.*

To solve this problem, we begin with the general expression for $\Omega(E)$ in a two-state system,

$$\Omega(E) = \frac{N!}{(E/b)!(N - E/b)!}$$

which arises from the number of ways of picking E/b particles in the up position. Each of systems 1 and 2 has this form for its density of states. With this formula, we can make a table for the number of microstates at different values of E_1: see Table 4.1.

Table 4.1 Numbers of microstates at different values of E_1 for Example 4.3

$\Delta E/b$	E_1/b	E_2/b	$\Omega_1(E_1)$	$\Omega_2(E_2)$	Ω'_T
>1	<0	>6	0	<924	0
1	0	6	$\dfrac{6!}{0!6!} = 1$	$\dfrac{12!}{6!6!} = 924$	$1 \times 924 = 924$
0	1	5	$\dfrac{6!}{1!5!} = 6$	$\dfrac{12!}{5!7!} = 792$	$6 \times 792 = 4{,}752$
−1	2	4	$\dfrac{6!}{2!4!} = 15$	$\dfrac{12!}{4!8!} = 495$	$15 \times 495 = 7{,}425$
−2	3	3	$\dfrac{6!}{3!3!} = 20$	$\dfrac{12!}{3!9!} = 220$	$20 \times 220 = 4{,}400$
−3	4	2	$\dfrac{6!}{4!2!} = 15$	$\dfrac{12!}{2!10!} = 66$	$15 \times 66 = 990$
−4	5	1	$\dfrac{6!}{5!1!} = 6$	$\dfrac{12!}{1!11!} = 12$	$6 \times 12 = 72$
−5	6	0	$\dfrac{6!}{6!0!} = 1$	$\dfrac{12!}{0!12!} = 1$	$1 \times 1 = 1$
<−5	>6	<0	0	0	$0 \times 0 = 0$

The total number of microstates summed over all values of E_1 is $924 + 4{,}752 + 7{,}425 + 4{,}400 + 990 + 72 + 1 = 18{,}564$. Because the supersystem is isolated, the principle of equal a priori probabilities demands that each of these is visited with equal frequency. However, more microstates lie at certain values of E_1 than others, making the observed distribution of E_1 non-uniform. For example, the fraction of time the system spends at $E_1 = 2b$ is

$$\wp(E_1 = 2b) = \frac{7{,}425}{18{,}564} = 40\%$$

On the other hand, the fraction of time the system spends at $E_1 = 6b$ is

$$\wp(E_1 = 6b) = \frac{1}{18{,}564} = 0.005\%$$

Returning to our combined supersystem, we can count the total number of microstates for all E_1 with a simple summation,

$$\Omega_T = \sum_{\text{all } E_1} \Omega_1(E_1, V_1, N_1)\Omega_2(E_T - E_1, V_2, N_2) \tag{4.13}$$

In principle, the limits of E_1 in the summation span from negative to positive infinity, but it is important to note that either Ω_1 or Ω_2 could be zero for some values of E_1 and E_2.

In other words, there may be energies for which there are no possible microstates. In any case, Ω_T is used to normalize the *macrostate* distribution of E_1, similar to the approach in Example 4.3,

$$\wp(E_1) = \frac{\Omega_1(E_1, V_1, N_1)\Omega_2(E_T - E_1, V_2, N_2)}{\Omega_T} \tag{4.14}$$

In many cases, the energy is continuous and not exchanged in discrete values. In such instances, the summation for Ω_T becomes an integral,

$$\Omega_T = \int \Omega_1(E_1, V_1, N_1)\Omega_2(E_T - E_1, V_2, N_2)dE_1 \tag{4.15}$$

In general, the density of states increases sharply with energy, which is a common feature of macroscopic systems. As a result, there is typically a maximum in $\wp(E_1)$ because Ω_1 decreases and Ω_2 increases as energy moves from system 1 to system 2. We find the location of the maximum by differentiating the summed term in Eqn. (4.13) with respect to E_1,

$$\frac{\partial}{\partial E_1}[\Omega_1(E_1, V_1, N_1)\Omega_2(E_T - E_1, V_2, N_2)]|_{V_1, N_1, V_2, N_2} = 0 \tag{4.16}$$

at the maximum. On suppressing for now the constant V and N variables and evaluating the derivative, we obtain

$$0 = \Omega_1(E_1^*)\frac{\partial\Omega_2(E_T - E_1^*)}{\partial E_1} + \Omega_2(E_T - E_1^*)\frac{\partial\Omega_1(E_1^*)}{\partial E_1}$$

$$= \Omega_1(E_1^*)\frac{\partial\Omega_2(E_2^*)}{\partial E_2}(-1) + \Omega_2(E_2^*)\frac{\partial\Omega_1(E_1^*)}{\partial E_1} \tag{4.17}$$

We have designated the energies at the maximum by E_1^* and E_2^*. By rearranging and simplifying, we have

$$\frac{\partial\ln\Omega_1(E_1^*)}{\partial E_1} = \frac{\partial\ln\Omega_2(E_2^*)}{\partial E_2} \tag{4.18}$$

In other words, the value of E_1 that maximizes Ω_T and $\wp(E_1)$ is found from the point at which the energy derivatives of $\ln\Omega$ are equal in the two systems. According to Eqn. (4.9), we can write the condition for the maximum using the entropy,

$$\frac{\partial S_1(E_1^*)}{\partial E_1} = \frac{\partial S_2(E_2^*)}{\partial E_2} \tag{4.19}$$

Alternatively, using the partial derivative definitions for S,

$$\frac{1}{T_1} = \frac{1}{T_2} \rightarrow T_1 = T_2 \tag{4.20}$$

We finally see that the number of microstates is a maximum when the temperatures of the two systems are equal. This is our major result. We have determined the

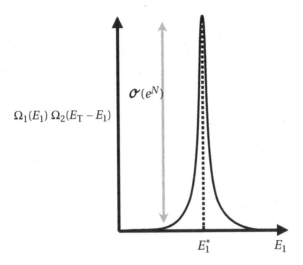

Figure 4.5. When two systems 1 and 2 are allowed to exchange energy, the number of combined (supersystem) microstates is very sharply peaked, with its relative height on the order of exp(N), where N is the number of molecules.

conditions for *thermal equilibrium*, the ultimate state of affairs when two systems can freely exchange energy, and we have proven that it results in the equality of temperatures.

What about energies besides the precise values E_1^* and E_2^* that give the maximum number of microstates? Should we not observe them as well, with a relative frequency in proportion to the number of microstates at each different value of E_1? The answer is two-fold: these microstates do contribute, but their impact is typically vanishingly small. There are overwhelmingly more microstates at E_1^* than at any other energy, and the maximum implied by Eqn. (4.16) is extraordinarily sharply peaked. If we were to change the value of E_1 away from E_1^* by just a few fractions of a percent, the number of microstates would drop precipitously, by orders of magnitude.

That the maximum is so sharp has nothing to do with the type of systems at hand, but rather the mere fact that they are macroscopic and composed of huge numbers of atomic degrees of freedom. The density of states grows exponentially with the system size; a simple scaling law is $\Omega = \exp[Ns(E/N)/k_B]$, where s is the intensive entropy, a function of the average energy per molecule. In macroscopic systems N is on the order of Avogadro's number, and tiny changes in the per-particle entropy are multiplied by the huge prefactor N. This causes the number of microstates to decrease dramatically as one moves even slightly away from the maximum, as illustrated in Fig. 4.5.

The sharpness of the maximum has major ramifications for the observed behavior of the two systems after coming into contact. It means that effectively they will spend nearly all of their time at the specific values E_1^* and E_2^*, simply because there are immensely more microstates there. If we were to measure the instantaneous value of E_1, we would always obtain a number nearly identical to E_1^*. From our macroscopic perspective it would then appear that E_1 and E_2 reached constant, specific values. It is for this very reason that we can speak of the systems as having well-defined energies at thermal equilibrium, even though in reality we know that their energies fluctuate by very small, undetectable amounts as their molecules continue to interact at the boundary.

By the same logic, we can approximate the summation in Eqn. (4.13) by the number of microstates at the maximum since the other terms are by comparison so much smaller,

$$\Omega_T \approx \Omega_1 \left(E_1^*, V_1, N_1\right)\Omega_2 \left(E_2^*, V_2, N_2\right) \tag{4.21}$$

This approximation is called the *maximum-term method*, and it occurs frequently in statistical mechanics due to the recurring presence of such sharply peaked distributions. Upon taking the logarithm, we find that the entropies are additive,

$$S_T = S_1 \left(E_1^*, V_1, N_1\right) + S_2 \left(E_2^*, V_2, N_2\right) \tag{4.22}$$

Compare this result with Eqn. (4.11) for the supersystem entropy *before* the two subsystems were brought into contact. The main difference is that, at thermal equilibrium, the values E_1 and E_2 cannot be independently controlled because the systems are no longer isolated; instead, they approach the maximum-microstate values given by the condition of Eqn. (4.19). We summarize these ideas below.

> When two large bodies are allowed to exchange energy and come to **thermal equilibrium**, the distribution of the exchanged energy implied by the principle of equal a priori probabilities is very sharply peaked. It is so sharply peaked that we can treat the final state as if there are definite, fixed energies of the bodies after heat exchange. These energies are the values that maximize the number of microstates in the combined supersystem, and correspond to the case when the temperatures of the two bodies are equal.

Example 4.4   *To an initial approximation, we can expand any entropy function to second order in energy: $S(E)/N = a + b(E/N) - c(E/N)^2$, where a, b, and c are all constants. Here, the factors of N are explicitly included to show that the entropy is extensive, i.e., that S grows linearly with the size of the system, as do E and N. The negative sign in the third term stems from the fact that the entropy is always a concave function of E.*

Assume that two systems 1 and 2, both with this form for their entropy functions, and with the same number of particles N, come into thermal contact. Compute the distribution of $\wp(E_1)$ at equilibrium if the total energy of the two systems equals $E_1 + E_2 = E_T$. How does the probability of seeing the value $E_1 = E_1^*$ compare with that of an energy that is just a small deviation from E_1^*?

We know according to the principle of equal a priori probabilities that the probability distribution of E_1 is given by

$$\wp(E_1) \propto \Omega_1(E_1)\Omega_2(E_2)$$
$$= \exp\left[aN + bE_1 - cN^{-1}E_1^2\right]\exp\left[aN + bE_2 - cN^{-1}E_2^2\right]$$
$$= \exp\left[aN + bE_1 - cN^{-1}E_1^2 + aN - b(E_T - E_1) - cN^{-1}(E_T - E_1)^2\right]$$
$$= \exp\left[2aN - bE_T - cN^{-1}\left(2E_1^2 - 2E_TE_1 + E_T^2\right)\right]$$

We can find the maximum of $\wp(E_1)$ by setting its derivative equal to zero. To make things simpler, we can take the derivative of its logarithm,

$$\frac{\partial}{\partial E_1} \ln \wp(E_1) = -cN^{-1}(4E_1^* - 2E_T) = 0$$

So we find that

$$E_1^* = E_T/2$$

We now compute the ratio of the probability at E_1^* to some other value that is a slight deviation away, $E_1 = \gamma E_1^*$, where γ is close to 1,

$$\frac{\wp(E_1^*)}{\wp(E_1 = \gamma E_1^*)} = \frac{\exp\left[2aN - bE_T - cN^{-1}\left(2E_T^2/4 - 2E_TE_T/2 + E_T^2\right)\right]}{\exp\left[2aN - bE_T - cN^{-1}\left(2\gamma^2 E_T^2/4 - 2E_T\gamma E_T/2 + E_T^2\right)\right]}$$

$$= \exp\left[\frac{cN^{-1}E_T^2(\gamma - 1)^2}{2}\right] = \exp\left[\frac{cN}{2}\left(\frac{E_T}{N}\right)^2(\gamma - 1)^2\right]$$

How does this ratio of probabilities change as the size of the systems grows? Notice that E_T scales as N since it is extensive. This means that the argument of the exponential in the last line above scales like N, for fixed γ,

$$\ln\left[\frac{\wp(E_1^*)}{\wp(E_1 = \gamma E_1^*)}\right] \sim N\left(\frac{N}{N}\right)^2(\gamma - 1)^2 = N(\gamma - 1)^2$$

For macroscopic systems in which $N \sim 10^{23}$, this prefactor becomes extremely large such that, for any value of γ except for the precise value of 1 (for which $E_1 = E_1^*$), the term in the exponential is extremely large. The probability at E_1^* is therefore overwhelmingly greater than that at any other energy.

Consider the case when $N = 10^{23}$ and there is a 0.01% deviation away from E_1^* such that $\gamma = 0.9999$. The probability at E_1^* is of the order of $\exp(10^{15})$ times greater than this seemingly tiny fluctuation in energy.

4.5   General forms for equilibrium and the principle of maximum entropy

So far we have considered the case of energy exchange, but certainly there are physically realistic situations in which two systems can exchange volume (e.g., the

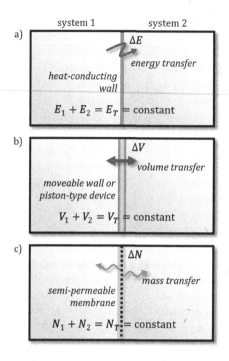

Figure 4.6. Two systems 1 and 2 can exchange different kinds of conserved physical quantities as they approach equilibrium. (a) Systems in thermal contact can exchange energy through a conductive wall. (b) Systems that share a common boundary that can be displaced will exchange volume. (c) Systems separated by a semi-permeable membrane can exchange molecules (mass) of a particular species.

wall between them is movable) or particles, molecules, or mass (e.g., they are separated by a semi-permeable membrane). Figure 4.6 compares these different modes of exchange. In all cases, the overall amount of a particular physical quantity is conserved, whether it is energy, volume, or mass. It is also important to note that, while Fig. 4.6 illustrates these different modes with *physical* separations between the two systems, there need not be any. For example, the systems can also be conceptual *control volumes* of our choosing, with imaginary boundaries that we impose for the sake of thermodynamic analysis.

For systems that exchange particles and volume, the conditions for which the super-system has a maximum number of microstates are derived in an identical manner to that before, by maximizing Ω_T as a function of the exchanged physical quantities. For example, consider two systems that can exchange both energy and volume. We seek to find the equilibrium values of both E_1^* and V_1^* that maximize the accessible microstates. The energy contribution to the maximum again gives us Eqn. (4.19) and hence $T_1 = T_2$. The volume part is

$$\frac{\partial}{\partial V_1}[\Omega_1(E_1, V_1, N_1)\Omega_2(E_T - E_1, V_T - V_1, N_2)]|_{E_1, N_1, E_2, N_2} = 0 \qquad (4.23)$$

which simplifies to

$$\frac{\partial \ln \Omega_1}{\partial V_1} = \frac{\partial \ln \Omega_2}{\partial V_2} \quad \rightarrow \quad \frac{\partial S_1}{\partial V_1} = \frac{\partial S_2}{\partial V_2} \quad \rightarrow \quad \frac{P_1}{T_1} = \frac{P_2}{T_2} \quad \rightarrow \quad P_1 = P_2 \tag{4.24}$$

In other words, the maximum occurs when the pressures of the two systems are identical, which we would have intuitively expected from a force balance. Such a condition is termed *mechanical equilibrium*.

Alternatively, we could have considered two systems that can exchange both energy and molecules of the same species. In this case, one of our maximum-microstate conditions becomes

$$\frac{\partial}{\partial N_1} [\Omega_1(E_1, V_1, N_1)\Omega_2(E_T - E_1, V_2, N_T - N_1)]_{E_1, V_1, E_2, V_2} = 0 \tag{4.25}$$

giving us

$$\frac{\partial \ln \Omega_1}{\partial N_1} = \frac{\partial \ln \Omega_2}{\partial N_2} \quad \rightarrow \quad \frac{\partial S_1}{\partial N_1} = \frac{\partial S_2}{\partial N_2} \quad \rightarrow \quad \frac{\mu_1}{T_1} = \frac{\mu_2}{T_2} \quad \rightarrow \quad \mu_1 = \mu_2 \tag{4.26}$$

In this case, particle exchange ensures equality of the chemical potentials in the two systems and, hence, *chemical equilibrium*. If more than one species were being exchanged, we would have an equality of chemical potentials between like species for each component allowed to transfer.

In summary, by allowing systems to exchange energy, volume, or mass, and by invoking the principle of equal a priori probabilities, we recover several different kinds of equilibrium.

The conditions for equilibrium between two single-component systems are

(1) $T_1 = T_2$ **thermal equilibrium**
(2) $P_1 = P_2$ **mechanical equilibrium**
(3) $\mu_1 = \mu_2$ **chemical equilibrium**

Again, each of these equalities originates from the maximization of the total number of microstates, $\Omega_T = \Omega_1 \times \Omega_2$, under conditions in which something is exchanged between two systems but conserved in total. The procedure is equivalent to maximizing the total entropy of the supersystem, $S_T = S_1 + S_2$. These arguments form the basis of one of the most central principles of thermodynamics.

The **second law of thermodynamics** prescribes that, in an *isolated* system, all spontaneous processes can only increase the total entropy of the system.

This statement merely means that systems spend most of their time at the states with the most microstates, effectively picking out conditions that maximize microstates. As we saw, systems do not really make any kind of specialized choice; they spend equal time in *every* microstate. However, because *we view* the system in terms of macroscopic rather than microscopic variables and because one *macro*state contains so many more microstates than any other, it seems to us as if systems select these conditions exclusively. The second law merely says that we do not observe the macroscopically rare states.

There is therefore nothing mysterious about the entropy. It is a tool to help find the conditions that are the most probable and that dominate our macroscopic view of a system. You may now wonder what would happen if the distribution of transferred energies were not sharply peaked such that there existed a noticeable probability for fluctuations away from the maximum-term energy. Would the second law still hold? Such cases actually do emerge in very small systems consisting of only a few molecules. While the maximum-term approximation of Eqn. (4.21) is no longer reasonable, the total number of microstates involving the full sum over energies is still given by Eqn. (4.13). In these cases, as long as the systems still obey the principle of equal a priori probabilities, they will experience energy fluctuations that maximize the number of microstates explored over time. In other words, the sum of microstates in Eqn. (4.13) for systems in thermal contact will always be greater than the number in Eqn. (4.10). Small-system cases simply mean that macroscopic simplifications, such as the maximum-term method, cannot be applied.

The second law also suggests the direction in which energy, volume, or particles will be exchanged as two coupled systems approach equilibrium. The entropy can only increase with respect to some exchange of these quantities because the combined supersystem will proceed to macroscopic conditions that maximize the number of microstates. For a small change in conditions on the road to equilibrium, we can examine the differential entropy of the supersystem, which must be greater than or equal to zero,

$$dS_T \geq 0 \tag{4.27}$$

From Eqn. (4.11),

$$dS_1 + dS_2 \geq 0 \tag{4.28}$$

Using the fundamental equation for each system,

$$\frac{1}{T_1} dE_1 + \frac{P_1}{T_1} dV_1 - \frac{\mu_1}{T_1} dN_1 + \frac{1}{T_2} dE_2 + \frac{P_2}{T_2} dV_2 - \frac{\mu_2}{T_2} dN_2 \geq 0 \tag{4.29}$$

Consider the specific case of thermal equilibrium with $dV = 0$ and $dN = 0$. The total energy $E_1 + E_2 = E_T$ is constant, giving $dE_1 + dE_2 = 0$ or $dE_2 = -dE_1$. Thus, upon approaching equilibrium, Eqn. (4.29) shows that

$$\left(\frac{1}{T_1} - \frac{1}{T_2} \right) dE_1 \geq 0 \tag{4.30}$$

This important result states that energy will transfer only from hotter to colder systems. If dE_1 is positive, system 1 has gained energy. The only way this inequality can be satisfied is then if $(1/T_1 - 1/T_2) \geq 0$, or $T_1 \leq T_2$. If dE_1 is negative, we must have $T_1 \geq T_2$. In either case, the system that gains energy is the one that has a lower temperature. An equivalent perspective is that heat energy will flow only opposite to the direction of, or downward along, a temperature *gradient*. Similar arguments can be made to develop constraints on the direction of volume and particle transfer for pressure and chemical-potential differences (gradients), respectively. These tasks are left as problems for the reader at the end of the chapter.

4.6   The second law and internal constraints

We derived the maximum-entropy condition from simple scenarios involving two systems that came to mutual equilibrium through different kinds of exchange. A perfectly equivalent way of viewing the same processes is in terms of a single large supersystem with an *internal constraint* that is removed. In this view, the internal constraint initially prevents part of the supersystem (system 1) from interacting with another part (system 2). Its removal allows a certain kind of exchange, be it energy, volume, or mass. Since the supersystem is isolated, this implies that an alternative way of writing the second law is

$$S(E, V, N | \text{internal constraint}) \leq S(E, V, N) \qquad (4.31)$$

where "internal constraint" on the left implies that the entropy must be calculated in the presence of the constraint. This would be equivalent to counting the number of microstates in the constrained case. Rewriting this in terms of the density of states gives

$$\Omega(E, V, N | \text{internal constraint}) \leq \Omega(E, V, N) \qquad (4.32)$$

Though it may seem obvious that a constraint can only operate to reduce the number of possible microstates, the implications of this result are quite broad. Any time we remove an internal constraint, the entropy must either remain the same or increase.

An equivalent formulation of the second law is that spontaneous processes in isolated systems resulting from the removal of an internal constraint can only increase the entropy.

The concept of internal constraints appears throughout thermodynamics. An internal constraint could be an insulating barrier that prevents heat flow or the presence of a wall or membrane that prevents mass transfer. As we will find in more detail later, an internal constraint can also exist naturally due to sluggish kinetics. It could be the observation that a system maintains a liquid state below its freezing point due to slow crystallization kinetics, at which point it would be supercooled. It could also be the fact that a macromolecule like a protein does not spontaneously decompose into its constituent amino acids due to slow reaction kinetics. Such examples show that an internal constraint need not be something that we actually physically impose. We must only recognize when one is present if we are to examine the ultimate behavior of a system when the constraint is removed, either deliberately or naturally.

An **internal constraint** is a statement of the fact that a system does not have full access to all possible internal configurations (microstates), irrespective of whether the constraint is physically imposed or naturally results from slow system kinetics.

An internal constraint also need not be a binary choice, such as the presence or absence of a partition. Some constraints are associated with a value or *internal degree of freedom*. In our previous example for thermal equilibrium, the energy E_1 was a super-system degree of freedom that changed as it came to equilibrium. Similarly, the cases of mechanical and chemical equilibrium involved degrees of freedom related to V_1 and N_1, respectively. In general, we can indicate the particular value of an internal degree of freedom as ξ. The entropy of the system is then

$$S(E, V, N, \xi) = k_B \ln \Omega(E, V, N, \xi) \tag{4.33}$$

We could fix the value of ξ at ξ_1 such that the system entropy will be $S(E, V, N, \xi_1)$. For example, we could insert an internal insulating partition that prevents energy exchange and hence fixes E_1. Instead, we could allow ξ to vary freely by removing the partition. Under these circumstances and presuming that the system remains isolated, the second law indicates that ξ will adopt a value ξ^* that maximizes $S(E, V, N, \xi)$:

$$S(E, V, N, \xi_1) \leq S(E, V, N, \xi^*) \tag{4.34}$$

where ξ^* maximizes $S(E, V, N, \xi^*)$. In other words, releasing any degrees of freedom associated with a constraint will always maximize and increase the entropy. Another way of thinking about an internal constraint is that it involves some *perturbation* away from equilibrium. For example, a transient increase in the density or energy in some spatial region can be a perturbation. Therefore, we can express the second law in a different but quite general way.

> The second law prescribes that an isolated system at equilibrium will maximize its entropy. That is,
>
> S is **maximized** at constant E, V, N conditions
>
> where the maximum is found with respect to any internal perturbations of the system, or unconstrained internal degrees of freedom.

4.7   Equivalence with the energy-minimum principle

With the idea of internal constraints, we can re-cast the second law in terms of an energy-minimization principle. Assume a system has an unconstrained internal degree of freedom ξ such that its entropy is $S(E, V, N, \xi)$. At equilibrium, the observed value of ξ is determined by

$$\left(\frac{\partial S}{\partial \xi}\right)_{E, V, N} = 0 \quad \text{and} \quad \left(\frac{\partial^2 S}{\partial \xi^2}\right)_{E, V, N} < 0 \tag{4.35}$$

Let us now imagine that instead we hold the system at constant S, V, N conditions. It may seem unusual to maintain constant entropy, but conceptually we can imagine a scenario in which we systematically add or remove energy so that the system's entropy remains at a fixed value. In effect, we vary E to compensate for changes in the value of ξ so as to keep S constant. This E variation is given by

$$\left(\frac{\partial E}{\partial \xi}\right)_S = -\left(\frac{\partial E}{\partial S}\right)_\xi \left(\frac{\partial S}{\partial \xi}\right)_E = -T\left(\frac{\partial S}{\partial \xi}\right)_E \tag{4.36}$$

where we omitted the subscripts V and N because these are constant throughout. The first equality entails a simple mathematical relation for derivatives called the "triple-product rule." In the second equality, we introduce the temperature through the fundamental equation. Equation (4.36) indicates that, when the system is at an entropy maximum and the right derivative is zero, it is also at an energy extremum. To determine whether this is a maximum or minimum, we examine the second derivative,

$$\left(\frac{\partial^2 E}{\partial \xi^2}\right)_S = -\left(\frac{\partial S}{\partial \xi}\right)_E \frac{\partial}{\partial \xi}\left[\left(\frac{\partial E}{\partial S}\right)_\xi\right]_S - \left(\frac{\partial E}{\partial S}\right)_\xi \frac{\partial}{\partial \xi}\left[\left(\frac{\partial S}{\partial \xi}\right)_E\right]_S$$

$$= -\left(\frac{\partial S}{\partial \xi}\right)_E \left(\frac{\partial^2 E}{\partial S \partial \xi}\right) - \left(\frac{\partial E}{\partial S}\right)_\xi \frac{\partial}{\partial \xi}\left[\left(\frac{\partial S}{\partial \xi}\right)_E\right]_S$$

$$= -T\frac{\partial}{\partial \xi}\left[\left(\frac{\partial S}{\partial \xi}\right)_E\right]_S \tag{4.37}$$

In going from the second to the third line, we used the fact that $(\partial S/\partial \xi)_E = 0$ at an entropy maximum. The remaining derivative is slightly more difficult because of the change of constant conditions from E to S. Consider a function $X(E, \xi)$. The total differential is

$$dX = \left(\frac{\partial X}{\partial E}\right)_\xi dE + \left(\frac{\partial X}{\partial \xi}\right)_E d\xi \tag{4.38}$$

Taking the $d\xi$ derivative at constant S conditions gives

$$\left(\frac{\partial X}{\partial \xi}\right)_S = \left(\frac{\partial X}{\partial E}\right)_\xi \left(\frac{\partial E}{\partial \xi}\right)_S + \left(\frac{\partial X}{\partial \xi}\right)_E \tag{4.39}$$

Letting X be equal to $(\partial S/\partial \xi)_E$, we can then fully evaluate (4.37),

$$\left(\frac{\partial^2 E}{\partial \xi^2}\right)_S = -T\left[\left(\frac{\partial^2 S}{\partial E \partial \xi}\right)\left(\frac{\partial E}{\partial \xi}\right)_S + \left(\frac{\partial^2 S}{\partial \xi^2}\right)_E\right]$$

$$= -T\left(\frac{\partial^2 S}{\partial \xi^2}\right)_E \tag{4.40}$$

The first term in brackets vanishes because of the extremum condition, $(\partial E/\partial \xi)_S = 0$. The second derivative of the entropy is negative at a maximum and therefore our final result implies

$$\left(\frac{\partial^2 E}{\partial \xi^2}\right)_S > 0 \tag{4.41}$$

Thus the system is at an energy minimum. The implication of these mathematical arguments is that a maximum in entropy at constant E, V, and N is equivalent to a minimum in energy at constant S, V, and N. Therefore, the second law can adopt an equivalent interpretation in terms of the internal energy.

> The second law prescribes that, at equilibrium,
>
> E is **minimized** at constant S, V, N conditions
>
> with respect to any internal perturbations of the system, or unconstrained internal degrees of freedom.

4.8   Ensemble averages and Liouville's theorem in classical systems

Earlier we mentioned the concept of microstate probabilities and described an example in which these could be constructed by monitoring a system at equilibrium over a very long period of time and making tallies of instantaneous microstates. A remarkable feature of equilibrium is that there is an equivalent way for determining these probabilities that involves viewing not a single system over a long period of time, but rather a very large number of identical systems, each at equilibrium at a single instant in time.

Consider a scenario in which there are many copies of the same classical, isolated system, each with identical macroscopic states. These systems all have the same values of E, V, and N, but they are at different microstates initially and evolve independently in time according to Newton's laws of motion. We call this collection of systems an *ensemble*. In fact, the term ensemble originally stems from this kind of scenario involving many macroscopically identical systems. Eventually we will let the number of such systems grow to infinity.

At any one moment in time, we could interrogate the ensemble and examine the current microstate in each system. Here, a microstate in one system is defined by the complete set of atomic positions $\mathbf{r}^N = (\mathbf{r}_1, \mathbf{r}_2, \ldots, \mathbf{r}_N)$ and momenta $\mathbf{p}^N = (\mathbf{p}_1, \mathbf{p}_2, \ldots, \mathbf{p}_N)$. We would like to characterize the *distribution* of microstates in this large ensemble. That is, we would like to compute the probability of finding systems within our ensemble that have any given microstate $(\mathbf{r}^N, \mathbf{p}^N)$. Since there are infinitely many microstates, we denote this probability using a continuous probability density,

$$\wp(\mathbf{r}^N, \mathbf{p}^N)d\mathbf{r}^N \, d\mathbf{p}^N \tag{4.42}$$

which gives the fraction of the systems in the ensemble that have their positions within the differential range $\mathbf{r}^N \pm d\mathbf{r}^N/2$ and their momenta within the differential range $\mathbf{p}^N \pm d\mathbf{p}^N/2$. In other words, we can think of $\wp(\mathbf{r}^N, \mathbf{p}^N)$ as the local density of systems in the ensemble at any moment in time with positions $\mathbf{r}^N$ and momenta $\mathbf{p}^N$. This density is not in three dimensions, but exists in a highly dimensional space where there is an axis for each position and momentum coordinate ($6N$ axes in total). We call the space of $(\mathbf{r}^N, \mathbf{p}^N)$ coordinates *phase space*.

Note that the distribution $\wp(\mathbf{r}^N, \mathbf{p}^N)$ can change with time because each system in the ensemble continuously evolves from one microstate to another. However, $\wp(\mathbf{r}^N, \mathbf{p}^N)$ acts like a density because the total probability must integrate to one, a constant. This means that we can apply a conservation law to the evolution of $\wp(\mathbf{r}^N, \mathbf{p}^N)$. In generalized form for a density $\rho(\mathbf{x})$, the *continuity equation* emerges from the constraint of the conserved total integrated ρ,

$$\frac{\partial \rho}{\partial t} + \nabla \cdot (\rho \mathbf{u}) = 0 \tag{4.43}$$

Here, $\mathbf{u}$ is the velocity of the system in the space of $\mathbf{x}$ due to flow, such that u_i is the flow along the axis x_i. Expanding the divergence operator,

$$\frac{\partial \rho}{\partial t} + \sum_i \left(u_i \frac{\partial \rho}{\partial x_i} + \rho \frac{\partial u_i}{\partial x_i} \right) = 0 \tag{4.44}$$

In the case of the ensemble, the degrees of freedom $\mathbf{x}$ are simply the combined sets of positions and velocities since both of these must be specified to define a microstate: $\mathbf{x} = (\mathbf{r}^N, \mathbf{p}^N)$. The density is given by $\rho = \wp(\mathbf{r}^N, \mathbf{p}^N)$.

We now must compute the flow velocity in the ensemble. That is, for systems at a given value of $(\mathbf{r}^N, \mathbf{p}^N)$, we need to know how fast they are moving to new values of $(\mathbf{r}^N, \mathbf{p}^N)$ and in what direction. In other words, we want to know the velocity vector

$$\mathbf{u} = \left(\frac{\partial \mathbf{r}^N}{\partial t}, \frac{\partial \mathbf{p}^N}{\partial t} \right) = (\dot{\mathbf{r}}^N, \dot{\mathbf{p}}^N) \tag{4.45}$$

as a function of the microstate $(\mathbf{r}^N, \mathbf{p}^N)$. Newtonian principles completely determine this. First, we are guaranteed that all systems that sit at the same values of $(\mathbf{r}^N, \mathbf{p}^N)$ will go to *exactly* the same place as they evolve in time since the atomic trajectory is deterministic. Second, Newton's equations tell us precisely how fast the coordinates $(\mathbf{r}^N, \mathbf{p}^N)$ will change. Recall that the total Hamiltonian is $H = U(\mathbf{r}_1, \mathbf{r}_2, \ldots) + \sum p_i^2 / 2m_i$. Now take note of the following H derivatives:

$$\frac{\partial H}{\partial \mathbf{r}_i} = \frac{\partial U}{\partial \mathbf{r}_i}$$

$$= \frac{\partial \, \mathbf{p}_i}{\partial t} \tag{4.46}$$

by Newton's Law, and

$$\frac{\partial H}{\partial \mathbf{p}_i} = \frac{\mathbf{p}_i}{m} = \frac{\partial \mathbf{r}_i}{\partial t} \tag{4.47}$$

Since $\mathbf{p}_i = m\mathbf{v}_i$. Therefore, the flow velocity is given in terms of the Hamiltonian as

$$\mathbf{u} = \left(\frac{\partial H}{\partial \mathbf{p}^N}, -\frac{\partial H}{\partial \mathbf{r}^N} \right) \tag{4.48}$$

With Eqns. (4.44) and (4.48), the continuity equation becomes

$$\frac{\partial \wp}{\partial t} + \sum_i \left(\frac{\partial H}{\partial \mathbf{p}_i} \cdot \frac{\partial \wp}{\partial \mathbf{r}_i} - \frac{\partial H}{\partial \mathbf{r}_i} \cdot \frac{\partial \wp}{\partial \mathbf{p}_i} + \wp \frac{\partial}{\partial \mathbf{r}_i} \cdot \frac{\partial H}{\partial \mathbf{p}_i} - \wp \frac{\partial}{\partial \mathbf{p}_i} \cdot \frac{\partial H}{\partial \mathbf{r}_i} \right) = 0 \tag{4.49}$$

The final two terms in this expression give the same second derivative of the Hamiltonian, only with the order of the derivative switched. These are equal,

$$\frac{\partial}{\partial \mathbf{r}_i} \cdot \frac{\partial H}{\partial \mathbf{p}_i} = \left(\frac{\partial^2 H}{\partial x_i \partial p_{x,i}}, \frac{\partial^2 H}{\partial y_i \partial p_{y,i}}, \frac{\partial^2 H}{\partial z_i \partial p_{z,i}} \right) = \frac{\partial}{\partial \mathbf{p}_i} \cdot \frac{\partial H}{\partial \mathbf{r}_i} \tag{4.50}$$

This means that the last two terms in the continuity equation cancel out, leaving finally

$$\frac{\partial \wp}{\partial t} + \sum_i \left(\frac{\partial H}{\partial \mathbf{p}_i} \cdot \frac{\partial \wp}{\partial \mathbf{r}_i} - \frac{\partial H}{\partial \mathbf{r}_i} \cdot \frac{\partial \wp}{\partial \mathbf{p}_i} \right) = 0 \tag{4.51}$$

This is known as *Liouville's equation*, after the French mathematician Joseph Liouville. It has important implications for the total derivative of $\wp$. In the above expression, $\partial \wp / \partial t$ gives the change in the density of systems in the ensemble at a fixed microstate. It would be what we would measure if we sat at a specific point in phase space and watched the density of systems change as they moved in and out of that region. On the other hand, the total (material) derivative would give us the change in the probability as we *moved with* the systems that are initially at a given value of $(\mathbf{r}^N, \mathbf{p}^N)$. It is defined by

$$\frac{d\wp}{dt} \equiv \frac{\partial \wp}{\partial t} + \sum_i \left(\frac{\partial \wp}{\partial \mathbf{r}_i} \cdot \frac{\partial \mathbf{r}_i}{\partial t} + \frac{\partial \wp}{\partial \mathbf{p}_i} \cdot \frac{\partial \mathbf{p}_i}{\partial t} \right)$$

$$= \frac{\partial \wp}{\partial t} + \sum_i \left(\frac{\partial \wp}{\partial \mathbf{r}_i} \cdot \frac{\partial H}{\partial \mathbf{p}_i} - \frac{\partial \wp}{\partial \mathbf{p}_i} \cdot \frac{\partial H}{\partial \mathbf{r}_i} \right)$$

$$= 0 \tag{4.52}$$

The last line is zero by virtue of Liouville's equation in (4.51). In other words, if we were to select a system in the ensemble and follow it with time through phase space, the local density of other systems around it would not change. We therefore say that phase space is *incompressible*.

A second consequence of Liouville's equation is that the microstate distribution $\wp(\mathbf{r}^N, \mathbf{p}^N)$ does not change with time if the initial distribution $\wp_0(\mathbf{r}^N, \mathbf{p}^N)$ is a function of the Hamiltonian only,

$$\wp_0(\mathbf{r}^N, \mathbf{p}^N) = f(H(\mathbf{r}^N, \mathbf{p}^N)) \tag{4.53}$$

This is readily shown by substituting $\wp_0$ into Liouville's equation,

$$\frac{\partial \wp}{\partial t} = -\sum_i \left(\frac{\partial H}{\partial \mathbf{p}_i} \cdot \frac{\partial \wp}{\partial \mathbf{r}_i} - \frac{\partial H}{\partial \mathbf{r}_i} \cdot \frac{\partial \wp}{\partial \mathbf{p}_i} \right)$$

$$= -\sum_i \left[\frac{\partial H}{\partial \mathbf{p}_i} \cdot \left(\frac{df}{dH} \frac{\partial H}{\partial \mathbf{r}_i} \right) - \frac{\partial H}{\partial \mathbf{r}_i} \cdot \left(\frac{df}{dH} \frac{\partial H}{\partial \mathbf{p}_i} \right) \right] = 0 \tag{4.54}$$

That is, the microstate probabilities described by the distribution $\wp(\mathbf{r}^N, \mathbf{p}^N)$ become time-invariant if the probability of each microstate depends only on its total energy $H(\mathbf{r}^N, \mathbf{p}^N)$. Note that this is very close to what the principle of equal a priori probabilities dictates: it says that the probability of a microstate depends only on its energy E_m and whether or not that energy equals the macroscopic energy, $E_m = E$. Therefore, this equation shows that, if a system obeys the principle, its microstate probabilities will remain constant with time, which is one manifestation of equilibrium. Note, however, that this derivation does not exclude the possibility of other forms of the microscopic probabilities that might also be independent of time. That is, the equal a priori assumption is more specific than the result of Eqn. (4.54).

Problems

Conceptual and thought problems

4.1. Write an expression for the probability that all of the air molecules in a 6 m × 6 m × 3 m room will suddenly all occupy a 1 m × 1 m × 1 m cube in the back corner. Assume that the room is filled with an ideal gas. Make sure your expression uses only *measurable* quantities like pressure P, temperature T, and volume V.

4.2. Two containers have identical volumes and numbers of particles but hold different ideal gas species, as depicted in Fig. 4.7. The containers are then joined and the particles allowed to mix. In one case, they are joined such that the final volume remains the same as each initial container. In another case, they are joined such that the pressure remains the same. Write an expression for the change in entropy upon mixing for each case.

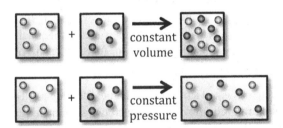

Figure 4.7. Two different ideal-gas mixing scenarios.

4.3. An insulated system contains two compartments 1 and 2, containing N_1 and N_2 molecules of an ideal gas, respectively. The compartments are linked by a movable, insulating, impermeable wall. Thus the compartments can exchange volume, while the total volume $V_T = V_1 + V_2$ is conserved.

(a) Assume that each box can be subdivided into very small cells of volume v; each cell serves as one particular location where one or more ideal gas particles can be placed. Find an expression for the density of states of the entire system in terms of v, N_1, N_2, V_T, and $x_1 = V_1/V_T$. Neglect energies in your microstate counting.

(b) Find an expression for the value of x_1 and hence V_1 that maximizes the density of states.

(c) What happens if the volume is slightly different than its value at the density-of-states maximum? Consider another value of V_1, given by $V_1 = 0.9999 V_1^*$, where

V_1^* is the value found in part (b). Determine the base-10 logarithm of the ratio of the number of microstates at the two volumes, $\log[\Omega(0.9999V_1^*)/\Omega(V_1^*)]$. Take N_1 and N_2 to be 10^{23}. Given the principle of equal a priori probabilities, what does this result imply for the frequency with which the volume $0.9999V_1^*$ will be seen, relative to V_1^*?

(d) Show thermodynamically that, if $P_1 > P_2$, the approach to equilibrium involves compartment 1 gaining and compartment 2 losing volume. Assume that the temperatures of the two systems are the same.

4.4. A volume V consists of equal numbers of moles of two ideal gases A and B. Find an expression for the probability that the gases will spontaneously demix into regions of pure A and pure B. Calculate the number of molecules for which the probability is 1%, 0.01%, and 0.0001%.

Fundamentals problems

4.5. Consider Fig. 4.4 involving two systems that come to equilibrium through energy exchange. Assume now that both systems, 1 and 2, are two-state models.

(a) What happens if the systems have initial macrostates $N_1 = 10$, $E_1 = 4b$, $N_2 = 20$, and $E_2 = 5b$? Find the total number of microstates after thermal equilibrium, and the distribution $\wp(E_1)$.

(b) Instead, the systems have initial macrostates $N_1 = 10\lambda$, $E_1 = 4b\lambda$, $N_2 = 20\lambda$, and $E_2 = 5b\lambda$, where λ is a large number. Find an expression for the relative probability $\wp(E_1{=}4\lambda)/\wp(E_1{=}E_1^*)$, where E_1^* maximizes the total number of microstates. What is this probability for $\lambda = 10$, 100, and 1000? You may wish to use Stirling's approximation.

4.6. Starting with the principle of equal a priori probabilities, develop a full derivation of the conditions of equilibrium for the cases in which two subsystems can exchange energy and (a) volume through a common movable wall, or (b) particles/mass through a membrane.

4.7. Consider two subsystems that can exchange volume as in Fig. 4.6(b). What happens if the movable wall separating the two systems is insulated, such that no heat can pass between the two systems? What are the equilibrium conditions in this case? Can the pressures on the opposite sides of the walls differ? Hint: even though no heat can pass through the walls, the systems will still be able to exchange energy. How?

4.8. Consider two subsystems that can exchange energy and mass as in Fig. 4.6(c). Imagine that there are two components in each subsystem, but only component A can pass through the membrane, while component B cannot. What are the conditions for equilibrium in this case?

4.9. A single particle sits on an infinite two-dimensional surface. Its potential energy is given by a harmonic expression: $U = ar^2$, where r is its distance from the origin and a is a constant. Count the number of molecular configurations, or

Table 4.2 Distinct microstates for problem 4.10

n_1	n_2	n_3	Shorthand
3	0	0	***||
0	3	0	|***|
0	0	3	||***
2	1	0	**|*|
2	0	1	**||*
1	2	0	*|**|
1	0	2	*||**
0	2	1	|**|*
0	1	2	|*|**
1	1	1	*|*|*

microstates, in this system for a given fixed total potential energy (neglecting kinetic energies). In order to count the infinite number of classical states, you will assume that the position variables of the particle are discretized into integer multiples of a distance δx. Write an expression for the number of microstates $\Omega(U)$ that have energy between U and $U + dU$, where dU is a small differential window of potential energy.

4.10. Earlier in this book we discussed the simple two-state system. Instead, consider an infinite-state system that consists of a number N of non-interacting "particles." Each particle can occupy one of an infinite number of energy levels that are evenly spaced. The energy of particle i is given by ϵn_i, where n_i is a positive integer starting at zero. The total energy of the system of N particles is given by

$$E = \sum_i \epsilon n_i = \epsilon \sum_i n_i$$

Consider, for example, $E = 3\epsilon$ and $N = 3$. The list of distinct microstates is given in Table 4.2. There are ten microstates in total in this case. In the right column, we have written a "stars and bars" shorthand for each state. A star "*" signifies one quantum of energy (ϵ) is assigned to a particle, and the partitions "|" divide the particles from each other. Thus, the number of distinct microstates here is equivalent to the number of ways of picking which three of the five possible characters in the shorthand notation are going to be bars versus stars.

(a) What is wrong with the expression $\Omega = N^{E/\epsilon} = 3^3$, which assumes that each quantum is assigned independently to a particle? What is wrong with the expression $\Omega = N^{E/\epsilon}/(E/\epsilon)!$, which presumably attempts to correct this problem?

(b) Instead, write a general expression for the density of states $\Omega(E, N)$ using the shorthand notation. With Stirling's approximation, develop an expression for the entropy $S(E, N)$.

(c) Find an expression for the constant-volume heat capacity, $C_V \equiv (\partial E/\partial T)_{V,N}$.

(d) As we will see in Chapter 14, the infinite-state model serves as a simple quantum-mechanical description of perfect monatomic crystals, in which solutions to Schrödinger's equation show that $\epsilon = h\nu$. For a real crystal, the equivalent number of infinite-state particles is three times the number of atoms due to three-dimensional degeneracies. For copper, $h\nu/k_B \approx 343$ K. Compute the heat capacity of copper at room temperature $(20\,°C)$ and compare your answer with the experimental value of 24.5 J/mol K.

Applied problems

4.11. Living cells are surrounded by a plasma membrane composed of amphipathic phospholipids. These molecules self-assemble into a bilayer in aqueous solvent so as to point their hydrophobic tails inwards, while their highly polar head groups are exposed to water. Many peptides – very small polymers of amino acids – are thought to be useful as therapeutics for their interactions with these cell membranes. For example, they may disrupt bacterial membranes and exhibit antibacterial activity. Consider N identical peptides that are membrane-associated. Assume that each can either sit at the extracellular interface in the hydrophilic region or insert itself into the hydrophobic bilayer, as in Fig. 4.8.

(a) You can view this problem as one in which two systems can freely exchange peptide molecules, one corresponding to the hydrophilic extracellular region and one corresponding to the hydrophobic bilayer interior. According to the entropy-maximization principle, what condition must be satisfied for equilibrium of peptide exchange?

(b) Consider a highly simplified model in which the change in energy for going from the extracellular interface to the bilayer region is given by ϵ. Find an expression for the fraction of the peptides that are in the bilayer as a function of temperature.

(c) An approximate energy change for moving a single glutamine amino acid from the bilayer interface to a hydrocarbon environment is 0.2 kcal/mol. Assume that a homopeptide made of n glutamines then has a transfer energy of $n \times$ (0.2 kcal/mol). What is the maximum length of such a peptide n for which at

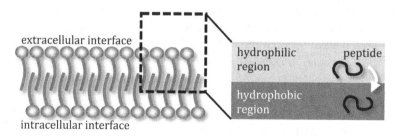

Figure 4.8. A peptide can partition in the exterior or interior of a lipid bilayer in a cell's plasma membrane.

least 10% of the peptides will be present inside the bilayer? Assume that the temperature is $T = 300$ K.

4.12. In order to be effective, many small-molecule drugs must pass through the cell membrane and into its interior, where they typically function by binding tightly to a particular protein implicated. Here, you will analyze this problem using simple molecular models.

(a) First consider the generic case without attention to specific numbers. The cell and its surroundings are two systems, C and S, that are isolated from the rest of the world but can exchange energy and drug molecules with each other. Show that, due to the principle of equal a priori probabilities, the most likely state of the two systems upon energy and molecule exchange corresponds to the point at which their temperatures and chemical potentials are equal.

(b) Show also that if $\mu_C > \mu_S$ then the approach to equilibrium corresponds to the cell losing and the surroundings gaining drug molecules. What does your physical intuition tell you about the general dependence of μ on concentration? Hint: think about diffusion.

4.13. Extending the previous problem, consider a specific model of protein–drug binding inside the cell. Let $N_D = c_D V_{cell}$ and $N_P = c_P V_{cell}$ denote the numbers of drug and protein molecules in the cell, respectively, where c_D and c_P are their concentrations in number per volume. For the sake of simplicity, consider the proteins to be immobile within the cell cytoskeleton. So that you can count microstates, discretize all of space into regions the size of the binding region in the protein, with volume v, such that the total number of "pixels" inside a cell is $M = V_{cell}/v$. Assume that the N_P protein molecules each occupy one pixel and cannot change their location. Drug molecules can occupy any of the pixels. When a drug sits on one of the protein-occupied pixels, there is a favorable binding energy of $-\epsilon$; otherwise, the energy is zero. Let x denote the fraction of the N_P sites with a bound drug molecule and y the fraction of the remaining $(M - N_P)$ sites with a free drug molecule.

(a) Write an expression for the entropy in terms of N_P, M, x, and y. Use Stirling's approximation.

(b) Assuming that N_D is less than the number of pixels, write an expression that relates x and y to each other, and to the values N_P, N_D, and M. Using the connection between the entropy and temperature, show that the equilibrium values of x and y satisfy the relation

$$\frac{y(1-x)}{x(1-y)} = e^{-\epsilon/k_B T}$$

Hint: Use the chain rule for the entropy derivative, e.g.,

$$(\partial S/\partial E) = (\partial S/\partial x)(\partial x/\partial E)$$

(c) A pharmaceutical company is developing a drug to target a specific protein that exists in intracellular concentrations of 10^{-7} mol/l. In order to be effective, the drug must bind to 20% of its protein target. The binding volume of this particular protein cavity is 200 $Å^3$/molecule. It is estimated that the drug can be delivered to the interior of the cell at concentrations of 10^{-6} mol/l. What must the binding energy ϵ be, in kcal/mol, at the physiological temperature of $T = 293$ K?

4.14. Resveratrol is a bioactive phytochemical produced in the skins of grapes that is hypothesized to contribute to the health benefits of red wine. This molecule can exist in one of two isomeric forms, *cis* and *trans*. While both forms display bioactivity, the overwhelming constituent found in grape skins is the *trans* form, and this particular isomer is thought to have the greatest positive impact on health. In what follows, you will create a very simplified model of resveratrol isomerization equilibrium. You will need to make several assumptions for a solution of N resveratrol molecules.

- Each molecule can freely interconvert between the two states: the *cis* state or the *trans* state. Neglect other possible conformations.
- Kinetic energies and interactions with the solvent are neglected.
- The solution is dilute such that resveratrol molecules do not interact with one another.
- The conformational internal energy of a molecule in the *trans* state is 0 and that of a molecule in the *cis* state is $-\epsilon$. Thus the total energy is given by $E = -n\epsilon$, where n is the number of molecules in the *cis* state.

(a) Develop an expression for the entropy of this simple model. Begin by finding the number of microstates as a function of the number of *cis* molecules, n, and hence the total energy, E. Simplify your answer using Stirling's approximation, and rewrite it in terms of the fraction of *cis* isomers $x \equiv n/N$ and the total number of particles N.

(b) Find the fraction of *cis* isomers x as a function of temperature.

(c) Fremont [*Life Sciences* **66**, 663 (2000)] estimates that the equilibrium fraction of *cis* isomers is around 91% at near-ambient temperature (300 K). What value of ϵ does this suggest? Express your answer in kcal/mol and compare it with the energy of a typical hydrogen bond in water (~2 kcal/mol).

(d) Given the numerical results in part (c), at what temperature will 60% of the molecules be in the *cis* isomer? What about 50%?

(e) Stable solutions of nearly pure *trans*-resveratrol can be prepared, so long as they are shielded from harsh lighting conditions. Why is this possible, in light of your analysis in part (d)?

FURTHER READING

H. Callen, *Thermodynamics and an Introduction to Thermostatistics*, 3rd edn. New York: Wiley (1985).

K. Denbigh, *The Principles of Chemical Equilibrium*, 4th edn. New York: Cambridge University Press (1981).

T. L. Hill, *An Introduction to Statistical Thermodynamics*. Reading, MA: Addison-Wesley (1960); New York: Dover (1986).

E. A. Jackson, *Equilibrium Statistical Mechanics*. Mineola, NY: Dover (1968).

A. I. Khinchin, *Mathematical Foundations of Statistical Mechanics*. New York: Dover (1949).

L. D. Landau and E. M. Lifshitz, *Statistical Physics*, 3rd edn. Oxford: Butterworth-Heinemann (1980).

D. A. McQuarrie, *Statistical Mechanics*. Sausalito, CA: University Science Books (2000).

J. W. Tester and M. Modell, *Thermodynamics and Its Applications*, 3rd edn. Upper Saddle River, NJ: Prentice Hall (1997).

R. C. Tolman, *The Principles of Statistical Mechanics*. New York: Dover (1979).

5 The fundamental equation

5.1 Equilibrium and derivatives of the entropy

We have now seen that the entropy is a tool to find the most likely macroscopic state of a system: the macroscopic conditions that have the greatest number of microstates. We also saw that the conditions for equilibrium between two bodies are intimately linked to derivatives of the entropy. For two single-component systems 1 and 2, there are three types of equilibrium:

thermal equilibrium $\qquad \dfrac{\partial S_1}{\partial E_1} = \dfrac{\partial S_2}{\partial E_2} \rightarrow \dfrac{1}{T_1} = \dfrac{1}{T_2} \rightarrow T_1 = T_2$

mechanical (and thermal) equilibrium $\qquad \dfrac{\partial S_1}{\partial V_1} = \dfrac{\partial S_2}{\partial V_2} \rightarrow \dfrac{P_1}{T_1} = \dfrac{P_2}{T_2} \rightarrow P_1 = P_2$

chemical (and thermal) equilibrium $\qquad \dfrac{\partial S_1}{\partial N_1} = \dfrac{\partial S_2}{\partial N_2} \rightarrow \dfrac{\mu_1}{T_1} = \dfrac{\mu_2}{T_2} \rightarrow \mu_1 = \mu_2$

How do we know that the derivatives of the entropy should involve the quantities T, P, and μ, and not something else? It is precisely because we use equilibrium to measure these quantities in reality. For example, when we measure temperature we allow a thermometer to come to thermal equilibrium with another body with which it exchanges energy. We then use changes in the equilibrium state and properties of the thermometer fluid to characterize the point at which thermal equilibrium is reached. We construct a scale called temperature that depends on these properties. Similarly, we measure pressure by placing a small measuring device in contact with another body such that the two can exchange volume, for instance, through a piston attached to a spring or through liquid that is displaced in a column. We can measure the displacement at equilibrium and calculate the net force exerted by both systems, which must be equal.

Why do the entropy derivatives entail the *specific* quantities $1/T$, P/T, and $-\mu/T$, in which the temperature and pressure are on an absolute scale? This will become clearer in our discussion of work and the first law in the next chapter, but consider for now the following qualitative argument. We could construct a temperature- or pressure-measuring device based on an ideal gas, for which we can determine an explicit, analytical form for the entropy (as we did roughly in Chapter 2 and will do more thoroughly in Chapter 9). We know that $(\partial S/\partial V)_{E,N} = Nk_B/V$ from this analysis, and experiments on ideal gas volumetric properties show that this must be equal to the quantity P/T which we measure in our defined units. If we bring the ideal-gas device to equilibrium with another body, the derivative relations above must still hold. In this sense the ideal gas can serve as a reference point for the thermodynamic quantities that determine equilibrium conditions. The fact that the derivatives of the entropy involve *inverse* temperatures, which may seem confusing at first, is merely an effect of how scales for measuring "hotness" developed historically. If we wanted, we could instead adopt a "coldness" scale based on $1/T$, where 0 represents an infinitely hot state and $+\infty$ is absolute zero.

One way in which we know that the E, V, and N derivatives of the entropy correspond to $1/T$, P/T, and $-\mu/T$, respectively, is because these are the quantities that become equal between two bodies at equilibrium, and because any measurement of T, P, or μ necessarily involves bringing a device into equilibrium with the system of interest. The fact that all of the derivatives involve inverse temperatures is due to the convention for measuring "hotness."

5.2   Differential and integrated versions of the fundamental equations

As we saw in Chapter 2, a convenient summarization of the function $S(E, V, N)$ is the differential form of the fundamental equation,

$$dS = \frac{1}{T}\, dE + \frac{P}{T}\, dV - \frac{\mu}{T}\, dN \tag{5.1}$$

This is of course the *entropy version* of the fundamental equation. An equivalent *energy version* can be made by rearranging the differentials,

$$dE = T\, dS - P\, dV + \mu\, dN \tag{5.2}$$

Equation (5.2) describes the function $E(S, V, N)$ and its partial derivatives:

$$\left(\frac{\partial E}{\partial S}\right)_{V,N} = T(S, V, N), \quad \left(\frac{\partial E}{\partial V}\right)_{S,N} = -P(S, V, N), \quad \left(\frac{\partial E}{\partial N}\right)_{S,V} = \mu(S, V, N) \tag{5.3}$$

All that we have done in going from the entropy version to the energy version is to switch the independent variable from E to S. It is as if we had transformed the equation $w = f(x; y, z)$ by inverting to solve for x as a function of w, that is, $x = g(w; y, z) = f^{-1}(w; y, z)$. Notice also that the E derivatives depend on the independent variables S, V, and N.

We must make a brief comment on our notation for the internal energy using the variable E. This convention is to keep consistency with the microscopic picture of matter, in which the internal energy is the sum of the atomic kinetic energy K and potential energy U. A frequently seen alternate convention designates the internal energy as U. In order to avoid confusion with the atomic potential energy, we will forgo this practice, and always write the total internal energy as E.

The reason for transforming to the energy fundamental equation is that it has a simpler connection to the physical quantities T, P, and μ. We will work more frequently with this version because of this simplicity. Note in particular that this equation involves both extensive and intensive quantities.

> An **extensive** quantity scales linearly with the size of the system. Some examples are E, S, V, and N. An **intensive** quantity is independent of the size of the system upon scaling whilst maintaining the same equilibrium state, such as T, P, and μ.

All of the independent variables S, V, N, and E are extensive, whereas the derivatives are intensive. This fact alone means that the function $E(S, V, N)$ has a very simple form according to a mathematical argument called Euler's theorem. That form is

$$E = TS - PV + \mu N \qquad (5.4)$$

Equation (5.4) gives the *integrated form* of the fundamental equation. It shows that the function $E(S, V, N)$ is given by the sum of the independent variables times the derivatives of E with respect to them. In this expression, it is important to keep in mind that T, P, and μ are not formally independent, but are functions of S, V, and N. It is as if we were to write

$$E = \left(\frac{\partial E}{\partial S}\right)_{V,N} S + \left(\frac{\partial E}{\partial V}\right)_{S,N} V + \left(\frac{\partial E}{\partial N}\right)_{S,V} N \qquad (5.5)$$

In other words, Euler's theorem says that the function $E(S, V, N)$ has the special property that its value for any set of S, V, and N is also given by the sum on the RHS of (5.5), involving derivatives of $E(S, V, N)$ evaluated at those values.

Euler's theorem applies in this case because of the extensive behavior of the energy. The proof is as follows. Since E scales linearly with system size,

$$\lambda E(S, V, N) = E(\lambda S, \lambda V, \lambda N) \qquad (5.6)$$

Taking the λ derivative of both sides gives

$$E(S, V, N) = \frac{\partial E(\lambda S, \lambda V, \lambda N)}{\partial(\lambda S)}\frac{\partial(\lambda S)}{\partial \lambda} + \frac{\partial E(\lambda S, \lambda V, \lambda N)}{\partial(\lambda V)}\frac{\partial(\lambda V)}{\partial \lambda} + \frac{\partial E(\lambda S, \lambda V, \lambda N)}{\partial(\lambda N)}\frac{\partial(\lambda N)}{\partial \lambda}$$

$$= \frac{\partial E(\lambda S, \lambda V, \lambda N)}{\partial(\lambda S)} S + \frac{\partial E(\lambda S, \lambda V, \lambda N)}{\partial(\lambda V)} V + \frac{\partial E(\lambda S, \lambda V, \lambda N)}{\partial(\lambda N)} N \qquad (5.7)$$

Since λ can be any arbitrary number, Eqn. (5.7) must hold for the case in which $\lambda = 1$,

$$E(S, V, N) = \frac{\partial E(S, V, N)}{\partial S} S + \frac{\partial E(S, V, N)}{\partial V} V + \frac{\partial E(S, V, N)}{\partial N} N$$

$$= T(S, V, N)S - P(S, V, N)V + \mu(S, V, N)N \qquad (5.8)$$

That concludes the proof. By an analogous argument, we can construct an integrated form for the entropy version of the fundamental equation. We can proceed by applying Euler's theorem to the extensivity property $\lambda S(E, V, N) = S(\lambda E, \lambda V, \lambda N)$, or we can simply rearrange the integrated energy form. Either returns the same result,

$$S = \frac{E}{T} + \frac{PV}{T} - \frac{\mu N}{T} \tag{5.9}$$

Example 5.1 *Show that the Sackur–Tetrode equation, the exact expression for the entropy of a monatomic ideal gas with molecular mass m, obeys Euler's theorem for the integrated form for the entropy. The Sackur–Tetrode equation is*

$$S = Nk_B \ln\left[\left(\frac{E}{N}\right)^{3/2}\left(\frac{V}{N}\right)\right] + Nk_Bc \quad \text{where } c = \frac{5}{2} + \frac{3}{2}\ln\left(\frac{4\pi m}{3h^2}\right)$$

Note that this entropy obeys extensive scaling, $S(\lambda E, \lambda V, \lambda N) = \lambda S(E, V, N)$. To check Euler's theorem, we must compute the three partial derivatives of S,

$$\left(\frac{\partial S}{\partial E}\right)_{V,N} = \frac{3Nk_B}{2E}; \quad \left(\frac{\partial S}{\partial V}\right)_{E,N} = \frac{Nk_B}{V}; \quad \left(\frac{\partial S}{\partial N}\right)_{E,V} = -\frac{5k_B}{2} + k_B \ln\left[\left(\frac{E}{N}\right)^{3/2}\left(\frac{V}{N}\right)\right] + k_Bc$$

Now we apply Euler's theorem,

$$S = \left(\frac{\partial S}{\partial E}\right)_{V,N} E + \left(\frac{\partial S}{\partial V}\right)_{E,N} V + \left(\frac{\partial S}{\partial N}\right)_{E,V} N$$

$$= \left(\frac{3Nk_B}{2E}\right)E + \left(\frac{Nk_B}{V}\right)V + \left(-\frac{5k_B}{2} + k_B \ln\left[\left(\frac{E}{N}\right)^{3/2}\left(\frac{V}{N}\right)\right] + k_Bc\right)N$$

$$= Nk_B \ln\left[\left(\frac{E}{N}\right)^{3/2}\left(\frac{V}{N}\right)\right] + Nk_Bc$$

The result is the same, as we expected.

5.3 Intensive forms and state functions

It is often convenient to make intensive versions of extensive variables. The recipe is simple: take the ratio of two extensive quantities, such as the energy per particle or the number of molecules per volume. The most common way to construct intensive variables is to use a per-molecule or per-mole basis, although a per-volume basis would be no less valid. We will define the energy per particle as

$$e \equiv \frac{E}{N} \tag{5.10}$$

Since the energy is extensive, we can let $\lambda = 1/N$ in (5.6) to find that

$$e = \frac{1}{N}E(S, V, N) = E\left(\frac{S}{N}, \frac{V}{N}, 1\right) \tag{5.11}$$

In other words, e depends only on the per-particle entropy $s = S/N$ and volume $v = V/N$. Therefore, we can write that e is a function of two independent variables,

$$e = e(s, v) \tag{5.12}$$

In going from an extensive to an intensive energy, we reduced the number of independent variables by one. This occurs because we give up information about the system, and that information is precisely the system size.

We now seek a fundamental differential equation for $e(s, v)$ by finding the partial derivatives. Note that

$$e(s, v) = \frac{E(sN, vN, N)}{N} \tag{5.13}$$

By taking the s derivative,

$$\left(\frac{\partial e}{\partial s}\right)_v = \frac{\partial}{\partial s}\frac{E(sN, vN, N)}{N}$$

$$= \frac{1}{N}\frac{d(sN)}{ds}\frac{\partial E(sN, vN, N)}{\partial(sN)}$$

$$= \frac{1}{N}N\frac{\partial E}{\partial S}$$

$$= T \tag{5.14}$$

and then the v derivative,

$$\left(\frac{\partial e}{\partial v}\right)_s = \frac{\partial}{\partial v}\frac{E(sN, vN, N)}{N}$$

$$= \frac{1}{N}\frac{d(vN)}{dv}\frac{\partial E(sN, vN, N)}{\partial(vN)}$$

$$= \frac{1}{N}N\frac{\partial E}{\partial V}$$

$$= -P \tag{5.15}$$

we find that we can write

$$de = T\,ds - P\,dv \tag{5.16}$$

What happened to the chemical potential when we changed to the intensive version? Because we used the variable N to make all quantities intensive, the chemical potential is no longer a partial of the energy function. Instead, we can recover it using the integrated form of the fundamental equation,

$$\frac{E}{N} = \frac{TS}{N} - \frac{PV}{N} + \frac{\mu N}{N} \tag{5.17}$$

(5.17)

Rearranging gives

$$\mu = e - Ts + Pv$$

$$= e - \left(\frac{\partial e}{\partial s}\right)_v s - \left(\frac{\partial e}{\partial v}\right)_s v \tag{5.18}$$

This shows that we can still compute the chemical potential from the function $e(s, v)$. Therefore, the intensive form of the fundamental equation retains a relationship to all intensive variables. This fact has significant implications.

Once s and v have been given, the values of all other intensive variables are fixed at equilibrium in an isolated, single-component system. e is found from $e(s, v)$, T and P stem from derivatives of $e(s, v)$, and the chemical potential is given by $\mu = e - Ts + Pv$. Thus we need to know only two pieces of intensive information in order to completely specify the **thermodynamic state** of the system, that is, the values of all other intensive variables.

For extensive variables, we need one additional piece of information, namely the size of the system N.

For systems with more than one component, there will be multiple chemical potentials ($\partial E/\partial N_1 = \mu_1$, $\partial E/\partial N_2 = \mu_2$, ...). Therefore, we would need $C - 1$ more pieces of information to specify the intensive state of a system of C components, for a total of $C + 1$ intensive variables.

The crux of these statements is very general and powerful: for C components we need $C + 1$ pieces of information (intensive variables) to specify the macroscopic, equilibrium state of a single-phase, non-reacting system. For multiphase systems, the mere fact that there exist two phases in equilibrium can serve as one piece of information. For reactive systems, equilibrium constants serve as additional information. We will learn more about such cases in later chapters.

Must the two pieces of information specifying the thermodynamic state of a system be v and s specifically? It turns out that we can use other sets of parameters by swapping these variables for their *conjugates*, the variables that relate to them in the partial derivatives of a fundamental equation. For example, we can swap s for T. This is because we can obtain the value of s by solving the following equation at fixed values of T and v,

$$\frac{\partial e(s, v)}{\partial s} = T \tag{5.19}$$

While one might be worried that there are multiple solutions for s that satisfy this equality, the concavity of the entropy and hence convexity of the energy guarantees that there is just one. Similarly, we can swap P for v, since we can solve

$$\frac{\partial e(s, v)}{\partial v} = -P \tag{5.20}$$

to find v as a function of s and P. Here, the concavity condition also guarantees a single solution. On the other hand, we cannot swap s for P because P is conjugate to v and not s.

In that case, there is no equation or concavity condition that can guarantee a unique solution for s. Thus, each piece of information used to specify the thermodynamic state must originate from different conjugate variable pairs. The same is true if we switch to other intensive versions of the fundamental equation, e.g., the energy per volume.

All thermodynamic variables described thus far are called *state functions*, meaning that their values in intensive form are exactly specified once one is given the $C + 1$ pieces of information required to specify the equilibrium state. For example, if one prepares a certain single-component system with specific values of s and v, there is one and only one pressure that system can exhibit at true equilibrium; no other value will be detected as long as s and v remain fixed, unless that system strays from equilibrium or, as we will see later, is metastable. It does not matter what processes a system experienced beforehand or how it was prepared, the values of state functions will always be the same at the same thermodynamic state.

The $C + 1$ intensive pieces of information specifying the thermodynamic state of a system must originate from different conjugate pairs. A **conjugate pair** appears as an independent variable and its derivative in a fundamental equation. Examples include (s, T) and (v, P), which stem from $e(s, v)$. In general, x and y are intensive conjugate pairs if $y = (\partial f/\partial x)_z$, where $f(x, z)$ is any intensive fundamental function.

The values of **state functions** are completely determined by the thermodynamic state of the system, regardless of its history or preparation. State functions include S, E, V, N, T, P, μ, and intensive versions thereof.

Example 5.2 *At constant V and N conditions, the energy of a perfect crystal at temperatures near absolute zero is found to obey $E = E_0 + \alpha T^4$, where E_0 and α are constants (both proportional to the system size in order to maintain extensivity). Compute $S(T)$ and $S(E)$.*

From the fundamental equation we have

$$dE = T\,dS - P\,dV + \mu\,dN$$

$$= T\,dS \quad \text{(constant } V \text{ and } N)$$

Taking the T derivative,

$$\left(\frac{\partial E}{\partial T}\right)_{V,N} = T\left(\frac{\partial S}{\partial T}\right)_{V,N}$$

Substituting the measured energy relation,

$$4\alpha T^3 = T\left(\frac{\partial S}{\partial T}\right)_{V,N}$$

Now we integrate,

$$\int dS = \int 4\alpha T^2 \, dT \quad \text{(constant } V \text{ and } N)$$

giving

$$S = S_0(V, N) + \frac{4}{3}\alpha T^3$$

where S_0 is an integration constant. It depends on V and N because those were held constant during the integration. This completes the determination of $S(T)$. To obtain $S(E)$, we back-substitute for T using the measured relation

$$T = \left(\frac{E - E_0}{\alpha}\right)^{1/4}$$

so that

$$S = S_0(V, N) + \frac{4}{3}\alpha^{1/4}(E - E_0)^{3/4}.$$

Alternatively, we could have substituted for T before integration using

$$\left(\frac{\partial S}{\partial E}\right)_{V,N} = \frac{1}{T}$$

$$= \left(\frac{E - E_0}{\alpha}\right)^{-1/4}$$

And then the integration would look like

$$\int dS = \int \left(\frac{E - E_0}{\alpha}\right)^{-1/4} dE$$

The final answer is the same as before.

Example 5.3   *Find the pressure from the above example as $P(T, V, N)$. This is the equation of state. In general, E_0 and α are V- and N-dependent.*

There are two approaches. The first is as follows:

$$P = T\left(\frac{\partial S}{\partial V}\right)_E$$

$$= T\frac{\partial}{\partial V}\left[S_0 + \frac{4}{3}\alpha^{1/4}(E - E_0)^{3/4}\right]_E$$

$$= T\frac{\partial S_0}{\partial V} + \frac{1}{3}T\alpha^{-3/4}(E - E_0)^{3/4}\frac{\partial \alpha}{\partial V} - T\alpha^{1/4}(E - E_0)^{-1/4}\frac{\partial E_0}{\partial V}$$

$$= T\frac{\partial S_0}{\partial V} + \frac{1}{3}T^4\frac{\partial \alpha}{\partial V} - \frac{\partial E_0}{\partial V}$$

The second approach is distinct but ultimately equivalent:

$$dE = T\,dS - P\,dV + \mu\,dN$$

Taking the V derivative at constant T and using constant N conditions gives

$$\left(\frac{\partial E}{\partial V}\right)_T = T\left(\frac{\partial S}{\partial V}\right)_T - P$$

On rearranging, we have

$$P = -\left(\frac{\partial E}{\partial V}\right)_T + T\left(\frac{\partial S}{\partial V}\right)_T$$

$$= -\left[\frac{\partial E_0}{\partial V} + \frac{\partial \alpha}{\partial V}T^4\right] + T\left[\frac{\partial S_0}{\partial V} + \frac{4}{3}T^3\frac{\partial \alpha}{\partial V}\right]$$

$$= T\frac{\partial S_0}{\partial V} + \frac{1}{3}T^4\frac{\partial \alpha}{\partial V} - \frac{\partial E_0}{\partial V}$$

which is the same as before.

Example 5.4   *At constant T and P, a liquid at its boiling point can also be a vapor or gas. This is similar to the situation in which there are two "systems" – two phases – that are in equilibrium with each other and can exchange energy, volume, and particles as the liquid and vapor phases interact. In terms of the intensive energies e^L and e^G and volumes v^L and v^G (L = liquid, G = gas), write an expression for the entropy difference $\Delta s = s^G - s^L$.*

Notice here that the two phases reach equilibrium with each other by sharing energy, volume, and particles. Thus the conditions for thermal, mechanical, and chemical equilibrium apply, meaning that $T^L = T^G$, $P^L = P^G$, and $\mu^L = \mu^G$. If these conditions were not met, phase coexistence would not be possible; one of the phases would shrink away and the system would become single-phase. Therefore, we can define the temperatures, pressures, and molar volumes without superscripts: $T^L = T^G = T$, $P^L = P^G = P$, and $\mu^L = \mu^G = \mu$.

We write out the integrated fundamental equation for E for each phase,

$$E^L = TS^L - PV^L + \mu N^L \quad \text{and} \quad E^G = TS^G - PV^G + \mu N^G$$

Dividing each side by the respective N variable gives

$$e^L = Ts^L - Pv^L + \mu \quad \text{and} \quad e^G = Ts^G - Pv^G + \mu$$

Now we can subtract the first of these equations from the second to find

$$\Delta e = T\,\Delta s - P\,\Delta v$$

Notice that the μ terms vanished since $\mu^L = \mu^G$. On rearranging the expression, we have

$$\Delta s = \frac{\Delta e + P\,\Delta v}{T}$$

We could also write this expression as

$$\Delta s = \frac{\Delta(e + Pv)}{T}$$

The quantity $e + Pv$ appears frequently in thermodynamics, and it is called the molar enthalpy or per-particle enthalpy, $h = H/N \equiv (E + PV)/N$. We will learn more about the enthalpy in later chapters, and why it emerges as useful in thermodynamics. For now we simply show that it relates conveniently to the entropy associated with a phase transition,

$$\Delta s = \frac{\Delta h}{T}$$

Problems

Conceptual and thought problems

5.1. Starting from the fundamental equation, show that T, P, and μ are all intensive and independent of the size of the system. Namely, use the relation $E(\lambda S, \lambda V, \lambda N) = \lambda E(S, V, N)$ to show that T, P, and μ are independent of λ.

5.2. A friend suggests that the intensive pair (e, T) is enough information to determine the state of a single-phase, single-component system. Is he or she right or wrong? Explain.

5.3. For a single-component system, prove that $(\partial V/\partial N)_{T,P} = v$. In general, prove that $(\partial X/\partial N)_{T,P} = X/N$, where X is any extensive quantity.

5.4. A liquid is heated at constant pressure in a smooth container to a temperature above its boiling point, at which it becomes superheated and metastable. Boiling chips are added and the liquid transforms irreversibly into a vapor phase. Why does the relationship $\Delta s = \Delta h/T$ not describe the change in entropy for this particular transition?

5.5. You are given the function $E(T, V, N)$ rather than $E(S, V, N)$. Show that you would not be able to obtain all thermodynamic properties from the former. Hint: consider how you would find the value of S.

5.6. Given T and v, prove that there is a unique solution for s in Eqn. (5.19). First show that $e(s, v)$ is convex if $s(e, v)$ is concave.

Fundamentals problems

5.7. A particular quantity W satisfies the extensivity relation $W(\lambda X, \lambda Y, z) = \lambda W(X, Y, z)$, where X and Y are extensive but z is intensive. Find an expression for the integrated form of W.

5.8. Using the extensivity property of the entropy, find the fundamental equation for the intensive quantity $s = S/N$ in both differential and integrated forms.

5.9. Consider a system of C components with internal energy $E(S, V, N_1, N_2, \ldots, N_C)$. Find the differential and integrated forms of the fundamental equation for $e = E/N$, where $N = \sum_i N_i$.

5.10. Find the fundamental equation for the intensive quantity $\hat{e} = E/V$ in differential form. Also give an expression for the pressure P in terms of $\hat{e}$ and $\hat{s} = S/V$. Define $\rho \equiv v^{-1} = N/V$.

5.11. An experimentalist measures the per-particle heat capacity of a system and finds the relation $c_v(T, v) = \alpha(v)T^2$. It is also known that the energy and entropy are both zero at absolute zero. Find an expression for $e(s, v)$. Then, write an expression for the equation of state $P(T, v)$.

5.12. The Sackur–Tetrode equation gives the absolute value of the entropy of the ideal gas, and has the form

$$S = Nk_B \ln\left[\left(\frac{E}{N}\right)^{3/2}\left(\frac{V}{N}\right)\right] + Nk_B c \text{ where } c = \frac{5}{2} + \frac{3}{2}\ln\left(\frac{4\pi m}{3h^2}\right)$$

(a) Find expressions for $T(e, v)$, $P(e, v)$, and $\mu(e, v)$, where e and v are the per-particle energy and volume, respectively. Which of these require a quantum description of the world in order to be defined in absolute terms?

(b) Find an expression for $\mu(T, P)$.

Applied problems

5.13. The term photon gas describes a system in which radiation inside a rigid container is maintained at a constant temperature T. In this particular case, the number of photons is *not* conserved since they can be created and destroyed. Therefore, it is impossible to fix the particle number N and the chemical potential is always zero. It is also found that the pressure obeys the relation $P = E/(3V)$. With these constraints only, show that the entropy and energy are given by the following, where c is a volume-dependent constant:

$$E = (3/4)cT^4$$
$$S = cT^3$$

FURTHER READING

H. Callen, *Thermodynamics and an Introduction to Thermostatistics*, 3rd edn. New York: Wiley (1985).

K. Denbigh, *The Principles of Chemical Equilibrium*, 4th edn. New York: Cambridge University Press (1981).

A. Z. Panagiotopoulos, *Essential Thermodynamics*. Princeton, NJ: Drios Press (2011).

J. W. Tester and M. Modell, *Thermodynamics and Its Applications*, 3rd edn. Upper Saddle River, NJ: Prentice Hall (1997).

6 The first law and reversibility

6.1 The first law for processes in closed systems

We have discussed the properties of equilibrium states and the relationship between the thermodynamic parameters that describe them. Let us now examine what happens when we *change* from one thermodynamic state to another. That is, let us consider *processes*.

An important concept is that of a *path*. A process can be defined by the path of changes that take a system from one state to another. The changes in state variables are independent of such paths. That is, they depend solely on the initial and final equilibrium states of the process, and not on the intervening steps. State variables include S, E, V, N, T, P, and μ.

The first law of thermodynamics describes the change in the internal energy of a system during a process. It accounts for these changes in terms of energy transfers into or out of the system due to interactions with the surroundings. For closed systems that do not exchange mass with their surroundings, its general differential form is

$$dE = \delta Q + \delta W \tag{6.1}$$

The first law is a statement of the conservation of energy. Recall from Newton's equations of motion for the time evolution of classical molecular systems that the total energy is conserved. Therefore, if the energy E of such a system changes then it must be explained by a mechanism of energy transfer with the surroundings.

The **first law of thermodynamics** defines two mechanisms by which energy can transfer between a system and its surroundings.

δW denotes differential energy transfer due to **work**. Work represents energy transfer that we can measure (and potentially use or perform) at the macroscopic level. This could be the movement of a piston to accomplish mechanical work, but it could also be a number of other measurable changes, like the production of an electric current (electrochemical work).

δQ denotes energy transfer due to **heat**. Heat is essentially the part of the change in energy that cannot be captured and measured directly at the macroscopic level. It is energy distributed to molecules in a random fashion that produces no net macroscopic motions or forces. In this sense, δQ represents a kind of microscopic component of energy transfer.

The first law simply states that changes in the energy of a system must exactly balance with any transfers of heat and work.

Before we continue, it is worthwhile noting that an alternate convention distinct from (6.1) employs a negative work, $-\delta W$, in the first law. This simply changes the interpretation of positive and negative values of work. We can rationalize signs on the basis of physical intuition about energy transfers. In our positive convention, "The system does work on its surroundings" indicates that some energy leaves the system and enters the surroundings in the form of work; thus, $\delta W < 0$ since the energy of the system decreases. Similarly, "Heat was added to the system" means that $\delta Q > 0$ since the system's energy is increased.

We write δQ and δW with the symbol δ rather than the total differential d as we did for dE. This special treatment indicates that δQ and δW are path-dependent. That is, heat and work are not state functions. To illustrate the idea, assume that we knew only the beginning and end states of a process, for example, $T_1, P_1 \rightarrow T_2, P_2$. It would be possible to determine the change in energy $\Delta E = E_2 - E_1$, and in fact we could use any process or path between these two state points to measure it. On the other hand, there would be no way to determine the work and heat associated with the process using that information alone. In fact there could be many processes that started at T_1, P_1 and ended at T_2, P_2 that had widely different values of work and heat. The first law states that the sum of these two quantities, however, would be independent of path.

Heat and work are **path-dependent** quantities, although their sum is not.

Because of these considerations, we express the first law in integrated form as

$$\Delta E = Q + W \tag{6.2}$$

Neither Q nor W is preceded by the symbol Δ. These quantities are not state functions, but rather variables that characterize the process itself. A statement of ΔQ would erroneously suggest that the system itself could be characterized by values of Q at the beginning and at the end of the process, which is impossible since heat represents a path-dependent energy transfer.

No matter how slow the process is or how careful we are, δQ and δW are always path-dependent. Mathematically they are *inexact differentials*, which underlies the use of the symbol δ. An inexact differential denotes a quantity of two or more independent variables whose integral cannot be written as a difference of a function at the two limits. For example, there is no function g that ever satisfies

$$\int_{T_1, P_1}^{T_2, P_2} \delta Q \neq g(T_2, P_2) - g(T_1, P_1) \tag{6.3}$$

The same is true of the work. Essentially, the value of Q found from summing all differential heat transfers δQ throughout the process depends on the path taken from $T_1, P_1 \rightarrow T_2, P_2$. On the other hand, there does exist such a function for the energy, since E is a state function,

$$\int_{T_1,P_1}^{T_2,P_2} dE = E(T_2, P_2) - E(T_1, P_1) \tag{6.4}$$

Consequently, we say that the energy and all other state functions have *exact differentials*.

6.2   The physical interpretation of work

At the most basic level, we can think of work as the energy associated with our system "doing something" observable. When we describe a system, we are implicitly indicating that it is subject to a number of *macroscopic* or external constraints. For example, if we are able to speak of a system volume V, there are constraints associated with the region of space in which the system resides. These may be physical constraints, like the walls of a container, or they may be conceptual ones, such as a theoretical control volume. Constraints may also involve, for example, areas of interfaces or imposed electromagnetic fields. Regardless, when the system "does something," these constraints change in a measurable way, and work is the associated energy.

Changing macroscopic constraints, and hence doing work, involves forces. We are almost always interested in the forces from the point of view of the surroundings, since those are what we measure, use, and control. Think of a gas in a piston. We may be interested in the force required on our part, by the surroundings, to compress the gas to half its volume. Or we may want to know the force felt by the surroundings as the gas expands, which we could put to use in an engine. With that in mind, a general differential definition of mechanical work that we can measure is

$$\delta W = \mathbf{F}_{\text{ext}} \cdot d\mathbf{r} \tag{6.5}$$

$\mathbf{F}_{\text{ext}}$ is the external force imposed or felt by the surroundings on the system, and $d\mathbf{r}$ is a differential distance over which it acts. The dot product shows that the work stems from displacements along the same direction as the force. If the force acts only along a single direction, say the x one, then it is easy to rewrite Eqn. (6.5) in terms of scalar quantities, $\delta W = F_{\text{ext}} \, dx$.

Consider the ideal, frictionless, gas-filled piston depicted in Fig. 6.1. Both the external force F_{ext} from the surroundings and the internal force from the gas F act at the piston interface. If the piston is stationary, then the two forces must balance exactly. If the forces are unequal, the piston will accelerate and move according to Newton's laws. In that case, let the differential distance over which the piston moves in a small amount of time be dx. By convention we take F_{ext} to be positive when the external force is directed downward, and the work to be positive when the gas is compressed. These conventions are adopted so that the work gives the net energy change from the system's point of view. Our expression for the work simplifies to

$$\delta W = -F_{\text{ext}} \, dx \tag{6.6}$$

For compression, $F_{\text{ext}} > 0$, $dx < 0$, and thus $\delta W > 0$, as we would expect for adding energy to the system in the form of work. For expansion, $F_{\text{ext}} > 0$, $dx > 0$, and $\delta W < 0$.

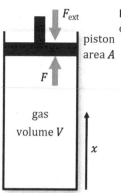

Figure 6.1. A simple piston scenario for the compression or expansion of a gas.

Note that the area is constant during either process, enabling us to rewrite the work in pressure–volume terms,

$$\delta W = -\left(\frac{F_{ext}}{A}\right)d(xA) = -P_{ext}\,dV \qquad (6.7)$$

Keep in mind that the pressure in Eqn. (6.7) is the external one, defined in terms of the external force, $P_{ext} = F_{ext}/A$. There is also an *internal* or *system* pressure, $P = F/A$. These two quantities are not necessarily equal, and, in fact, if they were exactly equal then the piston would never move and there would be no work. However, we can consider the limiting situation in which P_{ext} and P are nearly equal but one is infinitesimally larger so as to accomplish a net change in volume. Only in this particular circumstance do we have $P \approx P_{ext}$, and we are able to express the work as

$$\delta W = -P\,dV \qquad (6.8)$$

These considerations may seem obvious in retrospect, but there are important subtleties that have significant implications for connections to thermodynamic quantities. Namely, the work in which we are interested involves *external* pressures and forces, and only in the limit of mechanical equilibrium in which all forces balance can we replace these quantities with corresponding *internal* ones.

Such an analysis can be extended to many different kinds of work, beyond the pressure–volume variety. We will not discuss these in great detail, but merely mention a few for the purposes of illustration. For example, there are two- and one-dimensional analogues to volume work. If a system contains an interface that we stretch so as to increase its area $\mathcal{A}$, it can exert a force that resists deformation. The corresponding work is

$$\delta W = \gamma\,d\mathcal{A} \qquad (6.9)$$

where γ is the interfacial tension or surface free energy, with units of energy per area. Equation (6.9) lacks a negative sign because here expansion (i.e., stretching) results in an increase in system energy, unlike the volume expansion of a fluid at positive pressure. Similarly, if we stretch an elastic band or other one-dimensional system, the work is

$$\delta W = \tau\,dL \qquad (6.10)$$

where τ is the tension in the band, with units of energy per length, and L is its length. Another type of work involves the charging of a system in the presence of an electrostatic field. For a localized charge of magnitude and sign q,

$$\delta W = \Phi \, dq \tag{6.11}$$

where Φ is the local electrostatic potential. In all of these cases, the sign convention is such that δW is positive when the system gains energy. We also only indicated the *reversible* limit in which all of the external forces are equal to internal ones, in analogy with Eqn. (6.8). For example, there could be external tensions relevant to Eqns. (6.9) and (6.10) in nonequilibrium processes.

6.3   A classic example involving work and heat

Let us examine an example that readily demonstrates the path dependence of heat and work and the idea of internal and external forces. Imagine a gas held at constant temperature inside of an ideal, massless piston. On top of the piston is a mound of sand as shown in Fig. 6.2.

In state 1, the sand rests on top of the piston and the gas has volume V_1. In state 2, the sand has been removed except for a very small amount; the gas has expanded to $V_2 > V_1$ and the piston has been raised. Both states are in equilibrium at the same temperature, that of the surroundings. Note that the external force acting on the system is due to the mass of the sand. For the purposes of illustration, we will neglect any forces due to atmospheric pressure or piston mass.

We can now consider two scenarios to move from state 1 to state 2. In the first, the sand is swept off instantly and the system proceeds directly from 1 to 2. This case corresponds to negligible macroscopic work because the external force is removed prior to the piston changing its position, with the exception of the very small weight of the residual sand. The piston has not moved anything against a force,

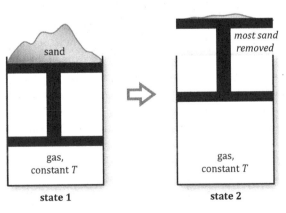

Figure 6.2. A gas-expansion process in which the external force, due to the weight of a pile of sand, can be removed either instantaneously or gradually.

sand

most sand removed

gas, constant T

gas, constant T

state 1

state 2

and the work W is essentially zero. As a result, the change in energy of the system equals the heat transferred from the environment, $\Delta E = Q$, per the integrated first law in (6.2).

In the second scenario the sand is slowly removed, piece by piece, and the piston inches upwards little by little. At any point in the process, the piston moves any remaining sand against gravity and is performing work while doing so. In this case $W < 0$ because energy is removed from the system and added to the surroundings by increasing the gravitational potential energy of the sand. At the end of this process, however, the state of the system and hence ΔE is identical to the case in which the sand is swept off at once. Since ΔE is the same for both processes, the slow removal of sand results in a greater heat transfer $Q = \Delta E - W$ from the environment than in the instantaneous case.

6.4   Special processes and relationships to the fundamental equation

Until now, we have imposed no general restrictions on the types of processes considered. We now specialize to a certain class of special processes called quasi-static.

> A **quasi-static process** is idealized in that it is performed infinitely slowly so that at each point in time, including the initial state, the system is infinitely close to a local equilibrium.

Quasi-static processes represent an important case because they suggest that the fundamental equation might hold at each point during them. Of course, since the fundamental equation is valid only for systems at equilibrium, it does not generally apply throughout a process when arbitrary changes to the state of a system are made. For closed systems, it takes the form

$$dE = T\,dS - P\,dV \tag{6.12}$$

Let us consider a subset of quasi-static processes called *reversible* processes that have two further characteristics. First, these processes lack dissipative or frictional forces that would lead to irreversibilities – specifically, effects preventing a return of the system and environment to their original states without expenditure of work. Second, no internal constraints are removed since that would imply that earlier parts of the process are not in equilibrium when compared with later ones; such features too represent irreversibilities. We will return shortly with a more formal definition of reversible processes. For the time being, these considerations necessarily require that internal and external forces are nearly in balance at all times. We therefore have $P = P_{ext}$ and, for pressure–volume work,

$$\delta W_{rev} = -P\,dV \tag{6.13}$$

We use the subscript "rev" to indicate that the work corresponds to a process without irreversibilities. The first law then becomes

$$dE = \delta Q_{\text{rev}} - P\,dV \tag{6.14}$$

On comparing this version of the first law with the fundamental equation in (6.12), we can now identify the reversible heat transfer with the $T\,dS$ term,

$$\delta Q_{\text{rev}} = T\,dS \tag{6.15}$$

Equivalently, we can express the entropy in terms of such heat transfers,

$$dS = \frac{\delta Q_{\text{rev}}}{T} \tag{6.16}$$

Importantly, the differential change in entropy for a reversible, quasi-static process is given by the heat transfer divided by the temperature. This key connection provides us with a way to *measure* changes in entropy using calorimetric experiments. Such measurements must be performed carefully and slowly, without frictional dissipation. In reality, reversible processes represent an ideal limit that can never quite fully be achieved.

What about processes that are not quasi-static and reversible? In such situations, the process occurs at a finite rate and the system is unable to maintain equilibrium at each point in time. When the system ultimately returns to equilibrium, therefore, there is an increase in entropy according to the second law. In general, there is always an increase in entropy in going from a nonequilibrium to an equilibrium state. As a result, processes that are not quasi-static generate more entropy.

In idealized **reversible**, quasi-static processes, $dS = \delta Q/T$. In real processes, $dS > \delta Q/T$.

This implies the possibility of entropy generation in quasi-static processes without heat transfer. Real processes are never performed infinitely slowly, and hence there is always some generation of entropy.

Consider a process for which we know the beginning and end states. We can easily compute the change in S using any arbitrary path, including a hypothetical reversible one, even if the process itself is irreversible. This is because entropy is a state function. In general, two differential isothermal processes that accomplish the same change of state – one reversible and equilibrium, and one irreversible and nonequilibrium – have identical differentials dS. According to the inequality above, then

$$\left(\frac{\delta Q}{T}\right)_{\text{rev}} > \left(\frac{\delta Q}{T}\right)_{\text{irrev}} \qquad \rightarrow \qquad \delta Q_{\text{rev}} > \delta Q_{\text{irrev}} \tag{6.17}$$

If heat is transferred to the environment, both δQ terms are negative, and (6.17) indicates that nonequilibrium processes add more heat to the environment than do reversible ones. If instead the process involves the addition of heat to the system and δQ is positive, then less heat is required to accomplish the same change of state in the nonequilibrium case.

We have as yet addressed only pressure–volume work, so it is natural to wonder how these results would change if other types of work were present, for example, those in Eqns. (6.9)–(6.11). In such cases, the heat δQ can always be identified with the differential entropy dS in reversible processes because the fundamental equation will simply incur additional work terms. A general formula for a single-component system is

$$dE = T\,dS + \sum_i \delta W_{i,\text{rev}} + \mu\,dN \tag{6.18}$$

where the summation proceeds over different types of work, each in the reversible case.

Example 6.1   *Returning to the sand–piston example, calculate the heat and work for the case in which the sand is removed so slowly that the process can be considered quasi-static. Assume the gas in the piston is ideal.*

For this problem, we will need to know that the energy of an ideal gas at constant temperature does not depend on volume or pressure: $E = E(T, N)$ only. We will prove this relation later. A brief physical explanation is that an ideal gas possesses only kinetic energy, which is independent of the system volume. This contrasts with the potential energy, which depends on the particle coordinates and hence the available space.

Since the entire process occurs at constant temperature, $E_1 = E_2$ and $\Delta E = 0$. Therefore,

$$Q = -W$$

To compute the work, we integrate,

$$W = \int \delta W$$

$$= -\int_{V_1}^{V_2} P\,dV$$

$$= -\int_{V_1}^{V_2} \frac{Nk_{\mathrm{B}}T}{V}\,dV$$

$$= -Nk_{\mathrm{B}}T \ln\left(\frac{V_2}{V_1}\right)$$

The work is negative since $V_2 > V_1$. This is consistent with the fact that the system does work on the surroundings, i.e., it increases the potential energy of the sand.

6.5   Baths as idealized environments

Consider the addition or subtraction of heat from a very large body. For small heat transfers, the temperature of such a body changes very little. In the limit that the body is infinitely large, we have a *bath* or *reservoir* for which the temperature is constant despite the process, and heat always appears to be added differentially and reversibly. For baths, we can therefore integrate the entropy–heat relation,

$$T_{bath} \int dS_{bath} = \int \delta Q_{bath} \tag{6.19}$$

On rearranging, we have

$$\Delta S_{bath} = \frac{Q_{bath}}{T_{bath}} \tag{6.20}$$

Baths are an important concept in thermodynamics because frequently they are the mechanism by which we manipulate or maintain the state of a system. In addition, we often treat the surroundings of our system as idealized baths.

6.6   Types of processes and implications from the second law

A process for which there is no heat transfer is called *adiabatic*. All processes are necessarily adiabatic in isolated systems because there is no interaction with the surroundings and hence no mechanism for energy exchange. If an adiabatic process is also quasi-static and there are no irreversibilities, the fundamental equation holds throughout the duration of the process, $\delta Q = T\, dS$, and hence $dS = 0$. That is, the entropy remains constant and the process is both adiabatic and *isentropic*. On the other hand, for irreversible processes in isolated systems, the second law of course shows that the entropy can only increase.

When dealing with the more general case of a system that can exchange heat and work with its surroundings, the isolated system that is relevant to the second law is the world: the system plus its environment. Therefore, a general statement of the second law for any process is

$$dS_{world} = dS + dS_{surr} \geq 0 \tag{6.21}$$

where dS is the entropy change for the system and dS_{surr} is that of the surroundings.

In the special case that the entropy of the world remains constant during the course of the process, $dS_{world} = 0$, the process is *reversible*. Reversible processes are so called because they can proceed in either direction according to the second law. They represent a class of idealized processes that cannot quite be obtained in reality. Reversible processes are also quasi-static since they maintain equilibrium at each point in time; if they were to move away from equilibrium, the subsequent return of the system to the equilibrium state would involve an increase in entropy and hence the process would not be reversible. In a real process not performed under such conditions, the world incurs an increase in entropy, $dS_{world} > 0$. The distinctions among these different kinds of processes are conveniently expressed mathematically:

$dS = \delta Q/T$ for **reversible quasi-static** processes
$dS_{world} = 0$ for **reversible quasi-static** processes
$dS_{world} > 0$ for **real** processes
$\delta Q = 0$ for **adiabatic** processes
$dS = 0$ for **isentropic** processes and **reversible adiabatic** processes

Often we are interested in extracting the maximum possible work produced by a process or the minimum work required for a process, since these govern the performance of macroscopic machines. To assess this limiting work, imagine a differential change of state that takes the system from one set of conditions to another. We are interested in the subset of all processes making this change that result in an extremum of work. Because E and S are state functions the differential changes dE and dS are the same for all processes that begin and end at the given conditions. We can write

$$dS_{world} = dS + dS_{surr} \tag{6.22}$$

If during the course of the process some heat is transferred from the system to the surroundings, the latter will increase in energy content by $\delta Q_{surr} = -\delta Q$. Considering the surroundings to act as an ideal heat bath at temperature T_{surr},

$$dS_{surr} = \frac{\delta Q_{surr}}{T_{surr}} = -\frac{\delta Q}{T_{surr}} \tag{6.23}$$

and hence

$$dS_{world} = dS - \frac{\delta Q}{T_{surr}} \tag{6.24}$$

We can finally substitute the first law for δQ,

$$dS_{world} = dS - \frac{dE - \delta W}{T_{surr}} \tag{6.25}$$

Note that nothing in this derivation has been specialized to any particular type of process; the expression for dS_{world} is quite general.

Now we consider the effects of reversibility on the above equation. We have that $dS_{world} \geq 0$, where the equality holds for reversible processes and the inequality for real processes. Therefore,

$$dS - \frac{dE - \delta W}{T_{surr}} \geq 0 \tag{6.26}$$

On rearranging, we find that

$$-\delta W \leq T_{surr}\, dS - dE \tag{6.27}$$

The significance of $-\delta W$ is that it measures the amount of work the system is able to do on the surroundings in positive units. That is, greater values of the LHS indicate that the system is able to perform more useful work. As is evident from this expression, the amount of useful work that can be extracted is bounded by a maximum. This limit occurs when the process is reversible, in which case the equality holds. For all other processes, less work is obtained. This is a very general statement about reversible processes.

> The maximum useful work that a system can perform on its surroundings can be obtained only with idealized, reversible processes for which $dS_{\mathrm{world}} = 0$.

In other words, we obtain the maximum work from a process that generates zero net entropy in the world. The same equation shows that, if we must instead perform work on the system and δW is positive, it is bounded by a minimum. In those cases, the least applied work required on our part also occurs when we do not generate any entropy; otherwise, we must apply additional work that ultimately is dissipated through irreversibilities. Putting everything together, reversible processes are always the most efficient. They represent a fundamental limit on the maximum work that can be performed by systems or the minimum work needed to apply to them. Of course, no practical process can be performed at an infinitely slow pace, and there is always a trade-off between efficiency and speed.

Example 6.2 *Returning again to the sand–piston example, consider the two processes for removing the sand: (a) removing it grain-by-grain in an infinitely slow manner, and (b) removing it at a finite rate. Find the corresponding change in entropy for each type of process.*

Part (a). In the case of slow removal, we found in the last example that

$$dE = 0 \text{ (ideal gas at constant } T)$$
$$\delta Q = -\delta W$$
$$W = -\int P \, dV = -Nk_{\mathrm{B}}T \ln \left(\frac{V_2}{V_1}\right)$$

Since $V_2 > V_1$, $W < 0$ and the system does work on the environment. We also have that $Q = -W > 0$, so that heat is added to the system. If we consider the removal to be performed infinitely slowly, then the process is quasi-static and we can write

$$dS = \frac{\delta Q}{T}$$

At constant temperature, we can integrate to find

$$\Delta S = \frac{Q}{T} = Nk_{\mathrm{B}} \ln \left(\frac{V_2}{V_1}\right)$$

The entropy change of the system is positive. In order to assess the reversibility of the process, we need to compute the entropy change of the world, which is the sum of that for the piston and for the surroundings. The latter are at constant temperature and we can write

$$\Delta S_{surr} = \frac{Q_{surr}}{T} = -\frac{Q}{T}$$

where we used the fact that any heat gained by the system is lost from the surroundings, $Q_{surr} = -Q$. Examining this result, we find that the gain in entropy of the system is exactly the loss in entropy of the surroundings, and so $\Delta S_{world} = 0$. Hence, the process is reversible.

Part (b). Now we consider the case in which the sand is swept off at some finite rate. Before we remove the sand, we have an external pressure exerted by its weight, $P_{ext} = mg/A$, where m is the sand mass, g the gravitational constant, and A the area of the piston. This external pressure is in balance with an internal one exerted by the gas, which is $P = Nk_BT/V$ at equilibrium.

When we remove the sand quickly, the external pressure suddenly drops and is no longer balanced by the internal one; therefore, a net force accelerates the piston upwards. In general, for a finite-speed expansion process, we have

$$P_{ext} < P$$

The work corresponds to the force exerted by the external pressure, which is given by

$$W = -\int P_{ext} \, dV$$

This work is necessarily less than that of an ideal quasi-static process:

$$-\int P_{ext} \, dV > -\int P \, dV$$

since $P_{ext} < P$, or

$$W > W_{rev}$$

where W_{rev} is what we found in part (a). Both work terms are negative but the work performed on the environment is less than that in the ideal reversible case, $|W_{rev}| > |W|$. We know for this process that ΔS is the same as in the reversible case since S is a state function. However, the value of Q is less than that in the reversible case because

$$Q = -W \quad \rightarrow \quad Q < Q_{rev}$$

Considering that ΔS is the same, this means that we must have

$$dS > \delta Q/T$$

which is consistent with our discussion of irreversible processes. The loss in entropy of the surroundings is correspondingly less,

$$\Delta S_{surr} = -\frac{Q}{T} > -\frac{Q_{rev}}{T}$$

Therefore, $\Delta S_{world} > 0$ because ΔS is the same as in the reversible case but ΔS_{surr} is greater than that in the reversible case.

6.7   Heat engines

A heat engine is a machine that uses the transfer of heat from a high-temperature heat bath, or *source*, to a low-temperature bath, or *sink*, to perform macroscopic mechanical work. For example, this could be achieved by bringing a piston filled with a gas into contact with a hot body. The gas expands, performing work (e.g., raising a weight). Finally, the piston is brought into contact with a cold body to bring it back to the original state so that it is ready to perform work again. An important aspect of heat engines is that they are *cycles*: they must return the system to the original state with each iteration of the process.

Let us use the second law to analyze the heat-engine process. For simplicity of understanding, we will consider the cycle as consisting of two basic steps: (a) the transfer of energy from the hot source at temperature T_H and (b) the transfer of energy to the cold sink at T_C. In reality engines can entail many substeps in the cycle; however, these can usually be lumped into the two basic process paths of heating and cooling, and thus the derivation here is actually quite general.

Consider the first step in an engine cycle, when heat is transferred from the hot reservoir to the system. We have in general

$$dS_{world} = dS + dS_{surr} \tag{6.28}$$

Integrating this and applying it to the first step gives

$$\Delta S^a_{world} = \Delta S^a + \Delta S^a_{surr}$$

$$= \Delta S^a - \frac{Q_H}{T_H} \tag{6.29}$$

Here Q_H is the amount of energy that is transferred from the heat bath to the system. By convention, we take it to have positive units. Note that $Q^a = Q_H$, where Q^a is the heat transfer from the point of view of the system. The amount of work associated with this step is given by the first law,

$$W^a = \Delta E^a - Q^a$$

$$= \Delta E^a - Q_H \tag{6.30}$$

With the second step, we have

$$\Delta S^b_{world} = \Delta S^b + \Delta S^b_{surr}$$

$$= \Delta S^b + \frac{Q_C}{T_C} \tag{6.31}$$

where Q_C is the amount of energy, in positive units, that is transferred from the system to the cold sink. $Q^b = -Q_C$ because the system loses energy. The work is

$$W^b = \Delta E^b - Q^b$$

$$= \Delta E^b + Q_C \tag{6.32}$$

The total entropy change and net work for the process sums the two steps:

$$\Delta S_{world} = \Delta S^a_{world} + \Delta S^b_{world}$$

$$= \Delta S^a - \frac{Q_H}{T_H} + \Delta S^b + \frac{Q_C}{T_C} \tag{6.33}$$

and

$$W = W^a + W^b$$

$$= \Delta E^a - Q_H + \Delta E^b + Q_C \tag{6.34}$$

At the end of step (b), the system is in the same state as it was in at the beginning of step (a) due to the cyclic nature of the process. The net change in any state function over a cycle *must* be zero and we can write

$$\Delta S = \Delta S^a + \Delta S^b = 0 \tag{6.35}$$

$$\Delta E = \Delta E^a + \Delta E^b = 0 \tag{6.36}$$

These constraints simplify our expressions for the world entropy and the work for the heat engine,

$$\Delta S_{world} = -\frac{Q_H}{T_H} + \frac{Q_C}{T_C} \tag{6.37}$$

$$W = -Q_H + Q_C \tag{6.38}$$

We are interested in the efficiency of this process. For a heat engine, the efficiency is defined as the fraction of the heat drawn from the hot source that can be converted into useful work,

$$\eta = \frac{|W|}{Q_H} \tag{6.39}$$

Using Eqn. (6.38), we find

$$\eta = \frac{|-Q_H + Q_C|}{Q_H} = 1 - \frac{Q_C}{Q_H} \tag{6.40}$$

since $W < 0$.

We now apply the second law to Eqn. (6.37), which will place a constraint on the efficiency:

$$\Delta S_{world} = -\frac{Q_H}{T_H} + \frac{Q_C}{T_C} \geq 0 \tag{6.41}$$

A little rearrangement gives

$$1 - \frac{Q_C}{Q_H} \leq 1 - \frac{T_C}{T_H} \tag{6.42}$$

Recognizing that the LHS gives the efficiency, we finally have

$$\eta \leq 1 - \frac{T_C}{T_H} \tag{6.43}$$

Equation (6.43) shows that the best efficiency any heat engine can achieve, even in the reversible limit, is bounded by the second law.

The maximum achievable efficiency of a heat engine is determined only by the temperatures of the hot source and cold sink, and is given by $\eta_{max} = 1 - T_C/T_H$. For processes that are irreversible and hence result in net entropy generation, the efficiency is less than this maximum attainable value.

While the second law predicts the efficiency in reversible process cycles, how do we predict the efficiencies and properties of processes that are not reversible? In short, we need to know the exact and detailed paths that these processes follow, which is more information than simple knowledge of the beginning and end points of a few subprocesses.

6.8 Thermodynamics of open, steady-flow systems

In the previous sections, we considered the form of the first law when there is no mass transfer between the system and the surroundings. It is possible to develop an analogous equation for systems in which one or more *flow streams* cross the system boundary. For the sake of brevity, we will consider only the case in which there is no change in the internal state of the system, that is, the steady-state scenario.

Figure 6.3 shows a schematic diagram of an open system involving three flow streams. Each stream is associated with a mass flow rate $\dot{m}$ with units of mass per time, a mean flow velocity u in distance per time, and a height z at which the stream enters the system. Because the system is at steady state, there can be no accumulation of mass within it and the flow rates must balance according to the conservation equation

$$\sum_{\text{inlets } i} \dot{m}_i = \sum_{\text{outlets } j} \dot{m}_j \tag{6.44}$$

Here the sums proceed over all streams of a given type, inlet or outlet. In the case of Fig. 6.3, we have $\dot{m}_1 + \dot{m}_2 = \dot{m}_3$.

To adapt the first law to open systems, we must consider all of the transfers of energies crossing the system boundary. Each flow stream provides four kinds of energy transfer. The first is that associated with the internal energy carried by the stream. Let $\hat{e}$ denote the internal energy *per mass* in a stream, which is determined by its thermodynamic state (i.e., the temperature, pressure, and composition). Because the streams can have different thermodynamic states, $\hat{e}$ in general takes on distinct values for each. The total rate of internal energy carried into the system boundary by inlets and outlets is thus

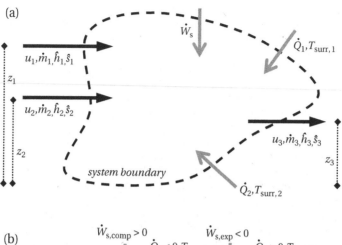

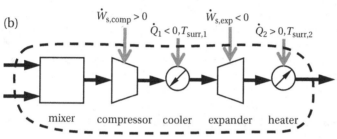

Figure 6.3. (a) A schematic diagram of an arbitrary process for an open system involving three flow streams (dark arrows). The system boundary is shown as the dashed line. Heat flows and net shaft work are indicated by gray arrows. (b) An actual process that might correspond to the schematic representation. The process involves mixing, compression, cooling, expansion, and heating of fluid streams.

$$\sum_{\text{inlets } i} \dot{m}_i \hat{e}_i - \sum_{\text{outlets } j} \dot{m}_j \hat{e}_j \qquad (6.45)$$

The second source of flow-stream energy is less immediately obvious. Because inlet streams must "push" fluid inside the system out of the way in order to flow, they actually perform pressure–volume work. Figure 6.4 shows a close-up of a stream flowing through a pipe that enters the system. The differential work done in moving a small volume element of fluid dV into the system is given by

$$\delta W_{\text{inlet}} = P \, dV \qquad (6.46)$$

where P is the pressure of the flow stream at the boundary. The work is positive because the flow accomplishes net work on the system. Moreover, the work is performed continuously and therefore the rate of work done by an inlet on a per-time basis is

$$\dot{W}_{\text{inlet}} = P\left(\frac{dV}{dt}\right) \qquad (6.47)$$

The derivative in Eqn. (6.47) indicates the rate of volume of fluid entering the system, which we can express in terms of the mass flow rate and the volume per mass of the fluid $\hat{v}$,

$$\dot{W}_{\text{inlet}} = P\hat{v}\dot{m} \qquad (6.48)$$

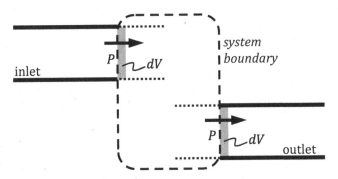

Figure 6.4. A schematic representation of the work performed by flow streams crossing the system boundary. A differential volume element dV (gray region) is acted upon by the pressure of the stream.

Similarly, fluid inside the system must do work to push fluid in the environment out of the way in outlet streams. By similar arguments, the work in this case is given by

$$\dot{W}_{\text{outlet}} = -P\hat{v}\dot{m} \tag{6.49}$$

where the negative sign appears because an outlet does work on the environment. In general, for Eqns. (6.48) and (6.49), the pressure and volume per mass are determined by the state of the stream at the boundary.

The final two energies associated with streams are the kinetic energy due to flow, which depends on the average fluid velocity, and the potential energy under gravity, which depends on the height at which a stream crosses the boundary. Both of these must be expressed as rates of energy per time, and their sum for any stream is given by

$$\frac{\dot{m}u^2}{2} + \dot{m}gz \tag{6.50}$$

where g is the gravitational constant.

On combining Eqns. (6.45), (6.48), (6.49), and (6.50) for all inlets and outlets, the total rate of energy transfer due to flowing streams is

$$\sum_{\text{inlets } i} \dot{m}_i \left(\hat{e}_i + P\hat{v}_i + \frac{u_i^2}{2} + gz_i \right) - \sum_{\text{outlets } j} \dot{m}_j \left(\hat{e}_j + P\hat{v}_j + \frac{u_j^2}{2} + gz_j \right) \tag{6.51}$$

The combination of the internal energy and pressure–volume terms is conveniently expressed using the thermodynamic quantity called the enthalpy. We briefly saw the enthalpy in Chapter 5, and we will learn more about its broader significance in the next chapter. For the time being, it is sufficient to say that the enthalpy is a state function and on a per-mass basis is defined by the following relation:

$$\hat{h} \equiv \hat{e} + P\hat{v} \tag{6.52}$$

Substitution into Eqn. (6.51) finally gives for the total rate of energy transfer due to flowing streams,

$$\sum_{\text{inlets } i} \dot{m}_i \left(\hat{h}_i + \frac{u_i^2}{2} + gz_i \right) - \sum_{\text{outlets } j} \dot{m}_j \left(\hat{h}_j + \frac{u_j^2}{2} + gz_j \right) \tag{6.53}$$

There are two additional mechanisms of energy transfer that are independent of flow streams. *Shaft work* is energy associated with continuous mechanical processes that drive internal changes to the system. For example, there may be a compressor, expander, turbine, mixer, or stirrer that resides within the system boundary. All would involve moving parts and would either perform work on the system (as in a compressor or stirrer) or capture work performed by the system for transfer to the surroundings (as in a turbine). We define the net rate of all shaft work by $\dot{W}_S$, which is a positive quantity if there is net work performed on the system and a negative one if the opposite applies. The second and final mechanism of energy transfer is heat exchange. The net rate of heat added to the system is described by $\dot{Q}$.

With these final considerations, we can now account for all energies crossing the system boundary. For steady-state conditions, we demand that they balance: the net rate of energy flow due to all of the streams, shaft work, and heat must be zero at steady state. The result is the open, steady-state version of the first law,

$$\sum_{\text{outlets } j} \dot{m}_j \left(\hat{h}_j + \frac{u_j^2}{2} + gz_j \right) - \sum_{\text{inlets } i} \dot{m}_i \left(\hat{h}_i + \frac{u_i^2}{2} + gz_i \right) = \dot{W}_S + \dot{Q} \tag{6.54}$$

Here we have separated shaft work and heat to the RHS of the balance, inverting the signs of the inlets and outlets on the LHS. Equation (6.54) forms the basis of any energy analysis of steady open systems involving mass flows.

In an analogous fashion, it is possible to adapt the second law to open systems. In this case, the net rate of entropy generation in the world must be greater than or equal to zero. The rate of entropy carried by inlet and outlet streams is given by $\dot{m}\hat{s}$, where $\hat{s}$ is the entropy per mass at the flow-stream conditions. If heat is exchanged with the surroundings, the rate of entropy change within them is $-\dot{Q}/T_{\text{surr}}$, where the negative sign accounts for the fact that $\dot{Q}_{\text{surr}} = -\dot{Q}$. In general, there may be more than one point along the boundary at which heat is exchanged; for each such case, we must have a separate term for the environment entropy that involves the local surroundings' temperature at that point. Therefore, the total rate of entropy generation in the world is

$$\dot{S}_{\text{world}} = \sum_{\text{outlets } j} \dot{m}_j \hat{s}_j - \sum_{\text{inlets } i} \dot{m}_i \hat{s}_i - \sum_{\text{heat transfers } k} \frac{\dot{Q}_k}{T_{\text{surr}, k}} \geq 0 \tag{6.55}$$

Per the second law, this quantity must never be negative or the process would be impossible. The most efficient processes occur when $\dot{S}_{\text{world}} = 0$.

Example 6.3 *A mysterious device takes as input a continuous stream of a monatomic ideal gas at initial temperature T_1 and pressure P_1. It then produces two outlet streams at reduced pressure $P_2 = P_3$, one at a higher temperature $T_2 = 1.10T_1$ and one at a lower temperature $T_3 = 0.95T_1$. The device is insulated and contains no moving parts. What is* the minimum pressure ratio P_1/P_2 for this process to be possible? Note: for a monatomic ideal gas, $(\partial h/\partial T)_P = c_P = 5k_B/2$ and both the internal energy e and the enthalpy h are functions of T only (i.e., they are P-independent).

A mass balance demands that

$$\dot{m}_1 = \dot{m}_2 + \dot{m}_3$$

The first law demands

$$\dot{m}_2\left(\hat{h}_2 + \frac{u_2^2}{2} + gz_2\right) + \dot{m}_3\left(\hat{h}_3 + \frac{u_3^2}{2} + gz_3\right) - \dot{m}_1\left(\hat{h}_1 + \frac{u_1^2}{2} + gz_1\right) = \dot{W}_S + \dot{Q}$$

In this case, there is no shaft work ($\dot{W}_S = 0$) and no heat transfer ($\dot{Q} = 0$). Moreover, changes due to the kinetic energies of fluid flow and gravitational potential energies are likely to be negligible. With these assumptions and using the mass conservation equation, the first law simplifies to

$$\dot{m}_2\hat{h}_2 + \dot{m}_3\hat{h}_3 - (\dot{m}_2 + \dot{m}_3)\hat{h}_1 = 0$$

Defining $f \equiv m_3/m_2$ and combining terms gives

$$(\hat{h}_2 - \hat{h}_1) + f(\hat{h}_3 - \hat{h}_1) = 0$$

Since there is only a single component, we can multiply by the molecular weight to convert per-mass to per-particle enthalpies,

$$(h_2 - h_1) + f(h_3 - h_1) = 0$$

This equation shows that we must calculate two changes in the per-particle enthalpy: one from state 1 to state 2, and one from state 1 to state 3. Because the enthalpy depends only on temperature for an ideal gas,

$$h_2 - h_1 = \int_{T_1}^{T_2}\left(\frac{\partial h}{\partial T}\right)_P dT = \frac{5}{2}k_B(T_2 - T_1)$$

and similar for the change $h_3 - h_1$. With these expressions, the first law becomes

$$\frac{5}{2}k_B(T_2 - T_1) + f\frac{5}{2}k_B(T_3 - T_1) = 0$$

Solving for f,

$$f = \frac{T_1 - T_3}{T_2 - T_1} = \frac{T_1 - 0.95T_1}{1.10T_1 - T_1} = 0.5$$

We now apply the second law to this problem. In a manner similar to what we found before, it simplifies to

$$(s_2 - s_1) + f(s_3 - s_1) \geq 0$$

This time, we must compute the corresponding entropy changes. We start with the fundamental relation

$$ds = \frac{1}{T}de + \frac{P}{T}dv$$

Its temperature derivative shows that

$$\left(\frac{\partial s}{\partial T}\right)_P = \frac{1}{T}\left(\frac{\partial e}{\partial T}\right)_P + \frac{P}{T}\left(\frac{\partial v}{\partial T}\right)_P$$

$$= \frac{1}{T}\left(\frac{\partial e}{\partial T}\right)_P + \frac{k_B}{T}$$

$$= \frac{1}{T}\frac{\partial}{\partial T}(e + k_B T)_P$$

$$= \frac{1}{T}\frac{\partial}{\partial T}(e + Pv)_P$$

$$= \frac{1}{T}\left(\frac{\partial h}{\partial T}\right)_P = \frac{5k_B}{2T}$$

and its pressure derivative gives

$$\left(\frac{\partial s}{\partial P}\right)_T = \frac{1}{T}\left(\frac{\partial e}{\partial P}\right)_T + \frac{P}{T}\left(\frac{\partial v}{\partial P}\right)_T$$

$$= 0 + \frac{P}{T}\left(\frac{-k_B T}{P^2}\right)$$

$$= -\frac{k_B}{P}$$

These suggest that the total differential of $s(T, P)$ is

$$ds = \frac{5k_B}{2T}dT - \frac{k_B}{P}dP$$

We can integrate this expression directly to find

$$s_2 - s_1 = \int_{T_1}^{T_2}\frac{5k_B}{2T}dT - \int_{P_1}^{P_2}\frac{k_B}{P}dP = \frac{5}{2}k_B \ln\left(\frac{T_2}{T_1}\right) - k_B \ln\left(\frac{P_2}{P_1}\right)$$

Substituting this result and an analogous one for the change from state 1 to 3 into the open second law gives

$$\frac{5}{2}k_B \ln\left(\frac{T_2}{T_1}\right) - k_B \ln\left(\frac{P_2}{P_1}\right) + f\left[\frac{5}{2}k_B \ln\left(\frac{T_3}{T_1}\right) - k_B \ln\left(\frac{P_2}{P_1}\right)\right] \geq 0$$

Solving for P_1/P_2,

$$\ln\left(\frac{P_1}{P_2}\right) \geq (1+f)^{-1}\left[-\frac{5}{2}\ln\left(\frac{T_2}{T_1}\right) - \frac{5}{2}f \ln\left(\frac{T_3}{T_1}\right)\right]$$

$$\geq (1+0.5)^{-1}\left[-\frac{5}{2}\ln(1.10) - \frac{5}{2}(0.5)\ln(0.95)\right]$$

$$\geq 0.116$$

So we must have that $P_1 > 1.12P_2$. If there were no pressure drop then the only function of the device would be to accomplish a net transfer of heat from a cold to a hot body without any work, which is impossible according to the second law.

Example 6.4   *Water is transported through a supply line from a mountain reservoir at elevation 240 m. The line crosses a pass at elevation 680 m before reaching a town at sea level. Pumps are distributed throughout to move the water, at an outlet delivery rate of 10 gal/s. The inlet and outlets are cylindrical pipes 0.5 m and 0.1 m in diameter, respectively. Heat losses during transit along the line amount to 3 kJ/kg. How much work is required of the pumps? Assume that the inlet and outlet streams are at roughly the same thermodynamic conditions (e.g., the same temperature and pressure), and that the density of water can be assumed to be constant at 1.0 g/cm³.*

First, find the mass flow rate of water in the line,

$$\dot{m} = (1{,}000 \text{ kg/m}^3)(10 \text{ gal/s})(\text{m}^3/264.17 \text{ gal}) = 37.9 \text{ kg/s}$$

The velocities of the inlet and outlet water streams are

$$u_2 = \frac{10 \text{ gal/s}}{\pi(0.25 \text{ m})^2}\left(\frac{\text{m}^3}{264.17 \text{ gal}}\right) = 0.193 \text{ m/s}$$

$$u_1 = \frac{10 \text{ gal/s}}{\pi(0.05 \text{ m})^2}\left(\frac{\text{m}^3}{264.17 \text{ gal}}\right) = 4.82 \text{ m/s}$$

For this problem, the open first law simplifies to

$$\dot{m}\left(\hat{h}_2 + \frac{u_2^2}{2} + gz_2\right) - \dot{m}\left(\hat{h}_1 + \frac{u_1^2}{2} + gz_1\right) = \dot{W}_S + \dot{Q}$$

Because the inlet and outlet are at the same state, $\hat{h}_1 = \hat{h}_2$, and the enthalpy terms drop out. Rearranging to solve for the shaft work gives

$$\dot{W}_S = \left[\frac{1}{2}(u_2^2 - u_1^2) + g(z_2 - z_1)\right]\dot{m} - \dot{Q}$$

$$= \left[\frac{(4.82 \text{ m/s})^2 - (0.193 \text{ m/s})^2}{2} + \left(9.8 \frac{\text{m}}{\text{s}^2}\right)(0 - 240 \text{ m})\right]$$

$$\times \left(37.9 \frac{\text{kg}}{\text{s}}\right) - \left(37.9 \frac{\text{kg}}{\text{s}}\right)\left(-3{,}000 \frac{\text{J}}{\text{kg}}\right)$$

$$= 25.0 \text{ kW}$$

Note that, if the outlet were found to be hotter than the inlet stream, we would have had $\hat{h}_2 > \hat{h}_1$. By virtue of the open energy balance, this would have increased $\dot{W}_S$.

Problems

Conceptual and thought problems

6.1. An object has a position in space denoted by **x**. It experiences a *conservative force* that can be written as the gradient of a potential energy,

$$\mathbf{F}_c = -\nabla U$$

If an external force $\mathbf{F}_{ext}$ acts on the object such that the total force experienced is $\mathbf{F} = \mathbf{F}_c + \mathbf{F}_{ext}$, prove that the total change in energy of the object is

$$\Delta E = \Delta K + \Delta U = \int \mathbf{F}_{ext}\, d\mathbf{x}$$

where K is the kinetic energy.

6.2. A liquid is slowly cooled at constant volume. Its energy decreases. The fundamental equation says that the change in entropy of the liquid, $dS = dE/T$, is negative. Is this in violation of the second law of thermodynamics? Why, or why not?

6.3. Your friend claims that, in any closed cyclic process, the change in entropy for the cycle must be zero because the entropy is a state function and because the beginning and end points of the cycle are the same. Therefore, all cyclic processes must be reversible. What is wrong with your friend's analysis? Explain.

6.4. An open glass of liquid water is placed into a household freezer maintained at atmospheric pressure. The water solidifies into ice. Is work done on or by the system? Find expressions for the work and heat per mole associated with the freezing process. Presume that the change in internal energy Δe is known or can be calculated. Note that water freezes from a density of ρ_L (liquid) to ρ_X (crystal) where $\rho_X < \rho_L$. Is the work per mole pressure-dependent?

6.5. In the famous Joule experiment, a mass attached to a pulley system falls a height h and powers a stirrer that mechanically agitates an insulated container of water. It is found that the temperature increase in the water ΔT is proportional to h. Is there any heat in this process? Explain with an analysis of the first law.

6.6. Which of the following processes are possible? Explain.
(a) a quasi-static but irreversible process
(b) a reversible isochoric process
(c) an irreversible, adiabatic process
(d) an irreversible, isentropic process

6.7. True or false? ΔS for a single-component system in going from T_1, P_1 to T_2, P_2 is higher for an irreversible than for a reversible process. Explain.

6.8. You need to estimate the energy flows in a slow, multi-step cycle involving a dense, nonideal gas. For example, you may model the gas using a cubic equation of state. For each of the following equations, indicate whether or not you can use them for your calculations:

(a) $Q + W = 0$ for an isothermal step

(b) $Q = \Delta H$ for an isobaric step

(c) $Q = \Delta E$ for an isochoric step

6.9. You need to estimate the work involved in a multi-step process cycle involving a dense, clearly nonideal gas. To a very good approximation, however, the heat capacity for this gas is constant with temperature. For each of the following equations, indicate whether or not you can use it for your calculations:

(a) $W/N = -k_B T \ln(V_2/V_1)$ for an isothermal step

(b) $\Delta E = C_V \Delta T$ for an isobaric step

(c) $\Delta H = \Delta E + V \Delta P$ for an isochoric step

(d) $Q = C_P \Delta T$ for an isobaric step

6.10. A well-insulated container consists of two halves of equal volumes V separated by a partition. On one half, there are N moles of a monatomic ideal gas at initial temperature T_i. On the other, there is a vacuum. The partition is suddenly removed, and the gas expands. Recall for a monatomic ideal gas that $PV = Nk_B T$ and $E = (3/2)Nk_B T$.

(a) Calculate the following quantities: the work done by the gas, the heat transferred to the gas, the change in energy, the change in entropy, the final temperature, and the final pressure.

(b) Consider the relationship discussed in this chapter, $\delta Q = T\, dS$. Why does an increase in entropy not correspond to a heat transfer in this case?

(c) Instead of being suddenly removed, the partition is gradually moved to the empty side of the container so that the gas expands reversibly against the force of the partition. Repeat part (a) for this scenario.

6.11. A rigid box contains two compartments separated by a partition. On one side is an ideal gas consisting entirely of green particles, while the other side holds an ideal gas of entirely red particles; the two particle types have identical masses. The partition is removed. Both Jane and Bob watch this process, but Bob is color blind and cannot tell the difference between green and red. Jane claims that the entropy of the world has increased and thus work must be done on the system to return it to its original state. Bob claims that there has been no entropy change, the process is reversible, and no work is required to return the system to its original state. Both claim that their analyses are consistent with the laws of thermodynamics. How can both of them be correct about the work? Explain.

Fundamentals problems

6.12. Show for any constant-pressure process that the first law reduces to $\Delta H = Q$, where H is the enthalpy, a state function defined by $H \equiv E + PV$. In contrast, show that it reduces to $\Delta E = Q$ for any constant-volume process.

6.13. Consider an ideal gas with constant $c_V = (\partial e/\partial T)_v = (3/2)k_B$ and $c_P = (\partial h/\partial T)_P = c_V + k_B$. Derive expressions for the work per molecule W/N, heat per molecule Q/N, and change in internal energy per molecule $\Delta e = \Delta E/N$, for each of the following processes:

(a) adiabatic expansion/compression

(b) isothermal expansion/compression

(c) isochoric heating/cooling

(d) isobaric heating/cooling

Put your answers in terms of the initial and final temperatures and pressures, T_1, P_1, T_2, and P_2.

6.14. Repeat the previous problem for the case in which the gas is still ideal but has $c_V = \alpha T$, where α is a constant.

6.15. Prove that the change in entropy per molecule for an ideal gas transitioning between two states, T_1, v_1, $P_1 \rightarrow T_2$, v_2, P_2, is given by the following two expressions:

$$\Delta s = \int_{T_1}^{T_2} \frac{c_v}{T} \, dT + k_B \ln\left(\frac{v_2}{v_1}\right)$$

$$= \int_{T_1}^{T_2} \frac{c_v + k_B}{T} \, dT - k_B \ln\left(\frac{P_2}{P_1}\right)$$

where v is the volume per molecule and $c_v \equiv (\partial e/\partial T)_v$ gives the intensive per-molecule constant-volume heat capacity.

6.16. A P-V diagram is a convenient way to illustrate different thermodynamic paths, with V along the x-axis and P along the y-axis. Starting at a common (T_1, P_1), illustrate the direction of the following quasi-static processes for an ideal monatomic gas: (a) isothermal compression to half the original volume, (b) adiabatic compression to half the original volume, (c) isochoric heating to twice the original temperature, and (d) isobaric heating to twice the original temperature.

6.17. Reconsider Example 6.2 involving the piston-and-sand scenario.

(a) Repeat the analysis for the reverse case in which sand is added either instantaneously or gradually to the piston and the ideal gas is compressed. Calculate the work W, heat transfer Q, and change in internal energy ΔE for these two scenarios.

(b) Consider the complete thermodynamic cycle in which sand is first removed and then put back. Draw the cycle on a P-V diagram, where the x-axis is V and the y-axis is P. Do this for both the instantaneous case and the gradual case.

(c) What are the net W, Q, and ΔE for the cycles in each scenario? Be sure to indicate the sign. Show that the net work W is related to an area in the cycle diagram.

6.18. Consider two containers that have volumes V_1 and $V_2 = 2V_1$ and that both contain the same number of molecules of an ideal gas. The containers initially have temperatures T_1 and T_2, but are then joined, allowing the gases to mix. Assume that the gases have constant heat capacities $c_V = (\partial e/\partial T)_V$ and $c_P = (\partial h/\partial T)_P = c_V + k_B$. Find the final temperature in the container T_f and the overall change in entropy for the process ΔS if

(a) the gases in the two containers are identical.

(b) the gases in the two containers are different species but have the same c_V and c_P.

6.19. An ideal gas in contact with a heat/temperature bath is suddenly and instantaneously compressed to ten times its original pressure. What is the minimum heat that the bath removes per mole of liquid?

6.20. If placed in very smooth, clean containers, many liquids can be supercooled to temperatures below their melting points without freezing, such that they remain in the liquid state. If left long enough, these systems will eventually attain equilibrium by spontaneously freezing. However, in the absence of nucleation sites like dirt or imperfections in the walls of the container, the time required for freezing can be long enough to permit meaningful investigations of the supercooled state.

Consider the spontaneous freezing of supercooled liquid water. For this problem you will need the following data taken at atmospheric pressure: $c_P = 38$ J/K mol for ice (at 0 °C, but relatively constant over a small T range); $c_P = 75$ J/K mol for liquid water (at 0 °C, but relatively constant); and $\Delta h_m = 6{,}026$ J/mol for liquid-ice phase equilibrium at 0 °C. Here Δx indicates $x_{\text{liquid}} - x_{\text{crystal}}$ for any property x. Recall that $H \equiv E + PV$, $h = H/N$, and $c_P = (\partial h/\partial T)_P = T(\partial s/\partial T)_P$, and that h and s are state functions.

(a) In an isolated system, when freezing finally occurs, does the entropy increase or decrease? What does this imply in terms of the relative numbers of microstates before and after freezing? Is "disorder" a useful qualitative description of entropy here?

(b) If freezing happens fast enough, to a good approximation the process can be considered adiabatic. If the entire process also occurs at constant pressure, show that freezing is a constant-enthalpy (isenthalpic) process.

(c) If one mole of supercooled water is initially at -10 °C (the temperature of a typical household freezer), what fraction freezes in an adiabatic process as it comes to equilibrium? Compute the net entropy change associated with this event. Note that this is not a quasi-static process.

(d) Consider instead the case when the freezing happens at constant T and P. Under these conditions, explain why the expression $\Delta s = \Delta h/T$ is valid when liquid water freezes at its melting point, but not when supercooled liquid water freezes.

(e) Compute ΔS_{water}, ΔS_{surr}, and ΔS_{world} when one mole of supercooled liquid water spontaneously freezes at -10 °C (e.g., in the freezer). Assume that the surroundings act as a bath.

6.21. A refrigerator is the opposite of a heat engine: it uses work supplied from the environment to move heat from a cold to a hot reservoir, effectively keeping the former at a lower temperature. The coefficient of performance measures the

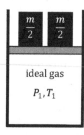

atmosphere   **Figure 6.5.** The piston setup for Problem 6.22.

efficiency of this process, defined by $\omega = |Q_C|/|W|$. Find an expression for the maximum value of ω that is possible according to the second law.

6.22. Consider the piston device depicted in Fig. 6.5 in which two weights each of mass $m/2$ sit on top of a chamber containing N molecules of an ideal gas. The piston itself is massless and exposed to atmospheric pressure on the outside of the device. The gas is maintained at a constant temperature T_1 through contact with the container walls. The initial pressure of the gas is P_1. An expansion process is performed according to the following two steps: one of the weights is suddenly removed and the gas expands to an intermediate volume with corresponding decreased pressure P_2. Subsequently the second weight is removed and the gas expands to a new pressure $P_3 = P_{atm}$. The temperature is maintained constant throughout.

(a) Write an expression for the work per molecule, W/N, performed by the gas in terms of k_B, T_1, and the pressures P_1, P_2, and P_3.

(b) Write an expression for the efficiency of the expansion process, η, in terms of k_B, T_1, and the pressures P_1, P_2, and P_3. The efficiency is given by the ratio of the actual work to that which would be obtained if the expansion had been performed in a continuous, slow, reversible manner.

6.23. How significant are interfacial contributions to the work during an isotropic expansion process? Consider a system consisting of water in a cubic volume $V = 1$ L and at $T = 100\,°C$ and $P = 1$ atm.

(a) If the water is entirely vapor, estimate the work done if the volume is increased isotropically and under isothermal conditions by 10%.

(b) Instead let the water be 50% vapor and 50% liquid on a volume basis, such that an interface perpendicular to the z direction is present. Estimate the work done for the same process. Note that, because the volume change is small and occurs isothermally, the final state still contains both liquid and vapor and is at the same pressure. Account for surface effects, noting that the liquid–vapor surface tension of water is about 72 mJ/m^2. What is the relative magnitude of bulk pressure–volume work relative to interfacial work?

6.24. A spherical droplet of diameter 1.0 cm rests on a superhydrophobic surface, such that it completely "beads up." Approximately how much work is required to split the droplet into two smaller ones of equal diameter? The surface tension of water is about 72 mJ/m^2.

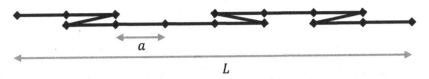

Figure 6.6. A simple elastic band consists of segments that can be oriented in one of two directions.

Applied problems

6.25. A simple model of a one-dimensional elastic or rubber band is shown in Fig. 6.6. The band consists of N different segments of length a. Each segment can point in either the positive or the negative x direction. The total end-to-end length of the band, for a given segment configuration, is L. By end to end, we mean the distance between the first point of the first segment and the last point of the last segment, regardless of the configuration of the intervening segments. Let the symbol τ denote the force exerted by the band or, equivalently, the negative force required to hold the band at a given length.

(a) What is the fundamental equation for this system, in differential entropy form ($dS =$)?

(b) Write an expression for L in terms of the fraction of segments pointing in the negative direction, x. Then determine the density of states and entropy as a function of x.

(c) Find the force τ that the band exerts as a function of L. What happens to τ as $L \to Na$?

(d) Compute the work per segment (W/N) required to stretch the band in a quasi-static manner from a resting state ($\tau = 0$) to fully extended, at $T = 300$ K. Make sure the sign of the work is appropriate. What is the corresponding heat transfer Q/N to or from the environment?

(e) When stretching DNA using optical tweezers, experiments suggest that the force experienced at a given L is linear in temperature. In light of the model above, what does this imply for the relative entropic and energetic contributions to DNA stretching?

6.26. One mole of argon is initially at T_1 and P_1. It undergoes the following cycle of quasi-static processes:

- $1 \to 2$: isothermal compression to half of its volume
- $2 \to 3$: adiabatic expansion back to its original volume
- $3 \to 1$: an isochoric process back to T_1 and P_1

Assume that argon is well described as an ideal gas with constant heat capacities $c_V = (3/2)k_B$ and $c_P = (5/2)k_B$.

(a) Sketch this process on a P–v diagram, where the x-axis is the molar volume v and the y-axis is the pressure P. Clearly label points 1, 2, and 3 and the direction of each process.

(b) Does this process do net work on the environment or require net work?

(c) If $T_1 = 25\,°C$, find T_3 in K.

(d) What are the signs of Q_{12} and Q_{31}, the heat transfers in the first and last steps, respectively? For $T_1 = 25\,°C$, find Q_{net} for the entire cycle.

6.27. One mole of nitrogen initially has temperature $T_1 = 25\,°C$ and pressure $P_1 = 1$ bar. It undergoes the following series of quasi-static processes:

- isothermal compression until $V_2 = (1/2)V_1$
- isobaric expansion at $P_2 = P_3$
- adiabatic expansion back to T_1 and P_1

Assume that nitrogen is well described as an ideal gas with constant heat capacity $c_V = (5/2)k_B$ and $c_P = (7/2)k_B$.

(a) Sketch this process on a P-v diagram, where the x-axis is the molar volume v and the y-axis is the pressure P. Clearly label points 1, 2, and 3 and the direction of each process.

(b) Find T_3 in K.

(c) Show that the net heat flow Q for the entire process is proportional to T_1.

6.28. A Carnot engine/cycle consists of the following reversible process steps between states 1, 2, 3, and 4:

- $1 \rightarrow 2$: adiabatic compression
- $2 \rightarrow 3$: isothermal expansion at T_H
- $3 \rightarrow 4$: adiabatic expansion
- $4 \rightarrow 1$: isothermal compression at T_C

Assume that the working fluid is an ideal monatomic gas with $c_V = (3/2)k_B$ and $c_P = (5/2)k_B$. Prove that the Carnot cycle achieves the maximum efficiency allowed by the second law for a heat engine, $\eta_{max} = 1 - T_C/T_H$.

6.29. The air standard Otto cycle is an idealized model of combustion engines not unlike that found in automobiles running on unleaded fuel. It consists of the following process steps between states 1, 2, 3, and 4 taking place inside a piston apparatus:

- $1 \rightarrow 2$: adiabatic compression
- $2 \rightarrow 3$: isochoric heating (due to fuel combustion)
- $3 \rightarrow 4$: adiabatic expansion
- $4 \rightarrow 1$: isochoric cooling (re-fueling the piston)

Assume that the working fluid in the piston is an ideal gas with temperature-independent heat capacities c_V and $c_P = c_V + k_B$.

(a) Show that the efficiency of this engine is given by $\eta = 1 - (T_4 - T_1)/(T_3 - T_2)$.

(b) Further show that the efficiency can be written as $\eta = 1 - r^{1-\gamma}$, where $r \equiv V_1/V_2$ is the so-called compression ratio and $\gamma \equiv c_P/c_V$. •

6.30. A fireman is using a high-pressure stream of water from a hose to combat a raging forest fire. At one end, the hose has an inside diameter of $d_1 = 5$ cm and is connected to a high-pressure water reservoir at $P_1 = 300$ psi. At the other end is a nozzle with diameter $d_2 = 3$ cm that is exposed to atmospheric pressure $P_2 = 1$ atm. To a good approximation, flowing water in this process can be modeled as having a constant temperature and constant mass density $\rho = 1 \text{ g/cm}^3 = 1{,}000 \text{ kg/m}^3$. Moreover, for an incompressible liquid, the following equation well describes isothermal enthalpy changes:

$$\Delta h \approx v \, \Delta P$$

where v is the volume per mole or molecule (depending on the basis of h).

(a) How much faster is the velocity of the exiting water at the nozzle than that of the entering stream at the reservoir? Express your answer as a factor (e.g., $u_2 = 1.5 \, u_1$).

(b) If the hose is well insulated, find the exit velocity of the stream of water, in m/s.

(c) Find the exit volumetric flow rate of water, in gal/s.

(d) Instead, you suspect that the hose is not well insulated because you find that the actual, measured exit velocity is 90% of what you calculated in part (b). The inlet velocity, however, remains the same. This suggests that frictional losses result in a dissipation of heat to the environment as the water maintains constant temperature. Find the heat lost in kJ per kg of water.

6.31. A blood vessel 1 mm in diameter and carrying a flow of blood at 20 cm/s splits into ten smaller vessels of equal flow and all 0.1 mm in diameter. The pressure of the blood in the larger vessel is about 60 mmHg. Assume that no blood crosses through the walls of the blood vessels, the flows are steady, and that blood can be approximated as a homogeneous, incompressible liquid with density $1{,}060 \text{ kg/m}^3$. Note that, for an incompressible liquid, the following equation well describes isothermal enthalpy changes:

$$\Delta h \approx v \, \Delta P$$

(a) What is the velocity of the blood in the smaller vessels, in cm/s?

(b) What should be the pressure of the blood in the smaller vessels, in mmHg? Assume there is no substantial heat transfer to or from the blood during the process of the flow splitting.

(c) If, instead, some heat is transferred to the blood from the surrounding tissue, would the pressure of the blood in the smaller vessels be higher or lower than your answer in part (b)? Be sure to justify your answer.

6.32. An impressive outdoor fountain emits a continuous, vertical jet of water into the air. The jet emerges from a cylindrical nozzle 3 cm in diameter and carries 2.3 gal/s of liquid. A pump is used to move fluid to the nozzle through a hose of diameter 6 cm that is connected to a reservoir (pool) of water. Assume that water is incompressible and that the pressure and temperature of the jet immediately outside of the nozzle are very close to those of the reservoir. Neglect any heat transfer.

(a) What is the velocity of the water jet leaving the nozzle, in m/s? What is the velocity of the water entering the hose from the reservoir, in m/s?

(b) How much power is required to operate the pump, in W?

(c) How high will the jet reach, in m? Assume that the returning water does not fall back on top of the jet.

FURTHER READING

H. Callen, *Thermodynamics and an Introduction to Thermostatistics*, 3rd edn. New York: Wiley (1985).

K. Denbigh, *The Principles of Chemical Equilibrium*, 4th edn. New York: Cambridge University Press (1981).

J. R. Elliot and C. T. Lira, *Introductory Chemical Engineering Thermodynamics*, 2nd edn. Upper Saddle River, NJ: Prentice Hall (2012).

L. D. Landau and E. M. Lifshitz, *Statistical Physics*, 3rd edn. Oxford: Butterworth-Heinemann (1980).

A. Z. Panagiotopoulos, *Essential Thermodynamics*. Princeton, NJ: Drios Press (2011).

J. M. Smith, H. V. Ness, and M. Abbott, *Introduction to Chemical Engineering Thermodynamics*, 7th edn. New York: McGraw-Hill (2005).

J. W. Tester and M. Modell, *Thermodynamics and Its Applications*, 3rd edn. Upper Saddle River, NJ: Prentice Hall (1997).

7   Legendre transforms and other potentials

7.1   New thermodynamic potentials from baths

We began our discussion of equilibrium by considering how the entropy emerges in isolated systems, finding that it is maximized under constant E, V, N conditions. In practical settings, however, systems are not often isolated and it is difficult to control their energy and volume in the presence of outside forces. Instead, it is easier to control so-called field parameters like temperature and pressure.

In this chapter, we will discuss the proper procedure for switching the independent variables of the fundamental equation for other thermodynamic quantities. In doing so, we will consider non-isolated systems that are held at constant temperature, pressure, and/or chemical potential through coupling to various kinds of baths. In such cases, we find that entropy maximization requires us to consider the entropy of both the system and its surroundings. Moreover, new thermodynamic quantities will naturally emerge in this analysis: additional so-called *thermodynamic potentials*.

To achieve conditions of constant temperature, pressure, or chemical potential, one couples a system to a bath. As discussed before, a bath is a large reservoir that can exchange energy, volume, or particles with the system of interest. While exchanging volume/energy/particles alters the system state, such changes are so minuscule for the bath that it is essentially always at the same equilibrium condition. Figure 7.1 gives a schematic representation of ways in which a system can be coupled to different kinds of baths.

As we saw in Chapter 4, systems that freely exchange $E/V/N$ ultimately reach thermal/mechanical/chemical equilibrium by finding the same $T/P/\mu$. Here, the main difference with these previous findings is that the bath is so large that it stays at the same $T/P/\mu$, and hence it *sets* the final equilibrium values of these parameters for the system. We consider the bath to be ideal in that it is infinitely large. Indeed, it must be infinitely sized in order to rigorously maintain the corresponding constant system conditions.

7.2   Constant-temperature coupling to an energy bath

Let us consider the specific case in which a system is held at a constant temperature by equilibrating with an ideal heat bath. In reaching equilibrium, the entropy of the world is maximized subject to energy conservation,

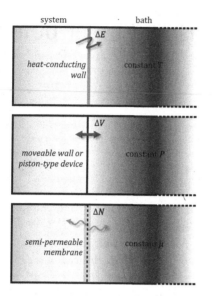

Figure 7.1. A system can be coupled to different kinds of baths. Because a bath is so large, its equilibrium state remains constant even though it can exchange different quantities (energy, volume, particles) with the system.

$$S_{world} = S + S_{bath} \quad \text{(maximum at equilibrium)}$$

$$E_{world} = E + E_{bath} \quad \text{(constant)} \tag{7.1}$$

Here, the unsubscripted variables are those pertaining to the system. Since the state of the bath does not change for any measurable energy transfer, we write the integrated form of its fundamental equation with the implicit approximation that all of its intensive state parameters are constant,

$$S_{bath} = \frac{E_{bath}}{T_{bath}} + \frac{P_{bath} V_{bath}}{T_{bath}} - \frac{\mu_{bath} N_{bath}}{T_{bath}}$$

$$= \frac{E_{bath}}{T_{bath}} + \text{(constants)} \tag{7.2}$$

where T_{bath} is fixed due to the constancy of state but E_{bath} depends on the exchange with the system. We also know that the energy of the bath is the balance of the total energy and that of the system; therefore,

$$S_{bath} = \frac{E_{world} - E}{T_{bath}} + \text{(constants)} \tag{7.3}$$

The energy of the world does not change, and we can absorb it into the constants,

$$S_{bath} = -\frac{E}{T_{bath}} + \text{(constants)} \tag{7.4}$$

Finally, we insert this expression into the entropy of the world,

$$S_{world} = S - \frac{E}{T_{bath}} + \text{(constants)} \tag{7.5}$$

To find the point of equilibrium between the two systems, we maximize S_{world} with respect to the energies of the system and bath. Note here that T_{bath} is *constant* and does not change with E. Because T_{bath} is the final equilibrium temperature of our system, we simply write it as T. Moreover, the constants group is unaffected by the energies and hence we can neglect it during maximization. On the other hand, S is dependent on E since changes to the system energy will affect its entropy. As a result, we might write the equilibrium expression as

$$\max \left[S_{world} \right] \rightarrow \max \left[S(E, V, N) - \frac{E}{T} \right] \tag{7.6}$$

That is, equilibrium occurs when the energy E of the system satisfies the expression in (7.6) with T treated as a constant. By convention, we define a quantity that entails this particular combination of variables in a slightly rearranged manner. The *Helmholtz free energy* is given by

$$A \equiv E - TS \tag{7.7}$$

Now the equilibrium condition becomes

$$\max \left[-\frac{A}{T} \right] \tag{7.8}$$

Because T is constant, this extremum is equivalent to

$$\min [A] \tag{7.9}$$

That is, the Helmholtz free energy attains a minimum at equilibrium. A is a function new to us that, importantly, is specific to the *system*. It does not rely on the detailed properties of the bath, and it is a distinct *thermodynamic potential* of the system. We have shown that maximizing the entropy of the world, which includes both the system and the bath, is equivalent to minimizing the Helmholtz free energy of the system alone.

Crucially, the equilibrium, minimized value of A is not a function of E, V, and N, the natural variables of the entropy, but is instead a function of T, V, and N. That is, we lose E as an independent variable but gain T as one. Physically, this makes sense. For one thing, we know that T is determined by the bath. It is something that we specify independently since we choose the particular kind of bath that we bring in contact with the system. Moreover, the value E of the system is determined by the entropy-maximization/Helmholtz-free-energy-minimization procedure. We cannot set E independently because its value is controlled by coupling to the bath; it becomes an *internal degree of freedom*.

At equilibrium for closed systems held at a constant temperature and volume, the **Helmholtz free energy** $A \equiv E - TS$ reaches a minimum. For pure systems, this equilibrium value of A has independent variables T, V, and N, so that we have the thermodynamic potential and function $A(T, V, N)$.

We note that the Helmholtz free energy is sometimes written with the variable F in the physics community. In this text, we use A instead throughout.

Example 7.1 *Find the Helmholtz free energy of a monatomic ideal gas, for which the entropy function is*

$$S(E, V, N) = k_B N \ln \left[\left(\frac{E}{N} \right)^{3/2} \frac{V}{N} \right] + N k_B c$$

where c is a constant.

Conceptually we put the gas in equilibrium with a heat bath with which it can exchange energy. The Helmholtz free energy is minimized, and its form is

$$A = E - TS$$

$$= E - N k_B T \ln \left[\left(\frac{E}{N} \right)^{3/2} \frac{V}{N} \right] - N k_B T c$$

We minimize with respect to E for fixed T, V, and N,

$$0 = \left(\frac{\partial A}{\partial E} \right) = 1 - \frac{3 N k_B T}{2E}$$

Solving for E gives

$$E = \frac{3}{2} N k_B T$$

which is consistent with the fact that $C_V = (3/2) N k_B$ for a monatomic ideal gas. Back-substitution into the expression for the Helmholtz free energy finally gives

$$A = \frac{3}{2} N k_B T - N k_B T \ln \left[\left(\frac{3 k_B T}{2} \right)^{3/2} \frac{V}{N} \right] - N k_B T c$$

$$= N k_B T \left\{ \frac{3}{2} - \ln \left[\left(\frac{3 k_B T}{2} \right)^{3/2} \frac{V}{N} \right] - c \right\}$$

which is only a function of T, V, and N.

7.3   Complete thermodynamic information and natural variables

Having introduced the function $A(T, V, N)$, it is worthwhile discussing the completeness of the thermodynamic information contained in it. Recall that only three pieces of information are required in order to completely specify the extensive thermodynamic state of a single-component system, after which all other thermodynamic variables have a unique value. The complete list of variables includes S, E, V, N, T, P, and μ. If we pick three of these to specify the state, then we seek a thermodynamic function that can provide the values of the other four. In the case of the entropy, the three variables to be specified are E, V, and N; the function $S(E, V, N)$ gives the entropy directly, and T, P, and μ extend from its partial derivatives. In that sense, $S(E, V, N)$ is a function that gives a complete set of thermodynamic information.

When we change from E to T in what we hold constant, we must switch from S to A as the relevant system potential. One might wonder why the function $S(T, V, N)$ is not as fundamental as $A(T, V, N)$. Indeed, $S(T, V, N)$ is a perfectly valid state function because we can always tabulate the entropy using the parameters T, V, and N, which come from different conjugate pairs. However, even though such a function can be constructed and is useful, it ultimately results in a loss of some information. That is, we cannot manipulate $S(T, V, N)$ so as to extract the *complete* set of thermodynamic variables at a given state point. For example, how could we find E from $S(T, V, N)$? We would take the following approach:

$$dS = \frac{1}{T}\, dE + \frac{P}{T}\, dV - \frac{\mu}{T}\, dN \quad \rightarrow \quad \left(\frac{\partial S}{\partial T}\right)_{V,N} = \frac{1}{T}\left(\frac{\partial E}{\partial T}\right)_{V,N} \tag{7.10}$$

By integrating, we obtain

$$E = \int T\left(\frac{\partial S}{\partial T}\right)_{V,N} dT + f(V, N) \tag{7.11}$$

In finding E, we end up with an unknown V- and N-dependent constant due to the fact that we need to perform an integration. We can never find the value of that constant from the function $S(T, V, N)$ alone. In contrast, if we knew $S(E, V, N)$ we would never incur any unknown constants in the determination of the other thermodynamic properties (T, P, μ) because all of these stem from its derivatives. In this sense, we call $S(E, V, N)$ *fundamental*, in that it contains all thermodynamic information, while $S(T, V, N)$ is not. We also say that the *natural variables* of the thermodynamic potential S are E, V, and N. These are general terms that apply to all potentials, as summarized in Table 7.1.

As alluded above, the Helmholtz function $A(T, V, N)$ is also fundamental. We do not lose any information in $A(T, V, N)$ despite the fact that it represents a function where T is now an independent variable rather than E. All other thermodynamic properties extend from derivatives of A. This is easily shown from its differential form,

Table 7.1 Natural variables

Thermodynamic potential	Natural variables	Fundamental form
Entropy S	E, V, N	$S(E, V, N)$
Internal energy E	S, V, N	$E(S, V, N)$
Helmholtz free energy A	T, V, N	$A(T, V, N)$

$$dA = d(E - TS)$$
$$= dE - T\, dS - S\, dT$$
$$= (T\, dS - P\, dV + \mu\, dN) - T\, dS - S\, dT \tag{7.12}$$

where we have substituted the differential form for the fundamental energy equation. On simplifying, we have

$$dA = -S\, dT - P\, dV + \mu\, dN \tag{7.13}$$

We could also take a total derivative of $A(T, V, N)$ to find

$$dA = \left(\frac{\partial A}{\partial T}\right)_{V,N} dT + \left(\frac{\partial A}{\partial V}\right)_{T,N} dV + \left(\frac{\partial A}{\partial N}\right)_{T,V} dN \tag{7.14}$$

The comparison of Eqns. (7.13) and (7.14) shows that the partial derivatives of the Helmholtz free energy are

$$\left(\frac{\partial A}{\partial T}\right)_{V,N} = -S, \qquad \left(\frac{\partial A}{\partial V}\right)_{T,N} = -P, \qquad \left(\frac{\partial A}{\partial N}\right)_{T,V} = \mu \tag{7.15}$$

We see that a complete set of thermodynamic information can be derived from A, given that the state conditions are specified by T, V, and N. Therefore, we conclude that $A(T, V, N)$ is the fundamental potential at constant T, V, N conditions.

7.4   Legendre transforms: mathematical convention

Above we derived A by considering the physical process in which a system is attached to a heat bath. Mathematically speaking, there is a deeper theoretical foundation for transforming the function $E(S, V, N)$ into $A(T, V, N)$. Per the considerations above in (7.6), we can find the value of A at any state point using the construction

$$A(T, V, N) = E^* - TS(E^*, V, N) \tag{7.16}$$

where E^* is the value of the internal energy E that minimizes the expression on the RHS, for given values of T, V, and N. Equivalently, we can switch to the energy version of the fundamental equation to show that

$$A(T, V, N) = E(S^*, V, N) - TS^* \tag{7.17}$$

where now S^* is the value of the entropy that minimizes A. In either case, we find the value of one free variable (an internal degree of freedom) that minimizes the Helmholtz free energy.

These constructions are closely related to a mathematical operation called the *Legendre transform*. Consider a convex function $f(x)$ and let its derivative be $p = df/dx$. A Legendre transform produces from $f(x)$ a new function $g(p)$, where the independent variable is p. This new function contains all of the same information as the original function so that it can be easily transformed back. The exact procedure is

$$\mathcal{L}[f(x)] = g(p)$$
$$= \max_x [px - f(x)]$$
$$= px^* - f(x^*) \text{ where } x^* \text{ solves } \frac{df(x^*)}{dx} = p \qquad (7.18)$$

In other words, the Legendre transform produces a new function equivalent to $px^* - f(x^*)$, where p is independently specified and x^* is the value of x that maximizes this expression at fixed p. This is easy to see: simply take the x derivative of $px - f(x)$ and set it equal to zero to obtain the third line in (7.18).

To show that the Legendre transform maintains all of the information in the original function, we merely need to show that $g(p)$ can be converted back to $f(x)$. The inverse of a Legendre transform turns out to be itself. We compute the derivative of $g(p)$ to demonstrate this,

$$\frac{dg(p)}{dp} = \frac{d}{dp}[px^* - f(x^*)]$$
$$= p\frac{dx^*(p)}{dp} + x^* - \frac{df(x^*)}{dx}\frac{dx^*(p)}{dp}$$
$$= p\frac{dx^*(p)}{dp} + x^* - p\frac{dx^*(p)}{dp} \quad \left(\text{since } \frac{df(x^*)}{dx} = p\right)$$
$$= x^* \qquad (7.19)$$

Then, computing $h(y) \equiv \mathcal{L}[g(p)]$,

$$\mathcal{L}[g(p)] = yp^* - g(p^*) \text{ where } \frac{dg(p^*)}{dp} = y \qquad (7.20)$$

Substituting for $g(p^*)$ from before, and using the fact that $dg(p^*)/dp = x^* = y$, gives

$$\mathcal{L}[g(p)] = yp^* - [p^*x^* - f(x^*)]$$
$$= (x^*)p^* - [p^*x^* - f(x^*)]$$
$$= f(y) \qquad (7.21)$$

We see that the original function is recovered when we apply the Legendre transform twice, $\mathcal{L}[\mathcal{L}[f]] = f$.

Example 7.2 *Compute the single and double Legendre transforms of $f(x) = e^x$.*

Starting with the definition of $\mathcal{L}[f(x)]$,

$$g(p) = px^* - f(x^*)$$
$$= px^* - e^{x^*}$$

The equation for x^* is

$$\frac{df(x^*)}{dx} = p \quad \rightarrow \quad e^{x^*} = p \quad \rightarrow \quad x^* = \ln p$$

On back-substituting for x^*, we find

$$g(p) = p \ln p - p$$

Now we compute the Legendre transform of $g(p)$,

$$h(y) = \mathcal{L}[g(p)]$$

$$= yp^* - g(p^*)$$

$$= yp^* - p^* \ln p^* + p^*$$

The condition for p^* gives

$$\frac{dg(p^*)}{dp} = y \quad \rightarrow \quad \ln p^* = y \quad \rightarrow \quad p^* = e^y$$

so that

$$h(y) = ye^y - e^y \ln e^y + e^y$$

$$= e^y$$

We see that we recover the same function, $h(y) = e^y$, as our original one, $f(x) = e^x$.

7.5   Legendre transforms: thermodynamic convention

Legendre transforms convert one thermodynamic potential to another, but there is a subtlety in that thermodynamics makes use of *negative* Legendre transforms, with a sign change in their construction. Moreover, thermodynamics entails multivariate functions, and one must specify which variable is the subject of the transform. Consequently, in a thermodynamic context we define for a convex function $f(x, y, z)$,

$$\mathcal{L}_x[f(x,y,z)] = g(p,y,z)$$
$$= \min_x[f(x,y,z) - px] \tag{7.22}$$
$$= f(x^*,y,z) - px^* \text{ where } x^* \text{ solves } \frac{\partial f(x^*,y,z)}{\partial x} = p$$

Here, the subscript x on the Legendre operator indicates that we perform the transform with respect to the independent variable x and not y or z. There is no need to be concerned about the difference with the mathematical definition; this is simply a convention-based sign change. It is important to note that the minimum operation in (7.22) applies when f is convex in x. If instead f is concave in x then the extremum operation is a maximum.

From now on, we will use the symbol $\mathcal{L}$ to denote the Legendre transform in the thermodynamic convention. In shortcut form, we write it as

$$\mathcal{L}_x[f] = f - x \cdot \left(\frac{\partial f}{\partial x}\right) \tag{7.23}$$

Consider as an example the expression for the Helmholtz free energy in (7.17),

$$A(T, V, N) = E(S^*, V, N) - TS^* \tag{7.24}$$

where S^* is the value of S that minimizes this expression,

$$\frac{\partial}{\partial S}[E(S, V, N) - TS]_{S=S^*} = 0 \tag{7.25}$$

or, equivalently,

$$\frac{\partial E(S^*, V, N)}{\partial S} = T \tag{7.26}$$

This set of equations is exactly a Legendre transform. By comparing (7.22) with (7.24) and (7.25) we see that $f = E$, $x = S$, and $p = T$. In fact, we can write the expression for the Helmholtz free energy explicitly as the entropy Legendre transform of the internal energy,

$$A(T, V, N) = \mathcal{L}_S[E(S, V, N)]$$

$$A = E - TS \quad \text{(minimum at constant } T, V, N \text{ equilibrium)} \tag{7.27}$$

Here, the subscript S indicates a transform with respect to the independent variable S.

In performing a Legendre transform of the internal energy, we swapped S for T as an independent variable to produce the Helmholtz free energy A. However, this is not the only Legendre transform that we can perform since there are three independent variables. One can consider transformations between any two conjugate variables. This means that we can also swap V for P and μ for N since Legendre transforms trade an independent variable for the derivative of the thermodynamic potential with respect to it. However, one cannot swap non-conjugate pairs, such as P for T. In general, Legendre transforms are the mathematical equivalent of connecting our system to various kinds of baths.

Legendre transforms enable one to switch an independent variable of a thermodynamic potential to its conjugate. The function produced corresponds to the relevant thermodynamic potential at constant conditions corresponding to the new independent variables.

7.6　　The Gibbs free energy

Let us examine this principle in the case of equilibrium at both constant T and P. In other words, consider a system that exchanges both energy and volume with a bath. We proceed using the second law as before,

$$
\begin{aligned}
S_{\text{world}} &= S + S_{\text{bath}} && (\text{maximum at equilibrium}) \\
E_{\text{world}} &= E + E_{\text{bath}} && (\text{constant}) \\
V_{\text{world}} &= V + V_{\text{bath}} && (\text{constant})
\end{aligned}
\tag{7.28}
$$

The very large bath admits the behavior

$$
S_{\text{bath}} = \frac{E_{\text{bath}}}{T_{\text{bath}}} + \frac{P_{\text{bath}} V_{\text{bath}}}{T_{\text{bath}}} + (\text{constants})
\tag{7.29}
$$

Expressing the bath quantities in terms of the difference between the world and the system gives

$$
S_{\text{world}} = S - \frac{E}{T_{\text{bath}}} - \frac{P_{\text{bath}} V}{T_{\text{bath}}} + (\text{constants})
\tag{7.30}
$$

The maximum-entropy equilibrium condition therefore becomes

$$
\max\left[S_{\text{world}}\right] \rightarrow \max\left[S - \frac{E}{T} - \frac{PV}{T}\right]
\tag{7.31}
$$

where we replaced T_{bath} with T and P_{bath} with P since it is through the bath that the system temperature and pressure are ultimately maintained. In multiplying the RHS by the constant $-T$, we find the equivalent equilibrium condition

$$
\min[E(S, V, N) - TS + PV]
\tag{7.32}
$$

The quantity in brackets is called the *Gibbs free energy*, $G \equiv E - TS + PV$. Here, equilibrium is obtained by minimizing (7.32) with respect to the values of E and V for given independent parameters T and P. If we proceed with the E minimization, we see from our earlier results that this is equivalent to

$$
\min[A(T, V, N) + PV]
\tag{7.33}
$$

We could also have tackled this problem using Legendre transforms. Since the system is connected to a bath with which it can exchange energy and volume, this means that we swap both $S \rightarrow T$ and $V \rightarrow P$ by performing two transforms. We can start with the Helmholtz free energy, which already entails a Legendre transform in the entropy, and perform an additional transform in the volume. The resulting potential is the Gibbs free energy,

$$
\begin{aligned}
G(T, P, N) &= \mathcal{L}_V[A(T, V, N)] \\
G &\equiv A + PV && (\text{minimum at constant } T, P, N \text{ equilibrium})
\end{aligned}
\tag{7.34}
$$

Notice that the Gibbs free energy is naturally a function of the variables T, P, and N. Moreover, like the Helmholtz free energy, it is a minimum at equilibrium, but it applies to systems at constant T, P, N conditions. We now examine the partial derivatives of G,

$$dG = dA + P\,dV - V\,dP$$
$$= (-S\,dT - P\,dV + \mu\,dN) + P\,dV + V\,dP \tag{7.35}$$

in which we substituted the differential expression for dA. Simplifying,

$$dG = -S\,dT + V\,dP + \mu\,dN \tag{7.36}$$

This result shows that the partials are in fact

$$\left(\frac{\partial G}{\partial T}\right)_{P,N} = -S, \qquad \left(\frac{\partial G}{\partial P}\right)_{T,N} = V, \qquad \left(\frac{\partial G}{\partial N}\right)_{T,P} = \mu \tag{7.37}$$

We can also relate the Gibbs free energy directly to the internal energy,

$$G(T, P, N) = \mathcal{L}_V[\mathcal{L}_S[E(S, V, N)]]$$
$$G = E - TS + PV \tag{7.38}$$

The two Legendre transforms provide the mathematical equivalent of connecting our system to energy and volume baths.

At equilibrium for closed systems held at a constant temperature and pressure, the **Gibbs free energy** $G \equiv E - TS + PV$ reaches a minimum. For pure systems, this equilibrium value of G has independent variables T, P, and N, so that we have the thermodynamic potential and function $G(T, P, N)$.

7.7 Physical rationale for Legendre transforms

Why do Legendre transforms naturally emerge when we consider alternate constant conditions? The reason stems from the fundamental form of the bath entropy. In general, we can express S_{bath} in terms of its extensive properties $X_{1,bath}$, $X_{2,bath}$,..., and the intensive conjugates of those, $y_{1,bath} = (\partial S_{bath}/\partial X_{1,bath})$, $y_{2,bath} = (\partial S_{bath}/\partial X_{2,bath})$, and so on. According to Euler's theorem,

$$S_{bath} = \sum_i y_{i,bath}\, X_{i,bath} \tag{7.39}$$

For example, if $X_{1,bath} = E_{bath}$ then $y_{1,bath} = 1/T_{bath}$. For a bath we take the intensive parameters to be constant, since its state is insensitive to small variations in the extensive quantities. Presume a bath is now coupled to a system such that it can exchange extensive quantity X_1 with $X_1 + X_{1,bath} = X_{1,world}$ constant. The second law then requires

$$S_{\text{world}} = S + S_{\text{bath}}$$

$$= S + y_{1,\text{bath}}\, X_{1,\text{bath}} + \sum_{i \neq 1} y_{i,\text{bath}}\, X_{i,\text{bath}}$$

$$= S + y_{1,\text{bath}}(X_{1,\text{world}} - X_1) + (\text{constants})$$

$$= S - y_{1,\text{bath}}\, X_1 + (\text{constants}) \tag{7.40}$$

We recognize that the intensive parameter $y_{1,\text{bath}}$ sets or determines the equivalent parameter in the system, y_1. Thus we may rewrite the expression above more formally as

$$\max\,[S_{\text{world}}] \rightarrow \max_{X_1}\,[S(X_1, X_2, \ldots) - y_1 X_1] \tag{7.41}$$

where maximization takes place with respect to the value X_1 at specified, fixed y_1. Note that the expression on the RHS is an entropy Legendre transform. At the maximum with respect to X_1,

$$\left(\frac{\partial S(X_1^*)}{\partial X_1}\right) = y_1 \tag{7.42}$$

In shortcut form, we can then write

$$\max\,[S_{\text{world}}] \rightarrow \max\,\left[S - \left(\frac{\partial S}{\partial X_1}\right)X_1\right] \tag{7.43}$$

In other words, connecting a system to a bath – regardless of the particular kind and what is exchanged – naturally leads to a Legendre transform of the entropy. Importantly, all variables on the RHS of (7.43) are strictly properties of the system; we have removed those of the bath by ignoring the constant terms. That is in fact the main feature of this approach. Rather than consider the second law directly for the combined system and bath, we convert to an equivalent form involving a different extremized quantity that depends only on system properties, and is independent of the bath.

7.8   Extremum principles with internal constraints

The entropy maximization described above was presented with respect to a system variable that is in exchange with the bath, such as E. In Chapter 4, we also discussed the idea of internal degrees of freedom and constraints. When an internal constraint is removed or when an internal degree of freedom exists, an isolated system will behave so as to maximize its entropy according to the second law. Using the notation of the previous section, we have

$$S(X_1, X_2, \ldots) = \max_{\xi}\, S(X_1, X_2, \ldots, \xi) \tag{7.44}$$

where ξ gives the value of the internal degree of freedom. It turns out that this kind of extremum principle applies to all thermodynamic potentials. In the heat-bath scenario, for example, the system energy E can be considered an internal degree of freedom and the Helmholtz free energy is minimized with respect to it. Alternatively, other degrees of freedom could be present, for example, due to an internal, freely moving partition that separates two compartments within the system boundary.

Let us generalize the way in which internal degrees of freedom behave when a system is connected to various kinds of baths. We need to consider the possibility that the system entropy can change due to both an internal degree of freedom and the bath-exchange variable. On substituting Eqn. (7.44) into (7.41) we find

$$\max [S_{\text{world}}] \rightarrow \max_{X_1} \left[\max_{\xi} \ [S(X_1, X_2, \ldots, \xi)] - y_1 X_1 \right] \qquad (7.45)$$

A reordering of the maximization conditions gives

$$\max [S_{\text{world}}] \rightarrow \max_{\xi} \left[\max_{X_1} \ [S(X_1, X_2, \ldots, \xi) - y_1 X_1] \right] \qquad (7.46)$$

Consider as an example a system connected to a thermal bath,

$$\max [S_{\text{world}}] \rightarrow \max_{\xi} \left[\max_{E} \ [S(E, V, N, \xi) - E/T] \right] \qquad (7.47)$$

Since T is constant, we can rewrite this as

$$\max [S_{\text{world}}] \rightarrow \min_{\xi} \left[\min_{E} [E - TS(E, V, N, \xi)] \right]$$

$$= \min_{\xi} [A(T, V, N, \xi)] \qquad (7.48)$$

The result shows that the Helmholtz free energy is minimized at constant T, V, and N with respect to any unconstrained internal degrees of freedom or internal *perturbations*. Similarly for the Gibbs free energy,

$$\max [S_{\text{world}}] \rightarrow \max_{\xi} \left[\max_{E, V} \ [S(E, V, N, \xi) - E/T - PV/T] \right]$$

$$\rightarrow \min_{\xi} \left[\min_{E, V} [E - TS(E, V, N, \xi) + PV] \right]$$

$$= \min_{\xi} [G(T, P, N, \xi)] \qquad (7.49)$$

The general message is the following.

The relevant potential at a set of constant conditions is extremized not only with respect to the variables describing what is being exchanged with baths, but also with respect to any general internal degree of freedom. For example,

$$A \text{ is } \textbf{minimized} \text{ at constant } T, V, N \text{ conditions}$$

and

$$G \text{ is } \textbf{minimized} \text{ at constant } T, P, N \text{ conditions,}$$

where the minimum is found with respect to any internal perturbations of the system, or unconstrained internal degrees of freedom.

7.9   The enthalpy and other potentials

Now we shift perspective slightly and consider Legendre transforms of the energy $E(S, V, N)$ rather than the entropy $S(E, V, N)$. What physical bath scenarios correspond to cases in which S remains constant? As we described in Chapter 4, conceptually we can hold a system at constant S by artificially and continuously adjusting its energy so that the entropy stays at a fixed value, even if its volume or composition changes. If we maintain constant S conditions in this way but allow a system to exchange other quantities like volume or mass, then,

$$S_{\text{world}} = S + S_{\text{bath}}$$

$$= S_{\text{bath}} + (\text{constants}) \tag{7.50}$$

Let us now assume that the system exchanges volume with the bath specifically, in order to maintain constant-pressure conditions. Continuing as we have before by expanding the bath entropy, we obtain

$$S_{\text{world}} = \frac{E_{\text{bath}}}{T_{\text{bath}}} + \frac{P_{\text{bath}} V_{\text{bath}}}{T_{\text{bath}}} + (\text{constants})$$

$$= \frac{E_{\text{world}} - E}{T_{\text{bath}}} + \frac{P_{\text{bath}} (V_{\text{world}} - V)}{T_{\text{bath}}} + (\text{constants})$$

$$= \frac{-E - PV}{T_{\text{bath}}} + (\text{constants}) \tag{7.51}$$

Since the temperature of the bath is constant, maximization of the world entropy can finally be written as

$$\max \left[S_{\text{world}} \right] \rightarrow \min_{V} [E(S, V, N) + PV]$$

$$= \min_{V}[H] \tag{7.52}$$

In the last line we introduced the enthalpy. This important thermodynamic quantity is a Legendre transform of the energy,

$$H(S, P, N) = \mathcal{L}_V \left[E(S, V, N) \right]$$

$$H \equiv E + PV \quad \text{(minimum at constant } S, P, N \text{ equilibrium)} \tag{7.53}$$

The enthalpy is the thermodynamic potential for constant S, P, N conditions, and it reaches a minimum at equilibrium. The differential form of H is found by differentiating (7.53),

$$dH = T\,dS + V\,dP + \mu\,dN \qquad (7.54)$$

which summarizes the partial derivatives,

$$\left(\frac{\partial H}{\partial S}\right)_{P,N} = T, \qquad \left(\frac{\partial H}{\partial P}\right)_{S,N} = V, \qquad \left(\frac{\partial H}{\partial N}\right)_{S,P} = \mu \qquad (7.55)$$

The enthalpy is not called a *free* energy for the reason that the temperature is not constant. The nomenclature "free" stems from the fact that A and G remove the term TS from the internal energy E, which loosely represents the part of the energy due to unorganized molecular heat motion. In this sense, the remaining parts represent energy available to do useful work.

The enthalpy may seem like an unusual potential in that systems infrequently achieve constant S, P, N conditions. However, it naturally emerges as a useful quantity in many scenarios, as we have already seen in Chapters 5 and 6, such as isobaric heating, steady-state flow, and phase transitions. Even though the constant conditions in many of these cases are not fixed S, P, N, it can be shown that the enthalpy nonetheless contains relevant information.

At equilibrium for closed systems held at a constant entropy and pressure, the **enthalpy $H \equiv E + PV$ is at a minimum.** For pure systems, this equilibrium value of H has independent variables S, P, N, so that we have the thermodynamic potential $H(S, P, N)$. The second law takes the following form:

H is **minimized** at constant S, P, N conditions,

where the minimum is found with respect to any internal perturbations of the system, or unconstrained internal degrees of freedom.

What about other thermodynamic potentials? We have pursued only Legendre transforms in energy and volume, but we can transform in the particle number as well. If there are multiple components, we will be able to transform each of the N_i separately, so the number of possible new thermodynamic potentials can be quite large. These potentials are just as valid as those we have discussed so far; it simply turns out that the specific ones given by S, E, A, G, and H are most commonly useful.

7.10   Integrated and derivative relations

The internal energy given in integrated form by $E = TS - PV + \mu N$ provides similar relations for the other potentials,

$$\begin{aligned} A &= E - TS = -PV + \mu N \\ G &= E - TS + PV = \mu N \\ H &= E + PV = TS + \mu N \end{aligned} \qquad (7.56)$$

We could have also arrived at these expressions by applying Euler's theorem directly to the new thermodynamic potentials themselves, using the extensivity relations

$$A(T, \lambda V, \lambda N) = \lambda A(T, V, N)$$
$$G(T, P, \lambda N) = \lambda G(T, P, N) \tag{7.57}$$
$$H(\lambda S, P, \lambda N) = \lambda H(S, P, N)$$

Here, only the extensive variables are scaled in each case. Notice in particular that the chemical potential in single-component systems is simply the per-particle Gibbs free energy, $\mu = G/N$. In general, for multicomponent systems, we have

$$E = TS - PV + \sum_i \mu_i N_i$$
$$A = -PV + \sum_i \mu_i N_i$$
$$G = \sum_i \mu_i N_i \tag{7.58}$$
$$H = TS + \sum_i \mu_i N_i$$

While these relations describe integrated thermodynamic potentials, there are also ways to interrelate them using derivatives. Consider, for example, the relationship between E and A,

$$E = A + TS$$
$$= A - T \left(\frac{\partial A}{\partial T} \right)_{V,N} \tag{7.59}$$

where we expressed S as a partial derivative of A. By dividing by T^2 and combining the two terms, we obtain

$$\frac{E}{T^2} = \frac{A}{T^2} - \frac{1}{T} \left(\frac{\partial A}{\partial T} \right)_{V,N}$$
$$= -\frac{\partial}{\partial T} \left(\frac{A}{T} \right)_{V,N} \tag{7.60}$$

Equation (7.60) shows that we can extract the function $E(T, V, N)$ from a derivative of $A(T, V, N)$. We could have also performed the transformation in the opposite direction, by expressing T as an S derivative of E,

$$A = E - \left(\frac{\partial E}{\partial S} \right)_{V,N} S \quad \rightarrow \quad \frac{A}{S^2} = -\frac{\partial}{\partial S} \left(\frac{E}{S} \right)_{V,N} \tag{7.61}$$

which gives a route to $A(S, V, N)$ from $E(S, V, N)$. Moreover, we can use all of the expressions $G = A + PV$, $G = H - TS$, and $H = E + PV$ to generate additional relations:

$$\frac{E}{T^2} = -\frac{\partial}{\partial T}\left(\frac{A}{T}\right)_{V,N}, \qquad \frac{E}{P^2} = -\frac{\partial}{\partial P}\left(\frac{H}{P}\right)_{S,N}$$

$$\frac{A}{S^2} = -\frac{\partial}{\partial S}\left(\frac{E}{S}\right)_{V,N}, \qquad \frac{A}{P^2} = -\frac{\partial}{\partial P}\left(\frac{G}{P}\right)_{T,N}$$

$$\frac{G}{S^2} = -\frac{\partial}{\partial S}\left(\frac{H}{S}\right)_{P,N}, \qquad \frac{G}{V^2} = -\frac{\partial}{\partial V}\left(\frac{A}{V}\right)_{T,N} \qquad (7.62)$$

$$\frac{H}{T^2} = -\frac{\partial}{\partial T}\left(\frac{G}{T}\right)_{P,N}, \qquad \frac{H}{V^2} = -\frac{\partial}{\partial V}\left(\frac{E}{V}\right)_{S,N}$$

Of course, there exist more such relations for the many other thermodynamic potentials that can be constructed by Legendre-transforming the particle-number variables in multicomponent systems. In general, it is not advised to memorize these; rather, it is important to understand the principles so such relations can be derived on the fly.

Example 7.3 *An experimentalist determines the heat capacity of a substance to obey the empirical relation $c_V(T, v) = \alpha T^2 v^2$, where α is a constant. The experimentalist also finds the entropy and energy to be zero at absolute zero, for all volumes. Compute the per-particle thermodynamic potentials $e(s, v)$, $h(s, P)$, $a(T, v)$, and $g(T, P)$.*

To solve this problem, we start with the definition of the constant-volume heat capacity,

$$c_V = \left(\frac{\partial e}{\partial T}\right)_V$$

The fundamental equation at constant volume shows $de = T\,ds$, such that

$$c_V = T\left(\frac{\partial s}{\partial T}\right)_V$$

We integrate these two expressions to compute the energy and entropy as functions of temperature,

$$e = \int c_V\,dT = \left(\frac{\alpha}{3}\right)T^3 v^2 + \text{constant}(v)$$

$$s = \int \frac{c_V}{T}\,dT = \left(\frac{\alpha}{2}\right)T^2 v^2 + \text{constant}(v)$$

Since we know the entropy and energy are both zero at $T = 0$, the constants are zero,

$$e = \left(\frac{\alpha}{3}\right)T^3 v^2, \qquad s = \left(\frac{\alpha}{2}\right)T^2 v^2$$

To find $e(s, v)$ we need to solve for the temperature as a function of entropy,

$$T = \left(\frac{2s}{\alpha v^2}\right)^{1/2}$$

giving

$$e(s, v) = \left(\frac{\alpha}{3}\right)\left(\frac{2s}{\alpha v^2}\right)^{3/2} v^2 = \left(\frac{8}{9\alpha}\right)^{1/2} s^{3/2} v^{-1}$$

To find the Helmholtz free energy, we use

$$a(T, v) = e - Ts$$

$$= \left(\frac{\alpha}{3}\right)T^3 v^2 - T\left[\left(\frac{\alpha}{2}\right)T^2 v^2\right]$$

$$= -\left(\frac{\alpha}{6}\right)T^3 v^2$$

For the enthalpy, we need to find the pressure. One way is to use the energy derivative,

$$P = -\left(\frac{\partial e}{\partial v}\right)_s = \left(\frac{8}{9\alpha}\right)^{1/2} s^{3/2} v^{-2}$$

Another way is to take the Helmholtz free energy derivative,

$$P = -\left(\frac{\partial a}{\partial v}\right)_T = \left(\frac{\alpha}{3}\right)T^3 v$$

It is easy to verify that the two pressure expressions are equal using $s = (\alpha/2)T^2 v^2$. Since the enthalpy depends on the entropy, and not on the temperature, we use the former. First we express volume in terms of pressure, one of the independent variables of the enthalpy:

$$v = \left(\frac{8}{9\alpha}\right)^{1/4} P^{-1/2} s^{3/4}$$

Then, the enthalpy takes the form

$$h(s, P) = e + Pv$$

$$= \left(\frac{8}{9\alpha}\right)^{1/2} s^{3/2}\left[\left(\frac{8}{9\alpha}\right)^{1/4} P^{-1/2} s^{3/4}\right]^{-1} + P\left[\left(\frac{8}{9\alpha}\right)^{1/4} P^{-1/2} s^{3/4}\right]$$

$$= \left(\frac{128}{9\alpha}\right)^{1/4} P^{1/2} s^{3/4}$$

Finally, we find the Gibbs free energy. We first express the volume as a function of temperature and pressure based on the pressure expression derived from the Helmholtz energy,

$$v = \left(\frac{3}{\alpha}\right)PT^{-3}$$

By constructing the Gibbs from the Helmholtz free energy, we obtain

$$g(T, P) = a + Pv$$

$$= -\left(\frac{a}{6}\right)T^3\left[\left(\frac{3}{a}\right)PT^{-3}\right]^2 + P\left[\left(\frac{3}{a}\right)PT^{-3}\right]$$

$$= \left(\frac{3}{2a}\right)P^2 T^{-3}$$

We could have also used $g = h - Ts$. The result would have been the same.

7.11 Multicomponent and intensive versions

The most general forms for each of the thermodynamic potentials involve multiple particle-number variables to indicate the amounts of different species. Let us denote the collection of these as $\{N\} = \{N_1, N_2, \ldots\}$. Table 7.2 summarizes the general potentials and their interrelationships.

An equivalent way of writing the sums involving chemical potentials is to use vectors, $\mathbf{N} = (N_1, N_2, \ldots)$ and $\boldsymbol{\mu} = (\mu_1, \mu_2, \ldots)$. Then we can take advantage of vector dot products as $\sum_i \mu_i \, dN_i = \boldsymbol{\mu} \cdot d\mathbf{N}$ and $\sum_i \mu_i \, dN_i = \boldsymbol{\mu} \cdot \mathbf{N}$. This alternate notation is frequently used.

When the system contains only a single component, we can define per-particle or molar versions of each of the thermodynamic potentials. The procedure is identical to what we did earlier for the energy fundamental equation in Chapter 5, and can be

Table 7.2 Differential and integrated forms of extensive thermodynamic potentials

Potential	Differential form	Integrated form
$S(E, V, \{N\})$	$dS = \frac{1}{T}\,dE + \frac{P}{T}\,dV - \sum_i \frac{\mu_i}{T}\,dN_i$	$S = \frac{E}{T} + \frac{PV}{T} - \sum_i \frac{\mu_i N_i}{T}$
$E(S, V, \{N\})$	$dE = T\,dS - P\,dV + \sum_i \mu_i \, dN_i$	$E = TS - PV + \sum_i \mu_i N_i$
$H(S, P, \{N\})$	$dH = T\,dS + V\,dP + \sum_i \mu_i \, dN_i$	$H = E + PV$ $= TS + \sum_i \mu_i N_i$
$A(T, V, \{N\})$	$dA = -S\,dT - P\,dV + \sum_i \mu_i \, dN_i$	$A = E - TS$ $= -PV + \sum_i \mu_i N_i$
$G(T, P, \{N\})$	$dG = -S\,dT + V\,dP + \sum_i \mu_i \, dN_i$	$G = E + PV - TS$ $= A + PV$ $= H - TS$ $= \sum_i \mu_i N_i$

Table 7.3 Differential and integrated forms of single-component intensive thermodynamic potentials on a per-particle or per-mole basis

Per-particle or per-mole potential	Differential form	Integrated relations
$s(e, v)$	$ds = \dfrac{1}{T} de + \dfrac{P}{T} dv$	$\dfrac{\mu}{T} = -s + \dfrac{e}{T} + \dfrac{Pv}{T}$
$e(s, v)$	$de = T ds - P dv$	$\mu = e - Ts + Pv$
$h(s, P)$	$dh = T ds + v dP$	$h = e + Pv$
		$\mu = h - Ts$
$a(T, v)$	$da = -s dT - P dv$	$a = e - Ts$
		$\mu = a + Pv$
$g(T, P)$	$dg = -s dT + v dP$	$g = e + Pv - Ts$
		$= a + Pv$
		$= h - Ts$
		$\mu = g$

applied both to the differential form and to the integrated form of the potentials. As before, we will use lower-case letters to indicate quantities that are normalized by N. Table 7.3 summarizes the intensive potentials for a single-component system. Notice that the per-particle Gibbs free energy is equal to the chemical potential. This is the case only for a single component.

7.12   Summary and look ahead

We have now learned an important and powerful generalization of the second law. Systems at conditions other than constant E, V, N require us to use a different thermodynamic potential beyond $S(E, V, N)$ or $E(S, V, N)$. That potential can be constructed by connecting the system to a bath of some kind and maximizing the world entropy. Equivalently, we obtain the appropriate potential by performing a Legendre transformation, which constructs a new function as $g = f - x(df/dx)$. Legendre-transformed potentials retain a complete set of thermodynamic information, and can be interrelated in both integrated and derivative forms.

Before closing this topic, we highlight one point that we will revisit in Chapter 17. Our current approach has been to couple a system to a bath and then maximize $S + S_{\text{surr}}$. Remember, however, that this maximization is really an approximation to find the most probable macrostate with the largest number of microstates. This approximation works well because the distribution of microstates is so sharply peaked around one value of what is transferred between the system and its surroundings.

Despite the utility of the maximum-term approximation, there are still *fluctuations* around the maximum as slightly different amounts of, say, energy are exchanged between the bath and the system. These fluctuations are small and rare, since they correspond to states of the system with greatly diminished numbers of microstates, but they nonetheless exist. That is, even though we cannot detect small variations in the amount of energy exchanged between the system and the bath, these variations are real, such that the system energy is never rigorously constant when the system is connected to a heat bath.

This leads to an important generality: only one of each conjugate pair of variables can be truly constant at a time. If the temperature is held constant, by using an infinite heat bath, then the energy of the system experiences slight fluctuations. If the volume is held constant, then the pressure fluctuates. This point is not obvious at the macroscopic level since the fluctuations are so small, but it is an essential feature at the microscopic scale. Moreover, we will find in Chapter 17 that the magnitude of fluctuations is connected to thermodynamic properties that we can actually measure.

Problems

Conceptual and thought problems

7.1. A single-component, constant-volume system is connected to a bath that regulates its temperature. The system can also exchange molecules with the bath. What is the fundamental function X for this system, both in differential form and in integrated form? What are its natural variables?

7.2. The chemical potential can be thought to correspond to a kind of force that generates a reversible work term in the fundamental equation. What is the nature of this force, and to what process does it correspond?

7.3. Let $f(x) = x^2$. Find the Legendre transform $g(y) = \mathcal{L}[f(x)]$ using the thermodynamic convention. Then find the double negative Legendre transform, $h(z) = -\mathcal{L}[-\mathcal{L}[f(x)]]$ and show that $f(x)$ and $h(z)$ are the same function.

7.4. Find the following Legendre transforms of the entropy: $\mathcal{L}_E[S]$, $\mathcal{L}_V[S]$, and $\mathcal{L}_V[\mathcal{L}_E[S]]$. To which thermodynamic potentials do these relate? Write both the integrated form and the differential form of each potential.

7.5. What is the fundamental function X of natural variables (S, P, μ), both in differential form and in integrated form?

7.6. What is the integrated form of the fundamental function Z of natural variables (T, P, μ), for a single-component system? Why does the integrated form have this special behavior?

7.7. Show that the function $S(E, P, N)$ is not fundamental, that is, it cannot be used to find all thermodynamic properties of a single-component system.

7.8. Consider a single-component, single-phase system. For each of the following, consider the uniqueness of the *intensive* thermodynamic state of the system.

Indicate whether the set of variables under-specifies the state (i.e., does not provide enough information), over-specifies the state (i.e., one or more variables are redundant), or exactly specifies the state. Explain.

(a) T, P, v

(b) $s(=S/N)$, T

(c) μ, P

7.9. Show that

$$\left(\frac{\partial\mu}{\partial P}\right)_T > 0$$

for all substances.

7.10. For a single-component system, classify the following as fundamental, not fundamental, or incomplete (meaning that not enough information is provided to completely specify the function):

(a) $H(T, P, N)$

(b) $E(S, T, N)$

(c) $A(T, V, N)$

7.11. For a binary system, classify the following as fundamental, not fundamental, or incomplete (meaning that not enough information is provided to completely specify the function):

(a) $X(E, P/T, N_1, N_2)$, where $X \equiv S - PV/T$

(b) $Y(T, P, N_1, \mu_2)$, where $Y \equiv A - \mu_2 N_2$

(c) $Z(E, V, \mu_1/T, N_2)$, where $Z \equiv S + \mu_1 N_1/T$

7.12. For a two-component system of A and B, classify the following as fundamental, not fundamental, or incomplete (meaning that not enough information is provided to specify the function).

(a) $H(S, P, \mu_A, \mu_B)$

(b) $E(S, T, N_A, N_B)$

(c) $a(T, v, x_A)$, where x_A is the mole fraction of A.

7.13. A friend suggests that the fundamental equation for the intensive Gibbs free energy in a binary mixture of A and B is given by

$$dg = -s\, dT + v\, dP + \mu_A\, dx_A + \mu_B\, dx_B$$

where $g = G/N = G/(N_A + N_B)$. What is wrong with your friend's statement? What should the differential form for dg really be? Derive this and explicitly show how μ_A and μ_B can be determined from the function g.

7.14. A particular substance is at vapor–liquid equilibrium at a given T and P, but under zero-gravity conditions. The liquid forms a *spherical* drop inside the vapor. Why? Hint: find a potential that is minimized and that can also account for the shape of the drop.

Fundamentals problems

7.15. For a single-phase, single-component system, consider the following list of intensive variables: T, P, μ, $s = S/N$, and $v = V/N$. Write down the complete list of each pair of these variables that uniquely specifies the state of the system.

7.16. A pot of N liquid molecules completely vaporizes at its boiling temperature T_b at pressure P. Assume the process is slow enough to be considered quasi-static. The latent heat of vaporization is Δh_{vap}. Write expressions for Q, W, ΔE, ΔS, and ΔG for the process, where ΔX indicates $X^G - X^L$. Assume ideal behavior for the vapor state, and that the liquid molar volume is much smaller than that of the gas.

7.17. Prove that

$$h = -T^2 \left(\frac{\partial(\mu/T)}{\partial T} \right)_P$$

starting from an intensive fundamental equation.

7.18. Let $\hat{e} = E/V$, $\hat{s} = S/V$, $\hat{a} = A/V$, and $\rho = N/V$. Show that

$$\hat{a} = -\hat{s}^2 \left(\frac{\partial(\hat{e}/\hat{s})}{\partial \hat{s}} \right)_\rho$$

7.19. Find the differential form for the fundamental equation of the energetic quantity $X = PV$. Indicate the independent variables of this quantity. Also indicate the integrated form of the relationship of X to E.

7.20. Consider a binary mixture of A and B and let $\hat{g} = G/N_A$. Find the fundamental form for the differential $d\hat{g}$. Show how μ_A and μ_B can be determined from $\hat{g}$ as functions of its natural variables.

7.21. A system exhibits a heat capacity C_P that is approximately constant with temperature. Show that, at fixed pressure, the Gibbs free energy follows a form like

$$G = G_0 + (C_P - S_0)(T - T_0) - TC_P \ln(T/T_0)$$

where T_0 is a reference temperature at which the entropy (S_0) and Gibbs free energy (G_0) are known. Keep in mind that the extensive heat capacity is defined by $C_p \equiv (\partial H/\partial T)_P$.

7.22. Gibb's approach to an interface between two phases was to treat it as a separate system governed by its own thermodynamic properties. He assumed that the interface has negligible volume such that its fundamental equation is

$$dE^s = T \, dS^s + \gamma \, dA + \sum_i \mu_i \, dN_i^s$$

where γ is the surface tension and A is the interfacial area. Note that no P-V term appears because the volume is always zero. The quantities E^s, S^s, and N_i^s are the *surface excess* energy, entropy, and number of particles, respectively.

(a) In reality, the interface is "fuzzy" in that thermodynamic variables like the density vary smoothly on traversing from one phase to another, even though the variation generally occurs over molecular distances. Consider a single-component system at liquid–vapor equilibrium. Sketch the density as a function of position z. Let $z \ll 0$ correspond to the liquid phase with bulk density ρ^L and $z \gg 0$ to the gas with ρ^G. Sketch the function $\rho(z)$.

(b) Generally, the total number of molecules in the system is given by $N = N^L + N^G + N^s$. However, demonstrate from your result in part (a) that an exact location for the interface, z_{int}, can be chosen so that N^s vanishes. In other words, show how one can pick z_{int} so that

$$\int_{-\infty}^{\infty} \rho(z) \mathcal{A} \, dz = \int_{-\infty}^{z_{int}} \rho^L \mathcal{A} \, dz + \int_{z_{int}}^{\infty} \rho^G \mathcal{A} \, dz$$

(c) Using part (b), show that the surface tension is the surface excess Helmholtz free energy per area of the interface.

7.23. Continuing the previous problem, instead consider a two-component system in which species 1 is a solvent and species 2 a dilute solute. Let z_{int} be chosen so that $N_1^s = 0$.

(a) Show that the chemical potential of species 2 obeys the *Gibbs adsorption isotherm*,

$$d\mu_2 = \frac{\mathcal{A}}{N_2^s} \, d\gamma \quad \text{(constant } T)$$

(b) Show that the surface tension generally decreases when solutes adsorb at the interface. Hint: find a suitable derivative that embodies this statement.

Applied problems

7.24. A simple model of a low-temperature solid is the following. Atoms interact with a repulsive pairwise potential energy of the form $u_{ij}(r_{ij}) = \text{constant} \times r_{ij}^{-n}$, where r_{ij} is the atom–atom separation distance and n is a constant, typically 9 or 12. For such solids, the Helmholtz free energy can be approximated as $A(T, V, N) = 3Nk_BT \ln[(N/V)^m \, b/T]$ [Shell et al., *J. Chem. Phys.* **118**, 8821 (2003)]. Here $m \equiv (n+2)/6$ and b is a constant with units of [temperature] · [volume]^m. Using this, find expressions for

(a) the thermodynamic potentials $E(S, V, N)$ and $\mu(T, P)$, and

(b) the intensive heat capacities $c_V(T, P)$ and $c_P(T, P)$, and the equation of state, $P(T, \rho)$, where $\rho = N/V$.

7.25. Prove that the maximum work attainable from a closed system undergoing a constant-temperature volume expansion is ΔA, the difference in Helmholtz free energy for the process. Similarly, prove that the minimum required work needed to compress a closed system at constant temperature is also ΔA.

7.26. In water, there is a driving force for the association of nonpolar solutes that arises because the overall free energy of the system can be reduced by minimizing the solute–water interface. A simple way of viewing this hydrophobic interaction is through the problem of the solvation or *hydration* of a nonpolar solute. We can think of this process as a "reaction" where one solute in vacuum plus a bath of liquid water, held at constant T and P, "react" to form a solute immersed in water:

$$\text{solute} + \text{water} \leftrightarrow \text{solvated solute}$$

(a) What thermodynamic quantity associated with this process will determine whether or not hydration is favorable?

(b) Let us say the quantity in the previous question is ΔX_{solv}. For large spherical solutes of radii greater than or equal to $R \approx 1$ nm, it can be shown that the approximation $\Delta X_{\text{solv}} \approx 4\pi\gamma R^2$ holds. Here γ is an effective surface tension. If two solvated spherical oil droplets of radius R associate into a single spherical droplet, show that the change in ΔX for this process is given by

$$\Delta X_{\text{assoc}} = 4\pi\gamma(2^{2/3} - 2)R^2$$

Is this process favorable or unfavorable?

(c) An order-of-magnitude estimate of γ is 55 dyne/cm. Assume that the density of oil droplets is 0.9 g/cm^3. What is ΔX_{assoc} for two single-molecule oil "droplets" with $R = 1$ nm, in kcal/mol of oil? Assume the molecular weight of the oil is equivalent to that of decane. How does the strength of this interaction compare with the van der Waals force?

7.27. When interfacial effects contribute to the properties of a system, one must modify the fundamental equation to include surface effects. Consider a spherical vapor bubble with volume V in coexistence with the surrounding liquid, all at temperature T. We might write the differential form of the bubble's Helmholtz free energy as

$$dA = -S\, dT - P\, dV + \mu\, dN + \gamma\, d\mathcal{A}$$

where γ is the surface tension, a positive quantity, and $\mathcal{A}$ is the area of the liquid–gas interface.

(a) If the liquid is considered to act as a "volume" bath at pressure P_b for the bubble, find an integrated expression for the quantity that is minimized when the bubble comes to an equilibrium volume V. Assume that both the vapor and the liquid are already thermostatted, $T_b = T$, and that no particle exchange occurs between the vapor and liquid (this is valid if the diffusion process is slow relative to pressure equilibration).

(b) Show that, at the equilibrium bubble radius R (or volume V), the pressure inside the bubble exceeds that on the outside by an amount equal to $2\gamma/R$. This is called *Young's law*.

(c) Assume that the pressure of the vapor inside the bubble behaves as an ideal gas. Note that the Gibbs phase rule demands that the bubble pressure is not an independent parameter, but determined by the other state conditions of the bubble. Show that bubbles described in this way – that can exchange volume but not particles with the liquid – are metastable. Namely, show that the second derivative of the relevant free energy is positive with respect to the bubble radius.

FURTHER READING

H. Callen, *Thermodynamics and an Introduction to Thermostatistics*, 3rd edn. New York: Wiley (1985).

K. Denbigh, *The Principles of Chemical Equilibrium*, 4th edn. New York: Cambridge University Press (1981).

J. W. Tester and M. Modell, *Thermodynamics and Its Applications*, 3rd edn. Upper Saddle River, NJ: Prentice Hall (1997).

8 Maxwell relations and measurable properties

8.1 Maxwell relations

The thermodynamic potentials emerging from Legendre transforms allow us to switch independent variables and give rise to alternate versions of the second law at different constant conditions:

$S(E, V, N)$	entropy	maximum at constant E, V, N equilibrium
$E(S, V, N)$	energy	minimum at constant S, V, N equilibrium
$H(S, P, N)$	enthalpy	minimum at constant S, P, N equilibrium
$A(T, V, N)$	Helmholtz free energy	minimum at constant T, V, N equilibrium
$G(T, P, N)$	Gibbs free energy	minimum at constant T, P, N equilibrium

Here we discuss some of the mathematical properties of these functions and their consequences for relationships between thermodynamic variables. These considerations will allow us to connect quantities that are difficult to measure directly – like the entropy and the chemical potential – to variables that we can easily access using experiments.

To start, note that all of the potentials above have multiple independent variables. A general feature of well-behaved multivariate functions is that the mixed partial derivatives do not depend on the order in which they are taken,

$$\frac{\partial^2 f}{\partial x \, \partial y} = \frac{\partial^2 f}{\partial y \, \partial x} \tag{8.1}$$

or, writing out the derivatives explicitly,

$$\frac{\partial}{\partial x}\left[\left(\frac{\partial f}{\partial y}\right)_x\right]_y = \frac{\partial}{\partial y}\left[\left(\frac{\partial f}{\partial x}\right)_y\right]_x \tag{8.2}$$

This fact has important consequences for potentials because their derivatives involve thermodynamic quantities. As an example, consider the energy equation

$$dE = T\,dS - P\,dV + \mu\,dN \tag{8.3}$$

Examine the second derivative with respect to S and V at constant N conditions:

$$\frac{\partial^2 E}{\partial S \, \partial V} = \frac{\partial^2 E}{\partial V \, \partial S} \quad \rightarrow \quad \frac{\partial}{\partial S}\left[\left(\frac{\partial E}{\partial V}\right)_{S,N}\right]_{V,N} = \frac{\partial}{\partial V}\left[\left(\frac{\partial E}{\partial S}\right)_{V,N}\right]_{S,N} \tag{8.4}$$

Substituting for the inner derivatives using the fundamental equation in (8.3) gives

$$\frac{\partial}{\partial S}[-P]_{V,N} = \frac{\partial}{\partial V}[T]_{S,N} \quad \rightarrow \quad -\left(\frac{\partial P}{\partial S}\right)_{V,N} = \left(\frac{\partial T}{\partial V}\right)_{S,N} \tag{8.5}$$

Therefore, a relationship between the derivatives of P and T emerges simply from the fact that these two quantities are related by a common second derivative of a thermodynamic potential. Such equalities based on potential second derivatives are called *Maxwell relations*, after James Maxwell, one of the founders of modern thermodynamics and electromagnetism. There are in fact many Maxwell relations, depending on which potential is used and which pair of independent variables is examined. Here is another example, based on the Gibbs free energy:

$$dG = -S \, dT + V \, dP + \mu \, dN$$

with

$$\frac{\partial^2 G}{\partial P \, \partial N} = \frac{\partial^2 G}{\partial N \, \partial P} \quad \rightarrow \quad \frac{\partial}{\partial P}\left[\left(\frac{\partial G}{\partial N}\right)_{T,P}\right]_{T,N} = \frac{\partial}{\partial N}\left[\left(\frac{\partial G}{\partial P}\right)_{T,N}\right]_{T,P} \tag{8.6}$$

On substituting for the inner derivatives, we find

$$\left(\frac{\partial \mu}{\partial P}\right)_{T,N} = \left(\frac{\partial V}{\partial N}\right)_{T,P} \tag{8.7}$$

which shows a relationship between the pressure dependence of the chemical potential and the N dependence of the system volume. In fact, the right-hand derivative is equal to v. These examples illustrate a basic recipe: pick a potential and a second derivative involving two of its independent variables, and then substitute first-derivative definitions to produce a Maxwell relation.

Maxwell relations connect two derivatives of thermodynamic variables, and emerge due to the equivalence of potential second derivatives under a change of operation order:

$$\frac{\partial^2 F}{\partial X \, \partial Y} = \frac{\partial^2 F}{\partial Y \, \partial X}$$

where F is a thermodynamic potential and X and Y are two of its natural independent variables.

Maxwell relations are quite important, for two reasons. First, they show us that derivatives of thermodynamic parameters are not completely independent. This can serve as a consistency check both in experiments and in pen-and-paper analysis. Second, they provide a method to express derivatives involving difficult-to-measure quantities in terms of ones that are readily accessible experimentally, as we will now see.

8.2   Measurable quantities

How does one measure the entropy or chemical potential from experiments? These kinds of quantities are not usually directly accessible in the lab. Typically what we can measure are mechanical quantities like pressure, bulk quantities like volume and density, and thermal properties like temperature and heat flow (e.g., by slow heat-exchange experiments in which we can measure temperature changes in a coupled reference body). Of the thermodynamic variables that we have discussed thus far, the following are considered *measurable*:

T	temperature
P	pressure
V	volume
N or m	number of particles or mass (related by the molecular weight)
x_i	mole fraction or other measure of composition
Δh_{latent}	enthalpy (latent heat) of phase change

There are also several readily measured material properties that depend on derivatives of thermodynamic variables. Because these quantities measure the change in a system parameter in response to an infinitesimal perturbation, they are termed thermodynamic *response functions*:

$$C_V \equiv \left(\frac{\partial E}{\partial T}\right)_{V,N} = T\left(\frac{\partial S}{\partial T}\right)_{V,N} \qquad \text{constant-volume heat capacity}$$

$$C_P \equiv \left(\frac{\partial H}{\partial T}\right)_{P,N} = T\left(\frac{\partial S}{\partial T}\right)_{P,N} \qquad \text{constant-pressure heat capacity}$$

$$\kappa_T \equiv -\frac{1}{V}\left(\frac{\partial V}{\partial P}\right)_{T,N} = -\left(\frac{\partial \ln V}{\partial P}\right)_{T,N} \qquad \text{isothermal compressibility}$$

$$\alpha_P \equiv \frac{1}{V}\left(\frac{\partial V}{\partial T}\right)_{P,N} = \left(\frac{\partial \ln V}{\partial T}\right)_{P,N} \qquad \text{thermal expansivity or expansion coefficient}$$

These response functions are defined such that they are positive in the case of "normal" systems, such as ideal gases and simple liquids. There are occasional exceptions in some of them; for example, the thermal expansivity of liquid water below 4 °C is negative. There are additional measurable response functions that we have not listed; some of them can be expressed as combinations of the above, and some of them emerge in systems with other thermodynamic variables, such as those involving electromagnetic or interfacial work. Note that the heat capacities as presented above are extensive, while more conventionally we might see intensive versions such as $c_V = C_V/N$ and $c_P = C_P/N$.

Maxwell relations enable us to express experimentally inaccessible quantities in terms of the measurable ones just listed. Consider the following example derivative,

$$\left(\frac{\partial S}{\partial P}\right)_{T,N} = -\frac{\partial^2 G}{\partial P\,\partial T}$$

$$= -\left(\frac{\partial V}{\partial T}\right)_{P,N} \tag{8.8}$$

This is reminiscent of the definition for α_P, so we can substitute

$$\left(\frac{\partial S}{\partial P}\right)_{T,N} = -V\alpha_P \tag{8.9}$$

Hence the entropy's dependence on pressure at constant temperature is related to the thermal expansivity. Many times, however, getting to measurable quantities is not so easy. Consider the Maxwell relation

$$\left(\frac{\partial P}{\partial T}\right)_{V,N} = -\frac{\partial^2 A}{\partial T\,\partial V}$$

$$= \left(\frac{\partial S}{\partial V}\right)_{T,N} \tag{8.10}$$

Here, we moved to something that is certainly not measurable. Fortunately, there are other ways to relate thermodynamic variables using the principles of multivariate calculus. We can use the so-called "triple-product rule," which shows that

$$\left(\frac{\partial P}{\partial T}\right)_{V,N}\left(\frac{\partial V}{\partial P}\right)_{T,N}\left(\frac{\partial T}{\partial V}\right)_{P,N} = -1 \tag{8.11}$$

On rearranging, we have

$$\left(\frac{\partial P}{\partial T}\right)_{V,N} = \frac{-1}{(\partial V/\partial P)_{T,N}(\partial T/\partial V)_{P,N}}$$

$$= -\frac{(\partial T/\partial V)_{P,N}}{(\partial V/\partial P)_{T,N}}$$

$$= \frac{V\alpha_P}{V\kappa_T} \tag{8.12}$$

which finally gives an expression involving only measurable quantities,

$$\left(\frac{\partial P}{\partial T}\right)_{V,N} = \frac{\alpha_P}{\kappa_T} \tag{8.13}$$

We now turn to a more complex example that relies on additional mathematical manipulations; let us find the relationship between C_V and C_P. From the definitions of these quantities we can infer a good place to start: both response functions depend on a derivative of the entropy with temperature. The difference between the two is the variable that is held constant in the derivative. We can relate a change in the constant conditions of a derivative in the following way. Construct the function $S(T, V)$ assuming constant N conditions throughout. This is not a fundamental potential because S is not a function of its natural variables. We can still, however, perform the construction since S is a state function and T and V stem from different conjugate pairs. We choose T as an independent variable because it is the temperature derivative that is relevant to the heat capacities. We now expand $S(T, V)$ in differential form as

$$dS = \left(\frac{\partial S}{\partial T}\right)_V dT + \left(\frac{\partial S}{\partial V}\right)_T dV$$

$$= \frac{C_V}{T} dT + \left(\frac{\partial S}{\partial V}\right)_T dV \tag{8.14}$$

To relate the constant-volume conditions to those at constant pressure, we take the T derivative of this expression at constant P. This is operationally equivalent to dividing by dT and applying constant P conditions to any complete derivative that is formed,

$$\left(\frac{\partial S}{\partial T}\right)_P = \frac{C_V}{T} + \left(\frac{\partial S}{\partial V}\right)_T \left(\frac{\partial V}{\partial T}\right)_P \tag{8.15}$$

We recognize that the LHS can be replaced using C_P and the rightmost derivative connects to α_P,

$$\frac{C_P}{T} = \frac{C_V}{T} + V\alpha_P \left(\frac{\partial S}{\partial V}\right)_T \tag{8.16}$$

Finally, we use the Maxwell relation of (8.10) to address the last term, giving

$$\frac{C_P}{T} = \frac{C_V}{T} + V\alpha_P \left(\frac{\partial P}{\partial T}\right)_V$$

$$= \frac{C_V}{T} + V\alpha_P \left(\frac{\alpha_P}{\kappa_T}\right) \tag{8.17}$$

from the previous example. On simplifying everything, we obtain

$$C_P = C_V + \frac{TV\alpha_P^2}{\kappa_T} \tag{8.18}$$

As a final example, we compute a quantity called the isentropic compressibility, which is defined by

$$\kappa_S = -\frac{1}{V} \left(\frac{\partial V}{\partial P}\right)_S \tag{8.19}$$

Experimentally, the isentropic compressibility measures the fractional change in volume of a system during a reversible adiabatic ($dS = 0$) compression. To proceed, we use the triple-product rule to remove the entropy from the constant condition,

$$\kappa_S = \frac{1}{V}\left[\left(\frac{\partial S}{\partial V}\right)_P \left(\frac{\partial P}{\partial S}\right)_V\right]^{-1}$$

$$= \frac{1}{V}\frac{(\partial S/\partial P)_V}{(\partial S/\partial V)_P} \tag{8.20}$$

We use a calculus rule called "addition of variable" to expand the numerator and denominator by including temperature. We want to add temperature because there are no measurable properties that are derivatives of V at constant P and vice versa. The rule produces

$$\kappa_S = \frac{1}{V}\frac{(\partial S/\partial T)_V (\partial T/\partial P)_V}{(\partial S/\partial T)_P (\partial T/\partial V)_P}$$

$$= \frac{C_V}{V C_P}\frac{(\partial T/\partial P)_V}{(\partial T/\partial V)_P} \tag{8.21}$$

Note that the remaining derivatives all involve permutations of the same three variables. This suggests the use of the triple-product rule again,

$$\kappa_S = \frac{C_V}{V C_P}\left[\left(\frac{\partial T}{\partial P}\right)_V \left(\frac{\partial V}{\partial T}\right)_P\right]$$

$$= \frac{C_V}{V C_P}\left[-\left(\frac{\partial V}{\partial P}\right)_T\right] \tag{8.22}$$

Finally, we see that we can insert the expression for the isothermal compressibility,

$$\kappa_S = \frac{C_V}{C_P}\kappa_T \tag{8.23}$$

8.3   General considerations for calculus manipulations

Many thermodynamic calculus manipulations can be derived from a relatively simple procedure. The first step is to construct a state function involving the variables of interest. Some examples are $H(T, P, N)$, $S(T, P, N)$, $\mu(V, P, N)$, and $V(T, P, N)$. Note that, if the state function is a potential, we do not necessarily need to use natural variables. We can choose any independent variables that we like so long as they form a complete set of thermodynamic information (i.e., all stem from different conjugate pairs).

The second step is to write out the full differential of the state function. As an example, take $P(T, V, N)$,

$$dP = \left(\frac{\partial P}{\partial T}\right)_{V,N} dT + \left(\frac{\partial P}{\partial V}\right)_{T,N} dV + \left(\frac{\partial P}{\partial N}\right)_{T,V} dN \tag{8.24}$$

The third step is to set the differentials to zero for any terms that are constant in the problem of interest. For example, if N is constant then $dN = 0$ and we have

$$dP = \left(\frac{\partial P}{\partial T}\right)_{V,N} dT + \left(\frac{\partial P}{\partial V}\right)_{T,N} dV \quad \text{(constant } N\text{)} \tag{8.25}$$

We must keep in mind that the current equation corresponds to constant N conditions. This is important because any new derivatives that are formed will acquire N as a constant variable.

The fourth step is to take the derivative of interest, which essentially means dividing by the appropriate differential. All new derivatives acquire any previously applied constant conditions. Moreover, we must specify additional constant conditions as necessary to meet one less than the number of independent variables. In our example, we can also specify S to be constant when taking the T derivative,

$$\left(\frac{\partial P}{\partial T}\right)_{S,N} = \left(\frac{\partial P}{\partial T}\right)_{V,N_*} + \left(\frac{\partial P}{\partial V}\right)_{T,N} \left(\frac{\partial V}{\partial T}\right)_{S,N} \tag{8.26}$$

In conjunction with this procedure, we can then begin to substitute various fundamental definitions, expressions for measurable quantities, and Maxwell relations.

As another example, consider the expression $(\partial H/\partial T)_{V,N}$. When derivatives involve a thermodynamic potential, we often begin with its differential form

$$dH = T\,dS + V\,dP + \mu\,dN \tag{8.27}$$

Having constant N conditions implies

$$dH = T\,dS + V\,dP \quad \text{(constant } N\text{)} \tag{8.28}$$

Taking the temperature derivative at constant volume gives

$$\left(\frac{\partial H}{\partial T}\right)_{V,N} = T\left(\frac{\partial S}{\partial T}\right)_{V,N} + V\left(\frac{\partial P}{\partial T}\right)_{V,N} \tag{8.29}$$

in which we can begin to see opportunities to substitute response functions.

Several common calculus manipulation techniques are useful in thermodynamic analysis. These are briefly summarized below:

inversion $\qquad \left(\dfrac{\partial X}{\partial Y}\right)_Z = 1 \bigg/ \left(\dfrac{\partial Y}{\partial X}\right)_Z$

triple-product rule $\qquad \left(\dfrac{\partial X}{\partial Y}\right)_Z \left(\dfrac{\partial Z}{\partial X}\right)_Y \left(\dfrac{\partial Y}{\partial Z}\right)_X = -1$

addition of variable $\qquad \left(\dfrac{\partial X}{\partial Y}\right)_Z = \left(\dfrac{\partial X}{\partial W}\right)_Z \bigg/ \left(\dfrac{\partial Y}{\partial W}\right)_Z$

non-natural derivative $\quad A(X,Y) \rightarrow \left(\dfrac{\partial A}{\partial Y}\right)_Z = \left(\dfrac{\partial A}{\partial X}\right)_Y \left(\dfrac{\partial X}{\partial Y}\right)_Z + \left(\dfrac{\partial A}{\partial Y}\right)_X$

As we have seen earlier, there are also rules that apply specifically to thermodynamic potentials:

Maxwell relations $\qquad \left(\dfrac{\partial^2 F}{\partial X\,\partial Y}\right) = \left(\dfrac{\partial^2 F}{\partial Y\,\partial X}\right) \quad \longrightarrow \quad \left(\dfrac{\partial A}{\partial X}\right)_Y = \left(\dfrac{\partial B}{\partial Y}\right)_X$

potential transformation $\qquad \dfrac{\partial (F_1/X)}{\partial X} = -\dfrac{F_2}{X^2}$

The combination of all of these techniques enables us to relate virtually any thermo-dynamic derivative to the set of measurable variables described earlier.

Problems

Conceptual and thought problems

8.1. Derive the triple-product rule.

8.2. Consider the response functions C_V and C_P.
(a) Explain how both are in fact measurable for non-reactive systems. Consider the first law in conjunction with slow heating at different constant conditions, where Q can be quantified by coupling to a reference body. Why does the constant-volume heat capacity naturally involve E, whereas the constant pressure one H?
(b) Explain why the heat capacities cannot be measured directly in this manner if a reversible reaction occurs within the system.

8.3. A quasi-static adiabatic process is performed on a system starting at state 1. Indicate the direction in which the system will proceed in the P–T plane (i.e., find dP/dT), in terms of measurable quantities.

8.4. Rigorously show that $(\partial V/\partial N)_{T,P}$ of Eqn. (8.7) is equal to V/N.

8.5. Prove that $\mu(T, P)$ is fundamental in a single-component system. That is, show that all of the following properties can be derived from it: s, e, h, a, v, c_V, c_P, κ_T, and α_P.

8.6. Consider a binary system of species 1 and 2. Assume that the chemical potentials of both species are known as functions of T, P, and the mole fraction x_1, that is, $\mu_1(T, P, x_1)$ and $\mu_2(T, P, x_1)$. Show that this provides a complete set of intensive thermodynamic information. Specifically, show how these two functions could be used to determine $g = G/N$, $s = S/N$, and $h = H/N$, where $N = N_1 + N_2$.

Fundamentals problems

8.7. Given *only* the ideal gas equation of state, $PV = Nk_BT$, prove the following properties using fundamental equations and Maxwell relations. In the first two parts, you are proving that several ideal gas properties are volume- or pressure-independent.

(a) $(\partial E/\partial V)_{T,N} = 0$

(b) $(\partial C_V/\partial V)_{T,N} = 0$ and $(\partial C_P/\partial P)_{T,N} = 0$

(c) $C_P = C_V + Nk_B$

8.8. Express the following thermodynamic derivatives in terms of measurable quantities for a single-component system at constant N conditions.

(a) $\left(\dfrac{\partial H}{\partial P}\right)_T$

(b) $\left(\dfrac{\partial E}{\partial V}\right)_T$

(c) $\left(\dfrac{\partial S}{\partial T}\right)_H$

(d) $\left(\dfrac{\partial E}{\partial P}\right)_T$

(e) $\left(\dfrac{\partial A}{\partial G}\right)_T$

(f) $\left(\dfrac{\partial S}{\partial P}\right)_V$

(g) $\left(\dfrac{\partial E}{\partial S}\right)_P$

(h) $\left(\dfrac{\partial V}{\partial T}\right)_E$

8.9. Indicate whether each of the following relations is true in general:

(a) $\left(\dfrac{\partial S}{\partial T}\right)_{V,N} = \left(\dfrac{\partial P}{\partial V}\right)_{T,N}$

(b) $\left(\dfrac{\partial S}{\partial \mu}\right)_{T,V} = \left(\dfrac{\partial N}{\partial T}\right)_{\mu,V}$

(c) $\left(\dfrac{\partial \mu}{\partial V}\right)_{T,N} = \left(\dfrac{\partial P}{\partial N}\right)_{T,V}$

8.10. Consider a Maxwell relation in a multicomponent system based on the identity

$$\left(\frac{\partial^2 G}{\partial N_1 \, \partial N_2}\right)_{T,P} = \left(\frac{\partial^2 G}{\partial N_2 \, \partial N_1}\right)_{T,P}$$

(a) For a binary mixture, show that this implies a relation in terms of the chemical potentials of the components and their dependence on the mole fractions.

(b) Repeat part (a) for an arbitrary system of C components.

8.11. The van der Waals equation of state can be written as $P = k_B T\rho/(1 - b\rho) - a\rho^2$, where $\rho = N/V$ and a and b are positive constants.

(a) Find the change in molar internal energy, Δe, upon isothermal expansion of such a fluid from molar volume v_1 to v_2.

(b) Is this change greater or less than that for an adiabatic expansion between the same two volumes?

8.12. For an incompressible liquid, show that c_P must be independent of pressure and that $c_V \approx c_P$.

8.13. The isothermal compressibility of a fluid is found to be independent of temperature. Prove that the fluid's thermal expansivity is also independent of pressure.

8.14. The speed of sound c of a fluid is given by $c^2 = (\partial P/\partial \hat{\rho})_s$, where $\hat{\rho} = M\rho$ is the mass density, M the molecular weight, and ρ the number density.

(a) Find an expression for the speed of sound in terms of measurable quantities.

(b) How does this expression simplify for liquids far from their critical points?

8.15. One can define an isothermal compressibility for a mixture in the same manner as for a single-component system, holding all mole numbers constant. Show that it then obeys

$$\frac{1}{\kappa_T} = \sum_i N_i \left(\frac{\partial P}{\partial N_i}\right)_{T,V,N_{j\neq i}}$$

8.16. In the absence of external fields like gravity, gradients in a species' chemical potential in a homogeneous phase generally indicate that a system has yet to reach equilibrium. Such a situation will lead to diffusive mass transport that drives the system towards a state of uniform densities and concentrations. For a single-component system, show that a density gradient along the z direction indicates an underlying chemical potential gradient along that direction per the relation

$$\left(\frac{\partial \rho}{\partial z}\right)_T = \kappa_T \rho^2 \left(\frac{\partial \mu}{\partial z}\right)_T$$

8.17. Two liquids 1 and 2 are mixed, and it is found that the molar volume of the mixture follows the relation

$$v = x_1 v_1^* + x_2 v_2^* + x_1 x_2 c$$

where the x terms are mole fractions, v_1^* and v_2^* are the molar volumes of pure 1 and 2, and c is a constant with units of volume per mole. Find expressions for the P dependence of the chemical potentials μ_1 and μ_2 in terms of the pure-species properties and the constant c.

8.18. The pressure of a fluid is increased slightly. Show that the decrease in volume is always less in the isentropic than in the isothermal case.

8.19. Show, for any system for which c_P is a constant, that its molar volume must be linear in temperature at constant pressure.

Applied problems

8.20. A Joule–Thomson process is one by which a gas is throttled (expanded) with no heat exchanged. Therefore, any change in internal energy E is due to internal changes in pressure–volume work, and the total enthalpy is conserved. Often, during such an expansion, the temperature of the gas will change. The Joule–Thomson coefficient measures how much the temperature changes for a given pressure drop: $\mu_{JT} = (\partial T/\partial P)_H$. Here μ does not indicate a relationship to the chemical potential, but instead simply comes from convention. Show that the Joule–Thomson coefficient is given by

$$\mu_{JT} = \frac{V}{C_P}(T\alpha_P - 1)$$

8.21. The work required to stretch an elastic band of length L is given by $\delta W = \tau \, dL$, where τ gives the tension in the band. Is it easier to cool a band from T_1 to $T_2 < T_1$ when it is held at constant length or at constant tension? In other words, which requires the removal of less heat?

8.22. Liquid water exhibits a wealth of properties that are considered anomalous when compared with simple liquid models such as the van der Waals fluid. One well-known anomaly is the existence of a temperature of maximum density (TMD) under constant-pressure conditions. At atmospheric pressure this temperature is around 4 °C. In general, however, one can construct a line in the T, P plane that gives the TMD for any pressure. Consider the properties of this line.

(a) Along the TMD line, what must be true about the thermal expansion coefficient? Prove also that $(\partial P/\partial T)_V = 0$ along this line. Be careful to address all of the terms in your proof.

(b) Show that the slope of the TMD line is given by

$$\left(\frac{dP}{dT}\right)_{TMD} = \left(\frac{\partial \alpha_P}{\partial T}\right)_P \bigg/ \left(\frac{\partial \kappa_T}{\partial T}\right)_P$$

(c) Water's isothermal compressibility increases upon cooling. What can be said about the slope of the TMD line?

8.23. A pump is a piece of continuously operating process equipment that increases the pressure of a liquid stream. Assume that the pressure change ΔP is small and that the liquid is incompressible.

(a) Show that the per-particle work required as input to an adiabatic (i.e., well-insulated) pump is given by $W/N = \Delta h$, where Δh is the change in per-particle enthalpy of the liquid upon passing through the pump. Note: the pump is not a closed system.

(b) Find an expression for the per-particle work for an ideal, maximally efficient pump in terms of measurable properties of the liquid.

(c) Find an expression for the increase in temperature of the liquid stream, ΔT, in terms of its measurable properties.

(d) Prove that the temperature increase is greater when the pump is not maximally efficient, even if it remains adiabatic.

8.24. In magnetic materials, the presence of a magnetic field $\mathcal{H}$ induces a magnetic moment in the same direction within the system, M. The latter is a measure of the net (extensive) magnetic dipole. The differential work associated with such a process is $\mathcal{H}\, dM$.

(a) Find an expression for the differential form of the fundamental equation for this system, $dE = ?$

(b) At constant T and P, the density of a given magnetic material is found to increase slightly with the strength of the field $\mathcal{H}$. What does this mean for the magnetic moment when the pressure is changed under constant-temperature and -field conditions?

FURTHER READING

H. Callen, *Thermodynamics and an Introduction to Thermostatistics*, 3rd edn. New York: Wiley (1985).

K. Denbigh, *The Principles of Chemical Equilibrium*, 4th edn. New York: Cambridge University Press (1981).

J. W. Tester and M. Modell, *Thermodynamics and Its Applications*, 3rd edn. Upper Saddle River, NJ: Prentice Hall (1997).

9 Gases

Until now, most of what we have discussed has involved general relationships among thermodynamic quantities that can be applied to any system, such as the fundamental equation, reversibility, Legendre transforms, and Maxwell equations. In this and the coming chapters, we begin to investigate properties of specific types of substances. We will mostly consider very simple models in which only the essential physics is included; these give insight into the basic behaviors of solids, liquids, and gases, and actually are sufficient to learn quite a bit about them. Of course, there are also many detailed theoretical and empirical models for specific systems, but very often these theories simply improve upon the accuracy of the approaches rather than introduce major new concepts and qualitative behaviors.

Statistical mechanics provides a systematic route to state- and substance-specific models. If one can postulate a sufficiently simple description of the relevant atomic interaction energetics, the entropy or free energy can be determined in fundamental form. Ultimately our strategy for most of these simple models will be to determine the chemical potential $\mu(T, P)$ in single-component systems or $\mu(T, P, \{x\})$ for multi-component ones, where $\{x\}$ gives the mole fractions. In both cases, knowledge of the chemical potentials does indeed give a fundamental perspective, allowing us to extract all of the intensive thermodynamic properties. Moreover, for problems involving phase equilibrium, the chemical potential is the natural starting point, as we will see in Chapter 10.

9.1 Microstates in monatomic ideal gases

The first simple model that we will consider is that of an ideal gas.

A single-component collection of **ideal gas** molecules is characterized by the following properties:

(1) Molecules do not interact. That is, there are no intermolecular potential energies. For monatomic gases, the only energies are therefore those due to kinetic energy.
(2) Each molecule has a mass m but zero volume. That is, the molecules are point particles.

Essentially all gases behave like ideal gases in the limit that molecules do not "see" each other. This primarily happens in the low-density limit $\rho = N/V \to 0$, or alternatively $P \to 0$. In this case molecules are separated by distances that are, on average, much greater than the molecularly sized range of the interaction potential. It also happens at very high temperatures, $T \gg 0$. Under such circumstances and provided that the density is low, the interactions are weak relative to the thermal energy scale, $k_B T$.

We begin to evaluate the properties of an ideal gas by asking how many microstates exist for a given macroscopic N, V, and E. To do that, we need to be able to enumerate the microstates themselves. Classically, we can specify any position or velocity for a given molecule, and hence the system has access to an infinite number of microstates and a continuous range of total energy. Here, however, we will pursue a quantum-mechanical treatment of these systems so that we can calculate the absolute entropy.

Consider the case of a single ideal gas molecule in a fixed volume V, the so-called *particle-in-a-box* problem. Quantum mechanics dictates that, in order to find the energy spectrum for the particle, we need to solve the time-independent Schrödinger equation,

$$-\left(\frac{h^2}{8\pi^2 m}\right)\nabla^2 \Psi(x, y, z) + U(x, y, z)\Psi(x, y, z) = \epsilon \Psi(x, y, z) \qquad (9.1)$$

Here, ϵ gives the energy of the single particle. We will reserve the variable E for a system of particles. For a monatomic ideal gas, the potential energy is such that $U(x, y, z) = 0$. Using separation of variables, this equation can then be split further into each of the x, y, and z components as $\Psi(x, y, z) = \Psi_x(x)\Psi_y(y)\Psi_z(z)$. One obtains for the x component, after some rearrangement,

$$\frac{d^2\Psi_x(x)}{dx^2} + K^2\Psi_x(x) = 0 \quad \text{where} \quad K^2 = \left(\frac{8\pi^2 m\epsilon}{h^2}\right) \qquad (9.2)$$

This second-order differential equation determines $\Psi_x(x)$ and depends on a term K that combines the constants. We also need boundary conditions. Because the particle must be inside the box, we demand that the wavefunction approach zero at its edges,

$$\Psi_x(x = 0) = 0 \quad \text{and} \quad \Psi_x(x = L) = 0 \qquad (9.3)$$

where L is the box width. The solution to this kind of differential equation is a sum of sine and cosine functions, $\Psi_x(x) = A \sin(Kx) + B \cos(Kx)$. The only way to satisfy the first boundary condition is with $B = 0$. For the second boundary condition, we then have

$$\Psi_x(x = L) = A \sin(KL) = 0 \qquad (9.4)$$

We cannot let $A = 0$, otherwise the wavefunction would be zero everywhere. Therefore, we must conclude that K can only take on values such that $\sin(KL) = 0$, or, alternatively, $KL = n_x\pi$, where n_x is any integer greater than zero. From the definition of K, we then see that the energy must be such that

$$\epsilon = \frac{(n_x h)^2}{8mL^2} \quad \text{with } n_x = 1, 2, 3, \ldots \tag{9.5}$$

where n_x is a *quantum number* characterizing the state of the particle. The *spectrum* in (9.5) shows that the energy of a one-dimensional ideal gas particle adopts discrete values. On solving the other equations for the y and z directions, the result for a three-dimensional particle in a cubic box is

$$\epsilon = \frac{h^2}{8mL^2} \left(n_x^2 + n_y^2 + n_z^2 \right) \tag{9.6}$$

What happens if more than one particle is present? Since the particles do not interact, we can write the total wavefunction as a product of individual wavefunctions for each particle,

$$\Psi(x_1, y_1, z_1, x_2, y_2, z_2, \ldots) = \Psi_1(x_1, y_1, z_1)\Psi_2(x_2, y_2, z_2) \cdots$$

The result is that the energy spectrum of the total system of particles is given by

$$E = \frac{h^2}{8mL^2} \left(n_{x,1}^2 + n_{y,1}^2 + n_{z,1}^2 + n_{x,2}^2 + n_{y,2}^2 + n_{z,2}^2 + \cdots \right) \tag{9.7}$$

where there are now $3N$ integer quantum numbers.

Equation (9.7) is the starting point for computing the number of microstates $\Omega(N, V, E)$. Namely, we want to know how many ways there are to pick combinations of all n_x, n_y, and n_z quantum numbers so that the system has a particular total energy E. To do so, we will use an algebraic trick to approximate Ω that is valid in the limit of large numbers of particles and high energies (i.e., high temperatures). We will do this by counting how many combinations of quantum numbers give energies less than or equal to E, which we will denote by Φ. To find Ω, we then simply need to examine how Φ changes with E.

To begin, notice that the sum over all integer values of n can be viewed as the squared length of a vector in $3N$-dimensional space. That is, let $\mathbf{n} = (n_{x,1}, n_{y,2}, \ldots, n_{z,N})$ such that $n_{x,1}^2 + n_{y,1}^2 + \cdots = \mathbf{n}^2$. The vectors corresponding to an energy less than or equal to the value E must satisfy

$$E \geq \frac{h^2}{8mL^2} \mathbf{n}^2 \quad \rightarrow \quad \mathbf{n}^2 \leq \frac{8mL^2 E}{h^2} \tag{9.8}$$

The second expression reveals a constraint on the length of $\mathbf{n}$. The vectors satisfying this constraint must fit within the positive quadrant of a $3N$-dimensional sphere of radius $R = \sqrt{8mL^2 E/h^2}$. The situation in two dimensions is illustrated in Fig. 9.1. Here the number of vector end points lying inside the circle is nearly equal to the area, $\pi R^2/4$, where the factor of $1/4$ restricts the counting to one quadrant of the circle of radius R. Importantly, this area approximation improves as we count microstates with increasingly high energies and larger values of R.

We now translate the two-dimensional ideas of Fig. 9.1 to a $3N$-dimensional setting, the true dimensionality of the vector $\mathbf{n}$. In this case, we must find the volume of a hypersphere of $3N$ dimensions with radius R. For a d-dimensional space, the volume of

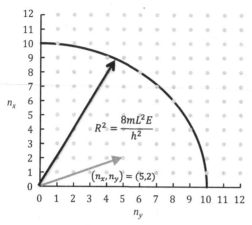

Figure 9.1. A schematic diagram for approximate counting of the number of two-dimensional integer vectors (n_x, n_y) that fit within a circle whose radius is constrained by the energy of the system. Adapted from Hill (1986).

a sphere of radius R is $C_d R^d$, where $C_d = \pi^{d/2}/\Gamma(d/2+1)$ and Γ is the gamma function. Therefore, the counting of the quantum microstates gives for Φ, the number of vectors with energy $\leq E$,

$$\Phi \approx \left(\frac{1}{2}\right)^{3N} \frac{\pi^{3N/2} R^{3N}}{\Gamma(3N/2+1)}$$

$$= \frac{1}{\Gamma(3N/2+1)} \left(\frac{2\pi m L^2 E}{h^2}\right)^{3N/2} \tag{9.9}$$

The factor of $(1/2)^{3N}$ is necessary in order to restrict the counting to the positive quadrant of the hypersphere since the quantum numbers must be positive.

Before proceeding, we must correct (9.9) to account for the fact that ideal gas particles of the same component are *indistinguishable* from each other. In a quantum sense, swapping the three quantum numbers (n_x, n_y, n_z) for one particle with those for another does not give rise to a new microstate. Indistinguishability stems from the mathematical properties of wavefunctions and is related to the Pauli exclusion principle. In our initial tabulation for Φ, we overcounted the number of indistinguishable microstates because we included all possible combinations of particle quantum numbers, regardless of whether or not two states differed by a swapping of quantum numbers. To correct for this, we need to remove from our counting the number of repeated sets of quantum numbers. To a very good approximation, at high energies where the vast majority of counted microstates assign distinct sets of quantum numbers to each particle, the over-counting is by a factor equal to the number of ways of swapping around quantum numbers between particles, $N!$. (This approximation breaks down at low temperatures and energies, where particles tend to have fewer quantum states available to them.) Therefore, a corrected version of Eqn. (9.9) becomes

$$\Phi = \frac{1}{N!\Gamma(3N/2+1)} \left(\frac{2\pi mL^2E}{h^2}\right)^{3N/2} \tag{9.10}$$

To compute Ω we need to find the number of states concentrated in a small energy range δE centered around a particular value of E. This is related to the derivative of Φ,

$$\Omega(E, V, N) = \frac{d\Phi}{dE}\delta E$$

$$= \frac{(3N/2)}{N!\Gamma(3N/2+1)} \left(\frac{2\pi mL^2}{h^2}\right)^{3N/2} E^{3N/2-1}\delta E \tag{9.11}$$

Finally, to obtain the entropy we take the logarithm and simplify using $\Gamma(x) = (x-1)!$, $\ln N! \approx N \ln N - N$, and $3N/2 - 1 \approx 3N/2$ for large values of N. The value of δE also becomes negligible relative to the magnitude of the other terms. The final result is

$$S(E, V, N) = k_\mathrm{B}N \ln\left[\left(\frac{E}{N}\right)^{3/2}\frac{V}{N}\right] + \frac{3}{2}k_\mathrm{B}N\left(\frac{5}{3} + \ln\left(\frac{4\pi m}{3h^2}\right)\right) \tag{9.12}$$

This famous result is the *Sackur–Tetrode equation*. It gives the exact functionality for the ideal gas entropy derived from quantum mechanics. It is an absolute entropy, and the zero of S depends on Planck's constant, a quantum parameter. Moreover, it is extensive, as we would expect; that is, we can check that (9.12) satisfies $S(\lambda E, \lambda V, \lambda N) = \lambda S(E, V, N)$.

It is important to remember that although the Sackur–Tetrode equation gives an absolute entropy, it is valid only at high energies due to the approximations made in its derivation. At low values of E, one must make a more detailed consideration of the number of microstates. As such, Eqn. (9.12) is inaccurate in the low-E, low-T regime.

9.2 Thermodynamic properties of ideal gases

Since Eqn. (9.12) gives the entropy in fundamental form, it contains all of the thermodynamic properties of an ideal, monatomic gas. For example, the relationship between energy and temperature is

$$\frac{1}{T} = \left(\frac{\partial S}{\partial E}\right)_{N,V} \rightarrow E = \frac{3}{2}Nk_\mathrm{B}T \tag{9.13}$$

The extensive heat capacity is thus $C_V = (3/2)Nk_\mathrm{B}$. Similarly, the equation of state follows

$$\frac{P}{T} = \left(\frac{\partial S}{\partial V}\right)_{N,E} \rightarrow P = Nk_\mathrm{B}T/V \tag{9.14}$$

This is the familiar ideal-gas law. Finally, the chemical potential is found from

$$\frac{\mu}{T} = -\left(\frac{\partial S}{\partial N}\right)_{V,E} \rightarrow \mu = -k_B T \ln\left[\left(\frac{2\pi m k_B T}{h^2}\right)^{3/2} \frac{V}{N}\right] \tag{9.15}$$

where we also made use of Eqn. (9.13) for the internal energy. We can rewrite (9.15) as

$$\mu = -k_B T \ln\left(\frac{V}{\Lambda(T)^3 N}\right) \tag{9.16}$$

Here we defined a new quantity $\Lambda(T)$ called the *thermal de Broglie wavelength* that has units of length,

$$\Lambda(T) \equiv \left(\frac{h^2}{2\pi m k_B T}\right)^{1/2} \tag{9.17}$$

Roughly speaking, the de Broglie wavelength gives a length scale above which particles behave classically. Namely, if the average distance between particles, $(V/N)^{1/3}$, becomes comparable to or smaller than Λ, quantum effects will be important. Notice that Λ increases at low temperatures, signaling the onset of quantum effects.

It is possible to calculate all other thermodynamic potentials for the ideal gas, such as the Helmholtz and Gibbs free energies A and G. These follow from simple manipulations using Legendre transforms. For example,

$$A(T, V, N) = E - TS$$

$$= \frac{3}{2} N k_B T - T\left\{ k_B N \ln\left[\left(\frac{\frac{3}{2} N k_B T}{N}\right)^{3/2} \frac{V}{N}\right] + \frac{3}{2} k_B N \left(\frac{5}{3} + \ln\left(\frac{4\pi m}{3h^2}\right)\right)\right\}$$

$$= -N k_B T \ln\left(\frac{V}{\Lambda(T)^3 N}\right) - N k_B T \tag{9.18}$$

The relevant thermodynamic potential at constant temperature and pressure is the Gibbs free energy, $G(T, P, N)$. For a single-component system, $G = \mu N$, so we can find the intensive Gibbs free energy as $\mu(T, P)$. Starting with Eqn. (9.16) and using (9.14) to substitute P for V, we find

$$\mu(T, P) = k_B T \ln P - k_B T \ln\left(\frac{k_B T}{\Lambda(T)^3}\right) = k_B T \ln P + \mu^\circ(T) \tag{9.19}$$

In the RHS, we are able to lump everything in the rightmost term into one function of temperature. Often μ° is called a *standard chemical potential*. That is, it contains the baseline chemical potential at a given temperature, while pressure effects are included separately through the first term. This division allows us to isolate the effect of pressure and shows a simplification in μ's functional dependence on T and P.

The keen reader might question the form of Eqn. (9.19). How can one take the log of the dimensional quantity pressure? Would not the choice of pressure units affect the result? Here, one must always be careful. The standard chemical potential involves the logarithm of a term with units of inverse pressure; when it is combined with $k_B T \ln P$, these units cancel out. Thus, whatever units are used for μ° must be used for P as well. In other words, we can think of the standard chemical potential as being associated with a certain unit of pressure, whether it is atmospheres, bars, or psi. One cannot have a standard chemical potential of this form that does not correspond to some particular kind of units. Alternatively, Eqn. (9.19) can be written so as to explicitly indicate the units,

$$\mu(T, P) = k_B T \ln \left(\frac{P}{P_0} \right) - k_B T \ln \left(\frac{k_B T}{\Lambda(T)^3 P_0} \right)$$

$$= k_B T \ln \left(\frac{P}{P_0} \right) + \mu^\circ(T) \tag{9.20}$$

where P_0 is a reference pressure that is incorporated into both terms on the RHS. In this case, the arguments of both logarithms are dimensionless, but the definition of the standard chemical potential is slightly different from that in (9.19).

Thus far, we have considered molecules to be point particles and monatomic, without structure. How does one treat ideal gases that have multiple atoms per molecule, such as O_2 and CO_2? In such cases, intramolecular bond-stretching and angle-bending potential energies exist in addition to kinetic energy. One can still assume that such molecules lack mutual interactions and occupy negligible volume in the $P \to 0$ limit. However, this problem requires a more detailed statistical-mechanical treatment. If one proceeds through the derivations, the equation of state remains the familiar ideal gas law and one can still write the chemical potential as $\mu(T, P) = k_B T \ln P + \mu^\circ(T)$, but the component $\mu^\circ(T)$ is distinct. Moreover, the additional intramolecular degrees of freedom contribute to the heat capacity. For a diatomic gas, C_V typically varies between $(5/2)Nk_B$ and $(7/2)Nk_B$ with T, compared with the temperature-independent value $(3/2)Nk_B$ for the monatomic case. The reader is referred to the texts by Hill and McQuarrie for detailed derivations of this case.

9.3   Ideal gas mixtures

Because ideal gas molecules are non-interacting, it is easy to determine the behavior of mixtures of different ideal gas species. Namely, the multicomponent density of states can simply be written as the product of those for each species,

$$\Omega(E, V, \{N\}) = \Omega_1(E_1, V, N_1) \times \Omega_2(E_2, V, N_2) \times \cdots \tag{9.21}$$

This also means that the entropy is additive,

$$S(E, V, \{N\}) = S_1(E_1, V, N_1) + S_2(E_2, V, N_2) + \cdots \tag{9.22}$$

The numbers N_i are fixed by the composition of the mixture. The energies E_i, on the other hand, are found by thermal equilibrium between the different species. How is this possible if the molecules are non-interacting? Here is a subtlety: for ideal gases we assume that the molecules are at the non-interacting limit but still have an infinitesimal amount of interaction in order to be able to exchange energy with one another and come to equilibrium.

The condition of thermal equilibrium between all gas species ensures that the energies E_i are found by the point at which the temperatures of the gases are equal, as we have seen more generally in Chapter 4. We can write this formally using the Helmholtz free energy. Simply multiply (9.22) by $-T$ and add to it the energy conservation equation, $E = E_1 + E_2 + \cdots = E$, to obtain

$$E - TS\,(E, V, \{N\}) = E_1 - TS_1\,(E_1, V, N_1) + E_2 - TS_2\,(E_2, V, N_2) + \cdots \quad (9.23)$$

or

$$A\,(T, V, \{N\}) = A_1\,(T, V, N_1) + A_2\,(T, V, N_2) + \cdots \quad (9.24)$$

Here we swapped E for T as independent variable. We can do this by demanding that the energies are found from the point at which all of the species temperatures are equal to each other and equal to the value T. Equivalently, we minimize (9.23) with respect to all E_i subject to the constraint of constant total energy. This is equivalent to performing a Legendre transform on each species to obtain the Helmholtz free energy.

To find the total pressure, we take the V derivative of Eqn. (9.24),

$$P = P_1 + P_2 + \cdots \quad (9.25)$$

We see that the total pressure is the sum of the *partial pressures* P_i of each of the gas species. This result is a distinctive property of ideal gas mixtures.

The **partial pressures** of a gaseous mixture are given by the expression $P_i = y_i P$, where $y_i = N_i/N$ is the mole fraction of component i.

In an ideal gas mixture, the partial pressures are identical to that which each component would experience at the same effective density in the pure state. Moreover, the sum of the partial pressures equals the total pressure, $P = N_1 k_B T/V + N_2 k_B T/V + \cdots = N k_B T/V$, and the mixture as a whole obeys the ideal gas law.

In nonideal gases, the partial pressures are *defined* by the relation $P_i = y_i P$. However, the partial pressure of each gas is no longer the same as that which it would experience in the pure-component state at the same effective density.

The chemical potential of each gas is the N_i derivative of $A(T, V, \{N\})$. In general, the chemical potential depends on the total composition of the mixture, denoted by the set of gas-phase mole fractions $\{y\}$. For an ideal gas,

$$\mu_i(T, P, \{y\}) = \frac{\partial A(T, V, \{N\})}{\partial N_i} = \frac{\partial A_i(T, V, N_i)}{\partial N_i}$$

$$= -k_B T \ln\left(\frac{V}{\Lambda(T)^3 N_i}\right) = \mu_i^\circ(T) - k_B T \ln\left(\frac{V}{N_i k_B T}\right) \quad (9.26)$$

After converting to P_i in favor of V using the ideal gas law, we obtain

$$\mu_i(T, P, \{y\}) = \mu_i^\circ(T) + k_B T \ln P_i$$
$$= \mu_i^\circ(T) + k_B T \ln P + k_B T \ln y_i \tag{9.27}$$

In other words, the chemical potential in an ideal gas mixture is decreased by the quantity $k_B T \ln y_i$ relative to the pure substance at the same pressure. Here $\mu_i^\circ(T)$ is the standard chemical potential of the pure gas i at temperature T. We can see this because we must recover the pure-component chemical potential in the limit $y_i \rightarrow 1$. Note also that the chemical potential of component i depends only on the composition of i, and none of the other compositions as in the general or nonideal case.

What is the free energy of mixing for ideal gases, at constant T and P? To determine this, we need to know the free energies before and after the mixing process,

$$G_{\text{before}} = \sum_i G_i(T, P, N_i) = \sum_i N_i \mu_i(T, P)$$
$$= \sum_i N_i \left[\mu_i^\circ(T) + k_B T \ln P \right] \tag{9.28}$$

and

$$G_{\text{after}} = G(T, P, \{N\}) = \sum_i N_i \mu_i(T, P, \{y\})$$
$$= \sum_i N_i \left[\mu_i^\circ(T) + k_B T \ln P + k_B T \ln y_i \right] \tag{9.29}$$

On combining these expressions we obtain

$$\Delta G_{\text{mix}} = G_{\text{after}} - G_{\text{before}}$$
$$= k_B T \sum_i N_i \ln y_i$$
$$= N k_B T \sum_i y_i \ln y_i \tag{9.30}$$

The negative temperature derivative of ΔG gives the entropy change,

$$\Delta S_{\text{mix}} = -\left(\frac{\partial \Delta G_{\text{mix}}}{\partial T} \right)_{P, \{N\}} = -N k_B \sum_i y_i \ln y_i \tag{9.31}$$

Since $\Delta G = \Delta H - T \Delta S$, this implies that $\Delta H = 0$ for the mixing process. In other words, the free energy change due to mixing is entirely entropic in nature. This is consistent with the lack of interactions in ideal gases. Moreover, when these gases mix at constant P, the total volume available to each species increases according to the ideal gas law. Since the volume increases, the entropy of each gas also increases. It is interesting to note that Eqn. (9.31) gives the *ideal entropy of mixing*, which one could derive using simple lattice statistics as in the problems at the end of Chapter 3.

9.4   Nonideal or "imperfect" gases

At modest pressures and temperatures, many gases exhibit deviations from ideal behavior. Gas molecules occupy a finite volume, do not behave like point particles, and exhibit subtle interactions with one another, such as van der Waals attractions. To account for nonideal behavior, it is common to introduce an "effective" pressure into the chemical potential in Eqn. (9.19). This corrected, fictitious pressure is called the *fugacity, f*. For a single component,

$$\mu(T, P) = \mu^\circ(T) + k_B T \ln f(T, P) \tag{9.32}$$

The fugacity is essentially *defined* by this equation; it gives a corrected pressure-like quantity that returns the *true* chemical potential for the nonideal gas. In the limit that gases behave in an ideal way – in the limit of low density and pressure – one has

$$f(T, P) \to P \text{ as } P \to 0 \tag{9.33}$$

An alternative formulation involves a *fugacity coefficient ϕ*,

$$\mu(T, P) = \mu^\circ(T) + k_B T \ln[\phi(T, P) P]$$

where

$$\phi(T, P) \equiv \frac{f(T, P)}{P} \tag{9.34}$$

It is clear from Eqn. (9.33) that

$$\phi(T, P) \to 1 \text{ as } P \to 0 \tag{9.35}$$

While its physical interpretation may seem vague, the fugacity is defined in this way because it is readily measured from pressure–volume experiments. Consider the fundamental relation for a single-component system,

$$\left(\frac{\partial \mu}{\partial P}\right)_T = v \tag{9.36}$$

Using (9.32),

$$k_B T \left(\frac{\partial \ln f}{\partial P}\right)_T = v \tag{9.37}$$

We can transform this into a form suitable for integration,

$$d\ln f = \frac{v}{k_B T} dP \quad (\text{constant } T) \tag{9.38}$$

In order to find f through integration, we must think carefully about the limits. One limit is that $f/P \to 1$ as $P \to 0$. We know the numerical value of the limit of f/P but not f itself, and need to re-express (9.38) in terms of f/P. Subtracting $d\ln P$ from both sides,

$$d\ln\left(\frac{f}{P}\right) = \left(\frac{v}{k_B T} - \frac{1}{P}\right) dP \quad (\text{constant } T) \tag{9.39}$$

Finally, we perform a definite integral from the state $P = 0$ to some pressure of interest $P = P'$,

$$\left[\ln\left(\frac{f}{P}\right)\right]_{P=P'} - \left[\ln\left(\frac{f}{P}\right)\right]_{P=0} = \int_0^{P'} \left(\frac{v}{k_B T} - \frac{1}{P}\right) dP \tag{9.40}$$

Using the limit of the fugacity to eliminate the second term on the LHS, we obtain

$$\ln\left(\frac{f(T, P')}{P'}\right) = \int_0^{P'} \left(\frac{v(T, P)}{k_B T} - \frac{1}{P}\right) dP \tag{9.41}$$

Here, we made explicit the dependences on temperature and pressure. Note that, to calculate the fugacity, one needs the variation of the molar volume of the gas between zero pressure and the pressure of interest. This behavior can be tabulated in experiments, and such data are often fitted to specific functional forms like the virial expansion, the van der Waals model, other cubic equations of state, or corresponding-states correlations. With data and models, it is typical to report the *compressibility factor*,

$$Z \equiv \frac{Pv}{k_B T} \tag{9.42}$$

Z is a convenient dimensionless number that in part measures the ideality of a gas, since ideal gases have $Z = 1$. By substituting for v in Eqn. (9.41) using Z, we obtain

$$\ln\left(\frac{f(T, P')}{P'}\right) = \int_0^{P'} \frac{Z(T, P) - 1}{P} dP \tag{9.43}$$

This equation shows that there is a very close relationship between the equation of state of a gas and its fugacity. In fact, many theoretical equation-of-state models have been used to suggest specific mathematical forms for fugacities.

9.5 Nonideal gas mixtures

In mixtures of nonideal gases, fugacities in general are not only functions of the pressure and temperature, but of composition as well,

$$\mu_i(T, P, \{y\}) = \mu_i^\circ(T) + k_B T \ln f_i(T, P, \{y\}) \tag{9.44}$$

Here it is also possible to calculate, using experiments on mixtures or guided by statistical-mechanical models, numerical values of the fugacities. For example, one might use Eqn. (9.43) at a given mixture composition. Often, however, it is reasonable to pursue a simpler approximation called the *Lewis–Randall rule*:

$$f_i(T, P, \{y\}) \approx y_i f_i(T, P) \tag{9.45}$$

Equation (9.45) presumes that the fugacity of a gas in a mixture is approximately equal to its mole fraction times the fugacity of the pure-component gas at the same T and P.

This approximation is motivated by the form of the ideal gas mixture chemical potential, in which $P_i = y_i P$. The Lewis–Randall rule simplifies the mixture equations because it eliminates the need to calculate the fugacities for different compositions, requiring only the pure-state quantities. Many nonideal gas mixtures can be described well by this approach at low to moderate pressures.

Problems

Conceptual and thought problems

9.1. Consider the results for ideal gases derived from quantum mechanics. Write an expression for the function $s(T, v)$ that includes the parameter $\Lambda(T)$.
(a) Show that the behavior of the entropy as $T \to 0$ is unrealistic in a quantum sense. What approximation in the derivation is responsible for this behavior?
(b) The Sackur–Tetrode equation is valid when the thermal de Broglie wavelength is much less than the average molecular separation distance, $\Lambda(T) < (V/N)^{1/3}$, which is generally the case at high temperatures. Calculate the temperature at which this inequality is first violated for an ideal gas at atmospheric pressure and with a mass of 6.6×10^{-26} kg/molecule (which is typical of argon).

9.2. At constant T, V, and N, does the entropy of an ideal gas increase or decrease with molecular weight? Explain why in physical terms.

9.3. An ideal gas mixture is pressurized at constant temperature. Indicate whether each of the following increases, decreases, or stays the same, or there is not enough information to tell.
(a) the total mixture heat capacity $C_P = (\partial H/\partial T)_{P,\{N\}}$
(b) the chemical potential μ_i of each component
(c) the entropy of mixing, ΔS_{mix}

9.4. An ideal gas mixture is heated at constant pressure. Indicate whether each of the following increases, decreases, or stays the same, or there is not enough information to tell.
(a) the partial pressure P_i of the dominant component
(b) the total mixture heat capacity $C_P = (\partial H/\partial T)_{P,\{N\}}$
(c) the chemical potential μ_i of each component
(d) the entropy of mixing, ΔS_{mix}

9.5. For a single-component system, indicate whether or not each of the following functions provides enough information to completely specify all of the intensive thermodynamic properties of a monatomic gas whose atomic mass is known:
(a) $s(T, P)$
(b) $f(T, P)$, where f is the fugacity
(c) $P(T, v)$ containing the limit $v \to \infty$

9.6. An ideal gas is throttled to a lower pressure. A throttling process lowers the pressure of a continuously flowing stream through a nozzle or an abrupt increase in the diameter of a pipe. Show first that throttling is a constant-enthalpy process. Then prove that the ideal gas entropy always increases in such processes.

9.7. Explain how a multicomponent gas system might constitute an ideal *mixture*, even if the individual species and the system as a whole do not behave like an ideal *gas*. Discuss the relationship of an ideal gas mixture to the Lewis–Randall rule.

9.8. A monatomic gas has a known equation of state of the form $v(T, P)$. Its fugacity is then determined by Eqn. (9.41). This in turn gives the gas' chemical potential and hence its entropy, $s = -(\partial \mu / \partial T)_P$. Is it possible that this entropy is an *absolute* entropy, even though the only information describing this particular gas is the equation of state, which does not necessarily contain quantum considerations? Explain.

Fundamentals problems

9.9. Prove that Eqns. (9.6) and (9.7), involving additivity of the energies for ideal gas particles, are correct. Hint: consider the separation-of-variables technique.

9.10. Show that the expression $\mu(T, P) = \mu°(T) + k_B T \ln P$ is valid for any ideal gas, regardless of its molecular structure (monatomic, diatomic, etc.). Hint: all ideal gases have the same equation of state.

9.11. The molar excess entropy for a real gas is defined as $s_{ex}(T, P) \equiv s(T, P) - s_{ig}(T, P)$, where s_{ig} is the entropy of an ideal gas at the same temperature and pressure. Show that the excess entropy is related to the fugacity by

$$s_{ex}(T, P) = -\left(\frac{\partial(k_B T \ln(f/P))}{\partial T}\right)_P$$

9.12. Consider the excess chemical potential of a gas, $\mu_{ex} = \mu - \mu_{ig}$, where μ_{ig} is the ideal contribution. Show that it can be calculated at a given number density ρ and temperature T by use of the relation

$$\frac{\mu_{ex}}{k_B T} = \int_0^\rho \frac{Z(T, \rho') - 1}{\rho'} \, d\rho' + Z(T, \rho) - 1 - \ln Z(T, \rho)$$

In addition, find μ_{ex} for a fluid described by a second-order virial expansion,

$$\frac{P}{k_B T} = \rho + B_2 \rho^2$$

9.13. The fugacity of a monatomic gas is found to obey the relation $f(T, P) = P \exp(bT)$, where b is a constant with units of inverse temperature. Find c_P and c_V for this gas.

9.14. A semi-permeable, immovable membrane separates two compartments of equal volumes V. Initially N_A molecules of component A are in the left-hand compartment and N_B molecules of component B are in the right-hand compartment; both are ideal gases. The membrane is impermeable to B but permeable to A. Find the fraction of A molecules on the LHS and the osmotic pressure Π (i.e., the pressure difference) across the membrane at equilibrium as functions of T, V, N_A, and N_B.

9.15. The fugacity of a monatomic imperfect gas might be expressed as a Taylor expansion of the pressure,

$$\frac{f(T, P)}{P} = 1 + z_1(T)P + z_2(T)P^2 + \cdots$$

where the coefficients $z_i(T)$ are, in general, dependent on the temperature.

(a) If the expansion is truncated at second order, show that z_1 and z_2 are constrained by

$$\frac{1 + 2z_1 P + 3z_2 P^2}{1 + z_1 P + z_2 P^2} > 0$$

(b) At low pressures, the expansion can be truncated at first order, and one can approximate $\ln(f/P) = \ln(1 + z_1(T)P) \approx z_1(T)P$. Assume that $z_1(T)$ can be expanded as $z_1(T) = b_0 + b_1 T$, where b_0 and b_1 are constants. Find the constant-pressure heat capacity c_P as a function of T and P.

(c) Show that as long as the quantity c_P/T is positive, b_1 must satisfy

$$b_1 < \frac{5}{4}\left(\frac{1}{TP}\right)$$

9.16. One kind of thermodynamic diagram that is useful in engineering problems is the P-H diagram, in which the x-axis is molar enthalpy and the y-axis is pressure.

(a) Several kinds of iso-lines can be drawn in such a diagram, including isotherms (constant temperature), isentrops (constant entropy), and isochores (constant molar volume). For each of these three cases, find expressions for the slope of these lines on a P-H diagram in terms of measurable properties.

(b) Find expressions for these slopes for the ideal monatomic gas, in terms of T and P. Explain on physical grounds any unusual results.

9.17. Prove that a mixture described by the Lewis–Randall rule, Eqn. (9.45), has the property that the enthalpy and volume of mixing are both zero.

Applied problems

9.18. In a continuous-flow process, a gas is throttled to half its density. The throttling is well described as a constant-enthalpy process. The constant-volume heat capacity of the gas is approximately $(5/2)k_B$ and its equation of state is $P = k_B T/v - a/v^2$, where a is a positive constant. Find an expression for the final temperature T_2 in terms of the initial temperature T_1 and molar volume v_1. Does the gas heat up or cool down due to throttling?

9.19. The equation of state for a certain gas is found to obey $P(T, \rho) = \rho k_B T/(1 - b\rho)$, where b is a positive constant. Find the compressibility factor, $Z(T, P)$, and the logarithm of the fugacity, $\ln f(T, P)$.

9.20. A certain gas obeys the equation of state $P = k_B T/(v - b)$ where b is a constant with units of volume per molecule. The applied pressure is increased by a small, differential amount ΔP in a quasi-static, adiabatic process. Show that the change in molar volume for this gas is given by

$$\Delta v = -\left(\frac{c_V}{c_V + k_B}\right)\left(\frac{v - b}{P}\right)\Delta P$$

where c_V is the molar heat capacity and v its original per-particle volume.

9.21. A monatomic gas is slowly and adiabatically compressed from initial conditions T_0 and P_0 to twice the original pressure. The equation of state of this gas is $P = k_B T/(v - b)$, where v is the volume per particle and b is a constant. Note that this is not an ideal gas and therefore you should not make any assumptions as such. What is the final temperature of the gas? How much work per molecule is required in order to accomplish this change of state?

9.22. Typically cooling can be accomplished by an expansion process. Consider a reversible, adiabatic expansion of an arbitrary system from total volume V_1 and temperature T_1 to $V_2 > V_1$ and T_2.

  (a) For small $\Delta V = V_2 - V_1$, show that the final temperature is given by an expression of the form $T_2 \approx T_1 \exp(K\,\Delta v)$, where K is a constant that depends on measurable quantities. Find K.

  (b) Is K always negative for all substances, such that an expansion produces a temperature drop? If not, under what conditions will it be positive? Be as specific as possible.

  (c) The fugacity of a gas is found to obey $f(T, P) = P \exp(bT)$, where b is a constant with units of inverse temperature. Show that the constant K for this gas is given by the following, where c_P^{IG} is the ideal gas heat capacity,

$$K = -\frac{P}{c_P^{IG}T - k_B T(2bT + 1)}$$

FURTHER READING

K. Denbigh, *The Principles of Chemical Equilibrium*, 4th edn. New York: Cambridge University Press (1981).

T. L. Hill, *An Introduction to Statistical Thermodynamics*. Reading, MA: Addison-Wesley (1960); New York: Dover (1986).

L. D. Landau and E. M. Lifshitz, *Statistical Physics*, 3rd edn. Oxford: Butterworth-Heinemann (1980).

D. A. McQuarrie, *Quantum Chemistry*. Mill Valley, CA: University Science Books (1983).

D. A. McQuarrie, *Statistical Mechanics*. Sausalito, CA: University Science Books (2000).

J. M. Smith, H. V. Ness, and M. Abbott, *Introduction to Chemical Engineering Thermodynamics*, 7th edn. New York: McGraw-Hill (2005).

10 Phase equilibrium

It is a familiar fact that pure substances tend to exist in one of three distinct states: solid, liquid, and gas. Take water, for example. As ice is heated at atmospheric pressure, it suddenly melts into liquid at a specific temperature. As the liquid continues to be heated, it eventually reaches a temperature at which it spontaneously vaporizes into a gas. These transitions are discontinuous; they occur at specific state conditions or particular combinations of T and P. At exactly those conditions, the system can exist in more than one form such that two (or more) phases are in equilibrium with each other.

Although we are typically familiar with phase behavior at atmospheric pressure, most substances experience a diverse set of phases over a broad range of pressures. Pure substances often have more than one crystalline phase, depending on the pressure. Figure 10.1 shows a schematic representation of a P–T phase diagram of water that illustrates the kind of complex behavior that can exist. In the case of mixtures, there are even more possibilities for phase equilibrium: for example, one can have equilibrium between two liquids of different compositions, or among multiple solid and liquid phases.

In this chapter, we address the two most important questions in the discussion of phases. First, what are the thermodynamic conditions for phase equilibrium? Second, why do phases change discontinuously? For example, why does water have a definite boiling-point temperature at ambient pressure? We will use both classical thermodynamics and statistical mechanics to tackle these questions.

10.1   Conditions for phase equilibrium

We must start by giving a precise definition to *phase*.

A **phase** is a homogeneous region of matter in which there is no spatial variation in average density, energy, composition, or other macroscopic properties.

Phases can also be distinct in their molecular structure. For example, water has multiple ice phases that differ in their crystallographic structure.

A phase can be considered a distinct "system" with boundaries that are interfaces with either container walls or other phases.

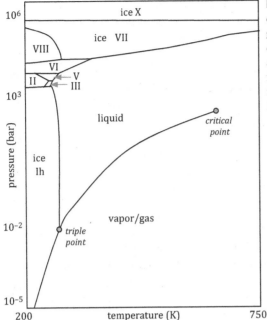

Figure 10.1. A schematic representation of the phase diagram of water. At greatly elevated pressures, several crystallographically distinct ice phases are seen, beyond the usual ice Ih formed at ambient pressure. Adapted from www1.lsbu.ac.uk/water/phase. html.

The notion of phase equilibrium means that there are two or more phase "systems" present. These systems exist spontaneously on their own without the use of partitions or membranes or other interventions, and, since they are in mutual equilibrium, they can exchange energy, volume, and particles. We have seen in Chapter 4 that, when such is the case, the conditions of equilibrium are given by maximization of the entropy subject to the constraints of constant total energy, volume, and number of particles. For two phases involving a single component, this becomes

$$\max[S^\alpha(E^\alpha, V^\alpha, N^\alpha) + S^\beta(E^\beta, V^\beta, N^\beta)] \tag{10.1}$$

Here we use superscripts α and β to indicate two phases, so that we can later reserve subscripts for chemical species within each phase. At an entropy maximum, we have,

$$dS^\alpha + dS^\beta = 0$$

and

$$\frac{1}{T^\alpha} dE^\alpha + \frac{P^\alpha}{T^\alpha} dV^\alpha - \frac{\mu^\alpha}{T^\alpha} dN^\alpha + \frac{1}{T^\beta} dE^\beta + \frac{P^\beta}{T^\beta} dV^\beta - \frac{\mu^\beta}{T^\beta} dN^\beta = 0 \tag{10.2}$$

where we employed the fundamental entropy equation. Because $dE^\alpha + dE^\beta = 0$, $dV^\alpha + dV^\beta = 0$, and $dN^\alpha + dN^\beta = 0$ by the constancy of the total energy, volume, and number of particles, this becomes

$$\left(\frac{1}{T^\alpha} - \frac{1}{T^\beta}\right) dE^\alpha + \left(\frac{P^\alpha}{T^\alpha} - \frac{P^\beta}{T^\beta}\right) dV^\alpha - \left(\frac{\mu^\alpha}{T^\alpha} - \frac{\mu^\beta}{T^\beta}\right) dN^\alpha = 0 \tag{10.3}$$

Recognizing that dE^α, dV^α, and dN^α can all vary independently, the conditions for equilibrium between two phases must then be given by

$$T^\alpha = T^\beta, \quad P^\alpha = P^\beta, \quad \mu^\alpha = \mu^\beta \tag{10.4}$$

It will be helpful to take the following viewpoint: imagine that we can manipulate each phase independently by changing its temperature, pressure, or chemical potential while retaining the same phase identity (e.g., liquid or vapor). As an example, consider liquid water and ice phases, each of which might be varied in temperature and pressure. To find the conditions of solid–liquid equilibrium, we must search (T, P) space for states satisfying the equalities in (10.4). The points at which these are satisfied then correspond to the melting line. If the equalities are not satisfied, just one phase will exist at those state conditions.

In fact, one of the variables in (10.4) is redundant. Recall that only two intensive variables are required in order to completely specify the thermodynamic state and hence all intensive properties of a single-component system. Therefore, if we choose T and P then μ for either phase is uniquely specified, and generally we can write $\mu = \mu(T, P)$. So, if we are looking for T and P combinations that enable equilibrium between liquid water and ice, we can simply solve the equation

$$\mu^\alpha(T, P) = \mu^\beta(T, P) \tag{10.5}$$

where the superscripts indicate each phase. Notice that we have already accounted for the thermal and mechanical equilibrium conditions by assuming that the temperatures and pressures are constant. This equality of chemical potentials constitutes an equation with two unknowns, T and P. If we specify the temperature, we can solve for the pressure corresponding to phase equilibrium, and vice versa.

Equation (10.5) indicates that just *one* intensive variable is needed in order to specify the state of the system when two phases are in equilibrium in a single-component system. If T is specified, P can be found by solving the chemical-potential equality, and vice versa. This reduction in the number of independent degrees of freedom occurs because the mere fact that two phases coexist provides a constraint on the chemical potentials. Figure 10.2 illustrates the procedure graphically, with the chemical potential in arbitrary units.

From Fig. 10.2, we can see that there is only one temperature at which the chemical potentials of the liquid and crystal are equal: the melting temperature T_m. Below T_m, the crystal has a lower chemical potential. Because the thermodynamic potential minimized at constant T and P is the Gibbs free energy G, and since $G = N\mu$ for a single-component system, the crystal is therefore the only phase present at $T < T_m$. Under those conditions, it is the most stable state of the system at equilibrium, since any liquid would increase the free energy. Similarly for $T > T_m$, the liquid is the only phase present.

If we consider variations in both temperature and pressure, we must examine the chemical potentials in a three-dimensional space as illustrated in Fig. 10.3. The intersection of the two chemical-potential surfaces defines a line in the T-P plane along which phase equilibrium can occur. This line is the melting line, and we can express it as either $P_m(T)$ or $T_m(P)$.

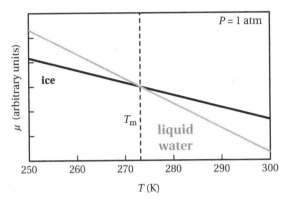

Figure 10.2. A schematic representation of the temperature dependence of the chemical potentials of liquid water and ice at ambient pressure. At 273 K, the chemical potentials are equal and phase equilibrium occurs. Above (or below) that temperature, liquid water (or ice) has the lower chemical potential and is the thermodynamically stable phase.

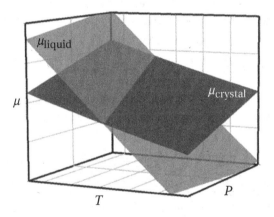

Figure 10.3. A schematic representation of the temperature and pressure dependences of the chemical potentials of ice and liquid water. The intersection of the two surfaces defines a melting line when projected onto the *T-P* plane (the floor of this graph).

What happens if three phases α, β, and γ are in equilibrium in a single-component system? In that case, two equalities of chemical potentials must be satisfied, $\mu^{\alpha} = \mu^{\beta}$ and $\mu^{\beta} = \mu^{\gamma}$ (the equation $\mu^{\alpha} = \mu^{\gamma}$ is clearly redundant). This amounts to two equations and two unknowns, *T* and *P*. Since there are no free variables, the temperature and pressure must be exactly specified in discrete solutions. This is why the triple point of a material – at which solid, liquid, and gas phases all coexist – is a point in the *T-P* plane, and not a line. These familiar examples illustrate some basic features of phase equilibrium.

At constant T and P in a single-component system, the phase that is observed is the one which has the lowest Gibbs free energy per particle, or, equivalently, the lowest chemical potential $\mu(T, P)$. At conditions where the chemical potentials of two or more phases are equal and at a minimum with respect to all other phases, the former can coexist in equilibrium with each other.

We can generalize these ideas in a mathematical framework that describes any kind of phase equilibrium, for arbitrary multicomponent systems. Consider the case in which there are n_{phase} phases and C components. There are C chemical potentials in each phase that depend not only on T and P, but on all of the compositions of their respective phases as well, $\{x\} = \{x_1, x_2, \ldots, x_C\}$, where x denotes mole fraction. Note that only $C - 1$ mole fractions are independent in each phase since they must always sum to unity. The conditions for equilibrium can then be cast as

$$\mu^{\text{phase } k}_{\text{component } i}(T, P, \{x\}^{\text{phase } k}) = \mu^{\text{phase } k+1}_{\text{component } i}(T, P, \{x\}^{\text{phase } k+1}) \qquad (10.6)$$

for all $i = 1, 2, \ldots, C$ and $k = 1, 2, \ldots, n_{\text{phase}} - 1$. There are $C(n_{\text{phase}} - 1)$ such equations in total and $2 + n_{\text{phase}}(C - 1)$ independent variables from T, P, and the $C - 1$ independent mole fractions in each phase. The number of degrees of freedom in this problem, n_{DOF}, is given by the total number of variables minus the number of chemical-potential constraint equations,

$$n_{\text{DOF}} = 2 + n_{\text{phase}}(C - 1) - C(n_{\text{phase}} - 1) = C - n_{\text{phase}} + 2 \qquad \textbf{(10.7)}$$

This is the famous *phase rule*, which was originally derived by Gibbs. Though we developed it on mathematical grounds, its physical interpretation is as follows.

Gibbs' phase rule asserts that the state conditions giving rise to equilibrium among n_{phase} phases in a C-component system are exactly characterized by n_{DOF} macroscopic, intensive degrees of freedom, with $n_{\text{DOF}} = C - n_{\text{phase}} + 2$. Only n_{DOF} pieces of information are required in order to uniquely specify the state of the entire system, knowing that n_{phase} phases are present. These pieces of information are typically drawn from T, P, and the phase compositions.

Consider the space of the variables T, P, and the set of compositions. Let us imagine the ways in which different values of n_{DOF} are manifested in this space. In the case $n_{\text{DOF}} = 0$, there are no free variables and phase equilibrium corresponds to single, discrete sets of conditions. These are points in the P–T diagram for single-component systems. For $n_{\text{DOF}} = 1$, state variables can move in concert with one another along one direction in state space, such as a line in the single-component P–T diagram. Finally, with $n_{\text{DOF}} = 2$, there is flexibility to move in two independent directions in the parameter space. For a single-component system, this corresponds to an area in the P–T plane.

The phase rule just described applies to non-reactive systems. If a set of components can interconvert by way of a reversible chemical reaction that also reaches equilibrium, then the rule is modified:

$$n_{\text{DOF}} = C - n_{\text{phase}} + 2 - n_{\text{reaction}} \qquad (10.8)$$

where n_{reaction} is the number of such equilibrium reactions. It might be easy to see why there is a reduction in the degrees of freedom: for each reaction there is an additional constraint equation that relates the concentrations of different components to an equilibrium constant. More technically, chemical reactions add additional chemical-potential equalities as discussed later in Chapter 20.

10.2 Implications for phase diagrams

A common way to depict the conditions giving rise to phase equilibrium is through a *phase diagram*. Indeed, there are many kinds of phase diagram, but two examples typical for a single-component system are shown in Fig. 10.4. In the *P–T* diagram, lines correspond to regions with two-phase equilibria. It is easy to see that $n_{\text{DOF}} = 1$ along these lines, since the indication of temperature automatically specifies the pressure and vice versa. On the other hand, $n_{\text{DOF}} = 0$ corresponds to a point, such as the triple point.

An alternative representation of the same behavior is shown in the *T–ρ* diagram in Fig. 10.4. This depiction differs in an important respect: two-phase regions correspond to areas rather than lines because the density changes discontinuously as the system moves from one phase to another. For example, if a liquid is heated at constant pressure,

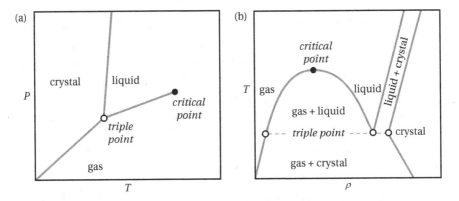

Figure 10.4. Schematic typical phase diagrams for simple, single-component systems like argon. (a) In the *P–T* diagram, single phases appear as areas, while phase equilibrium occurs along lines and at points. (b) In the *T–ρ* diagram, both single phases and two-phase equilibria appear as areas. In two-phase regions, the density ρ corresponds to the overall system density, the total number of moles divided by the total volume of both phases. At a given temperature, therefore, a horizontal line intersecting the boundaries surrounding a two-phase region gives the individual densities of the coexisting phases.

it will eventually begin to vaporize at the boiling temperature. For a small amount of heat added, the system consists of a minuscule amount of gas at low density and a large amount of liquid at high density, for an *overall* system density that remains close to that of the liquid. However, as more heat is added, the number of moles in the gas phase increases at the expense of that in the liquid; while the individual densities of these phases remains the same, the overall density is thus shifted towards that of the gas. It is this overall density that is represented on the x-axis of T-ρ diagrams within the two-phase areas, while the boundary lines give the densities of the individual coexisting phases.

The slopes of boundaries in phase diagrams are tightly connected to thermodynamic properties. Consider the liquid-gas boundary in the P-T diagram. Along this line the chemical potentials of both phases must be equal; otherwise, two-phase equilibrium would not be possible. Therefore, if we move a short distance along the liquid-gas boundary, the chemical potentials of the liquid and gas must change by exactly the same amount since they must continue to be equal at the ending point. In the limit of an infinitely small change,

$$d\mu^L(T, P) = d\mu^G(T, P) \quad \text{(along a phase boundary)} \tag{10.9}$$

Recognizing that $\mu = g$ for a single component and using the full differential of g, we have

$$-s^L \, dT + v^L \, dP = -s^G \, dT + v^G \, dP \quad \text{(along a phase boundary)} \tag{10.10}$$

On rearranging and completing the derivative, we obtain

$$\left(\frac{dP}{dT}\right)_{\text{phase boundary}} = \frac{s^G - s^L}{v^G - v^L} = \frac{\Delta s}{\Delta v} \tag{10.11}$$

Equation (10.11) asserts that the slope of the liquid-gas boundary in the P-T phase diagram is related to the molar entropy and volume differences of the two phases. Actually, this equation is valid for *any* phase boundary because we have done nothing to introduce gas- or liquid-specific behavior. One can go further to replace Δs using the enthalpy difference, $\Delta h = \Delta \mu + T \, \Delta s$. Since $\Delta \mu = 0$ along a phase boundary, we finally have

$$\left(\frac{dP}{dT}\right)_{\text{phase boundary}} = \frac{\Delta h}{T \, \Delta v} \tag{10.12}$$

This is the *Clapeyron equation*. It shows that the differences in enthalpies and densities of two phases at equilibrium allow us to predict how their equilibrium pressure will change as a function of temperature, or vice versa. For liquid-gas equilibrium, this pressure is in fact the *vapor pressure* of the liquid, defined as that of vapor-liquid equilibrium at a given temperature. Alternatively, Eqn. (10.12) shows how the *boiling temperature* changes with the applied pressure.

To a good approximation, we can simplify the Clapeyron equation further for the specific case of liquid-gas equilibrium by assuming the gas is ideal with $v^G = k_B T/P$.

Moreover, typically $v^G \gg v^L$ and so we can approximate $\Delta v \approx v^G \approx k_B T/P$. With these simplifications, we obtain the *Clausius–Clapeyron equation*:

$$\left(\frac{d\ln P}{dT}\right)_{\text{phase boundary}} = \frac{\Delta h}{k_B T^2} \tag{10.13}$$

The reader is encouraged to perform the full derivation of Eqn. (10.13). Finally, in cases where the enthalpy difference does not vary significantly with temperature, we can treat Δh as constant and integrate to obtain

$$\ln\left(\frac{P_2}{P_1}\right) = -\frac{\Delta h}{k_B}\left(\frac{1}{T_2} - \frac{1}{T_1}\right) \tag{10.14}$$

where (T_1, P_1) and (T_2, P_2) are two points on the liquid–gas phase boundary. If we treat T_1 and P_1 as reference "constants," this equation could also be written in a form that gives the saturation or vaporization pressure at any temperature,

$$\ln P^{\text{vap}}(T) = c_1 - \frac{c_2}{T} \tag{10.15}$$

where $c_1 = \Delta h_{\text{vap}}/(k_B T_1) + \ln P_1$ and $c_2 = \Delta h_{\text{vap}}/k_B$ are assumed constant. The form of this equation is remarkably similar to empirically fitted correlations used to predict the saturation pressure, such as the *Antoine equation*.

$$\ln P^{\text{vap}}(T) = c_1 - \frac{c_2}{c_3 + T} \tag{10.16}$$

The additional constant c_3, relative to (10.15), compensates for the assumption of a constant enthalpy of vaporization. Equation (10.16) is widely used in practice, and values for c_1, c_2, and c_3 for a large number of liquids are available in many reference texts and publications.

For equilibrium between solids and gases – sublimation – we can also use the Clausius–Clapeyron equation because the gas volume remains much greater than that of the crystal. However, for solid–liquid or solid–solid equilibrium, one must return to the original Clapeyron equation since this assumption becomes poor.

Example 10.1   *A pressure cooker is able to prepare food rapidly by enabling water to remain in a liquid state at temperatures higher than possible at atmospheric conditions. Calculate the boiling temperature of water in a pressure cooker rated at 15 psi above atmospheric pressure. The enthalpy of vaporization of water at ambient conditions is $\Delta h = 40.7$ kJ/mol.*

To address this problem, we use the integrated Clausius–Clapeyron equation of (10.14):

$$\frac{14.7 \text{ psi} + 15 \text{ psi}}{14.7 \text{ psi}} = \exp\left[-\frac{40{,}700 \text{ J/mol}}{8.31 \text{ J/mol K}}\left(\frac{1}{T_2} - \frac{1}{373 \text{ K}}\right)\right]$$

Solving for T_2,

$$T_2 = 393\ \text{K} = 120\,°\text{C}$$

10.3 Other thermodynamic behaviors at a phase transition

At a phase transition the second law demands that the chemical potentials, temperatures, and pressures of the phases present are equal. What about differences in other properties such as the entropy, energy, and volume? In general, these variables change discontinuously across a phase transition. Figure 10.5 illustrates this behavior for the liquid–vapor transition in water at atmospheric pressure. Clearly, at $T = 100\ °\text{C}$, there are sharp changes in the volume, entropy, and energy. In other words, Δv, Δs, and Δe can take on nonzero values that are characteristic of the boiling point.

Similarly, the change in enthalpy, $\Delta h = \Delta e + P\,\Delta v$, is also discontinuous. This is a particularly important quantity because the first law at constant pressure gives $Q = \Delta H$, which implies that heat added to a system at phase equilibrium is directly related to

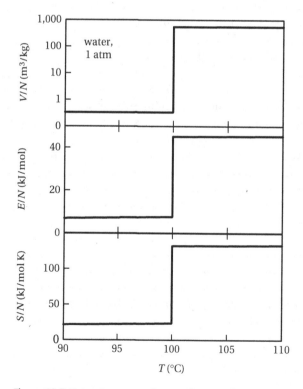

Figure 10.5. Extensive properties – and per mole or per-mass versions thereof – change discontinuously at the liquid–vapor phase transition. Illustrated here is the behavior of the molar volume, internal energy, and entropy for water under atmospheric pressure.

the system enthalpy change. We typically term the quantity Δh the *latent heat* or *latent enthalpy* of a given phase transition; for boiling it is the latent heat of vaporization.

How then does the intensive heat capacity, $c_P = (\partial h / \partial T)_P$, behave at a phase transition? If the enthalpy changes discontinuously at the boiling point, then c_P diverges. The heat and hence enthalpy added to the system at the boiling temperature convert liquid into vapor instead of increasing the system temperature. In other words, the enthalpy of vaporization is absorbed with zero change in temperature, which suggests the lack of a well-defined heat capacity.

The intensive state properties of the phases alone are insufficient to determine the amount of each present. Instead, one needs to make a connection to extensive variables. Any total extensive property of the entire two-phase system is obtained by summing over the respective contributions in each phase, neglecting boundary effects. For liquid and vapor or gas equilibrium,

$$
\begin{aligned}
V &= v^L N^L + v^G N^G \\
E &= e^L N^L + e^G N^G \\
S &= s^L N^L + s^G N^G \\
H &= h^L N^L + h^G N^G
\end{aligned}
\tag{10.17}
$$

where N^L and N^G denote the numbers of molecules in the liquid and gas phases, respectively. Here, the superscripts on the intensive properties v, e, s, and h indicate their values in the respective phases. If we divide Eqns. (10.17) by the total molecules present and let $f = N^G / N$ be the fraction of the system in the vapor phase and $1 - f = N^L / N$ that in the liquid, we obtain

$$
\begin{aligned}
V/N &= (1 - f)v^L + f v^G = v^L + f\,\Delta v \\
E/N &= (1-f)e^L + f e^G = e^L + f\,\Delta e \\
S/N &= (1-f)s^L + f s^G = s^L + f\,\Delta s \\
H/N &= (1-f)h^L + f h^G = h^L + f\,\Delta h
\end{aligned}
\tag{10.18}
$$

The terms on the LHS of (10.18) give the overall extensive quantities per molecule, for example, the total system volume normalized by the total number of molecules (liquid plus gas). The expressions on the RHS therefore relate these to the amount of the vapor phase f because the phase-specific properties v^L, v^G, e^L, e^G, s^L, s^G, h^L, and h^G all remain fixed for a given pressure and temperature – they are state variables. That is, the overall quantities (liquid plus gas) vary during a phase transition because the amounts of each phase change, not because the phases' intensive properties vary.

Figure 10.6 illustrates the relationship of system to individual phase properties using water and its enthalpy as an example. For a constant-pressure process, the first law $Q = \Delta H$ shows that heating water below its boiling point increases its enthalpy and raises the liquid temperature until it reaches 100 °C. At that point, the system is pure liquid water with an infinitesimally small amount of vapor. As heat is further added, the temperature remains constant but the amount of vapor grows until eventually the entire system consists of vapor and an infinitesimally small amount of liquid. As might be obvious, this occurs when heat in the amount of the enthalpy of vaporization is added. During vaporization, all of the properties of the liquid and gas phases remain

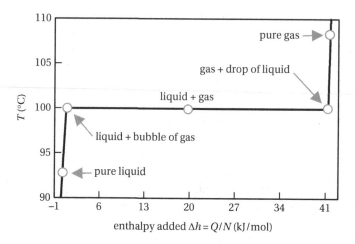

Figure 10.6. The constant-pressure addition of heat to liquid water eventually converts it into a vapor phase. The x-axis gives the overall change in system enthalpy $\Delta H/N$. Added heat continuously increases the enthalpy, but at the boiling point the amount of gas phase increases at the expense of the liquid, at constant temperature. This is in contrast to the pure liquid and gas phases, in which the temperature rises with increases in enthalpy.

constant: their densities, molar energies and entropies, etc. Finally, after the last drop of liquid has vanished, additional heat raises the temperature and increases the enthalpy of the homogeneous gas phase.

Example 10.2 *A small microwave oven adds heat to a glass of water at the rate of 500 W. Assume that the rate of heating is slow enough that the water can maintain a near-equilibrium state. If the water is initially at 20 °C (room temperature), how long will it take for the liquid to completely boil off?*

Assuming constant atmospheric pressure for the process, we have for the water

$$dE = \delta Q + \delta W = \delta Q - P\,dV$$

and

$$\delta Q = dE + P\,dV \quad \rightarrow \quad Q = \Delta E + P\,\Delta V = \Delta H$$

The work is nonzero because the water will expand against the pressure of the atmosphere as it is heated and changed into a gas. Thus, the heat added from the microwave oven increases the water's enthalpy. Because enthalpy is a state function, we can calculate the total change in H between the liquid and gas using a cycle by which we heat the liquid and then boil it. The total enthalpy is calculated as follows, assuming a constant heat capacity for the liquid,

$$\Delta H = Nc_P(T_H - T_C) + N\,\Delta h_{vap}$$
$$= N(75 \text{ J/mol K} \times 80 \text{ K} + 40{,}600 \text{ J/mol}) = N(46{,}600 \text{ J/mol})$$

For a typical glass of water, $N = 13$ mol. Thus,

$$t = \frac{\Delta H}{\dot{Q}} = \frac{(13 \text{ mol})(46{,}600 \text{ J/mol})}{500 \text{ J/s}} = 1{,}060 \text{ s} = 12 \text{ min}$$

10.4 Types of phase equilibrium

The considerations above show that a phase change often involves a discontinuity in enthalpy and volume. There are, however, special phase transitions where the changes in enthalpy and volume are exactly zero. These are often called *second-order phase transitions*, an old nomenclature originated by Paul Ehrenfest. In this classification scheme, phase transitions are characterized by the free energy derivatives that change discontinuously. Consider the chemical potential in a single-component system, for which $\mu = g$. *First-order phase transitions* occur when discontinuities arise in its first derivatives,

$$\partial\mu/\partial T = -s \quad \text{and} \quad \partial\mu/\partial P = v \qquad (10.19)$$

These are the most common and familiar phase transitions such as melting, boiling, sublimation, and liquid–liquid separation. On the other hand, second-order phase transitions have first derivatives (s and v) that remain continuous, but entail discontinuities in the second derivatives,

$$\frac{\partial^2 \mu}{\partial T^2} = -\frac{\partial s}{\partial T} = -\frac{c_P}{T}$$
$$\frac{\partial^2 \mu}{\partial P^2} = \frac{\partial v}{\partial P} = -v\kappa_T \qquad (10.20)$$
$$\frac{\partial^2 \mu}{\partial T \partial P} = \frac{\partial v}{\partial T} = v\alpha_P$$

In other words, second-order phase transitions involve discontinuities in the response functions c_P, κ_T, and α_P.

The Ehrenfest scheme is generally useful for describing a large range of different kinds of phase transitions. However, the modern notion of the order of a phase transition has a more subtle definition. Generally, a first-order transition is one that involves a nonzero latent heat Δh. Second-order transitions, on the other hand, are characterized by the behavior of molecular-scale correlations and how they react as one approaches the transition. Such analyses are the province of modern research in statistical mechanics, and are thus beyond the scope of this book. However, it is useful to recognize that such transitions indeed exist, occurring in magnetic materials, superfluids, and liquid crystals, for example. In addition, the liquid–gas transition at the critical point becomes a second-order transition because differences between the gas and liquid vanish and it becomes possible to continuously change between them, with no discontinuities.

10.5   Microscopic view of phase equilibrium

So far we have said nothing about why phase transitions exist. After all, a system of molecules does not "know" that it needs to change abruptly to a gas at a well-defined boiling temperature. The system instead knows only the fundamental atomic interactions governing the time evolution of its constituent molecules. Therefore, the observation that a system can spontaneously phase-separate into two distinct regions is a *consequence* of the nature of molecular interactions, not a separate feature of the molecular world. Phase transitions, it turns out, emerge in the limit of very large systems of molecules.

To illustrate this point, it helps to examine a model of molecular interactions. One of the simplest we can study is the so-called *lattice-gas model*, a very basic approach to the liquid–gas transition. The model postulates that the volume available to molecules is subdivided into small cells that form, for instance, a cubic lattice. Each cell is the size of a single molecule and can either contain one or not. When two molecules sit adjacent to each other in neighboring cells, they experience an attractive potential energy of $-\epsilon$. The approach is illustrated in two dimensions in Fig. 10.7.

Though it does not describe the atomic world in quantitative detail, the lattice gas captures the two most important features of all liquids and gases. First, molecules experience an attractive interaction upon close approach, typically due to van der Waals forces that are present in every molecular system. Second, molecules have some space in which to move around, depending on the density of the system, here in terms of the number of molecules relative to the number of lattice sites. Although this model seems very crude, these features capture the main physical driving forces behind the liquid–gas transition. In fact, near the critical point very many substances behave exactly like the lattice gas. This surprising result is called *universality*, and it was discovered and developed in the 1960s and 1970s.

The lattice gas is useful for describing other phase transitions. In particular, if instead of particles in the lattice one considers magnetic spins and makes a few variable substitutions, the model reproduces the ferromagnetic behavior of materials like iron. In fact, the model was originally introduced for this purpose in 1925 by Ernst Ising in his Ph.D. thesis, and hence is often called the *Ising model*. This is one of the most famous systems in statistical mechanics.

$$N = 17$$
$$V = 48$$
$$E = -6\epsilon$$

Figure 10.7. A schematic diagram of the lattice-gas model in two dimensions. Molecules cannot overlap and those that are nearest neighbors experience an attractive interaction of magnitude $-\epsilon$. A central molecule has four (six) nearest neighbors in two (three) dimensions.

To understand the origins of liquid–vapor phase behavior, we must develop the thermodynamic properties of this model by determining a potential that is a function of its natural variables, like $S(E, V, N)$, $E(S, V, N)$, or $G(T, P, N)$. For one and two dimensions, the lattice gas is exactly solvable. The one-dimensional case is fairly easy to derive and can be found in standard statistical-mechanical texts. The two-dimensional case is far more complicated. It was not until 1944 that Lars Onsager determined the mathematical solution to the two-dimensional Ising model that later won him the Nobel Prize. For three dimensions, it is quite possible that there are no analytical solutions.

Rather than seek an exact solution, here we find an approximate one that has the same basic features as the true one. We compute the Helmholtz free energy using the equation

$$A = E - TS \tag{10.21}$$

First, we determine the average energy expected for N lattice-gas molecules randomly distributed in a volume V. The random part is an approximation, since attractive interactions certainly could coax molecules into very non-random, clustered configurations. On the lattice, each molecule has z neighbors, with $z = 6$ in three dimensions. If we focus on one molecule, we can approximate the number of neighbors that it has, on average, by assuming that the local density is the same as the bulk density. In other words, we assume that the average number of neighbors is $z \times (N/V)$. The total energy contributed by the central molecule is then

$$E_{molecule} \approx -\frac{1}{2}\left(z\frac{N}{V}\right)\epsilon \tag{10.22}$$

The half stems from the fact that each pair interaction must be shared between two molecules, with $-\epsilon/2$ given to the central one and $-\epsilon/2$ to the neighbor. This particular simplification is called the *mean-field approximation* because we have assumed an average density and energy field surrounding each molecule, rather than performed a detailed enumeration of specific configurations. The mean-field approach is an extremely common way of evaluating statistical-mechanical models because it obviates the need to consider configurations explicitly. Finally, the total system energy is N times that of (10.22),

$$E = NE_{molecule} = -\frac{\epsilon z}{2}\left(\frac{N^2}{V}\right) = -\frac{\epsilon z}{2}\rho^2 V \tag{10.23}$$

where the number density is $\rho = N/V$. The next task is to compute the entropy term of (10.21) by counting configurations. We have no need to consider energies because the mean-field approximation assumes that all configurations have the same energy, that of (10.23). For V lattice sites and N molecules, the number of ways to arrange them is

$$\Omega = \frac{V!}{N!(V - N)!} \tag{10.24}$$

By taking the logarithm and applying Stirling's approximation, we obtain

$$S = k_B \ln \Omega = -k_B V[\rho \ln \rho + (1 - \rho)\ln(1-\rho)] \tag{10.25}$$

Putting Eqns. (10.23) and (10.25) together allows us to determine the free energy per particle,

$$\frac{A}{N} = -\frac{\epsilon z}{2}\rho + k_{\mathrm{B}}T\rho^{-1}[\rho \ln \rho + (1 - \rho)\ln(1 - \rho)] \tag{10.26}$$

To understand the behavior of this model as a function of temperature and pressure, we construct the per-particle Gibbs free energy, $g(T, P) = G(T, P, N)/N$. This is given by the Legendre transform,

$$g(T,P) = \frac{A}{N} + \frac{PV}{N}$$

$$= -\frac{\epsilon z}{2}\rho + k_{\mathrm{B}}T\rho^{-1}[\rho \ln \rho + (1 - \rho)\ln(1 - \rho)] + P\rho^{-1} \tag{10.27}$$

Recall that the Gibbs free energy is at a minimum at equilibrium for constant T and P conditions. Therefore, the equilibrium density for a particular T and P is given by the value of ρ that minimizes (10.27). This is equivalent to the statement that we find a value of ρ that minimizes $g(T, P)$ according to the recipe for a Legendre transform.

The density-dependence of Eqn. (10.27) at constant pressure and for three dimensions ($z = 6$) is illustrated in Fig. 10.7. It is clear that there are two local minima in the free energy with respect to density, but their behavior is highly temperature-dependent. At low T the minimum at higher densities is lowest, implying that the liquid is the dominant, minimum free energy at equilibrium. In contrast, at high T, a low-density minimum is the global extremum; this is the gas. At a particular intermediate temperature, the depths of the two minima are equal and the system can experience both states at equilibrium. This special state is the boiling temperature, T_{b}.

These considerations explain why a liquid changes discontinuously into a gas when heated at constant pressure: the gas state suddenly becomes the global free energy minimum. Why are there two distinct minima in the free energy curve? Qualitatively it is because distinct physical interactions dominate the behavior of each state. Liquids experience favorable energetic interactions because molecules are in close contact, and van der Waals forces contribute attractive (negative) energies. Gases have large entropies since they afford molecules a large volume in which to roam. As the liquid is heated, the drive to gain entropy overwhelms the favorable energetic interactions. Consider that the Gibbs free energy, $G = H - TS$, shows that as T increases the entropic term eventually becomes more important than the enthalpic one.

Of course, the free energy curves and hence boiling temperature in Fig. 10.8 will change with pressure. In general, if we were to examine these curves across the entire P-T phase diagram, we would see behavior reminiscent of Fig. 10.9. Notice that the locus of points for which the two minima have equal depth forms a line in the P-T diagram. This boundary could be called the boiling line, the saturation curve, or the liquid–vapor equilibrium line. As one moves from low to high temperatures along this boundary, the barrier between the two minima decreases. Eventually, at the *critical point*, the barrier disappears altogether. Above P_{c} and T_{c}, there is only one minimum in the free energy,

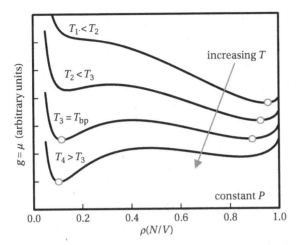

Figure 10.8. The behavior of the per-particle Gibbs free energy of the lattice-gas model, for different temperatures at constant pressure. Two minima appear with respect to density: at low temperatures the global minimum corresponds to liquid-like densities, while at high temperatures it occurs at gas-like ones. Each curve is shifted vertically by an arbitrary amount for clarity.

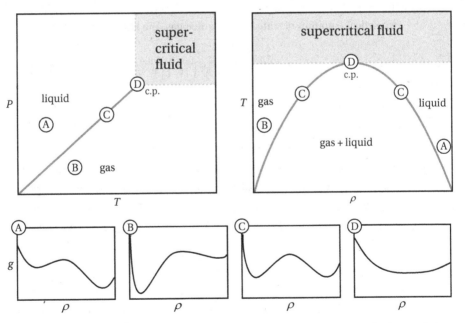

Figure 10.9. The free energy curves for several representative state points in the P–T and T–ρ phase diagrams of the top two panels are illustrated in the bottom ones. At coexistence, the liquid and gas minima have equal depth, but the barrier between them shrinks and eventually disappears as one approaches the critical point (c.p.). Above this point, the system is a supercritical fluid with no distinction between liquid and gas.

corresponding to a supercritical fluid, and there is no longer a distinction between liquid and gas.

Why is there a free energy barrier that separates the two minima in the free-energy function? Physically, the barrier is related to the penalty paid for the existence of an interface that separates the two phases present. It is therefore an unfavorable increase in free energy due to a *surface tension* relevant to intermediate densities that lie between the liquid and gas. At coexistence, the two phases attempt to minimize the surface area to minimize the free energy.

For a single-component liquid–gas transition, there are two minima in the Gibbs free energy as a function of density. The stable phase corresponds to the lowest-lying minimum at a given T and P. The boiling or saturation line corresponds to the locus of (T, P) points along which the two minima are of equal depth, separated by a free energy barrier. The critical point is the temperature and pressure at which the barrier vanishes, above which there is only one free energy minimum for the supercritical fluid.

It is relatively easy to compute the coexistence densities from the model. One approach is to simply tune T (or P) at fixed P (or T) until the minima predicted by Eqn. (10.27) are of equal depth. Frequently, however, we do not have direct access to this function, but rather must rely on some simpler experimentally-tuned model such as an analytical equation of state. In the lattice gas, the latter is given by

$$P = -\left(\frac{\partial A}{\partial V}\right)_{T,N} = -N\left(\frac{\partial \rho}{\partial V}\right)\left(\frac{\partial (A/N)}{\partial \rho}\right) = \rho^2\left(\frac{\partial (A/N)}{\partial \rho}\right)$$

$$= -\frac{\epsilon z}{2}\rho^2 - k_B T \ln(1 - \rho) \tag{10.28}$$

One can determine whether or not an equation of state predicts coexistence by examining its dependence on molar volume, $v = 1/\rho$, at constant temperature. Figure 10.10 illustrates this in the lattice gas. We see that the pressure displays a kind of non-monotonic behavior that is an artifact of the mean-field approximation. In reality, the pressure should display a flat line at the densities in the coexistence region, i.e., in between the liquid and gas volumes, v^L and v^G. Such behavior would correspond to the discontinuous change in molar volume as the pressure crossed the melting line at a given temperature. This would have been the case if we had performed an exact evaluation of the model in the large-system limit rather than employed the mean-field approximation.

The non-monotonic behavior of the kind in Fig. 10.10 is termed a *van der Waals loop*, and it is frequently encountered in approximate or empirical equations of state. To find the coexistence pressure and densities, one uses the *equal-area construction* as illustrated. We shall now derive this rule. The chemical potentials of the liquid and gas must be equal at coexistence, $\mu^G = \mu^L$; therefore, the chemical-potential difference as

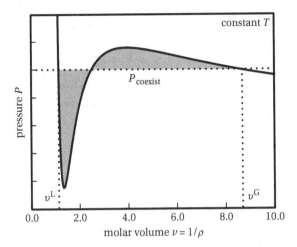

Figure 10.10. The equation of state $P(T, v)$ shows a van der Waals loop with molar volume at fixed temperature, indicating the presence of liquid–vapor equilibrium. The equal-area construction is used to determine the coexistence pressure and molar volumes (and hence densities) at this temperature.

we integrate across molar volume at constant temperature from gas to liquid must be zero, $\Delta\mu = 0$. Recalling that $(\partial\mu/\partial P)_T = v$ for a single-component system, we can write

$$\Delta\mu = 0 = \int_{P^L}^{P^G} v \, dP \tag{10.29}$$

Using the chain rule, $d(Pv) = v \, dP + P \, dv$,

$$0 = \Delta(Pv) - \int_{v^L}^{v^G} P \, dv \tag{10.30}$$

Since the integration limits are at the same coexistence pressure, we finally have

$$P_{\text{coexist}}(v^G - v^L) = \int_{v^L}^{v^G} P \, dv \tag{10.31}$$

Equation (10.31) is simply another way of saying that the shaded areas above and below the coexistence pressure in Fig. 10.10 must be equal: the total area under the curve should equal the area of the box formed between the x-axis end points v^G and v^L, and between 0 and P_{coexist} on the y-axis. This can happen only if the shaded areas exactly balance.

The critical temperature is reached when the pressure no longer exhibits the non-monotonic behavior shown in Fig. 10.10. As the temperature is increased, the pressure minimum and maximum become less pronounced until, at T_c, they converge at the same density and transform into an inflection point. Thus, at the critical point

$$\left(\frac{\partial P}{\partial \rho}\right)_T = 0 \quad \text{and} \quad \left(\frac{\partial^2 P}{\partial \rho^2}\right)_T = 0 \tag{10.32}$$

Substituting Eqn. (10.28) into these expressions gives

$$-\epsilon z \rho + \frac{k_B T}{1-\rho} = 0 \quad \text{and} \quad -\epsilon z + \frac{k_B T}{(1-\rho)^2} = 0 \qquad (10.33)$$

These two equations can be solved simultaneously for ρ and T, giving, for the critical density and temperature,

$$\rho_c = \frac{1}{2} \quad \text{and} \quad T_c = \frac{1}{4}\frac{\epsilon z}{k_B} \qquad (10.34)$$

In turn, (10.34) and (10.28) give the critical pressure,

$$P_c = \epsilon z \left(\frac{\ln 2}{4} - \frac{1}{8}\right) \qquad (10.35)$$

In three dimensions $z = 6$ and the critical temperature is $T_c = 1.5\epsilon/k_B$. The mean-field approximation actually overestimates the true critical parameters of this model. Detailed computer simulations have instead established $T_c \approx 1.12\epsilon/k_B$.

10.6   Order parameters and general features of phase equilibrium

We examined phase equilibrium in the liquid–vapor case using a Gibbs free energy function dependent on the density. More generally, for any kind of phase equilibrium, one needs to consider the behavior of a free energy as a function of some internal degree of freedom or *order parameter* that distinguishes between the two phases. For the liquid–vapor transition, a natural order parameter is the density. However, this is not always the most useful metric to choose. For example, in solid–liquid transitions the density difference can be very small, or even zero, and thus we might need a quite different order parameter, such as a crystalline lattice metric that quantifies the structural behavior of the molecules. For multicomponent mixtures, often the compositions (e.g., mole fractions) are useful as order parameters since the mixture might phase-separate into regions enriched or depleted in various components.

Once one or more appropriate order parameters are identified, the procedure is to examine a free energy expressed as a function of them in addition to its natural variables. Then, the stable phase corresponds to the values of the order parameters that minimize the free energy. If we were to find conditions for which more than one global minimum existed, then these conditions would correspond to phase equilibrium. It is possible that more than two global minima may be present. This occurs in a single-component system at its triple point. In multicomponent systems the possibilities for multiple minima increase per Gibbs' phase rule. For example, two immiscible liquid phases can be in equilibrium with a vapor phase; in this case, we would seek three global free energy minima in the multidimensional space of both density and composition.

Problems

Conceptual and thought problems

10.1. A liquid mixture of 1-propanol and 1-pentanol is prepared under standard conditions. Your labmate suggests that you determine a mixture boiling point by heating the solution (at constant pressure) and then noting the temperature at which it completely changes from all liquid to all vapor. What is wrong with your labmate's analysis? Explain.

10.2. The change in molar enthalpy of a particular phase transition at a given T and P is zero. For each of the following, indicate whether the statement is true or false, or there is not enough information to tell.
(a) The entropy of phase transition is also zero.
(b) A differential increase in P will always convert the system completely to the phase of higher density, even if T changes differentially as well.
(c) It is impossible for this system to be a single-component system.

10.3. A particular system consists of gaseous components A, B, and C that freely undergo the reversible reaction A $\leftrightarrow$ 2B + C. The reaction equilibrium is well described by an equilibrium constant $K_C = \rho_B^2 \rho_C / \rho_A$, where ρ gives the number density.
(a) In general, how many macroscopic degrees of freedom are there for this system?
(b) If the system is prepared as pure A with no B or C, how many degrees of freedom are there? Explain why the answer is different from that for part (a).

10.4. Consider a piston apparatus maintained at constant pressure $P = 1$ atm that is initially filled with pure liquid water. Upon heating it boils entirely at $T = 100\,°C$, after which the piston will have expanded significantly to accommodate the (pure) water vapor. On the other hand, consider liquid water heated on a stove that is exposed to the atmosphere. Upon heating it also boils at $100\,°C$, even though it is in equilibrium with a vapor phase containing inert species (mainly N_2 and O_2) during the entire process. How can the two cases be similar? You may want to assume that the inert species have negligible solubility in the liquid and that the vapor phase behaves ideally.

10.5. Critical points can be found in liquid–vapor and liquid–liquid phase equilibria, but are almost never present in equilibria involving solid crystalline phases. Explain why. Hint: at a critical point, the properties of the phases involved become identical.

10.6. What thermodynamic potential will exhibit two minima of equal depth for a system of components A and B that displays liquid–liquid phase separation? Give the potential in integrated and differential form.

Fundamentals problems

10.7. At atmospheric pressure, liquid water expands upon freezing. Explain what this means for the ice–water phase boundary in the P-T plane.

10.8. Consider heating a system across a first-order phase transition at constant pressure. Prove that the entropy of the lower-temperature phase must be less than that of the higher-temperature phase.

10.9. Consider equilibrium between two phases in a single-component system. Show that the change in chemical potential with pressure along the phase boundary is given by

$$\left(\frac{d\mu}{dP}\right)_{\text{phase boundary}} = \frac{\Delta(1/\hat{s})}{\Delta(1/s)}$$

where $s \equiv S/N$ and $\hat{s} \equiv S/V$.

10.10. Consider equilibrium between two phases in a single-component system. Show that the change in chemical potential with temperature along the phase boundary is given by

$$\left(\frac{d\mu}{dT}\right)_{\text{phase boundary}} = -\frac{\Delta(\rho s)}{\Delta\rho}$$

10.11. Originally, second-order phase transitions were defined as those for which the volumes and entropies did not change as one moved from one phase to another, i.e., $\Delta v = 0$ and $\Delta s = 0$. Show that movements along the phase boundary for such transitions must obey the following so-called Ehrenfest relationships:

$$\left(\frac{dP}{dT}\right)_{\text{phase boundary}} = \frac{\Delta\alpha_P}{\Delta\kappa_T}$$

$$\left(\frac{dP}{dT}\right)_{\text{phase boundary}} = \frac{\Delta c_P}{Tv\,\Delta\alpha_P}$$

10.12. By Gibbs' phase rule the latent heat of a first-order phase change depends on one thermodynamic variable, say T. For a single-component system, show that its state-dependence is related to measurable properties by

$$\left(\frac{d\Delta h}{dT}\right) = \Delta c_P + \Delta h\left[\frac{1}{T} - \left(\frac{\partial \ln \Delta v}{\partial T}\right)_P\right]$$

where $\Delta X \equiv X^\alpha - X^\beta$ denotes a difference in properties between the two phases α and β. Find also an approximate expression for this derivative in the case of solid–vapor equilibrium.

10.13. A Mollier diagram gives fluid properties as a function of the molar entropy (x-axis) and molar enthalpy (y-axis). In particular, it is highly useful in process calculations involving steam. Various iso-lines in this kind of diagram trace out paths of constant pressure, temperature, and quality (the fraction of vapor in the two-phase region).

(a) Find an expression for the slope of an isotherm in a Mollier diagram in terms of measurable quantities, in the single-phase region. Then, find an expression for the slope of an isobar.

(b) What is the functional relationship $H = f(S)$ for an ideal gas along an isobar, assuming a temperature-independent heat capacity? Assume the absolute entropy S_0 and enthalpy H_0 are known at some reference temperature T_0. What is the relationship for an isotherm?

(c) Where might an isotherm be vertical and have an infinite slope? Name an example.

10.14. A semi-permeable, immovable membrane separates two compartments of equal volumes V. Initially N_A molecules of ideal gas A are in the left-hand compartment and N_B molecules of ideal gas B are in the right-hand compartment. The membrane is impermeable to B but permeable to A. At equilibrium, what is the fraction of A molecules on the LHS? What is the pressure drop across the membrane (i.e., that the membrane must resist)? The temperature is constant at T.

10.15. If interfacial properties contribute to the bulk thermodynamic behavior of a system, the fundamental equation must be modified to include their effects. Generally, one can write for a single-component single-phase system

$$dG = -S\, dT + V\, dP + \mu\, dN + \gamma\, d\mathcal{A}$$

where $\mathcal{A}$ is the surface area of the system and γ is the surface tension. Note that γ is defined by the change in Gibbs free energy with area at constant T, P, and N. Generally, the surface tension is positive, which drives systems towards the minimization of interfacial areas by way of minimizing the Gibbs free energy.

(a) Show that the entropy of the system is independent of its area if the surface tension is constant with temperature.

(b) If the area is extensive (e.g., scales linearly with N), show that $(\partial \mu / \partial \gamma)_{T,P} = -\mathcal{A}/N$.

10.16. Consider the system described in the previous problem to be a single-component spherical droplet of radius R in a vacuum, neglecting the possibility of evaporation.

(a) Write an expression for the differential Helmholtz free energy of the system. Use this to find an expression relating the equilibrium value of R to the liquid pressure P and surface tension γ, at constant temperature and mole number.

(b) Instead of existing in a vacuum, the droplet sits inside a vapor phase of the same component that is metastable with respect to the liquid (i.e., it is in the process of condensing into the droplet). The droplet and vapor are in thermal and mechanical equilibrium, and are held at constant temperature and pressure. For this system, the Gibbs free energy at constant T and P is given by

$$dG = \mu^L\, dN^L + \mu^G\, dN^G + \gamma_{LG}\, d\mathcal{A}$$

where the superscripts L and G denote the liquid droplet and gas phases, respectively. As liquid condenses into the droplet, show that the free energy change with droplet radius is given by

$$\frac{dG}{dR} = 4\pi \left(\frac{\Delta\mu \, R^2}{v} + 2\gamma_{LG}R \right)$$

where $\Delta\mu = \mu^L - \mu^G$ and v is the liquid molar volume (which is approximately constant).

(c) What are the signs of each of the two terms in the expression for dG/dR? Which term dominates at small R and which at large R? Sketch $G(R)$, the free energy as a function of the droplet radius, setting $G(R = 0) = 0$. What is the size of the droplet at the free energy barrier? This approach might be used to find a critical nucleus size and to determine the kinetics of nucleation.

Applied problems

10.17. The saturation pressure for vapor–liquid equilibria for a given species is fitted to the Antoine equation, $\ln (P^{vap}/P_{ref}) = c_1 - c_2/(T + c_3)$, where c_1, c_2, and c_3 are material-specific constants. Assume the vapor obeys the equation of state $P = Nk_BT/(V - Nb)$, where b is another constant that modifies the ideal gas expression to account for molecular volume.

(a) Write an expression for the enthalpy of vaporization as a function of temperature.

(b) For water, $P_{ref} = 1$ kPa, $c_1 = 16.3872$, $c_2 = 3,885.70$ K, $c_3 = -42.980$K, and $b \approx 3.05 \times 10^{-5}$ m^3/mol. How does your estimate of $\Delta h_{vap}(T = 100 \,^\circ\text{C})$ compare with the experimental value of 40.7 kJ/mol? What is the most likely source of error?

10.18. At high pressures, water can form a number of ice phases differing in crystallographic structure. Using Fig. 10.1, rank order liquid water and its ices from lowest to highest density to the best of your ability.

10.19. A Kauzmann point is defined as a location in (T, P) space where the entropy difference between two phases equals zero. Figure 10.11 gives a schematic phase diagram of the isotope ^{3}He.

(a) Indicate any Kauzmann points on this diagram.

(b) In general for any substance, consider the locus of all points in the (T, P) plane for which $\Delta S = 0$ between two phases. This will form a line that extends

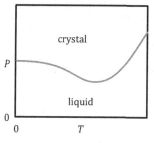

Figure 10.11. A schematic illustration of the solid–liquid phase diagram of the isotope ^{3}He [adapted from Stillinger, Debenedetti, and Truskett, *J. Phys. Chem.* B **109**, 11809 (2001)].

into the metastable regions of the two phases. What is the slope of this line, $(dP/dT)_{\Delta S=0}$, in measurable quantities?

10.20. A sealed, rigid container initially holds equal numbers of moles of pure liquid X and its vapor at temperature T_1 and pressure P_1. The container is then heated at constant volume to a temperature T_2, at which point it still contains liquid and vapor X. Assume that throughout this process the vapor behaves ideally and that the liquid has a constant molar volume v^L. Moreover, assume that the heat capacities of the liquid and vapor, c_P^L and c_P^G, as well as the vaporization enthalpy, Δh_{vap}, are independent of temperature and pressure.

(a) Find an expression for the final pressure P_2.

(b) Find an expression for the fraction f_2^L (on a molar basis) that is liquid at the end of the process. Your answer can include P_2 from above.

(c) Find an expression for the heat required relative to the total number of moles of substance X. Your answer can include P_2 and f_2^L from above.

10.21. Proteins are biopolymers that can generally exist in two different states: an unfolded state consisting of many unstructured conformations (e.g., the protein "flops" around a lot) and a folded state that is a highly ordered and stable three-dimensional structure. Typically, as one increases the temperature, a protein transitions from folded to unfolded. The quantity that determines which state will be dominant at any one temperature is the Gibbs free energy. In this problem, assume that the heat capacities of each state, $C_{P,folded}$ and $C_{P,unfolded}$, are constant.

(a) Show that the free energy of folding, $\Delta G_{fold} = G_{folded} - G_{unfolded}$, can be written as

$$\Delta G_{fold} = \Delta H_{fold, T_f}\left(1 - \frac{T}{T_f}\right) + \Delta C_P\left[T - T_f - T\ln\left(\frac{T}{T_f}\right)\right]$$

where T_f is the folding temperature (i.e., the temperature at which the protein switches from folded to unfolded), $\Delta H_{fold, T_f} = (H_{folded} - H_{unfolded})_{T_f}$ is the enthalpy of folding at the folding temperature, and $\Delta C_P = C_{P,folded} - C_{P,unfolded}$.

(b) The numbers of folded and unfolded protein molecules in a given solution are proportional to $\exp(-G_{folded}/k_B T)$ and $\exp(-G_{unfolded}/k_B T)$, respectively. Experiments can measure the fraction of protein molecules that are folded, f, as a function of temperature T. If the heat capacities of the unfolded and folded states are roughly equivalent, how would you estimate $\Delta H_{fold, T_f}$ from such data?

10.22. An insulated piston-type device contains pure 1-propanol that is compressed in a slow manner from 1 atm to 2 atm. Some useful properties of 1-propanol are listed below. You may want to assume ideal behaviors in the gas and liquid to facilitate your solutions.

$$T_b(1\text{ atm}) = 98\,^\circ\text{C}$$
$$\Delta h_{vap} = 47.5\text{ kJ/mol}$$
$$c_P^L = 144.4\text{ J/mol}\,^\circ\text{C}$$
$$c_P^G = 85.6\text{ J/mol}\,^\circ\text{C}$$
$$\rho^L = 13.8\text{ mol/l}$$

(a) If the initial state of the system consists of pure saturated vapor, prove that the final state is pure gas.

(b) Estimate the work per mole required to accomplish the compression in part (a).

(c) If the initial state instead contains pure saturated liquid, what is the final state of the system? Prove this. How does the magnitude of the work required in this case compare with that in the one above?

FURTHER READING

H. Callen, *Thermodynamics and an Introduction to Thermostatistics*, 3rd edn. New York: Wiley (1985).

K. Denbigh, *The Principles of Chemical Equilibrium*, 4th edn. New York: Cambridge University Press (1981).

K. Dill and S. Bromberg, *Molecular Driving Forces: Statistical Thermodynamics in Biology, Chemistry, Physics, and Nanoscience*, 2nd edn. New York: Garland Science (2010).

T. L. Hill, *An Introduction to Statistical Thermodynamics*. Reading, MA: Addison-Wesley (1960); New York: Dover (1986).

D. A. McQuarrie, *Statistical Mechanics*. Sausalito, CA: University Science Books (2000).

J. M. Smith, H. V. Ness, and M. Abbott, *Introduction to Chemical Engineering Thermodynamics*, 7th edn. New York: McGraw-Hill (2005).

J. W. Tester and M. Modell, *Thermodynamics and Its Applications*, 3rd edn. Upper Saddle River, NJ: Prentice Hall (1997).

11 Stability

11.1 Metastability

For single-component systems, we have seen that the stable, equilibrium phase at a given T and P is the one that minimizes the Gibbs free energy. Consider the liquid–gas transition, and the dependence of the Gibbs free energy on density. Above the boiling point, this function may resemble Fig. 11.1. The gas is the stable phase at these conditions since it corresponds to the global free energy minimum. Thus, if we were to wait long enough for true thermodynamic equilibrium to occur, the system at these conditions would eventually become a gas, no matter what its initial state is.

However, at shorter times it is possible that other states might be stable. We could prepare a system at the higher-density minimum ρ^L of Fig. 11.1 by heating a liquid from a lower temperature to one above its boiling point. At this point, the liquid would be *superheated* and could remain in that state for times long enough for us to make meaningful investigations of it. When it finally vaporizes, however, the liquid transforms irreversibly to the stable gas phase, and there is a net decrease in Gibbs free energy of the system and a corresponding increase in entropy in the world.

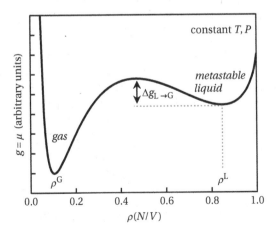

Figure 11.1. The dependence of the Gibbs free energy on density for the lattice-gas model above the boiling temperature for a pressure below the critical point. A liquid can exist in a metastable state even though it is not the global free energy minimum because a barrier separates it from the gas.

How does a superheated liquid exist for finite periods of time? Figure 11.1 shows that there is a barrier in free energy between the liquid and gas states. If the system is at ρ^L, it must temporarily increase its free energy by an amount $\Delta g_{L \to G}$ in order to travel to the gas state at ρ^G. This free energy barrier gives rise to a finite rate at which the liquid can turn to gas. If the barrier is high enough, the rate of spontaneous gas formation will be so slow that the superheated liquid will persist for observation time scales, or longer.

The key to forming a superheated liquid is to heat it in a clean, unagitated container. Particulates or disturbances can kick the system over to the gaseous state by lowering the free energy barrier between the two regions, by way of providing nucleation sites for bubbles, and thus greatly increasing the rate of liquid-to-gas conversion. Superheating is actually quite common; it is relatively easy to superheat liquid water in microwave ovens where energy can be added to the liquid without much disturbance. Adding sugar to, or suddenly agitating, a clean container of water superheated in a microwave is dangerous and can cause *explosive boiling*. In general, superheated liquids are part of a more general class of *metastable* systems.

A system that resides at a local but not global free energy minimum is **metastable**. The existence of a free energy barrier between a metastable and the globally stable state of the system enables the former to exist for periods of time that permit investigations of it. Ultimately, however, metastable systems will evolve to global equilibrium in the rigorous limit of infinite time.

There are many different types of metastable systems. We saw that *superheated liquids* exist above the boiling point and will eventually vaporize. Here the system exists at a local minimum in free energy with respect to the gas phase, and the free energy order parameter is the density. On the other hand, *supercooled liquids* exist below the melting line and will eventually freeze. In this case the system exists at a local minimum with respect to the crystal phase, and the relevant order parameter describes the degree of crystalline lattice structure.

Metastable phases are certainly not limited to pure liquids. *Metastable solids* include both superheated crystals that will eventually melt, and crystals with metastable lattice structures that will eventually transform to a different crystalline phase (i.e., a more stable lattice structure). In these cases, the local free energy minimum depends on one or more crystalline order parameters. Furthermore, *supersaturated solutions* include systems in which the amount of dissolved solute exceeds the equilibrium solubility, and they will eventually phase-separate by precipitation. Here the system exists at a local minimum with respect to concentration.

11.2   Common tangent line perspective on phase equilibrium

We now derive general thermodynamic conditions for metastability in terms of properties of the free energy. We will continue with the example of a superheated liquid, but the basic concepts discussed here apply generally to all of the metastable systems described above.

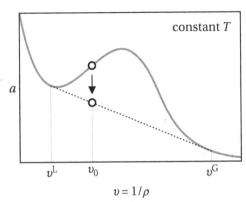

Figure 11.2. The common tangent line for the molar Helmholtz free energy determines the molar volumes of the liquid and gas phases, at a given temperature. A system at intermediate volumes can lower its free energy by jumping down to the common tangent line.

We first take a slightly different view of the free energy by examining the constant-volume rather than constant-pressure case. At constant volume, a system can phase-separate by spontaneously forming two separate regions within its boundaries, for a range of total system volumes. In contrast, at constant pressure, the total volume adjusts to the corresponding volume of the stable phase, leaving only a single phase except at the exact conditions of phase equilibrium. By switching to constant-volume conditions, therefore, we will be able to examine how far we can push a metastable system before it spontaneously phase-separates.

In particular, let us examine how much we can "stretch" or expand a superheated liquid before it spontaneously decomposes into an equilibrium gas and liquid phase. At a given temperature, the molar Helmholtz free energy looks similar to that shown in Fig. 11.2. This curve differs in a substantial way from Fig. 11.1; in this case we can specify the per-particle volume v of the system exactly since we are considering the case of constant T and V conditions, the natural variables of A. In Fig. 11.1 we specified T and P, and the volume (or density) was determined by the global free energy minimum. In other words, we can prepare a system with an exact location on the x-axis in Fig. 11.2, but the second law determines this in Fig. 11.1.

This Helmholtz free energy curve has some notable features. There are two ranges of density corresponding to the liquid and gas regimes. At intermediate volumes, the free energy increases, and, in this region, the system can phase-separate in order to lower its free energy. By doing so, it moves from the original curve to a *common tangent line* between the liquid and gas regimes. Note that the end points of the tangent line do not correspond to minima, but rather the points at which the slope of the line equals that of the free energy curve.

The interpretation of any common tangent line is the following. At any point on it, the line gives the free energy of a system that has phase-separated into two subsystems, one liquid with per-particle volume v^L and one gas with v^G. These two values come from the *end points* of the tangent line. The overall molar volume v_0 shown in Fig. 11.1 accounts for the total volume of the system, including both liquid and gas phases, such that

$$Nv_0 = N^G v^G + N^L v^L \qquad (11.1)$$

where N^G and N^L are the numbers of molecules in the gas and vapor phases, constrained by $N^G + N^L = N$. As one moves along the tangent line from liquid to gas, the amount of the gas phase (N^G) increases at the expense of the liquid (N^L) because v^L and v^G remain the same for a given temperature.

   Why does the tangent line represent the free energy of the phase-separated state? First, we show that any line connecting two points on the curve gives the free energy that would be obtained if the system phase separated into the two molar volumes at the line end points. Consider an intermediate volume v_0 corresponding to total volume $V = Nv_0$. If the system phase-separates, the volumes of the phases must sum to the total volume as in Eqn. (11.1). On dividing that expression by N and defining $f \equiv N^G/N = 1 - N^L/N$, this volume balance can be inverted to determine the mole fraction in the gas phase,

$$f = \frac{v_0 - v^L}{v^G - v^L} \tag{11.2}$$

We can also write the total free energy of the phase-separated system, $A_{PS} = Na_{PS}$, as the sum over the two phases,

$$Na_{PS} = N^G a(v^G) + N^L a(v^L) \tag{11.3}$$

where $a(v^G)$ and $a(v^L)$ are the free energies evaluated at the gas and liquid molar volumes. Dividing by N gives

$$a_{PS} = f[a(v^G) - a(v^L)] + a(v^L) \tag{11.4}$$

By substituting (11.2) into (11.4), we arrive at

$$a_{PS} = \left[\frac{a(v^G) - a(v^L)}{v^G - v^L}\right](v_0 - v^L) + a(v^L) \tag{11.5}$$

Equation (11.5) gives a straight line where the x coordinate is v_0 and the y coordinate is a_{PS}. The line connects the points $(v^G, a(v^G))$ and $(v^L, a(v^L))$ in the representation of Fig. 11.2, and it shows the free energy change for a phase-separated system as one moves along values of v_0. Higher values of v_0 near v^G indicate that there is more gas phase present according to (11.2). Similarly, lower values of v_0 near v^L imply more liquid. At any value of v_0 along the line, there are only two molar volumes present in the system, v^G and v^L. Only the relative amounts of liquid and gas change with v_0.

Any straight line connecting two points in the Helmholtz free energy diagram gives the free energy for the case in which the system is phase-separated into two subsystems at the line end points. The relative amounts of each phase for a specified overall system molar volume v_0 relate to the proximity of v_0 to each of the end points.

   We now consider how one picks the values v^G and v^L that are used in the determination of the equilibrium line. These are the molar volumes of the gas and liquid phases at the equilibrium state. Since phase separation occurs so as to minimize the free energy, the lowest line we can draw on this graph that still connects two points along the original free energy curve is the line that is tangent to the two minima. This can also be shown mathematically by minimizing a_{PS} as a function of v^G and v^L.

The tangent condition requires that the slope of the free energy be equal at v^G and v^L,

$$\frac{\partial a(v^G)}{\partial v} = \frac{\partial a(v^L)}{\partial v} \tag{11.6}$$

Using the fundamental relation $(\partial a/\partial v)_T = -P$, this condition readily shows that the pressures of the two phases are equal at equilibrium,

$$P^G = P^L \tag{11.7}$$

Notice that the temperatures of the phases are already identical because the Helmholtz free energy is considered at a given T. The final phase-equilibrium equality, that of chemical potentials, can also be extracted from the common tangent line. The slope at either tangent point must also be equal to the slope of the line connecting them,

$$\frac{\partial a(v^G)}{\partial v} = \frac{\partial a(v^L)}{\partial v} = -P = \frac{a(v^G) - a(v^L)}{v^G - v^L} \tag{11.8}$$

Simplifying the rightmost equality gives

$$-P(v^G - v^L) = a(v^G) - a(v^L) \tag{11.9}$$

Using the integrated relation $\mu = g = a + Pv$, we find

$$\mu^G = \mu^L \tag{11.10}$$

as we would expect from the equilibrium condition.

A **common tangent line** separating two minima in the Helmholtz free energy, with respect to molar volume v or density ρ, gives the free energy of a state in which the system has phase-separated into two regions. The properties of those regions are given by the intersection of the tangent line with the free energy curve. The tangency of this line ensures equality of pressures and chemical potentials between the two phases.

11.3   Limits of metastability

Let us now consider several distinct overall system volumes v_0 in the Helmholtz free energy curve, as illustrated in Fig. 11.3. At point A, the liquid is the stable state; there is no way to draw a line between any two points per the phase-separated expression in (11.5) that has the same global density but results in a new state lower in free energy than A. Therefore, only a single phase exists at A.

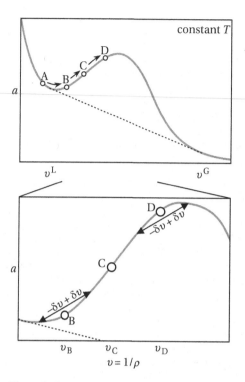

Figure 11.3. As a liquid is stretched, it explores qualitatively different regimes of the free energy curve that can be characterized by the initial behavior upon phase separation. Specifically, one considers a homogeneous system that just begins to form a higher-density phase of $v - \delta v$ (eventually the liquid) and a lower-density one of $v + \delta v$ (eventually the gas). At point B, the free energy must initially increase, whereas at point D it always decreases with respect to any perturbation in density. The transition occurs at the inflection point C, called the spinodal.

Consider expanding the liquid by increasing its volume such that the system moves from point A to B. At point B, the system could attain a lower-free-energy state by splitting into two phases with densities v^{L} and v^{G}, which would drop the free energy to the dotted common tangent line. If one is careful enough, however, one can maintain the system at the metastable state B without phase separation. To see this, we need to carefully consider the immediate region of the free energy, as shown in the bottom panel of Fig. 11.3. Imagine that the system is initially a homogeneous liquid at B and begins to phase-separate. During the early steps of this process, some regions in the volume begin to decrease in density, moving towards the gas phase, while others condense towards a denser liquid. If we evaluate the free energy of this initial part of the phase-separation process by drawing a local phase-separation line as per Eqn. (11.5), it lies *above* the value for the homogeneous liquid at point B. Only as the liquid and gas densities evolve to their final states, which are given by the *common tangent* line, is the system able to reach a state of net lower free energy. That is, the system must pass through a free energy barrier to phase-separate. This makes point

B *stable* with respect to small perturbations in density, even though it is metastable in a global sense.

Not all states can be metastable, however. When a system at point D in Fig. 11.3 begins to phase-separate, initial small perturbations to molar volume in the two phases will always lower the free energy, as can be seen by noting a local phase-separation line. Therefore, systems at D will always *spontaneously* phase-separate because there is no free energy barrier to overcome in order to form a gas phase. That is, D is *unstable*.

> **Unstable** states correspond to conditions at which infinitesimally small perturbations result in a decrease in free energy. Thus, for these states, no barrier prevents the evolution towards equilibrium.

Considering the curvature in the graph above, it is clear that the condition for stability is

$$\left(\frac{\partial^2 a}{\partial v^2}\right)_T > 0 \tag{11.11}$$

But, since $P = -\partial a/\partial v$,

$$\left(\frac{\partial P}{\partial v}\right)_T < 0 \tag{11.12}$$

for stable phases. This is a general *stability condition*. Systems that do not obey this relation are not at equilibrium and are not stable; they will spontaneously decompose into one or more stable phases. From the free energy curve in Fig. 11.3, we see that the stability condition is first violated at point C, where

$$\left(\frac{\partial^2 a}{\partial v^2}\right)_T = -\left(\frac{\partial P}{\partial v}\right)_T = 0 \tag{11.13}$$

This limit of stability lies within the metastable region of the liquid and is called the *spinodal*. A spinodal also exists for the gas. That is, we can find a distinct spinodal density by moving from the gas side to lower molar volumes until the second derivative of the free, energy reaches zero. This represents the highest density that the metastable gas can sustain before a liquid phase spontaneously forms.

Our analysis so far has considered a single temperature. If we were to examine multiple temperatures, we would find behavior reminiscent of that of Fig. 11.4. At each temperature, we find four relevant molar volumes: two that correspond to the stable gas and liquid at two-phase equilibrium and two that occur at spinodals. All four points merge at the critical temperature when the free energy loses its regime of negative curvature.

The molar volumes extracted from curves like those in Fig. 11.4 can be used to construct a detailed phase diagram of the fluid in T-ρ space, as shown in Fig. 11.5. Here, the *binodal* is the liquid–gas equilibrium line. If we were to wait forever, until global equilibrium, any point inside the binodal would eventually lead to a phase-separated

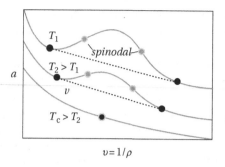

$v = 1/\rho$

Figure 11.4. The Helmholtz free energy varies with temperature. The presence of a region of negative curvature indicates the possibility of phase equilibrium, with gas and liquid phases indicated by the black dots (binodals) and their respective limits of stability (spinodals) by gray circles. At the critical temperature, the curvature vanishes, and all four binodal and spinodal points merge.

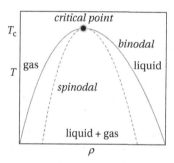

Figure 11.5. The T-ρ phase diagram gives a convenient depiction of the binodal and spinodal boundaries for vapor–liquid equilibrium.

system. However, if we are interested in the conditions that permit metastable states such as a superheated liquid, we must consider the dotted *spinodal* lines. At a given temperature, this boundary gives the maximum density to which we can compress a metastable gas and the minimum density to which we can expand a metastable liquid, before either spontaneously phase-separates. Metastable states are possible only between the binodal and spinodal lines. Any point inside the spinodal is unstable as a single homogeneous phase and will always spontaneously decompose into two phases.

A **spinodal** is a locus of state conditions that represent the limit of metastability. Inside a spinodal envelope, a system is unstable and will spontaneously phase-separate. A **binodal** gives the locus of conditions for the stable phase-separated state.

Figures 11.4 and 11.5 reveal some interesting features of the critical point. The liquid and gas phases become indistinguishable above T_c as the free energy barrier separating them vanishes. Moreover, the critical point lies both on the spinodal line and on the binodal line, and is therefore both an equilibrium coexistence point and a limit of stability. Because both boundaries end there, the critical point can be called a *limit of a limit of stability*. In mathematical terms, the points of inflection and the curvature in the free energy curve vanish at T_c,

$$\left(\frac{\partial^2 a}{\partial v^2}\right)_T = 0 \quad \text{and} \quad \left(\frac{\partial^3 a}{\partial v^3}\right)_T = 0 \tag{11.14}$$

From the fundamental equation, these can also be written as

$$\left(\frac{\partial P}{\partial v}\right)_T = 0 \quad \text{and} \quad \left(\frac{\partial^2 P}{\partial v^2}\right)_T = 0 \tag{11.15}$$

One way to find the critical point is to search for the conditions at which both relations in (11.15) hold.

It turns out that the critical point represents a kind of singularity in the thermodynamic functions of liquid–gas systems. As a result it is *non-analytic*, meaning that it is impossible to correctly Taylor-expand thermodynamic properties around this point since their derivatives can have discontinuities. Instead, a detailed understanding of behavior near the critical point requires a molecular model. Surprisingly, the lattice gas discussed in the previous chapter quantitatively captures the properties of most substances at their vapor–liquid critical points. This simple model works well because the critical point is a location in the phase diagram where substances have a kind of universal behavior, when rescaled in temperature and density. This observation was a great achievement of research in statistical mechanics in the 1970s.

A **critical point** lies at the terminus of both a binodal and a spinodal curve, where two phases in equilibrium become indistinguishable. It represents a non-analytic point in the phase diagram where the properties of substances can be scaled (appropriately non-dimensionalized) in a manner that shows universal behavior.

11.4 Generalized stability criteria

There are in fact many criteria for the stability of homogeneous phases. Equation (11.12) addressed only the variation of free energy with volume, but similar arguments for other extensive variables can be used to show that stable phases must obey a number of stability relations. In fact a *stability criterion* exists for the second derivatives of each of the thermodynamic potentials. The general form is

$$\frac{\partial^2 F}{\partial X^2} > 0 \tag{11.16}$$

where F is any extensive thermodynamic potential that stems from one or more of the Legendre transforms of E. X is then one of its *extensive* natural independent variables. It is also important that F has more than one natural extensive variable.

In simple terms, the positive second derivative in (11.16) ensures that thermodynamic potentials are at a minimum with respect to local perturbations to the extensive variables. The connection to extensive rather than intensive variables emerges because the former can be perturbed in opposite directions within two regions of the system; for example, it is the extensive variables that can be split between two phases. It turns out that if X were replaced by an intensive variable, the inequality would be reversed. Moreover, Eqn. (11.16) does not apply to potentials with only one extensive variable because the state of the system is already fixed by the intensive parameters. For example,

$$\left(\frac{\partial^2 G}{\partial N^2}\right)_{T,P} = \left(\frac{\partial^2 [Ng(T,P)]}{\partial N^2}\right)_{T,P} = 0 \tag{11.17}$$

The application of (11.16) to the standard thermodynamic potentials leads to a number of general stability criteria. The energy fundamental equation gives

$$\left(\frac{\partial^2 E}{\partial S^2}\right)_{V,N} > 0 \quad \rightarrow \quad \left(\frac{\partial T}{\partial S}\right)_{V,N} > 0$$

$$\left(\frac{\partial^2 E}{\partial V^2}\right)_{S,N} > 0 \quad \rightarrow \quad \left(\frac{\partial P}{\partial V}\right)_{S,N} < 0 \tag{11.18}$$

$$\left(\frac{\partial^2 E}{\partial N^2}\right)_{S,V} > 0 \quad \rightarrow \quad \left(\frac{\partial \mu}{\partial N}\right)_{S,V} > 0$$

From the enthalpy, we have

$$\left(\frac{\partial^2 H}{\partial S^2}\right)_{P,N} > 0 \quad \rightarrow \quad \left(\frac{\partial T}{\partial S}\right)_{P,N} > 0$$

$$\left(\frac{\partial^2 H}{\partial N^2}\right)_{S,P} > 0 \quad \rightarrow \quad \left(\frac{\partial \mu}{\partial N}\right)_{S,P} > 0 \tag{11.19}$$

And finally, the Helmholtz free energy gives

$$\left(\frac{\partial^2 A}{\partial V^2}\right)_{T,N} > 0 \quad \rightarrow \quad \left(\frac{\partial P}{\partial V}\right)_{T,N} < 0$$

$$\left(\frac{\partial^2 A}{\partial N^2}\right)_{T,V} > 0 \quad \rightarrow \quad \left(\frac{\partial \mu}{\partial N}\right)_{T,V} > 0 \tag{11.20}$$

These examples all originate from named thermodynamic potentials. However, we can create other potentials by performing Legendre transforms in the particle numbers as well. One such example is

$$Z = E - \mu N$$

with

$$dZ = T\,dS - P\,dV - N\,d\mu \tag{11.21}$$

The potential Z suggests the stability criterion

$$\left(\frac{\partial^2 Z}{\partial V^2}\right)_{S,\mu} > 0 \quad \rightarrow \quad \left(\frac{\partial P}{\partial V}\right)_{S,\mu} < 0 \tag{11.22}$$

Though (11.22) looks similar to the criterion in (11.20) developed from the Helmholtz free energy, in this case the derivative is at constant S and μ rather than T and N.

All stability criteria can be traced back to the concavity property of the entropy. Stability relations can also be developed by demanding that a system remain at a local potential minimum with respect to simultaneous perturbations in two extensive variables. One such example is

$$\left(\frac{\partial^2 E}{\partial S^2}\right)_{V,N} \left(\frac{\partial^2 E}{\partial V^2}\right)_{S,N} - \left(\frac{\partial^2 E}{\partial V\, dS}\right)_N^2 > 0 \tag{11.23}$$

Higher-order stability relations can also be considered if the derivatives described by Eqn. (11.16) are zero.

As is now apparent, there are many possible stability criteria. If one is violated, are all of the other criteria also violated simultaneously? In other words, what is the relationship between them? A detailed analysis shows that some stability criteria are always violated first, before others. For example, a system cannot violate $(\partial T/\partial S)_{V,N} > 0$ without first violating $(\partial P/\partial V)_{T,N} < 0$. In other words, if we expand a metastable liquid at constant temperature, we will first reach a density at which $(\partial P/\partial V)_{T,N} = 0$ before $(\partial T/\partial S)_{V,N}$ becomes negative. In general, the stability criteria that are the first to be violated have only a single extensive variable among the constant terms in their derivatives.

In a single-component system, the **stability criteria** to be violated first are the set of six relations

$$\left(\frac{\partial T}{\partial S}\right)_{P,N} > 0, \quad \left(\frac{\partial P}{\partial V}\right)_{T,N} < 0, \quad \left(\frac{\partial \mu}{\partial N}\right)_{T,V} > 0$$

$$\left(\frac{\partial T}{\partial S}\right)_{V,\mu} > 0, \quad \left(\frac{\partial P}{\partial V}\right)_{S,\mu} < 0, \quad \left(\frac{\partial \mu}{\partial N}\right)_{S,P} > 0$$

One important point of note is that all of these six criteria are violated simultaneously. Thus, at the spinodal, all six derivatives reach zero. We will not prove these facts here, but they can be shown using detailed mathematical manipulations to interrelate thermodynamic derivatives. The interested reader is referred to the excellent analysis in the texts by Tester and Modell and by Debenedetti.

Example 11.1 *Show that the isothermal compressibility, κ_T, is always positive.*

The definition of κ_T is

$$\kappa_T = -\frac{1}{V}\left(\frac{\partial V}{\partial P}\right)_T$$

From the stability criterion of (11.16), we know that

$$\left(\frac{\partial^2 A}{\partial V^2}\right)_T = -\left(\frac{\partial P}{\partial V}\right)_T > 0$$

If $\partial P/\partial V$ is always negative, then $\partial V/\partial P$ must be as well. Therefore, the isothermal compressibility must always be positive.

Problems

Conceptual and thought problems

11.1. Prove that the following stability relation must be true:

$$\frac{\partial^2 F}{\partial y^2} < 0$$

where y is an *intensive*, independent variable of the potential F.

11.2. In Fig. 11.3, there is a region in the free energy diagram for which $(\partial^2 a/\partial v^2)_T < 0$. In reality, this region is purely hypothetical, cannot be measured, and requires a mean-field assumption or other approximation to be represented. Explain why.

11.3. Show that c_V and c_P are always positive as long as T is positive.

11.4. What would happen if $(\partial P/\partial V)_T$ were positive? Consider two systems in mechanical equilibrium for which this is the case. If $P_1 > P_2$ initially, what happens as the system volume changes? Do the two systems become closer to or further away from equilibrium?

11.5. Consider the gray line drawn in the Helmholtz free energy diagram for a single-component system shown in Fig. 11.6. Is each of the following statements true or false, or is there not enough information to tell? Explain.

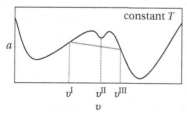

constant T

v^{I} v^{II} v^{III}

v

Figure 11.6. A schematic representation of a Helmholtz free energy diagram at constant temperature with multiple extrema.

(a) The gray line gives the free energy which would be obtained if the system phase separated into two phases of molar volumes v^{I} and v^{III}, as a function of the overall molar volume v.

(b) The pressures of the system at v^{I} and v^{III} are equal.

(c) It is impossible to maintain the system at molar volume v^{II}.

11.6. Indicate whether or not each of the diagrams in Fig. 11.7 is thermodynamically valid, and explain your reasoning.

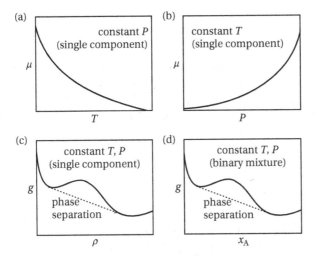

Figure 11.7. Several thermodynamic diagrams of different variables. In (d) x_A gives the mole fraction of component A in a binary A–B mixture.

11.7. Consider a binary solution of components A and B. Is each of the following quantities positive, negative, or zero, or is there not enough information to tell? Prove your results.

(a) $\left(\dfrac{\partial \mu_B}{\partial N_B}\right)_{T,V,\mu_A}$

(b) $\left(\dfrac{\partial \mu_B}{\partial P}\right)_{T,\mu_A}$

11.8. Find all stability criteria that are first violated for a ternary system of components 1, 2, and 3. In general, how many such stability criteria exist for a C-component system?

11.9. A solution of a polymer and a solvent can exhibit two critical points: a lower critical solution temperature (LCST) and an upper critical solution temperature (UCST). Above the LCST and below the UCST, the system can phase-separate into polymer-rich and solvent-rich regions. In between the two temperatures, the species are fully miscible.

(a) Sketch a phase diagram for such a system in the temperature (y-axis) and composition (x-axis) plane, at constant pressure. Include both the binodals and the spinodals.

(b) What conditions describe the behavior of thermodynamic properties at the LCST and UCST?

Fundamentals problems

11.10. The van der Waals equation of state is $P = k_B T/(v - b) - a/v^2$, where a and b are constants. This equation describes both a liquid and vapor phase, and therefore produces a vapor–liquid phase envelope.

(a) At constant temperature, there is a single equilibrium P, v^G, and v^L. Write three algebraic equations that can be solved simultaneously to give these values for a given T.

(b) Write an equation that can be solved to determine the spinodal volume, v^{spin}. Explain how the resultant equation can give both $v^{G,spin}$ and $v^{L,spin}$.

(c) Find the critical temperature T_c, volume v_c, and pressure P_c in terms of the constants a and b.

11.11. Consider the liquid and vapor spinodal lines of a pure substance in a P–T diagram.

(a) Draw a schematic diagram showing the binodal and two spinodal lines, including the critical point.

(b) Show that the slope of a spinodal is given by

$$\left(\frac{dP}{dT}\right)_{spin} = \frac{\alpha_P}{\kappa_T}$$

Hint: consider the function $P(T, v)$.

(c) Show that the slope can also be expressed as

$$\left(\frac{dP}{dT}\right)_{spin} = \frac{C_P}{TV\alpha_P}$$

11.12. Liquids can exist at negative pressures, i.e., under tension. In such cases, attractive interactions with the container walls cause the liquid to resist expansion and exert a restoring force such that $P < 0$. Such cases actually exist in nature, for example, in the ascent of sap in plants.

(a) Show that liquids under negative pressure can be (locally) stable.

(b) Prove that a liquid at negative pressure is always metastable with respect to coexistence with a vapor phase.

11.13. A thin liquid film is supported by a wire frame, such as a soap film. The work required to stretch the film at constant total liquid volume is given by $\delta W = 2\gamma \, d\mathcal{A}$, where γ is the surface tension and $\mathcal{A}$ is the area of the film. The factor of 2 stems from the fact that the film has two interfaces in contact with air, one on each side. Assume that the liquid is pure.

(a) Find an expression for the differential form of the fundamental equation for this system, $dE = ?$.

(b) Show that the surface tension relates to the excess Gibbs free energy (that beyond the bulk) per area of the film.

(c) The usual condition under which a film is stretched is at constant temperature and pressure. Show that the film volume will not change upon stretching in this case if the surface tension is pressure-independent.

(d) At constant T and P, under what conditions for γ will the film be unstable and rupture?

11.14. Consider the isothermal compressibility κ_T, the constant-volume heat capacity c_V, and the constant-pressure heat capacity c_P.

(a) What happens to the values of these quantities at a limit of stability?

(b) Prove that both κ_T and c_P approach a limit of stability at the same time.

(c) Prove that c_P always reaches a limit of stability before c_V. Hint: express the latter in terms of the former and consider the points at which each changes sign.

Hint: it may be useful in the above problems to examine the relevant inverse derivatives $(\partial T/\partial S)$ and $(\partial P/\partial V)$ rather than the response functions themselves.

Applied problems

11.15. Consider the Redlich–Kwong equation of state,

$$P = \frac{RT}{v - b} - \frac{a}{v(v + b)\sqrt{T}}$$

Here, a and b are positive constants, with $b < v$. Liquid water exhibits a temperature of maximum density of around 4 °C at ambient pressure. Is this equation of state capable of describing such behavior? Prove this rigorously.

11.16. Proteins in aqueous solution fold to unique three-dimensional structures at ambient conditions. However, they also have a first-order-like transition to a state of less-structured, floppy configurations that can be accomplished in a number of ways. The most common is by increasing T past the folding temperature at ambient pressure. They can also be unfolded by extreme compression, cooling, and expansion into the negative-pressure regime, at least in theory. Figure 11.8 gives a schematic of the "phase" behavior and these four mechanisms of unfolding.

(a) For each mechanism, indicate the signs of the volume and entropy changes associated with the folded-to-unfolded transition, namely $\Delta V = V_{\text{unfolded}} - V_{\text{folded}}$ and $\Delta S = S_{\text{unfolded}} - S_{\text{folded}}$.

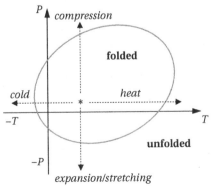

Figure 11.8. Proteins can exist in folded and unfolded states depending on the pressure and temperature. The gray line denotes the boundary between the two, called the *folding curve*. The * indicates ambient conditions.

(b) Explain why the cooling and stretching mechanisms might be difficult or impossible to realize in an experiment. Hint: what would happen if the solution contained no proteins?

11.17. At atmospheric pressure, water can be supercooled to about $-42\,^{\circ}\mathrm{C}$ before the rate of crystal nucleation becomes so fast that ice forms spontaneously. Calculate the difference in Gibbs free energy between the metastable liquid and ice phases at this temperature. At $T = 0\,^{\circ}\mathrm{C}$ their constant-pressure heat capacities are roughly 75 J/K mol and 38 J/K mol, respectively, and the enthalpy of melting is 6.0 kJ/mol.

FURTHER READING

H. Callen, *Thermodynamics and an Introduction to Thermostatistics*, 3rd edn. New York: Wiley (1985).

P. Debenedetti, *Metastable Liquids*. Princeton, NJ: Princeton University Press (1996).

K. Denbigh, *The Principles of Chemical Equilibrium*, 4th edn. New York: Cambridge University Press (1981).

K. Dill and S. Bromberg, *Molecular Driving Forces: Statistical Thermodynamics in Biology, Chemistry, Physics, and Nanoscience*, 2nd edn. New York: Garland Science (2010).

J. W. Tester and M. Modell, *Thermodynamics and Its Applications*, 3rd edn. Upper Saddle River, NJ: Prentice Hall (1997).

12 Solutions: fundamentals

12.1 Ideal solutions

Solutions are among the most important kinds of systems in the chemical and biochemical sciences and technologies, spanning an enormous range of different phenomena and scales. The simplest model of a solution is the so-called ideal one.

An **ideal solution** is defined by the following properties.

(1) Components are non-interacting. In other words, there are no explicit energetic interactions between molecules of different species.
(2) At constant temperature and pressure, an **ideal entropy of mixing** contributes to the total entropy due to the presence of different species.
(3) Ideal solutions are usually excellent approximations to real ones in the near-pure limit, in which there is one major component (e.g., a solvent) and the concentrations of all other species (solutes) are very small.

To develop a thermodynamic model of ideal solutions, we first evaluate the entropy of mixing using a simple model. Figure 12.1 illustrates a lattice mixing model for a binary system. Though the lattice certainly does not represent molecular-structural features of actual liquids, the final behavior of the mixing entropy is surprisingly insensitive to its use. Prior to mixing the particles are completely separated into different regions of the system volume. One configuration exists for this state, and thus the entropy is zero. If the particles are then allowed to mix randomly, the number of arrangements of N_A A particles and N_B B particles in the $N = N_A + N_B$ sites is given by the combinations formula

$$\Omega_{\text{mix}}(N_A, N_B) = \frac{(N_A + N_B)!}{N_A! N_B!} \approx \frac{(N_A + N_B)^{N_A + N_B}}{N_A^{N_A} N_B^{N_B}}$$

$$= \left(\frac{N_A}{N_A + N_B}\right)^{-N_A} \left(\frac{N_B}{N_A + N_B}\right)^{-N_B} \tag{12.1}$$

In the second line, we applied Stirling's approximation. By taking the logarithm to obtain the entropy, we find

N_A "A" particles N_B "B" particles

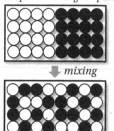

Figure 12.1. A simple lattice model can be used to determine the ideal entropy of mixing in terms of component mole fractions.

mixing

$$\Delta S_{\text{mix}}(N_A, N_B) = -N_A k_B \ln x_A - N_B k_B \ln x_B$$
$$= -N k_B (x_A \ln x_A + x_B \ln x_B) \tag{12.2}$$

where x_A and x_B are the mole fractions of A and B particles, respectively. We use x_i, rather than y_i as we did for gases, to indicate the solution phase. Equation (12.2) for the ideal entropy of binary mixing is easily generalized by analogous arguments to a system of C components,

$$\Delta S_{\text{mix}}(N_1, N_2, \ldots, N_C) = -N k_B \sum_i x_i \ln x_i \tag{12.3}$$

Note that Eqn. (12.3) gives the same mixing entropy as that which we found for ideal gases in Chapter 9. As in that case, an ideal solution augments the Gibbs free energy at constant temperature and pressure conditions relative to the pure-component free energies,

$$G(T, P, N_1, N_2, \ldots, N_c) = \sum_i G_i(T, P, N_i) - T \Delta S_{\text{mix}}(N_1, N_2, \ldots, N_c)$$
$$= \sum_i G_i(T, P, N_i) + N k_B T \sum_i x_i \ln x_i \tag{12.4}$$

The first term on the RHS of (12.4) sums the Gibbs free energies of each of the components in their *pure* states at the same temperature and pressure as the mixture; the direct sum is possible because the components do not interact. The second term is constructed so that the actual mixture entropy is an amount ΔS_{mix} greater than the contributions of the pure components. The temperature derivative of (12.4) shows this,

$$S = -\left(\frac{\partial G}{\partial T}\right)_{P, \{N\}}$$
$$= -\sum_i \left(\frac{\partial G_i}{\partial T}\right)_{P, N_i} + \Delta S_{\text{mix}}(N_1, N_2, \ldots, N_c)$$
$$= \sum_i S_i(T, P, N_i) - k_B N \sum_i x_i \ln x_i \tag{12.5}$$

To derive the chemical potentials, we take the N_i derivative of the free energy. Note that in general these depend on the entire composition of the system, $\{x\} = \{x_1, x_2, \ldots, x_c\}$,

$$\mu_i(T, P, \{x\}) = \left(\frac{\partial G}{\partial N_i}\right)_{T, P, N_{j \neq i}}$$

$$= \left(\frac{\partial G_i}{\partial N_i}\right)_{T, P, N_{j \neq i}} - T\left(\frac{\partial \Delta S_{\text{mix}}}{\partial N_i}\right)_{T, P, N_{j \neq i}} \tag{12.6}$$

After some simplification, we arrive at the most important equation describing ideal solution behavior,

$$\mu_i(T, P, \{x\}) = \mu_i^*(T, P) + k_B T \ln x_i \tag{12.7}$$

The function $\mu_i^*(T, P)$ gives the chemical potential of *pure* component i. Therefore, the chemical potential of each species in an ideal solution is decreased relative to that of the pure phase because the quantity $k_B T \ln x_i$ is negative. Moreover, the chemical potential of an ideal species depends only on its mole fraction and not on the other compositions in the system.

Equation (12.7), which will be the basis of nearly all of our subsequent ideal solution results, stems from a simple model that captures only mixing entropy effects. Some real systems obey Eqn. (12.4) and hence (12.7) over the full range of compositions for all species; these typically have molecular interactions that are similar among different components. Most others are found to follow the chemical potential of (12.7) only in the limit that one component is nearly pure. That is to say, it becomes an excellent approximation for mixtures of a solvent with $x \approx 1$ and other species that are present in very dilute amounts. Systems have this limit because such conditions lead to rare interactions between different molecular species, and mixing entropy rather than energy dominates solution behavior. However, in these cases the quantity μ_i^* in Eqn. (12.7) for a dilute component no longer represents the chemical potential of pure i at the same T and P. Instead, when applying the ideal chemical potential to nonideal solutions in the near-pure limit, the quantity μ_i^* takes on a different interpretation for species with small mole fractions.

In applying the ideal solution model to real systems at the near-pure limit, the character of $\mu_i^*(T, P)$ in the ideal chemical potential depends on the concentration of the species i.

If component i is nearly pure ($x_i \rightarrow 1$), μ_i^* gives the usual chemical potential one would expect for a single-component system of i at T and P.

If component i is dilute ($x_i \approx 0$), μ_i^* gives the chemical potential of a hypothetical state of infinitely dilute i in the solvent at that T and P. In this case, μ_i^* depends on the nature of the solvent, but importantly it remains independent of mole fraction. One can think of μ_i^* as the T- and P-dependent limit as $x_i \rightarrow 0$ of the quantity $\mu_i(T, P, \{x\}) - k_B T \ln x_i$.

It can be shown that the second, dilute convention is a consequence of the first, which is proved later in Example 12.1. Of course, if the solution behaves ideally over the whole range of concentrations, then the second convention is unnecessary and every component adheres to the first convention. In most of what follows we use the first convention, either by treating the dominant component or by assuming ideality through-out all compositions.

The chemical-potential expression of Eqn. (12.7) is the starting point for nearly all problems involving solutions that involve phase coexistence. At equilibrium the chemical potential of a component in one phase must be equal to its chemical potential in another, provided that it can freely transfer between the two. Therefore, a general strategy for solving problems involving solutions and phase equilibrium is to equate the relevant chemical potentials in each phase. We now discuss several applications of this approach.

12.2   Ideal vapor–liquid equilibrium and Raoult's law

Consider a mixture that is at a coexistence temperature and pressure such that there is equilibrium between a vapor and liquid phase, as in Fig. 12.2. For a species i that is present in both phases, the simplest approximation is that it behaves ideally in each. At equilibrium, its chemical potential in the solution (liquid) and vapor (gas) phases must be the same,

$$\mu_i^G = \mu_i^L \tag{12.8}$$

Using the ideal gas chemical potential from Chapter 9 and Eqn. (12.7) for the ideal solution phase, this becomes

$$\mu_i^\circ(T) + k_B T \ln P_i = \mu_i^*(T, P) + k_B T \ln x_i \tag{12.9}$$

where $P_i = y_i P$ is the partial pressure of i in the gas phase. A little rearrangement gives

$$y_i P = x_i \exp\left(\frac{\mu_i^* - \mu_i^\circ}{k_B T}\right) \tag{12.10}$$

We can use the fact that we know one limit of this equation, the case in which the mixture approaches pure species i. Equation (12.10) then implies vapor–liquid equilibrium of pure i. In that case, there exists a single macroscopic degree of freedom as per Gibbs' phase rule and thus the pressure and temperature must be related by

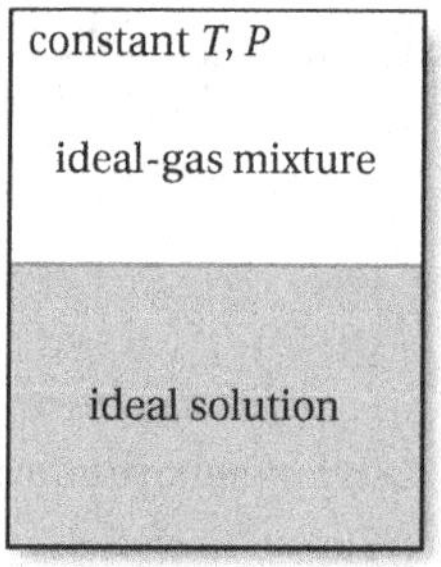

constant T, P

ideal-gas mixture

ideal solution

Figure 12.2. Raoult's law predicts vapor–liquid equilibrium for multicomponent systems in which both gas and solution phases behave ideally.

$P = P_i^{\text{vap}}(T)$, where P_i^{vap} is the vapor or saturation pressure of pure i. On taking the limit of $x_i \to 1$ and $P_i \to P_i^{\text{vap}}(T)$, we find

$$P_i^{\text{vap}} = \exp\left(\frac{\mu_i^* - \mu_i^\circ}{k_B T}\right) \tag{12.11}$$

If we move away from $x_i = 1$, then, to a good approximation, the exponential term involving chemical potentials remains pressure-independent. Therefore, we can rewrite a general relationship between composition and partial pressure by substituting (12.11) into (12.10) to remove the standard and pure chemical-potential terms,

$$y_i P = x_i P_i^{\text{vap}} \tag{12.12}$$

This important relationship is known as *Raoult's law*, and it shows that the partial pressure of a near-pure component in an ideal gas mixture above an ideal solution is given by the liquid mole fraction times the vapor pressure. Because it is based on the ideal solution model, Raoult's law gives accurate behavior in the vicinity of $x_i = 1$, but deviations far away from the pure state can be significant. Chapter 13 considers ideal, multicomponent vapor–liquid equilibrium and Raoult's law in greater detail.

12.3   Boiling-point elevation

The ideal solution model also predicts that the boiling-point temperature of a pure liquid will be increased when a nonvolatile solute is dissolved in it. Consider a liquid in equilibrium with its vapor phase at its boiling temperature, as shown in Fig. 12.3. When a small amount of solute is added and dissolves in the liquid phase, the temperature must be raised to recover liquid–vapor equilibrium at the same pressure. Here, we show why this behavior emerges and develop an expression for the change in the boiling temperature T_b. Since we will assume an ideal solution, our results will be valid asymptotically in the limit of an infinitely dilute amount of solute. We will also assume

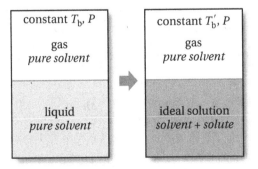

Figure 12.3. A nonvolatile solute that dissolves in a liquid raises its boiling point. At a given pressure P, vapor–liquid coexistence moves from temperature T_b for the pure case to temperature T_b' for the solution.

that the solute has a negligible vapor pressure such that it is absent in the vapor phase. Consider, for example, that the solute might be table salt and the solvent water.

At phase equilibrium, the chemical potentials of a component that can move between two phases must be equal, which is only the case for the solvent. When no solute is present, we have

$$\mu_{\text{solvent}}^{*G}(T_b, P) = \mu_{\text{solvent}}^{*L}(T_b, P) \tag{12.13}$$

where the * superscripts indicate pure-phase chemical potentials. Upon addition of solute, we have instead

$$\mu_{\text{solvent}}^{*G}(T_b', P) = \mu_{\text{solvent}}^{L}(T_b', P, x_{\text{solvent}}) \tag{12.14}$$

where T_b' is the new temperature at which the chemical potentials are equal, i.e., the new boiling temperature. From here on out, we will drop the "solvent" subscript; it will be understood that the un-subscripted chemical potentials and mole fractions are those of the solvent and not the solute. Using the ideal chemical potential of (12.7),

$$\mu^{*G}(T_b', P) = \mu^{*L}(T_b', P) + k_B T_b' \ln x \tag{12.15}$$

A more convenient way to write condition (12.15) is by introducing a quantity $\Delta\mu$ that gives the difference in solvent chemical potential between its pure-gas and pure-liquid phases,

$$\Delta\mu(T_b', P) = k_B T_b' \ln x \quad \text{where} \quad \Delta\mu(T_b', P) \equiv \mu^{*G}(T_b', P) - \mu^{*L}(T_b', P) \tag{12.16}$$

Clearly, $\Delta\mu$ is zero at the original boiling temperature. It is finite, however, at T_b'.

We will take a moment to find a more informative expression for $\Delta\mu$ that can be substituted into Eqn. (12.16). We begin by relating the chemical potentials at T_b' to those at T_b. For pure phases, $\mu = g$ and we have the thermodynamic identity

$$\left(\frac{\partial(\mu/T)}{\partial T}\right)_P = -\frac{h}{T^2} \tag{12.17}$$

Taking two instances of this equation for each of the liquid and gas phases and subtracting one from the other gives

$$\left(\frac{\partial(\Delta\mu/T)}{\partial T}\right)_P = -\frac{\Delta h}{T^2} \tag{12.18}$$

where Δh gives the enthalpy difference between the pure-gas and pure-liquid phases, or equivalently the heat of vaporization Δh_{vap}. For small changes in temperature, Δh_{vap} can be considered constant and we can integrate (12.18) between the limits T_b and T_b' to find the change in $\Delta\mu$,

$$\frac{\Delta\mu(T'_b, P)}{T'_b} - \frac{\Delta\mu(T_b, P)}{T_b} = \Delta h_{vap} \left(\frac{1}{T'_b} - \frac{1}{T_b} \right) \tag{12.19}$$

Since $\Delta\mu(T_b, P) = 0$ due to the pure-phase equilibrium at T_b, we have

$$\Delta\mu(T'_b, P) = \Delta h_{vap} \left(1 - \frac{T'_b}{T_b} \right) \tag{12.20}$$

We now have a suitable expression for $\Delta\mu$ involving measurable quantities. Substituting (12.20) into (12.16),

$$\Delta h_{vap} \left(1 - \frac{T'_b}{T_b} \right) = k_B T'_b \ln x \tag{12.21}$$

After rearrangement, we obtain a closed-form expression for the new boiling temperature,

$$T'_b = T_b \left(1 + \frac{k_B T_b}{\Delta h_{vap}} \ln x \right)^{-1} \tag{12.22}$$

This equation always predicts that small amounts of solute will increase the boiling point. When x is slightly less than one due to the addition of solute, the second term inside the parentheses is negative and the overall effect is to multiply T_b by a number greater than unity. Of course, for $x \to 1$, we recover the original solvent boiling temperature.

Since the ideal solution model underlying Eqn. (12.22) is generally valid only in the near-pure limit, there is no reason we should not further simplify the expression on the basis that $x \approx 1$. To do so, we use Taylor expansions. We can approximate

$$\left(1 + \frac{k_B T_b}{\Delta h_{vap}} \ln x \right)^{-1} \approx 1 - \frac{k_B T_b}{\Delta h_{vap}} \ln x \tag{12.23}$$

and

$$\ln x = \ln(1 - x_{solute}) \approx -x_{solute} \tag{12.24}$$

Our final expression for the elevated boiling temperature thus becomes

$$T'_b \approx T_b \left(1 + \frac{k_B T_b}{\Delta h_{vap}} x_{solute} \right) \tag{12.25}$$

Again Eqn. (12.25) shows that, as solute is added, the boiling temperature increases since Δh_{vap} is positive and the term in parentheses is greater than one. It is interesting to note that the change in boiling point is entirely independent of the nature of the solute; it depends only on the solvent heat of vaporization and the mole fraction. Such solute-unspecific behavior is generally termed a *colligative property*, and it emerges because the solute merely acts to increase the entropy of the solution phase, stabilizing it and thus requiring higher temperatures for vaporization. In fact, Eqn. (12.25) could be used to estimate the value of Δh_{vap} by measuring changes in the boiling temperature of a liquid with small but increasing additions of a solute.

12.4   Freezing-point depression

It is no surprise that ideal solutions also predict freezing-point depression. Consider a liquid solution in equilibrium with a pure crystalline (solid) phase of the solvent. For example, liquid salt water can coexist with pure solid ice. The derivation for the change in melting temperature T_m with solute concentration proceeds almost identically to the one above, starting with the equality of solvent chemical potentials,

$$-\Delta\mu(T'_m, P) = k_B T'_m \ln x$$

where

$$\Delta\mu(T'_m, P) \equiv \mu^{*L}(T'_m, P) - \mu^{*X}(T'_m, P) \tag{12.26}$$

Here, the superscript X indicates the crystal phase. Note that Eqn. (12.26) differs from (12.16) in that it contains a negative sign. As before, we can relate $\Delta\mu$ to the difference in enthalpies of the pure phases through (12.18). By integrating, we find

$$\Delta\mu(T'_m, P) = \Delta h_{fus}\left(1 - \frac{T'_m}{T_m}\right) \tag{12.27}$$

where Δh_{fus} is the latent heat of fusion, which is a positive quantity. By substitution of (12.27) into (12.26) and working through the same approximations as in Section 12.3, we finally obtain

$$T'_m \approx T_m\left(1 - \frac{k_B T_m}{\Delta h_{fus}}x_{solute}\right) \tag{12.28}$$

We therefore find that the addition of solute always lowers the melting temperature, since the term in parentheses is less than one. Why do solutes *raise* the boiling temperature but *lower* the melting temperature? It is because they always increase the entropy and decrease the free energy of the solution through the mixing entropy. This has a stabilizing effect on the liquid phase that broadens its range in temperature, in both directions.

12.5   Osmotic pressure

Osmotic pressures emerge when a solute cannot move past a physical barrier in a system, and are due to ideal mixing entropies. Consider an experiment similar to that depicted in Fig. 12.4. A solution is divided into two cells by a semi-permeable membrane that allows solvent to pass but not solute. Initially, the left cell contains pure solvent and the right one a dilute solution with both solute and solvent. At equilibrium the temperatures of the two sides must be the same since energy can freely move between the two systems. Each side is also at equilibrium with the environment temperature T. Moreover, the solvent can freely move from one side to another, so its chemical potential on the left must equal that on the right.

At equilibrium one finds that the liquid phase in the right cell has a greater height. While the pressure at the top of each compartment must be the same as that of the

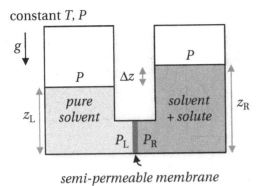

constant T, P

semi-permeable membrane

Figure 12.4. The ideal solution model predicts osmotic pressures. In the diagram above, solute is confined to the right compartment and cannot pass through the membrane, but solvent can. This gives rise to an osmotic pressure drop $\Pi = P_R - P_L = \rho g \Delta z$ across the membrane, where ρ is the density of the solution, g is the gravitational constant, and Δz is the height difference of the liquid compartments. Both P_L and P_R are greater than the atmospheric pressure P.

environment, P, the pressures at the bottom can be calculated from $P + \rho g z$, where g is the gravitational constant and z is the liquid height. As a result, an *osmotic* pressure manifests across the membrane, $\Pi = P_R - P_L = \rho g \Delta z$, where P_L and P_R are the respective left and right pressures at the bottom of the cells. Such a pressure drop can exist because the two sides cannot exchange volume and approach mechanical equilibrium; the membrane is fixed in position and must be strong enough to resist the osmotic forces.

To show the origins of the osmotic pressure, we again begin by considering the equality of solvent chemical potentials on each side,

$$\mu^*(T, P_L) = \mu(T, P_R, x) \tag{12.29}$$

where subscripts for the solvent have been omitted for simplicity. Substituting the ideal chemical potential of (12.7) into (12.29) gives

$$\mu^*(T, P_L) = \mu^*(T, P_R) + k_B T \ln x \tag{12.30}$$

The two pure-solvent chemical potentials differ only in the pressures at which they are evaluated. This enables us to simplify (12.30) using the single-component thermodynamic relation

$$\left(\frac{\partial \mu}{\partial P}\right)_T = v \tag{12.31}$$

For liquids the molar volume v changes very little with pressure. Thus, assuming incompressibility, we integrate to obtain

$$\mu^*(T, P_L) - \mu^*(T, P_R) \approx v(P_L - P_R) \tag{12.32}$$

Equation (12.32) can now be used to simplify the equilibrium condition of (12.30),

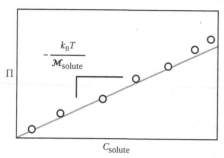

Figure 12.5. The variation of osmotic pressure with concentration of added solute provides a way to measure its molecular weight. Namely, the data in the dilute regime should be linear, with a slope given by $k_B T/\mathcal{M}_{solute}$.

$$-v\Pi = k_B T \ln x \qquad \rightarrow \qquad \Pi = -\frac{k_B T}{v} \ln x \qquad (12.33)$$

Here we substituted the osmotic pressure, $\Pi \equiv P_R - P_L$. Indeed, Eqn. (12.33) predicts that the bottom pressure and height will be higher in the right cell because $-\ln x$ will be a positive quantity. In fact, we can Taylor-expand the logarithm as $\ln x = \ln(1 - x_{solute}) \approx -x_{solute}$, so that our expression is in terms of the solute mole fraction,

$$\Pi = \frac{k_B T}{v} x_{solute} \qquad (12.34)$$

In experimental settings the solute's mole fraction is not often directly accessible; however, dilute solutions allow us to re-express (12.34) in terms of concentrations that can be computed from the mass of solute added and the system volume. Expanding v and x_{solute} and recognizing that $N_{solute} \ll N_{solvent}$,

$$\Pi = k_B T \frac{N_{solvent}}{V} \left(\frac{N_{solute}}{N_{solute} + N_{solvent}} \right)$$

$$\approx k_B T \frac{N_{solvent}}{V} \left(\frac{N_{solute}}{N_{solvent}} \right)$$

$$= k_B T \rho_{solute} \qquad (12.35)$$

where $\rho_{solute} = N_{solute}/V$. We have also $\rho_{solute} = c_{solute}/\mathcal{M}_{solute}$, where c_{solute} is the solute's concentration in the right cell in mass per volume and $\mathcal{M}_{solute}$ is its molecular weight on a per-particle basis. We thus arrive at our key result,

$$\Pi = \frac{k_B T c_{solute}}{\mathcal{M}_{solute}} \qquad (12.36)$$

The remarkable significance of the osmotic-pressure equation is that it provides a way to measure the molecular weight of arbitrary soluble solutes. The procedure is simple: a known mass of solute is dissolved into the solvent and the osmotic pressure is measured from the height difference Δz. The volume of the right cell is then used to determine c_{solute}, and, using (12.36), $\mathcal{M}_{solute}$ can be computed. Typically the osmotic pressure is measured at different values of c_{solute} and the slope at low dilutions is used to estimate the molecular weight, as illustrated in Fig. 12.5. Deviations from a linear relationship at higher concentrations can occur due to the breakdown of the ideal solution model and the mathematical approximations used to obtain the relationship of (12.36).

12.6   Binary mixing with interactions

What happens if we cannot treat a solution as consisting of non-interacting components and the interactions become important? Here, we derive the simplest model of interacting solutions from the lattice approach illustrated in Fig. 12.1. This slight extension of the ideal solution will accommodate an approximate energy of mixing in addition to ΔS_{mix} and is called the *regular-solution model*. We again consider the case in which A and B particles mix. Now, however, we assume that, when two A particles are next to each other, they experience a pairwise energy of w_{AA}. Similarly, two B particles interact with an energy w_{BB}, and an A particle and a B particle interact with energy w_{AB}. In general, each w parameter can either be negative (attractive) or positive (repulsive).

If we knew exactly the lattice configuration of A and B particles, we could sum all of the interactions to find the exact energy of the system. Here, instead, we use the *mean-field approximation* to compute an average energy of the randomly mixed state that will make the problem tractable. This is the same basic technique that we used to model the lattice gas in Chapter 10. The core of the approximation is that the particles are assumed to be well mixed such that the composition of A and B molecules around any one lattice site is equal to the overall composition of those components. In reality, differences in energetic interactions will give rise to differences in local compositions, but we will ignore those possibilities.

With the mean-field simplification, the local environment of any one lattice site is readily determined. The respective average numbers of A and B neighbors are zN_A/N and zN_B/N, where z is the total number of nearest neighbors to a lattice site (six for three dimensions). This allows us to compute the numbers of different types of contacts. If we examine a central A particle, it has (zN_A/N) A–A contacts. There are N_A such centers, so the total energy due to A–A interactions becomes

$$E_{mix, AA} = \frac{1}{2} \times w_{AA} \times N_A \times \frac{zN_A}{N}$$

$$= \frac{w_{AA}}{2} \frac{zN_A^2}{N} \tag{12.37}$$

The half in (12.37) accounts for the fact that each A–A interaction must be shared between the particle pair. By similar arguments for a central B particle, the total B–B interaction energy is

$$E_{mix, BB} = \frac{1}{2} \times w_{BB} \times N_B \times \frac{zN_B}{N}$$

$$= \frac{w_{BB}}{2} \frac{zN_B^2}{N} \tag{12.38}$$

Finally, we must account for the A–B cross-interactions. There are two contributions, one in which we consider an A particle at the center and B neighbors, and one for a B particle at the center with A neighbors. The result is

$$E_{mix, AB} = \frac{1}{2} \times w_{AB} \times N_A \times \frac{zN_B}{N} + \frac{1}{2} \times w_{AB} \times N_B \times \frac{zN_A}{N}$$

$$= w_{AB} \frac{zN_A N_B}{N} \tag{12.39}$$

By putting together Eqns. (12.37)–(12.39), we develop an expression for the average energy of the randomly mixed state, namely

$$E_{mix} = zN \left(\frac{w_{AA}}{2} x_A^2 + \frac{w_{BB}}{2} x_B^2 + w_{AB} x_A x_B \right) \tag{12.40}$$

where we have made the substitutions $N_A = x_A N$ and $N_B = x_B N$ in order to express the equation in terms of mole fractions. It is useful to reference this energy relative to the unmixed state of pure A and pure B in separate containers. In that case, the total number of A–A contacts is simply proportional to the number of molecules and neighbors, for $zN_A/2$ interactions in total. A similar argument exists for the pure-B state, such that the total unmixed energy is

$$E_{unmixed} = \frac{w_{AA} zN_A}{2} + \frac{w_{BB} zN_B}{2}$$

$$= zN \left(\frac{w_{AA}}{2} x_A + \frac{w_{BB}}{2} x_B \right) \tag{12.41}$$

We now subtract (12.41) from (12.40) and simplify using $x_B = 1 - x_A$ to find the change in energy of mixing. After some algebra,

$$\Delta E_{mix} = E_{mix} - E_{unmixed} = Nx_A(1 - x_A)k_B T \chi_{AB} \tag{12.42}$$

Here, we define a new quantity χ_{AB} called the *exchange parameter* or *chi parameter*,

$$\chi_{AB} \equiv \frac{z}{k_B T} \left(w_{AB} - \frac{w_{AA} + w_{BB}}{2} \right) \tag{12.43}$$

In effect χ_{AB} is a dimensionless parameter that summarizes the net effect of the three distinct pair interactions: it measures how much A–B interactions differ from the average of the A–A and B–B interactions. According to this definition, it is inversely proportional to temperature.

Equation (12.42) shows that there can be two ways a solution might lack an energy of mixing. The first, trivial case occurs if all interactions among the components are the same, $w_{AA} = w_{BB} = w_{AB}$. However, a second scenario is possible when differences in the interactions do exist but $w_{AB} = (w_{AA} + w_{BB})/2$. The two cases have identical physical behavior within the regular-solution model because only the single parameter χ_{AB} affects the mixing energy, not the three individual interactions.

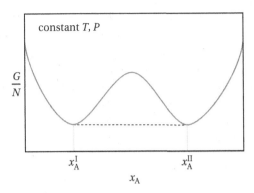

Figure 12.6. A schematic representation of the Gibbs free energy for the regular-solution model in the case $\chi_{AB} > 0$. Because the free energy increases at intermediate compositions, the system can spontaneously separate into two solution phases I and II that are depleted and enriched in species A, respectively. The common tangent line can be constructed to give the compositions and free energy of the phase-separated state.

Armed with Eqn. (12.42), we can now construct a solution free energy for the binary mixture. Accounting for both the energy and entropy of mixing gives

$$G(T, P, N_A, N_B) = G_A(T, P, N_A) + G_B(T, P, N_B)$$
$$- T\,\Delta S_{\text{mix}}(N_A, N_B) + \Delta E_{\text{mix}}(N_A, N_B) \qquad (12.44)$$

On simplifying, we obtain

$$G = G_A + G_B + Nk_B T[x_A \ln x_A + (1 - x_A)\ln(1 - x_A)] + x_A(1 - x_A)\,Nk_B T\,\chi_{AB}$$

$$(12.45)$$

Here, G_A and G_B give the pure-component free energies. Relative to the ideal solution of Eqn. (12.4), the new term on the RHS can have substantial effects on the thermodynamic properties of the solution.

Indeed, several distinct behaviors can occur, depending on the sign of the exchange parameter. Clearly, if $\chi_{AB} = 0$, the model simply recapitulates the ideal solution, and the energetics are irrelevant. However, if $\chi_{AB} < 0$, A–B cross-interactions are stronger than the self-interactions, and mixing becomes more favorable. As expected, in that case, the mixing energy is negative and lowers the free energy. On the other hand, if $\chi_{AB} > 0$ it is energetically more favorable for A particles to be in contact with other A particles and for B particles to be near other B particles. If ΔE_{mix} is large enough compared with $T\,\Delta S_{\text{mix}}$, the free energy can increase upon mixing. The system will then *phase-separate* into two solution phases, one rich in A and one lean in A, in order to reduce the total free energy.

Figure 12.6 shows how the regular-solution free energy might lead to liquid–liquid equilibrium. In the region between the compositions x_A^{I} and x_A^{II}, G increases with x_A. If the overall concentration lies between these two values, it is always favorable for the system to spontaneously phase-separate into two regions, one with composition x_A^{I} and another with composition x_A^{II}. The common tangent line gives the free energy of the phase-separated state. In this particular model, the compositions of the two phases are symmetric around $x_A = 1/2$, but in general this is not the case for real solutions displaying liquid–liquid separation. A detailed analysis of phase separation and the regular solution is left to the reader as a problem at the end of this chapter.

The exchange parameter χ_{AB} stemming from this model is a simple metric describing the mixing compatibility of two substances A and B. It is frequently used in the study of polymer mixtures, in which effective exchange parameters can be measured by fitting the predictions of the regular-solution and related models to experimental data, such as the conditions and concentrations associated with phase separation.

12.7   Nonideal solutions in general

For solutions whose properties deviate substantially from ideality, corrections are usually treated by introducing an *activity coefficient* γ into the expression for the component chemical potentials,

$$\mu_i(T, P, \{x\}) = \mu_i^*(T, P) + k_B T \ln[\gamma_i(T, P, \{x\})x_i] \qquad (12.46)$$

This approach is reminiscent of the technique used for imperfect gases, where a fugacity coefficient is introduced as a correction to the pressure. Again μ_i^* is the pure-phase chemical potential such that γ_i accounts for all nonidealities in the system. In general, the activity coefficient for each component depends on the state condition: the temperature, pressure, and all of the compositions. This can substantially complicate the analysis of solution phase behavior.

To completely specify γ_i one needs a convention for its limiting behavior. Just as in the ideal solution model, we must consider whether a species is nearly pure or dilute.

For nonideal solutions, two conventions determine the limiting behavior of the activity coefficient, depending on the concentration of a given species i.

If component i is nearly pure, we have that $\gamma_i \to 1$ as $x_i \to 1$. That is, as component i becomes the dominant component, the activity returns ideal solution behavior. Here μ_i^* gives the pure-phase, single-component chemical potential for component i.

If component i is dilute, we have instead that $\gamma_i \to 1$ as $x_i \to 0$. In this case, μ_i^* gives the chemical potential of a hypothetical state of infinitely dilute i solvated in the dominant component at that T and P.

These limits on the activity coefficients extend from the fact that any solution that becomes nearly pure should recover ideal behavior. The two conventions are closely related and must have these particular properties in order to be thermodynamically consistent. Namely, different species' chemical potentials are derivatives of a common free energy.

Nonideal activity coefficient models often stem from an approach to solution thermodynamics that is based on the idea of an excess mixture Gibbs free energy. The general approach is to express the solution free energy as an ideal part, including the entropy of mixing, plus a correction that accounts for deviations due to energetic, enthalpic, or other entropic components,

$$G(T, P, \{N\}) = \sum_i G_i(T, P, N_i) + Nk_BT \sum_i x_i \ln x_i + G_{ex}(T, P, \{N\}) \qquad (12.47)$$

where G_{ex} is the excess, nonideal contribution that in general depends on all of the properties of the mixture. It is not too difficult to see the relationship between this approach and the activity coefficient in the chemical potential; indeed, once a form for G_{ex} has been specified, the chemical potential and hence activity coefficient of a species i naturally follow from the N_i derivative of (12.47).

Many models have been developed to predict forms for G_{ex} and the activity coefficients that account for different molecular interactions and physics. These are frequently derived from basic molecular models and statistical mechanics, often with simplifications like the mean-field approximation. For example, the Margules, van Laar, Redlich-Kister, NRTL, UNIQUAC, UNIFAC, and Wilson models are frequently encountered theoretical frameworks in solution-phase thermodynamics. We will not discuss any particular activity model in greater detail in this book, but the interested reader is referred to the text by Smith, van Ness, and Abbott.

12.8   The Gibbs–Duhem relation

We found that relationships exist between different thermodynamic variables and their derivatives, such as Maxwell relations. Here we show that similar relationships exist between composition variables in mixtures. To begin, the Gibbs free energy for a mixture of C components in integrated form is

$$G(T, P, \{N\}) = \sum_i N_i \mu_i(T, P, \{x\}) \qquad (12.48)$$

where $\{x_i\} = \{x_1, x_2, \ldots, x_N\} = \{N_1/N, N_2/N, \ldots, N_C/N\}$ and N is the total number of particles. Remember that this integrated form of G stems from its extensivity in the values N_i and Euler's theorem. Taking the total differential,

$$dG = \sum_i N_i \, d\mu_i + \mu_i \, dN_i \qquad (12.49)$$

We now compare (12.49) with the differential form of the Gibbs free energy that extends from the Legendre transform of the energy,

$$dG = -S \, dT + V \, dP + \sum_i \mu_i \, dN_i \qquad (12.50)$$

On subtracting (12.49) from (12.50), we find

$$-S \, dT + V \, dP = \sum_i N_i \, d\mu_i \qquad (12.51)$$

This important result is called the *Gibbs–Duhem relation*. It shows that changes in the temperature, pressure, and chemical potentials of a mixture are interrelated. Essentially the Gibbs–Duhem relation is a re-expression of Gibbs' phase rule, which shows that $C + 1$ intensive variables completely specify the state of a single-phase system. Equation (12.51) involves $C + 2$ intensive variables given by T, P, and the μ_i. It shows that independent perturbations to $C + 1$ of these variables necessarily determine the change in the remaining one. It is worthwhile to note that Eqn. (12.51) can also be derived by Legendre-transforming the energy fundamental equation $E(S, V, \{N\})$ in each one of its independent variables.

Of particular interest are conditions of constant temperature and pressure, for which $dT = 0$ and $dP = 0$,

$$\sum_i N_i \, d\mu_i = 0 \qquad (\text{constant } T, P) \tag{12.52}$$

For a binary mixture of species 1 and 2, Eqn. (12.52) can be expressed in a particularly insightful form. We begin with

$$N_1 \, d\mu_1 + N_2 \, d\mu_2 = 0 \tag{12.53}$$

On taking the derivatives with respect to the mole fraction of the first component, we obtain

$$N_1 \left(\frac{\partial \mu_1}{\partial x_1} \right)_{T,P} + N_2 \left(\frac{\partial \mu_2}{\partial x_2} \right)_{T,P} \left(\frac{dx_2}{dx_1} \right) = 0 \tag{12.54}$$

The normalization $x_1 + x_2 = 1$ leads to $(dx_2/dx_1) = -1$. Dividing by N then gives an important constraint on the composition variation of the chemical potentials, namely

$$x_1 \left(\frac{\partial \mu_1}{\partial x_1} \right)_{T,P} = x_2 \left(\frac{\partial \mu_2}{\partial x_2} \right)_{T,P} \tag{12.55}$$

Clearly this relation is satisfied by an ideal solution, for which $\partial \mu_i / \partial x_i = k_B T / x_i$. For a nonideal solution, substitution of Eqn. (12.46) gives

$$x_1 \left(\frac{\partial \ln \gamma_1}{\partial x_1} \right)_{T,P} = x_2 \left(\frac{\partial \ln \gamma_2}{\partial x_2} \right)_{T,P} \tag{12.56}$$

Equation (12.56) can be useful in an experimental setting. If the activity of one component in a binary solution is known as a function of composition, it provides a mechanism to compute the activity of the other, by integrating from a reference state such as infinite dilution or pure phase.

The **Gibbs–Duhem relation** places constraints on the way in which the temperature, pressure, and chemical potentials in a multicomponent system can vary with respect to each other. It is a specific, mathematical statement of Gibbs' phase rule for a single phase.

Example 12.1 *Consider a binary mixture of 1 and 2. If the chemical potential of 1 is described by the ideal expression in the first convention for μ^*, then show that the chemical potential of 2 must be described by the ideal expression in either the first or the second convention. Repeat the exercise for a nonideal binary solution.*

Since species 1 is ideal, it follows Eqn. (12.7),

$$\mu_1 = \mu_1^*(T, P) + k_B T \ln x_1$$

where μ_1^* gives the chemical potential of pure 1 at the same temperature and pressure. The Gibbs–Duhem relation of (12.55) can be rewritten at constant T and P as

$$d\mu_2 = \frac{x_1}{x_2}\left(\frac{\partial \mu_1}{\partial x_1}\right) dx_2$$

$$= \frac{k_B T}{x_2}\, dx_2$$

We perform an integration, which incurs a temperature- and pressure-dependent constant $f(T, P)$,

$$\mu_2 = k_B T \ln x_2 + f(T, P)$$

This equation has the same form as Eqn. (12.7). Namely, the composition dependence of μ_2 is fully contained in the term $k_B T \ln x_2$. The integration constant $f(T, P)$ can be regarded as $\mu^*(T, P)$ in either convention for the ideal chemical potential. If the solution is ideal over the entire range of compositions, then the first convention applies to f. The second convention is relevant otherwise.

For a *nonideal* solution, the chemical potential of species 1 is described with an activity coefficient via Eqn. (12.46), such that $\gamma_1 \to 1$ as $x_1 \to 1$, as is required in order to recover its pure chemical potential μ_1^*. This means that the ideal expression applies to μ_1 in this limit and thus, by the derivation above, the only composition dependence that μ_2 can have is $k_B T \ln x_2$. In general, this means that we could adopt the convention that $k_B T \ln \gamma_2 \to c(T, P)$ as $x_2 \to 0$, where c is composition-independent. However, in this case we might as well absorb $c(T, P)$ into the function $f(T, P)$, which equals $\mu^*(T, P)$ in the second convention. Thus we instead adopt the convention that $k_B T \ln \gamma_2 \to 0$ and $\gamma_2 \to 1$ as $x_2 \to 0$.

12.9 Partial molar quantities

Often we are interested in the way in which solution properties change with the amount of each component present. The way we express changes of properties with composition is through *partial molar* quantities. For any extensive quantity X like the entropy, volume, or free energy, the partial molar quantity for component i is defined by

$$\overline{X}_i = \left(\frac{\partial X}{\partial N_i}\right)_{T, P, N_{j \neq i}} \tag{12.57}$$

A partial molar quantity for species i measures the change in the property X with the number of moles N_i, holding fixed the amounts of all of the other components. Note that $\overline{X}_i$ is *intensive*, since both X and N_i are extensive, and it is a function of T, P, and the compositions $\{x\}$. Moreover, the derivative in (12.57) is taken at constant $N_j \neq N_i$, which is not equivalent to a derivative at constant composition; namely, changing N_i affects all other mole fractions.

Because the quantity X is extensive, it obeys the relationship

$$X(T,P, \lambda N_1, \lambda N_2, \ldots) = \lambda X(T,P,N_1,N_2, \ldots) \tag{12.58}$$

We can therefore use Euler's theorem to show that the total quantity X is given by the particle-weighted sum of the partial molar quantities,

$$X = \sum_i N_i \left(\frac{\partial X}{\partial N_i} \right)_{T,P,N_{j \neq i}} = \sum_i N_i \overline{X}_i \tag{12.59}$$

Let us examine a few specific partial molar quantities. By virtue of its fundamental equation, the partial molar Gibbs free energy is simply the chemical potential,

$$\overline{G}_i = \left(\frac{\partial G}{\partial N_i} \right)_{T,P,N_{j \neq i}} = \mu_i \tag{12.60}$$

Next we consider partial molar entropies and volumes,

$$\overline{S}_i = \left(\frac{\partial S}{\partial N_i} \right)_{T,P,N_{j \neq i}}, \qquad \overline{V}_i = \left(\frac{\partial V}{\partial N_i} \right)_{T,P,N_{j \neq i}} \tag{12.61}$$

Referring to Eqn. (12.50), we see that Maxwell relations readily convert the entropy and volume derivatives into chemical-potential derivatives,

$$\overline{S}_i = -\left(\frac{\partial \mu_i}{\partial T} \right)_{P,\,\text{all } N_j}, \qquad \overline{V}_i = \left(\frac{\partial \mu_i}{\partial P} \right)_{T,\,\text{all } N_j} \tag{12.62}$$

Notice that the derivatives are now taken with *all* mole amounts held constant, i.e., at constant composition. These relationships easily lead to the partial molar enthalpy,

$$\overline{H}_i = \left(\frac{\partial H}{\partial N_i} \right)_{T,P,N_{j \neq i}} = \left(\frac{\partial (G + TS)}{\partial N_i} \right)_{T,P,N_{j \neq i}}$$

$$= \left(\frac{\partial G}{\partial N_i} \right)_{T,P,\,\text{all } N_{j \neq i}} + T \left(\frac{\partial S}{\partial N_i} \right)_{T,P,N_{j \neq i}}$$

$$= \mu_i - T \left(\frac{\partial \mu_i}{\partial T} \right)_{P,\,\text{all } N_j} \tag{12.63}$$

We can re-express the final two terms as a single temperature derivative, using the approach introduced in Chapter 7,

$$\overline{H}_i = -T^2 \left(\frac{\partial (\mu_i / T)}{\partial T} \right)_{P,\,\text{all } N_j} \tag{12.64}$$

At this point, you may notice similarities between the derivatives of the partial molar quantities in mixtures and analogous derivatives of intensive thermodynamic variables in pure systems. The single-component versions are

$$s = -\left(\frac{\partial \mu}{\partial T}\right)_P, \quad v = \left(\frac{\partial \mu}{\partial P}\right)_T, \quad h = -T^2\left(\frac{\partial(\mu/T)}{\partial T}\right)_P \tag{12.65}$$

Compare these relationships with Eqns. (12.62) and (12.64). The analogy with pure systems can be useful in remembering the partial molar relationships, but it also allows us to write a general expression for the chemical potential of one component in a *mixture*. Consider the state function $\mu_i(T, P, x_1, \ldots, x_{C-1})$. Its differential is given by

$$d\mu_i = \left(\frac{\partial \mu_i}{\partial T}\right)_{P,\text{ all } x_j} dT + \left(\frac{\partial \mu_i}{\partial P}\right)_{T,\text{ all } x_j} dP + \sum_{k=1}^{C-1}\left(\frac{\partial \mu_i}{\partial x_k}\right)_{T,P,\ x_{j\neq k}} dx_k$$

$$= -\overline{S}_i\, dT + \overline{V}_i\, dP + \sum_{k=1}^{C-1}\left(\frac{\partial \mu_i}{\partial x_k}\right)_{T,P,\ x_{j\neq k}} dx_k \tag{12.66}$$

That is, changes in the chemical potential can be expressed as functions of T, P, and $(C-1)$ mole fractions. The last mole fraction is redundant because of the constraint $\sum_i x_i = 1$. All of these independent variables are intensive.

Example 12.2 *Calculate the partial molar volume of a nonideal solution modeled with an activity coefficient.*

For this problem recall that

$$\mu_i(T, P, x_1, x_2, \ldots) = \mu_i^*(T, P) + k_B T \ln[\gamma_i(T, P, x_1, x_2, \ldots)x_i]$$

Using the expression for $\overline{V}_i$ in Eqn. (12.62),

$$\overline{V}_i = \left(\frac{\partial \mu_i^*}{\partial P}\right)_T + k_B T\left(\frac{\partial \ln \gamma_i}{\partial P}\right)_{T,\text{ all } x_j}$$

By the fundamental equation, the first term on the RHS gives the molar volume of pure i at the same temperature and pressure,

$$\overline{V}_i = v_i + k_B T\left(\frac{\partial \ln \gamma_i}{\partial P}\right)_{T,\text{ all } x_j}$$

Interestingly, this result can be used to find the pressure dependence of the activity coefficient. One can measure the volume of the solution V as a function of N_i and take a derivative to compute $\overline{V}_i$. The v_i term is simply the volume per particle of pure i at the same temperature and pressure. Therefore, the activity coefficient derivative is given by

$$\left(\frac{\partial \ln \gamma_i}{\partial P}\right)_{T,\text{ all } x_j} = \frac{\overline{V}_i - v_i}{k_B T}$$

In principle, one could use this expression to reconstruct the isothermal change in the activity coefficient between two different pressures.

Problems

Conceptual and thought problems

12.1. Explicitly derive Eqn. (12.3) for the ideal mixing entropy of a C-component system.

12.2. A tiny amount of salt is added to liquid water at standard conditions. Indicate whether the given quantity increases, decreases, or stays the same, or there is not enough information to specify the answer without knowing more about the system. Consider this to be an ideal solution.

(a) The pressure at which the water will freeze, at $0\,°C$.

(b) The chemical potential of the water.

(c) The partial molar volume of the water.

12.3. For an arbitrary binary mixture of species 1 and 2, indicate whether or not each of the following relations is true in general. Explain why or why not.

(a) $\left(\dfrac{\partial S}{\partial T}\right)_{V,N_1,N_2} = -\left(\dfrac{\partial V}{\partial P}\right)_{T,N_1,N_2}$

(b) $\left(\dfrac{\partial N_1}{\partial T}\right)_{V,\mu_1,\mu_2} = \left(\dfrac{\partial S}{\partial \mu_1}\right)_{T,V,\mu_2}$

(c) $\Delta V_{\text{mix}} < 0$

(d) $\left(\dfrac{\partial S}{\partial \mu_1}\right)_{T,V,\mu_2} = -\left(\dfrac{\partial N_1}{\partial T}\right)_{V,\mu_1,\mu_2}$

(e) $\left(\dfrac{\partial \mu_1}{\partial x_1}\right)_{T,P} > 0$

(f) $x_1\left(\dfrac{\partial \mu_1}{\partial T}\right)_{P} = x_2\left(\dfrac{\partial \mu_2}{\partial T}\right)_{P}$

12.4. Consider an ideal binary solution of components A and B. Is each of the following quantities positive, negative, or zero, or is there not enough information to tell?

(a) $\left(\dfrac{\partial \mu_A}{\partial N_A}\right)_{S,V,\mu_B}$

(b) $\Delta V_{\text{mix}} = V_{\text{soln}} - \displaystyle\sum_i V_{i,\text{pure}}$

(c) $\left(\dfrac{\partial S}{\partial N_A}\right)_{T,P,N_B}$

(d) $\left(\dfrac{\partial \mu_B}{\partial N_B}\right)_{T,P,\mu_A}$

(e) $\left(\dfrac{\partial \mu_A}{\partial \mu_B}\right)_{T,P}$

12.5. Indicate whether or not each of the following statements is always true.

(a) On average, particles will transfer from regions of lower to higher chemical potential at constant temperature and pressure in a single-component system.

(b) In a binary mixture of A and B,

$$\left(\frac{\partial \mu_B}{\partial N_B}\right)_{\mu_A, T, P} > 0$$

12.6. Consider the change in the Gibbs free energy of a binary mixture as the composition x_1 is varied, for constant N_2. A friend suggests that this quantity is given by the following, and is the same for the change at constant number of total moles $N = N_1 + N_2$,

$$\left(\frac{\partial G}{\partial x_1}\right)_{T,P,N_2} = \left(\frac{\partial G}{\partial x_1}\right)_{T,P,N} = N\mu_1$$

Show that your friend is wrong and provide the correct expressions for these two distinct derivatives.

12.7. Consider the osmotic-pressure apparatus as shown in Fig. 12.4. The membrane is permeable only to water, the solvent, and a number of moles of NaCl is added to the right compartment. For each of the following, indicate whether the height difference Δz will increase, decrease, or stay the same.

(a) The temperature is increased.

(b) Additional pure water is added to the left compartment.

(c) The solution is prepared with the same number of moles of $MgCl_2$ instead of NaCl.

12.8. A room-temperature glass of liquid water exposed to the atmosphere eventually evaporates, even though its temperature is well below the boiling point.

(a) Explain why this happens. What is the difference between evaporation and boiling?

(b) For what level of humidity, in mass of water per volume of total gas, will the water *not* evaporate at 25 °C? You may find the following equation for the vaporization pressure of pure water useful:

$$\log_{10}(P^{vap}) \approx 4.6543 - \frac{1435.264}{T/K - 64.848}$$

where P^{vap} is in bars and T is in K.

12.9. In a particular binary mixture, the excess Gibbs free energy of mixing is found to be $G_{ex}/Nk_BT = f(x_1)$, where f is a function with the behavior $f \to 0$ as either $x_1 \to 0$ or $x_1 \to 1$. It is suggested that the chemical potential of species 1 is given by

$$\mu_1(T, P, x_1) = \mu_1^*(T, P) + k_BT \ln x_1 + k_BT\left(\frac{df}{dx_1}\right)$$

What is wrong with this analysis? Find expressions for the chemical potentials μ_1 and μ_2.

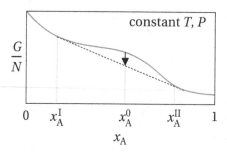

Figure 12.7. The overall Gibbs free energy per particle in a specific binary system of A and B predicts liquid–liquid phase separation. A system with overall composition at x_A^0 will spontaneously separate according to the common-tangent-line construction.

12.10. Prove that the composition dependence of the partial molar enthalpies in a binary solution obeys the relation

$$\left(\frac{\partial \overline{H}_1}{\partial \overline{H}_2}\right)_{T,P} = -\frac{x_2}{x_1}$$

12.11. Consider a binary system of A and B components. Under conditions of phase separation at constant temperature and pressure, the per-particle Gibbs free energy might look like Fig. 12.7. Since this is a two-component system, there are three degrees of freedom and one can fix the overall composition x_A^0 in addition to T and P.

For some overall compositions, the system can lower its free energy by phase-separating into A-rich and B-rich phases. This reduces the free energy by moving from the gray curve to the dotted line that is tangent at the two compositions x_A^I and x_A^{II}, which are the compositions of the two phases. Note that there are no minima or maxima in Fig. 12.7.

(a) For any solution in general, show that a line between two points in the gray free energy curve gives the free energy of a system with overall composition x_A^0 that is phase-separated into two phases with compositions specified at the line end points:

$$g_{PS}(x_A^0) = g(x_A^I) + (x_A^0 - x_A^I)\frac{g(x_A^{II}) - g(x_A^I)}{x_A^{II} - x_A^I}$$

(b) Show that, when this line is tangent to the gray curve at its end points, the chemical potentials in each phase are equal, $\mu_A^I = \mu_A^{II}$ and $\mu_B^I = \mu_B^{II}$. Hint: use an integrated relation to express the Gibbs free energy as a function of the chemical potentials and x_A, and then find the slope of the tangent line in three ways.

(c) As the temperature increases, the region of negative curvature in this diagram shrinks. At the critical temperature and composition, what two equations describe the properties of the free energy?

(d) For the regular-solution model specifically, find the critical temperature T_c above which complete mixing always occurs. Note that χ is temperature-dependent and given by $\chi = w/(k_B T)$ where w is shorthand for the expression $z(w_{AB} - (w_{AA} + w_{BB})/2)$. Express your answer in terms of w.

12.12. A container is prepared with equal numbers of moles of two immiscible liquids A and B (e.g., oil and water). A solute C will dissolve in both liquids, and a tiny amount of it is added to the overall system, totaling 0.01 moles. At equilibrium, it is found that the mole fraction of C in A is 0.001 and that that in B is 0.002. An additional 0.01 moles of C is then added. What are the new mole fractions? Provide a formal proof of your reasoning.

12.13. Consider a liquid solvent (species 1) with an amount of solute (species 2) dissolved in it. Explain how the activity coefficient γ_1 can be determined in the following cases and with the indicated experimental data.

(a) *The pure solute is a liquid at the same conditions.* The entire solution is brought to liquid–vapor equilibrium and the compositions of the phases as well as the partial pressures are measured as a function of temperature or pressure. Assume the vapor behaves ideally.

(b) *The pure solute is a solid at the same conditions.* The freezing-point temperature is measured as a function of the solution composition.

(c) In turn, prove that $\gamma_2(T, P, x_2)$ can be determined given knowledge of $\gamma_1(T, P, x_1)$.

12.14. Find an expression for the temperature dependence of the activity coefficient of a component i in solution,

$$\left(\frac{\partial \ln \gamma_1}{\partial T}\right)_{P, \text{ all } x_j}$$

In light of your result and Example 12.2, which exerts a greater influence on the activity coefficient, the temperature or the pressure?

Fundamentals problems

12.15. Explicitly derive Eqn. (12.7) from Eqn. (12.4). You may want to recall the chain rule

$$\left(\frac{\partial G}{\partial N_i}\right)_{N_{j \neq i}} = \left(\frac{\partial G}{\partial N}\right)_{\{x\}} \left(\frac{\partial N}{\partial N_i}\right)_{N_{j \neq i}} + \sum_k \left(\frac{\partial G}{\partial x_k}\right)_{N, x_{j \neq k}} \left(\frac{\partial x_k}{\partial N_i}\right)_{N_{j \neq i}}$$

12.16. Consider a solution of two components A and B at vapor–liquid equilibrium. Henry's law is an approximation that states that, if component A is very dilute then $P_A = K x_A$. Here, P_A is the partial pressure of A in the gas phase, x_A is the solution-phase mole fraction of A, and K is Henry's constant, specific to the two components involved but independent of concentration. Assuming ideal gases and solutions, express K in terms of standard chemical potentials. What makes Henry's law different from Raoult's law $(P_A = P_A^{\text{vap}} x_A)$?

12.17. Equation (12.36) for the osmotic pressure looks remarkably similar to the ideal gas law. Explain why this is so.

12.18. Prove that the Gibbs–Duhem relation also emerges when the energy fundamental equation is Legendre-transformed in all of its independent variables,

$$\mathcal{L}_S \mathcal{L}_V \prod_{i=1}^{C} \mathcal{L}_{N_i}[E(S, V, \{N\})]$$

12.19. Chapter 5 showed that the single-component differential form for the per-particle energy follows the relation $de = T\,ds - P\,dv$. Derive this in a different way, starting with the integrated form, $E = TS - PV + \mu N$. Hint: first divide by N.

12.20. Show that the following relation describing activity coefficients must hold for a binary solution,

$$\int_0^1 \ln\left(\frac{\gamma_1}{\gamma_2}\right) dx_1 = 0$$

12.21. In working with a binary solution, you decide to account for deviations from nonideality with a second-order Taylor expansion of the chemical potentials for components A and B:

$$\mu_A = \mu_A^* + k_B T \ln x_A + k_B T(C_{A0} + C_{A1}x_A + C_{A2}x_A^2)$$

$$\mu_B = \mu_B^* + k_B T \ln x_B + k_B T(C_{B0} + C_{B1}x_B + C_{B2}x_B^2)$$

Here, C_{A0}, C_{A1}, C_{A2}, C_{B0}, C_{B1}, and C_{B2} are all constants of the expansion, which are, in principle, dependent on T and P. The potentials μ^* indicate the usual pure-phase chemical potentials of each of the components at the same temperature and pressure. Show that the *only* thermodynamically valid form of these equations occurs when the constants take on values such that

$$\mu_A = \mu_A^* + k_B T \ln x_A + k_B TC(1 - x_A)^2$$

$$\mu_B = \mu_B^* + k_B T \ln x_B + k_B TC(1 - x_B)^2$$

where $C = C_{A2} = C_{B2}$. Compare these expressions with the chemical potentials in the regular-solution model.

12.22. Consider a binary mixture of two species 1 and 2 that is at vapor–liquid equilibrium. Both species are present in the liquid and gas phases. One might define a partition coefficient for each species between the two phases according to

$$K_1 \equiv \frac{y_1}{x_1}, \quad K_2 \equiv \frac{y_2}{x_2}$$

where the x and y terms denote mole fractions in the liquid and vapor phases, respectively. Assume that the temperature-independent enthalpies of vaporization and boiling temperatures for pure species 1 and 2, namely Δh_1, Δh_2, T_{b1}, and

T_{b2}, are known. Further assume that both phases behave as ideal mixtures, although the vapor phase is not necessarily an ideal gas (only an ideal mixture).

(a) Develop expressions for the partition coefficients as a function of temperature and the pure heats of vaporization and boiling temperatures.

(b) Is this model capable of predicting an azeotrope, for which $x_1 = y_1$? Why or why not? Justify your answer with an equation or two.

(c) Express x_1 and y_1 in terms of the partition coefficients K_1 and K_2.

(d) A single-phase liquid mixture is well below its boiling point and has an overall composition of species 1 of z_1. It is then heated to a temperature in the two-phase regime. Express the molar fraction of the solution that is in the vapor phase f as a function just of z_1 and the partition coefficients K_1 and K_2.

(e) Show that the temperature at which the entire solution in (d) vaporizes is given by an equation of the form

$$1 = C_1 \exp\left(\frac{\Delta h_1}{k_B T}\right) + C_2 \exp\left(\frac{\Delta h_2}{k_B T}\right)$$

Find C_1 and C_2.

12.23. The total Gibbs free energy of a multicomponent, liquid-phase mixture can be expressed as the ideal solution part plus an excess, $G = G_{\text{ideal}} + G_{\text{ex}}$. In the case of a binary regular solution of components 1 and 2 at constant pressure, one can write the excess as

$$\frac{G_{\text{ex}}}{Nk_B T} = x_1(1 - x_1)\chi_{\text{AB}}$$

where χ_{AB} is the exchange parameter. Here, assume that χ_{AB} is not constant but can have a first-order composition dependence of the form $\chi_{\text{AB}} = (b + cx_1)/T$, where b and c are constants. In the following, assume constant atmospheric pressure throughout.

(a) Find expressions for the chemical potentials $\mu_1(T, P, x_1)$ and $\mu_2(T, P, x_2)$.

(b) A mixture is known to exhibit liquid–liquid phase separation at low temperatures, terminating in a critical point at T_c where the critical composition is $x_{1,c}$. If the mixture is well described by this model, express b and c in terms of the critical temperature and composition.

(c) Consider a specific system with the overall composition of component 1 given by z_1 and with $c = 0$. Outline a simple strategy by which you could determine whether or not three-phase equilibrium between an ideal gas mixture and two solution phases would exist. That is, for a given value of the parameter b, how would you determine whether three-phase behavior is possible?

12.24. Consider the solubility of a gas (e.g., N_2) in a liquid well below its boiling point such that its vapor pressure is low. Does the gas' solubility typically increase or decrease with T at constant P? You may want to consider the equilibrium between a pure-gas phase and a liquid mixture. Provide a brief derivation of your result using ideal solution models.

12.25. A mixture containing species 1 with a dilute amount of species 2 is found to obey Raoult's law, $y_1 P_1 = x_1 P_1^{vap}$, near $x_A \approx 1$ but not at other compositions. Show that the solution then also obeys Henry's law for species 2 whenever it obeys Raoult's law for species 1, with the former being given by

$$y_2 P_2 = K x_2$$

where Henry's constant K depends on temperature and pressure but not on composition. That is, show that Raoult's law and Henry's law imply each other.

12.26. Define the overall per-particle volumes and enthalpies of a mixture as $v = V/N$ and $h = H/N$ with $N = \sum_i N_i$. Prove the following two relations governing their temperature and pressure dependence:

$$\left(\frac{\partial v}{\partial T}\right)_{P,\{x\}} dT + \left(\frac{\partial v}{\partial P}\right)_{T,\{x\}} dP = \sum_i x_i \, d\overline{V}_i,$$

$$\left(\frac{\partial h}{\partial T}\right)_{P,\{x\}} dT + \left(\frac{\partial h}{\partial P}\right)_{T,\{x\}} dP = \sum_i x_i \, d\overline{H}_i$$

Applied problems

12.27. How strong must cell walls be? Assume that a typical cell is spherical (radius $R \approx 7$ μm) and is 70% water on a volume basis, with the remaining fraction being largely due to macromolecules like proteins. Assume that there are about 10^8 such molecules in the cell.

(a) In general, the pressure inside the cell will be higher than that outside, and the cell wall will sustain a surface tension that balances this force difference. Show that the free energy is minimized, and thus the forces balance, when the surface tension satisfies $2\gamma = R(P_{in} - P_{out})$. Hint: first construct an integrated form for the total Helmholtz free energy that includes two pressure–volume terms and an interface term, $\gamma \times 4\pi R^2$.

(b) Compute the pressure difference that the cell wall would experience if placed in a container of pure water at 300 K. If a cell ruptures at a surface tension of $\gamma = 0.001$ N/m, what happens when it is placed in pure water?

(c) What minimum concentration of sucrose (molecular weight 342 g/mol) dissolved in the water will prevent lysis? Assume that the sucrose molecule is too big to transport inside of the cell.

12.28. Consider a binary nonideal solution of two liquids 1 and 2. The activity coefficient of species 1 is well described by $\ln \gamma_1 = 2Cx_1 x_2^2$, where C is a constant.

(a) Find an expression for $\ln \gamma_2(x_2)$.

(b) For what values of C will the solution be capable of liquid–liquid phase separation? Explain in detail.

12.29. Mixtures of methanol and 1-octanol are prepared, and their temperature is raised to a level above the boiling point of methanol but below that of 1-octanol, at constant atmospheric pressure. As a result, a two-phase liquid–vapor system is observed. Table 12.1 gives the mole fraction of methanol in each phase for two different temperatures.

(a) Using only these data, estimate the enthalpies of vaporization of pure methanol and pure 1-octanol in this temperature regime, $\Delta h_{vap,M}$ and $\Delta h_{vap,O}$, respectively. Express your answer in kJ/mol. Hint: you may want to consider ideal mixture models.

(b) Estimate the boiling temperatures of pure methanol and pure 1-octanol, $T_{b,M}$ and $T_{b,O}$, respectively. Express your answer in K.

Table 12.1 Mole fractions of methanol

T (K)	Liquid mole fraction of methanol	Vapor mole fraction of methanol
340.0	0.8978	0.9990
450.0	0.0127	0.4072

12.30. The Margules model for a binary mixture of two liquids 1 and 2 gives a total mixture Gibbs free energy that augments the ideal solution case by

$$\frac{G_{ex}}{Nk_BT} = (c_1x_1 + c_2x_2)x_1x_2$$

where c_1 and c_2 are constants. This model well correlates with a wide range of binary solution behavior.

(a) Show that this model returns ideal solution behavior for the appropriate chemical potentials in the near-pure limits, $x_1 \to 1$ and $x_2 \to 1$.

(b) If $c_1 = c$ and $c_2 = 2c$, for what values of c will the Margules model give a system that can phase-separate into two liquid phases? If c has the temperature dependence $c = wT$, what are the critical temperature and composition?

12.31. A modified Margules model for a particular binary mixture of liquids 1 and 2 postulates an excess free energy given by

$$G_{ex} = Nc[(x_1 + 3x_2)x_1x_2]$$

where c is a negative constant with units of energy.

(a) What form for the activity coefficient γ_1 does this equation imply?

(b) How does the entropy of mixing compare between this model and the ideal solution?

(c) Consider the case in which pure species 1 and 2 are completely mixed quasi-statically at constant pressure and temperature. An amount of heat Q (a negative quantity) is removed during this process to maintain the system at constant temperature. Find an expression for Q/N, where N is the total number of molecules, in terms of the model constants and mole fractions.

(d) Will your answer in part (c) be different if mixing is performed at constant total volume instead of constant pressure, i.e., if a partition between species 1 and 2 in a rigid container is slowly removed?

12.32. Consider a container of liquid that is held at its boiling temperature T at pressure P. A small amount of solute is added, but the temperature and pressure are maintained at their original values. Since the solute has elevated the boiling temperature, the system will condense entirely into a liquid.

One way to return the mixture to boiling is to reduce the pressure while maintaining the temperature at the original T. Find an equation that can be solved to find the new vaporization pressure P' as a function of solute mole fraction x_{solute} and liquid number density $\rho_L = v_L^{-1}$. Assume ideal solution behavior for the solvent species, and that a negligible amount of solute is in the vapor phase. To a fair approximation, the solvent vapor at these conditions can be described by the equation of state

$$v_G = \frac{k_B T}{P} + b + cP$$

where b and c are constants.

12.33. The solubility of nonpolar gases in liquid water is typically very low. Consider nitrogen in particular. Its solubility is characterized by the equilibrium mole fraction x_N of dissolved nitrogen in the liquid phase (with water mole fraction $x_W = 1 - x_N$) when the system is in equilibrium with a vapor phase (with corresponding mole fractions y_N and y_W). The mole fraction x_N is typically of the order 10^{-5}. The vapor pressure of water at 300 K is $P_W^{vap} = 3.5$ kPa, and you can assume that the impact of its temperature dependence on solubility is relatively weak. Use ideal models in what follows.

(a) At 1 bar and 300 K, estimate the mole fraction of water that is present in the vapor phase, y_W.

(b) Show that the solubility of nitrogen is given approximately by

$$x_N = C(T, P)(1 - y_W)P$$

where $C(T, P)$ is a constant that is independent of concentrations. Find an expression for $C(T, P)$ in terms of standard and pure chemical potentials.

(c) Do you expect the solubility to increase or decrease with temperature? Explain by finding the temperature dependence of $C(T, P)$.

12.34. At standard conditions, water and butanol phase-separate so that a mixture of them forms two liquid phases. To such a mixture, acetic acid is slowly added; with each increase in the amount added its mole fraction in each phase is determined experimentally.

(a) It is found for small amounts of added acetic acid that the mole fractions in the two phases nearly always have the same ratio and can be described by a *distribution* or *partition coefficient* $K = x_A^B/x_A^W$ that is approximately 3.4. Here the mole fractions pertain to acetic acid and the superscripts B and W to the butanol and water phases. Explain why this might be the case and find an expression for K in terms of thermodynamic quantities.

(b) A total of 0.5 moles of acetic acid is added to a mixture of 1 L each of water and butanol. Find the mole fractions of acetic acid in each phase. The densities of butanol and water are 0.81 g/ml and 1.0 g/ml, respectively. Assume that you can neglect any water that enters the butanol phase and vice versa.

12.35. The solubility of n-butanol in water is 7.7% by mass at $20\,°C$ and 7.1% at $30\,°C$. The solubility of water in n-butanol, on the other hand, is 20.1% at $20\,°C$ and 20.6% at $30\,°C$. The heat of solution is the enthalpy change of preparing a mixture from pure components, ΔH_{mix}. Estimate the heat of solution per mole of n-butanol in water when the former is dilute. Similarly, estimate the heat of solution per mole of water in n-butanol when water is dilute. To begin, first prove that the enthalpy of mixing for a solvent 1 with a dilute amount of species 2 is given by

$$\frac{\Delta H_{mix}}{N_2} \approx \overline{H}_2 - h_2^* = h_2^{**} - h_2^*$$

where $\overline{H}_2$ is the partial molar enthalpy of species 2 in the mixture, h_2^* is the molar enthalpy of pure liquid 2, and h_2^{**} is the molar enthalpy of liquid 2 in solvent when liquid 2 is in an infinitely dilute state. Then, consider equilibrium between water- and n-butanol-rich phases. Note that the partial molar enthalpies can be related to the T dependence of the chemical potentials.

FURTHER READING

K. Denbigh, *The Principles of Chemical Equilibrium*, 4th edn. New York: Cambridge University Press (1981).

K. Dill and S. Bromberg, *Molecular Driving Forces: Statistical Thermodynamics in Biology, Chemistry, Physics, and Nanoscience*, 2nd edn. New York: Garland Science (2010).

T. L. Hill, *An Introduction to Statistical Thermodynamics*. Reading, MA: Addison-Wesley (1960); New York: Dover (1986).

D. A. McQuarrie, *Statistical Mechanics*. Sausalito, CA: University Science Books (2000).

A. Z. Panagiotopoulos, *Essential Thermodynamics*. Princeton, NJ: Drios Press (2011).

J. M. Smith, H. V. Ness, and M. Abbott, *Introduction to Chemical Engineering Thermodynamics*, 7th edn. New York: McGraw-Hill (2005).

J. W. Tester and M. Modell, *Thermodynamics and Its Applications*, 3rd edn. Upper Saddle River, NJ: Prentice Hall (1997).

13 Solutions: advanced and special cases

In this chapter we extend the fundamental properties of solutions introduced in Chapter 12 to a variety of cases in which a more advanced analysis is required. The uninterested reader may wish to skip these topics. Here we first examine in detail the phenomenology and mathematical description of liquid–vapor equilibrium in multi-component systems, including nonideal ones. Subsequently, we consider two important classes of systems – polymers and strong electrolytes – for which the ideal solution never provides a reasonable model. We provide the basic models and conceptual underpinnings in these analyses; however, excellent elaborations on these topics can be found in the references noted at the end of the chapter.

13.1 Phenomenology of multicomponent vapor–liquid equilibrium

Because multicomponent vapor–liquid equilibrium is so important to the chemical process industries, we elaborate on the ideas initially developed in Chapter 12. Before we consider the mathematics of this problem, let us begin by describing some essential phenomenology. The behavior upon heating a pure liquid at constant pressure P is familiar: its temperature increases until the boiling temperature is reached, at which $P^{\text{vap}}(T_b) = P$. The liquid vaporizes entirely at T_b and then the temperature subsequently increases again as the vapor is heated.

In the case of liquid mixtures, however, vaporization does not occur entirely at a single temperature. Consider a binary mixture that is heated in the same manner at constant pressure. The T-x-y diagram in Fig. 13.1 illustrates the basic behavior for a liquid of initial mole fraction $z_1 = 0.4$ for species 1. At some point as the liquid is heated, a tiny amount of vapor will be formed at an initial temperature T_{bub} called the *bubble point*. The system will consist of nearly all liquid of composition $x_1 \approx 0.4$ with an infinitesimally small amount of vapor. The composition of the vapor will be richer in species 1 because it is the more volatile component, here with $y_1 \approx 0.95$.

As the temperature is raised further to T_2, the total number of moles of vapor increases at the expense of the liquid and the compositions of both phases change. Compared with the bubble point, both the vapor and liquid decrease in the fraction of species 1. Still, the vapor remains enriched in species 1 compared with the original mixture, $y_1 > z_1$, while the liquid remains depleted, $x_1 < z_1$. At T_2 in Fig. 13.1, we have $x_1 \approx 0.18$ and $y_1 \approx 0.88$.

Ultimately at some higher temperature T_{dew} called the *dew point*, there is only an infinitesimally small amount of liquid remaining, the system is greatly increased in

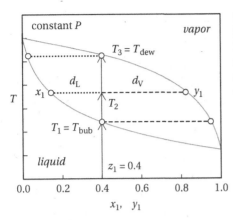

Figure 13.1. A T–x–y diagram illustrates vapor–liquid equilibrium for a binary mixture. Illustrated here is the behavior of a liquid mixture initially with mole fraction 0.4 of species 1 that is heated through the bubble and dew points.

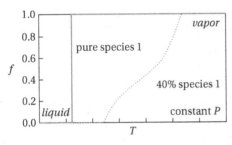

Figure 13.2. Pure liquids vaporize at a single temperature upon constant-pressure heating, while mixtures do so over a range of temperatures. Here, f gives the fraction of the system in the vapor phase on a mole basis.

volume, and the majority of it is vapor. This single drop of liquid has a composition enriched in species 2 (here, $x_1 \approx 0.03$), while the vapor necessarily has a composition nearly identical to that of the original mixture, $y_1 = z_1$. With further heating, the system no longer contains two phases but is entirely vapor. Figure 13.2 illustrates the basic difference between single- and multiple-component vapor–liquid equilibrium through the temperature dependence of the fraction of the system that is vapor, f.

As you might have surmised, the T–x–y diagram gives a convenient way to find the bubble- and dew-point temperatures and the mole fractions of the vapor and liquid phases. The recipe is simple: at a given temperature draw a horizontal *tie line* and find where it intersects the liquid and vapor phase boundaries. Tie lines also give the fraction of the system that is vapor or liquid. To see this, perform a mole balance on species 1,

$$z_1 N = x_1 N^L + y_1 N^G \tag{13.1}$$

where z_1 gives the overall or initial system composition, N the total number of moles of any species, and N^G and N^L those of the gas and liquid phases, respectively. Dividing both sides by N gives

$$z_1 = x_1(1-f) + y_1 f \tag{13.2}$$

where $f \equiv N^G/N$ gives the overall fraction of the system in the vapor phase. On solving for f we find

$$f = \frac{z_1 - x_1}{y_1 - x_1} \tag{13.3}$$

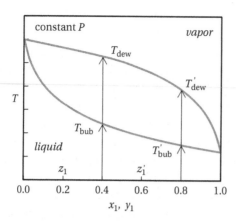

Figure 13.3. Mixtures of differing initial compositions lead to distinct bubble- and dew-point temperatures.

On comparing Eqn. (13.3) with Fig. 13.1, we see that the lengths of the tie lines $d_L = z_1 - x_1$ and $d_V = y_1 - z_1$ are exactly proportional to the amounts of vapor and liquid, respectively. That is,

$$f = \frac{d_L}{d_L + d_V} \tag{13.4}$$

In other words, the closer one is to the liquid boundary along a horizontal (constant-temperature) line, the more liquid the system will contain. It makes sense, therefore, that $d_L = 0$ and $f = 0$ at T_{bub}, while $d_V = 0$ and $f = 1$ at T_{dew}.

If we had started with a different initial mole fraction z_1 for the subcooled liquid mixture, we would have obtained different values for T_{bub} and T_{dew}. This is illustrated in Fig. 13.3, where one can see that both temperatures decrease if z_1 increases, that is, if the original liquid mixture has a higher fraction of the more volatile component. The behavior is also P-dependent. If we heat at a different pressure, we obtain distinct liquid and vapor boundaries in the diagram and hence new bubble and dew points.

The T-x-y diagram also contains the familiar behavior we would expect for the pure component cases. For $x_1 = 1$ both the vapor and liquid boundaries converge at the boiling point of pure species 1, $T_{b,1}$. On the other hand, for $x_1 = 0$, both converge at that of species 2, $T_{b,2}$. Notice that $T_{b,1} < T_{b,2}$, which is consistent with the fact that species 1 is the more volatile of the two and thus will be enriched in the vapor phase.

In contrast to pure liquids, multicomponent liquid solutions generally do not vaporize at a single temperature when heated isobarically. Instead, a first bubble of vapor appears at the bubble point, and the last drop of liquid disappears at the dew point. In between, the temperature controls both the compositions and the relative amounts of the liquid and vapor phases.

13.2   Models of multicomponent vapor–liquid equilibrium

A mathematical description of vapor–liquid equilibrium allows one to predict the phase behavior described above. Specifically, we can develop a system of equations that can be solved to give the equilibrium temperatures, pressures, compositions, and vapor fractions. We begin with the equality of chemical potentials for each species,

$$\mu_i^L(T, P, \{x\}) = \mu_i^G(T, P, \{y\}) \quad (C \text{ equations}) \tag{13.5}$$

for all components i. The initial or overall composition of the mixture can be specified, $\{z\} = \{z_1, z_2, \ldots, z_C\}$. Then, additional equations constrain the vapor and liquid compositions due to the mole balances, as with (13.2). Namely, we have

$$z_i = x_i(1 - f) + y_i f \quad (C \text{ equations}) \tag{13.6}$$

Of course, the mole fractions in each phase must all also sum to unity,

$$\sum_i x_i = 1, \quad \sum_i y_i = 1 \quad (\text{two equations}) \tag{13.7}$$

The summation $\Sigma_i z_i = 1$ does not provide an additional relation because it is naturally a consequence of Eqns. (13.6) and (13.7). To see this, simply sum (13.6) over the components.

How many variables are there in this problem? T, P, and f provide three, while the compositions $\{x\}$, $\{y\}$, and $\{z\}$ give $3C$. The total number of independent variables is thus $3 + 3C$, but we also have $2C + 2$ equalities: C from the chemical potentials in (13.5), C from the mole balances in (13.6), and two from the mole-fraction sums in (13.7). Therefore, we have a total of $3C + 3 - 2C - 2 = C + 1$ degrees of freedom (DOFs) that can be specified to constrain the values of the remaining variables. For example, we could specify the overall composition $\{z\}$, which is $C - 1$ DOFs, as well as T and P. Then we could solve Eqns. (13.5)–(13.7) to find the liquid and vapor compositions and the vapor fraction.

Think about these equations in relation to the T-x-y diagram in Fig. 13.1. In this case the pressure is fixed at P, which gives a particular set of T-x-y phase boundaries, and the temperature is specified as a location on the y-axis. Knowing the original composition of the mixture, z_1, then completes the specification of three DOFs and completely determines all equilibrium conditions according to this analysis. It is easy to see from the diagram that this would give enough information to locate a point within the vapor-liquid region and to find the values of x_1, y_1, and f.

To make progress, we need specific models for the chemical potentials in (13.5). If the gas and solution phases both behave ideally, the equality of chemical potentials for each species leads to Raoult's law as we saw in Chapter 12. Thus we can replace (13.5) with

$$y_i P = x_i P_i^{\text{vap}}(T) \quad (C \text{ equations}) \tag{13.8}$$

Note that the vapor pressure of species i is for its pure state. It is a function of temperature, and it is often modeled successfully by the Antoine equation,

$$\ln P^{\text{vap}}(T) = c_1 - \frac{c_2}{T + c_3} \tag{13.9}$$

where c_1, c_2, and c_3 are substance-dependent constants that can be found in any number of reference texts and databases.

Equations (13.6)–(13.9) now provide a practical, closed set of relations that can be used to predict vapor–liquid phase behavior for systems in the ideal limit. The variables in this system of equations are $T, P, f, \{x\}, \{y\}$, and $\{z\}$. We need to provide values for $C+1$ of them in order to arrive at a solution. Below we discuss several distinct solution scenarios involving different choices for the specified DOFs.

13.3   Bubble- and dew-point calculations at constant pressure

A *bubble-point calculation* predicts the temperature at which a liquid mixture will first begin to vaporize. It also gives the composition of the vapor phase that would be in equilibrium with a liquid of known composition. In a binary mixture, for example, it would predict y_1 given x_1. As we saw from the T-x-y diagram, the temperature T_{bub} corresponds to the case when a minuscule amount of vapor first appears. Thus,

$$f = 0 \text{ and } \{x\} = \{z\} \tag{13.10}$$

at the bubble point. Specifying f counts as one DOF. We also specify the pressure P and mixture composition $\{z\}$ as the remaining C DOFs. Equations (13.6)–(13.8) then become

$$y_i P = z_i P_i^{vap}(T_{bub}) \tag{13.11}$$

for all i, and

$$\sum_i y_i = 1 \tag{13.12}$$

Here we notate $T = T_{bub}$, since the temperature sought is the bubble-point temperature. To make use of these equations, we solve (13.11) for y_i and substitute into (13.12), giving

$$\sum_i \frac{z_i P_i^{vap}(T_{bub})}{P} = 1 \tag{13.13}$$

Note that P and all of the z_i are specified. P_i^{vap} depends only on the temperature through, for example, the Antoine equation. Therefore, we simply need to find the temperature that satisfies Eqn. (13.13). In general, we cannot solve for T_{bub} explicitly because of the complex way it enters into the vapor pressure. Instead, we can use numerical methods such as the iterative or root-finding techniques common in many mathematical software packages.

Once T_{bub} has been determined from the solution to (13.13), it is easy to find the composition of the first bubble of vapor that is formed using (13.11),

$$y_i = \frac{z_i P_i^{vap}(T_{bub})}{P} \tag{13.14}$$

Equation (13.14) shows that this approach can be used to find the set of vapor compositions $\{y\}$ that would exist in equilibrium with a liquid phase of known composition $\{x\}$.

In a manner analogous to the bubble point, we can find the temperature at which a vapor mixture will first form a minuscule drop of liquid upon cooling. This is a *dew-point calculation*. In this case,

$$f = 1 \text{ and } \{y\} = \{z\} \tag{13.15}$$

at the dew point. Then Eqns. (13.6)–(13.8) become

$$z_i P = x_i P_i^{\text{vap}}(T_{\text{dew}}) \quad \text{for all } i \tag{13.16}$$

and

$$\sum_i x_i = 1 \tag{13.17}$$

Solving for x_i using (13.16) and substituting into (13.17) gives a final equation that we can solve to find T_{dew},

$$\sum_i \frac{z_i P}{P_i^{\text{vap}}(T_{\text{dew}})} = 1 \tag{13.18}$$

As before, we typically need numerical methods to solve Eqn. (13.18). Once T_{dew} has been determined, however, the liquid composition follows directly from (13.16),

$$x_i = \frac{z_i P}{P_i^{\text{vap}}(T_{\text{dew}})} \tag{13.19}$$

Thus, dew-point calculations provide the set of liquid compositions $\{x\}$ that correspond to equilibrium with a vapor phase of specified composition $\{y\}$.

Example 13.1 *A mixture of 30% mole fraction n-hexane and 70% iso-heptane is prepared. What are the bubble- and dew-point temperatures at 2 atm?*

The vapor pressure of these two species is well described by the equation $\log_{10} P^{\text{vap}} = c_1 - c_2/(c_3 + T)$, where P^{vap} has units of mmHg, T has units of °C, and the constants c_1, c_2, and c_3 are $c_{1N} = 6.87024$, $c_{2N} = 1168.72$, and $c_{3N} = 224.210$ for n-hexane (N); and $c_{1I} = 6.87318$, $c_{2I} = 1236.03$, and $c_{3I} = 219.545$ for iso-heptane (I).

The bubble temperature is found by solving Eqn. (13.13), which in this case becomes

$$\frac{0.3 \times 10^{c_{1N}-c_{2N}/(c_{3N}+T_{\text{bub}})}}{2 \times 760 \text{ mmHg}} + \frac{0.7 \times 10^{c_{1I}-c_{2I}/(c_{3I}+T_{\text{bub}})}}{2 \times 760 \text{ mmHg}} = 1$$

The solution to this nonlinear equation is found numerically to be

$$T_{\text{bub}} = 81.8 \,°\text{C}$$

The dew point can be found with the same approach as applied to Eqn. (13.18),

$$\frac{0.3 \times 2 \times 760 \text{ mmHg}}{10^{c_{1N}-c_{2N}/(c_{3N}+T_{\text{dew}})}} + \frac{0.7 \times 2 \times 760 \text{ mmHg}}{10^{c_{1I}-c_{2I}/(c_{3I}+T_{\text{dew}})}} = 1$$

The solution is

$$T_{\text{dew}} = 84.6 \,°\text{C}$$

The small difference in temperature between the bubble and dew points is due to the fact that the species are so similar chemically and in terms of their pure-phase boiling points.

13.4   Flash calculations at constant pressure and temperature

If we operate in between the bubble and dew points, we need to perform what is called a *flash calculation*. Here we specify both the temperature T and the pressure P as two degrees of freedom, and the overall composition $\{z\}$ as the remaining one. We then solve for the vapor fraction and the liquid and vapor compositions. Note that we must make sure that our specified T is in between T_{bub} and T_{dew} at this pressure; otherwise, the system will have only a single phase and we will not be able to find a physically realistic solution to Eqns. (13.6)–(13.8).

The Rachford–Rice method offers a convenient and numerically well-behaved way to address flash calculations. It begins by re-expressing Raoult's law for each species in a slightly different form,

$$y_i = K_i x_i \quad \text{for all } i \tag{13.20}$$

with the definition

$$K_i \equiv \frac{y_i}{x_i} = \frac{P_i^{vap}(T)}{P} \tag{13.21}$$

Here, K_i is a *vapor–liquid distribution* or *partition coefficient* for species i. It is important to remember that it is both temperature- and pressure-dependent, as (13.21) shows explicitly, even though for simplicity we will not fully write out $K_i(T, P)$. Once T and P have been specified, we can easily compute all of the K_i terms through the Antoine equation.

The Rachford–Rice method proceeds as follows. First, the composition sums in Eqn. (13.7) are combined to give

$$\sum_i (y_i - x_i) = 0 \tag{13.22}$$

Using (13.20) with (13.6) to solve for x_i in terms of the distribution coefficients, we obtain

$$x_i = \frac{z_i}{1 + f(K_i - 1)} \tag{13.23}$$

Finally, substituting both (13.20) and (13.23) into (13.22) gives

$$\sum_i \frac{z_i (K_i - 1)}{1 + f(K_i - 1)} = 0 \tag{13.24}$$

Given T, P, and $\{z\}$, everything in Eqn. (13.24) is specified except for f, which is the variable for which we seek a solution. Again, we usually cannot do this explicitly, and need root-finding methods. However, this formulation is particularly numerically convenient because the LHS of Eqn. (13.24) is a monotonic function of f. Once we have solved for f, all other unknowns are then easy to find through

$$x_i = \frac{z_i}{1+f(K_i-1)} \quad \text{and} \quad y_i = K_i x_i \qquad (13.25)$$

It is important to keep in mind that we can find a physically realistic root of (13.24) only if we are in the two-phase region. Two simple checks let us know whether the conditions are instead such that the system is entirely liquid or vapor,

$$\sum_i z_i K_i < 1 \quad \text{if } T < T_{bub} \quad \text{(all liquid)} \qquad (13.26)$$

$$\sum_i \frac{z_i}{K_i} < 1 \text{ if } T > T_{dew} \quad \text{(all vapor)} \qquad (13.27)$$

As a reminder, the bubble- and dew-point temperatures depend on pressure and overall composition.

Example 13.2 *Calculate the liquid and vapor mole fractions for the mixture in Example 13.2 at $T = 83\,°C$ and $P = 2$ atm.*

First we need to calculate the distribution coefficients for the two species at this pressure:

$$K_N = \frac{10^{c_{1N}-c_{2N}/(c_{3N}+83\,°C)}}{2 \times 760 \text{ mmHg}} = 1.552$$

$$K_I = \frac{10^{c_{1I}-c_{2I}/(c_{3I}+83\,°C)}}{2 \times 760 \text{ mmHg}} = 0.8178$$

Now we can solve for f using Eqn. (13.24), which in this case becomes

$$\frac{0.3(1.552-1)}{1+f(1.552-1)} + \frac{0.7(0.8178-1)}{1+f(0.8178-1)} = 0$$

The solution is

$$f = 0.278$$

which gives the fraction of the system in the vapor phase. In turn, Eqn. (13.23) gives

$$x_N = \frac{0.3}{1+0.278(1.552-1)} = 0.248, \quad x_I = \frac{0.7}{1+0.278(0.8178-1)} = 0.752$$

Finally, the mole fractions in the vapor phase are given by Eqn. (13.20),

$$y_N = 1.552 \times 0.248 = 0.385, \quad y_I = 0.8178 \times 0.752 = 0.615$$

As expected, the mole fraction of the more volatile n-hexane in the liquid phase is lower than its overall mole fraction, while that in the vapor phase is higher.

13.5 Relative volatility formulation

Are there further assumptions that can simplify vapor–liquid phase calculations? We might be tempted to assume that the distribution coefficients K_i remain constant with temperature, but this would be too drastic a simplification and would discard key physics. Indeed, it is easy to see from the T-x-y diagram that these coefficients vary significantly with temperature. Instead, a more reasonable simplification is to assume that while the K_i change with temperature, they change in exactly the same proportions for all species. Let us define a *relative volatility* by comparing the K_i of a given component with that of some reference species in the mixture K_r,

$$\alpha_i \equiv \frac{K_i}{K_r} \tag{13.28}$$

Often the reference species is taken as the one with the highest boiling point, i.e., the least volatile component such that α_i is typically greater than unity. In principle, α_i depends on T because the K terms do as well. However, in many cases its temperature dependence is quite weak, and one can assume that it is constant. This is a particularly good approximation if the vapor phase behaves ideally and the heats of vaporization of all species are roughly identical.

In the relative-volatility formulation, the equations relating the liquid and vapor compositions can be simplified. We take Eqn. (13.20) for an arbitrary species i and divide it by the same equation for the reference species r,

$$\frac{y_i}{y_r} = \frac{K_i x_i}{K_r x_r} = \alpha_i \frac{x_i}{x_r} \tag{13.29}$$

On summing (13.29) over all i, we obtain

$$\frac{1}{y_r} \sum_i y_i = \frac{1}{x_r} \sum_i \alpha_i x_i \tag{13.30}$$

The sum on the LHS evaluates to unity. Subsequently dividing (13.30) into (13.29) gives

$$y_i = \frac{\alpha_i x_i}{\sum_j \alpha_j x_j} \tag{13.31}$$

Equation (13.31) is an important result because it gives a simple way to predict the vapor-phase mole fractions given the liquid-phase ones and the relative volatilities. If Raoult's law is valid, the α_i can be found using (13.21),

$$\alpha_i = \frac{K_i}{K_j} = \frac{P_i^{\text{vap}}(T)}{P_j^{\text{vap}}(T)} \tag{13.32}$$

Importantly, this shows that the relative volatilities are rigorously independent of pressure for ideal systems and can be calculated directly, for example, using the Antoine equation. Further, if α_i varies little with temperature, one can assume it is constant by taking an average value over the range of temperatures expected for liquid–vapor equilibrium.

Similar mathematical manipulations can be used to show the dependence of the liquid mole fractions on the vapor ones,

$$x_i = \frac{y_i/\alpha_i}{\sum_j y_j/\alpha_j}$$

(13.33)

Equations (13.31) and (13.33) do not provide distinct information; they are simply different formulations of the same underlying relationships. If you know the liquid compositions, use (13.31) to find the vapor one. If the situation is reversed and you know the vapor compositions, use (13.33). These relationships are often useful in modeling distillation processes, when one phase composition is known and the other must be predicted.

For a binary mixture, Eqns. (13.31) and (13.33) take on a particularly simple form. Assume that species 2 is the least volatile species and the reference. Then,

$$y_1 = \frac{\alpha_1 x_1}{1 + (\alpha_1 - 1)x_1} \quad \text{and} \quad x_1 = \frac{y_1/\alpha_1}{1 + (1/\alpha_1 - 1)y_1}$$

(13.34)

These relationships, as well as (13.31) and (13.33), hold for *any* mixture, since they are derived from the definitions of the relative volatilities. We can use them even when a system is highly nonideal. On the other hand, ideal mixtures allow us to use the simple formula given by (13.32) to compute the α values.

If we further assume that α_1 is constant with temperature, we can immediately draw an *x-y* diagram using Eqns. (13.34). (If α were not constant, but had a marked temperature dependence, we would instead have to do the full flash calculation at every temperature.) Figure 13.4 illustrates this behavior for a range of relative volatilities. Each point on an *x-y* curve represents a different temperature, going from $T_{b,1}$ at $x_1, y_1 = 1$ to $T_{b,2}$ at $x_1, y_1 = 0$. In other words, the relationship is a parametric plot of *x*, *y* pairs along temperature tie lines in Fig. 13.1.

13.6 Nonideal mixtures

For systems that deviate from ideality, we cannot use the simple expressions for the ideal gas and ideal solution chemical potentials that lead to Raoult's law. Instead, we typically model the gas phase using a fugacity coefficient and the liquid phase with an activity coefficient,

$$\mu_i^G(T, P, \{y\}) = \mu_i^\circ(T) + k_B T \ln[\phi_i(T, P, \{y\})y_i P]$$

(13.35)

$$\mu_i^L(T, P, \{x\}) = \mu_i^*(T, P) + k_B T \ln [\gamma_i(T, P, \{x\})x_i]$$

(13.36)

Note that both coefficients depend on T, P, and the respective phase compositions, $\{x\}$ or $\{y\}$. Of the two ways in which a system can deviate from ideality, it is usually the liquid phase that requires a more complex treatment. In solution, molecules are in close contact and experience many intricate interactions with each other, which can lead to

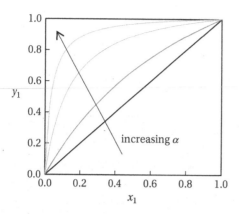

Figure 13.4. An x–y diagram illustrates the relationship between liquid- and vapor-phase compositions in a binary mixture. In this case, the relative volatility α has been assumed constant, and the curves illustrate the effect of its magnitude. As one moves along each line from $(0,0)$ to $(1,1)$, the associated temperature of vapor–liquid equilibrium decreases.

substantial enthalpies of mixing. In contrast, the gas can often be assumed near-ideal because intermolecular interactions are less significant.

In the general case, we use Eqns. (13.35) and (13.36) with the chemical potential equality of (13.5) to develop an expression that replaces Raoult's law for nonideal systems. As we did in Chapter 12, we take the limit $x_i, y_i \rightarrow 1$ to eliminate the standard and pure-phase chemical potentials by relating them to the saturation pressure. The final result is

$$\frac{\phi_i(T, P, \{y\})}{\phi_i^*(T, P)} y_i P = \gamma_i(T, P, \{x\}) x_i P_i^{\mathrm{vap}}(T) \tag{13.37}$$

where $\phi_i^*(T, P)$ gives the fugacity coefficient for pure i. Just as before, there are C of these relations that, combined with the mole balances and the composition summations, yield a system of equations that we can solve given $C + 1$ specified variables. Now, however, the numerical solution is significantly more complicated because all of the K_i and α_i have a composition dependence,

$$K_i \equiv \frac{y_i}{x_i} = \frac{\gamma_i(T, P, \{x\}) \phi_i^*(T, P)}{\phi_i(T, P, \{y\})} \frac{P_i^{\mathrm{vap}}(T)}{P} \quad \text{and} \quad \alpha_{ij} = \frac{K_i}{K_j} \tag{13.38}$$

For ideal systems, these variables are composition-independent, which makes solving the vapor–liquid-equilibrium equations more straightforward. Certainly numerical solutions to (13.37) are possible; however, they require careful consideration of the behavior of γ_i and ϕ_i in order to guarantee physically reasonable and converged solutions. This requires knowledge of the specific forms of these coefficients, which depend on the models chosen.

Let us consider a specific example in which the vapor phase remains ideal but the liquid is modeled by an excess Gibbs free energy,

$$G^{\mathrm{L}} = \sum_i G_i(T, P, N_i) + N^{\mathrm{L}} k_{\mathrm{B}} T \sum_i x_i \ln x_i + G_{\mathrm{ex}}(T, P, N^{\mathrm{L}}, \{x\}) \tag{13.39}$$

For illustration, we will use the binary *regular-solution model* of Chapter 12 that posits $G_{ex} = N^L c x_1 x_2$, where c is a constant with units of energy. For this mixture,

$$G^L = G_1(T, P, N_1^L) + G_2(T, P, N_2^L) + N^L k_B T(x_1 \ln x_1 + x_2 \ln x_2) + N^L c x_1 x_2 \quad (13.40)$$

The chemical potential of species 1 is then given by

$$\mu_1^L = \left(\frac{\partial G^L}{\partial N_1^L}\right)_{T, P, N_2^L}$$

$$= \left(\frac{\partial G_1}{\partial N_1^L}\right)_{T, P} + \frac{\partial}{\partial N_1^L}[N^L k_B T(x_1 \ln x_1 + x_2 \ln x_2) + N^L c x_1 x_2]_{T, P, N_2^L}$$

$$= \mu_1^*(T, P) + k_B T \ln x_1 + c(1 - x_1)^2 \quad (13.41)$$

To take the derivative we had to keep in mind that there are implicit dependences on N^L that might not be obvious. Namely, $N^L = N_1^L + N_2^L$, $x_1 = N_1^L/(N_1^L + N_2^L)$, and $x_2 = N_2^L/(N_1^L + N_2^L)$. An analogous derivation for species 2 gives

$$\mu_2^L = \mu_2^*(T, P) + k_B T \ln x_2 + c(1 - x_2)^2 \quad (13.42)$$

By comparing Eqns. (13.41) and (13.42) with the general nonideal chemical potential expression of (13.36), it is easy to see that the activity coefficients have the form

$$k_B T \ln \gamma_1 = c(1 - x_1)^2, \qquad k_B T \ln \gamma_2 = c(1 - x_2)^2 \quad (13.43)$$

It should be noted that these activity coefficients satisfy the Gibbs–Duhem relation, as must be the case. With these expressions for the distribution coefficients of (13.38) and assuming that the fugacity coefficients are unity,

$$K_1 = \exp\left[\frac{c(1 - x_1)^2}{k_B T}\right] \frac{P_1^{vap}(T)}{P}, \qquad K_2 = \exp\left[\frac{c(1 - x_2)^2}{k_B T}\right] \frac{P_2^{vap}(T)}{P} \quad (13.44)$$

which implies for the relative volatility

$$\alpha_1 = \frac{K_1}{K_2} = \frac{P_1^{vap}(T)}{P_2^{vap}(T)} \exp\left[\frac{c(x_2 - x_1)}{k_B T}\right] \quad (13.45)$$

These results could be used to perform a flash calculation for the regular solution. The equations to be solved are

$$y_1 = \exp\left[\frac{c(1 - x_1)^2}{k_B T}\right] \frac{P_1^{vap}(T)}{P} x_1$$

$$y_2 = \exp\left[\frac{c(1 - x_2)^2}{k_B T}\right] \frac{P_2^{vap}(T)}{P} x_2 \quad (13.46)$$

$$z_1 = x_1(1 - f) + y_1 f, \qquad z_2 = x_2(1 - f) + y_2 f \quad (13.47)$$

$$x_1 + x_2 = 1, \qquad y_1 + y_2 = 1 \quad (13.48)$$

If we specify z_1, T, and P, then we have unknowns y_1, y_2, x_1, x_2, z_2, and f. This is exactly six variables, and we have six equations in (13.46)–(13.48). Unfortunately, however, the

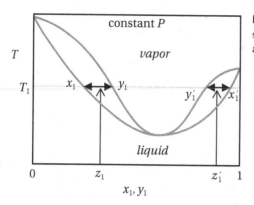

Figure 13.5. A schematic T-x-y diagram shows the presence of a minimum boiling azeotrope around $x_1 = y_1 = z_1 = 0.6$.

problem is highly nonlinear because of the way the compositions enter (13.46). We are not able to eliminate them to obtain a single equation to solve for f as with the Rachford–Rice method. Instead, we might use other numerical techniques. A naïve approach might involve calculating the K_i in (13.44) with "guess" compositions, then solving for the actual compositions using the Rachford–Rice method, and finally recomputing the K_i and iterating the process until $\{x\}$ converges. In reality, this is not the best numerical approach, but it illustrates the complexities involved for nonideal systems.

In some cases, the phase behavior of nonideal systems can be especially complicated, particularly if an *azeotrope* exists. An azeotrope occurs when the compositions of the liquid and vapor phases are identical. Figure 13.5 illustrates such behavior for a binary system using a T-x-y diagram. One can see that there are two qualitatively different regimes, depending on the initial composition of the mixture. If a mixture with initial composition z_1 is heated to T_1, a vapor that is enriched in component 1 will be produced. On the other hand, if the initial composition is z_1', the vapor will be depleted of species 1. There exists an intermediate composition near a mole fraction of 0.6 at which the vapor and liquid will have exactly the same composition. Such systems pose challenges for separations because it is never possible to move beyond an azeotropic composition using distillation alone.

The azeotrope in Fig. 13.5 has a *minimum boiling temperature* that is lower than either of the pure-phase boiling points. A liquid mixture at the azeotropic composition that is heated would vaporize entirely at this temperature. Alternatively, but less frequently, an azeotrope can have a *maximum boiling temperature*, in which case the mixture vaporizes at a higher temperature than either of the pure species.

13.7   Constraints along mixture vapor–liquid phase boundaries

The shapes of the phase boundaries in T-x-y diagrams are tightly connected to other thermodynamic properties. For the situation in which two mixture phases are in equilibrium, we can derive relationships analogous to the Clapeyron equation, showing how the thermodynamic state changes as we move along a boundary.

Consider a binary system of components 1 and 2 at vapor–liquid equilibrium. As we change state conditions along a boundary, the chemical potentials of each component in the gas and liquid phases must remain equal, $\mu_1^L = \mu_1^G$ and $\mu_2^L = \mu_2^G$. Thus changes in the chemical potential should satisfy

$$d\mu_1^L = d\mu_1^G$$
$$d\mu_2^L = d\mu_2^G \tag{13.49}$$

Using the general differential expression for the chemical potential developed in Chapter 12, Eqn. (13.49) becomes

$$-S_1^L \, dT + V_1^L \, dP + \left(\frac{\partial \mu_1^L}{\partial x_1}\right)_{T,P} dx_1 = -S_1^G \, dT + V_1^G \, dP + \left(\frac{\partial \mu_1^G}{\partial y_1}\right)_{T,P} dy_1$$

$$-S_2^L \, dT + V_2^L \, dP + \left(\frac{\partial \mu_2^L}{\partial x_2}\right)_{T,P} dx_2 = -S_2^G \, dT + V_2^G \, dP + \left(\frac{\partial \mu_2^G}{\partial y_2}\right)_{T,P} dy_2 \tag{13.50}$$

Here, the quantities S_i and V_i give the *partial molar* entropy and volume of component i in a given phase. We omit the overbars normally associated with partial molar quantities in order to maintain clarity. We now multiply the top equation in (13.50) by x_1 and the bottom one by x_2, and add them. We also use the fact that $dx_2 = -dx_1$ and $dy_2 = -dy_1$. After some simplification, we obtain

$$-[x_1 \, \Delta S_1 + x_2 \, \Delta S_2] dT + [x_1 \, \Delta V_1 + x_2 \, \Delta V_2] dP$$

$$+ \left[x_1 \left(\frac{\partial \mu_1^G}{\partial y_1}\right)_{T,P} - x_2 \left(\frac{\partial \mu_2^G}{\partial y_2}\right)_{T,P} \right] dy_1$$

$$- \left[x_1 \left(\frac{\partial \mu_1^L}{\partial x_1}\right)_{T,P} - x_2 \left(\frac{\partial \mu_2^L}{\partial x_2}\right)_{T,P} \right] dx_1 = 0 \tag{13.51}$$

where we have abbreviated $\Delta S_i \equiv S_i^G - S_i^L$ and $\Delta V_i \equiv V_i^G - V_i^L$. The derivatives of the chemical potentials with respect to composition in each phase are not independent, but are connected by Gibbs–Duhem relations,

$$y_1 \left(\frac{\partial \mu_1^G}{\partial y_1}\right)_{T,P} = y_2 \left(\frac{\partial \mu_2^G}{\partial y_2}\right)_{T,P}, \qquad x_1 \left(\frac{\partial \mu_1^L}{\partial x_1}\right)_{T,P} = x_2 \left(\frac{\partial \mu_2^L}{\partial x_2}\right)_{T,P} \tag{13.52}$$

On substituting (13.52) into (13.51), we find that the last bracketed term vanishes. This is why we multiplied Eqns. (13.50) by the mole fractions before adding them. Finally Eqn. (13.51) becomes

$$-[x_1 \, \Delta S_1 + x_2 \, \Delta S_2] dT + [x_1 \, \Delta V_1 + x_2 \, \Delta V_2] dP + \left[x_1 - x_2 \frac{y_1}{y_2} \right] \left(\frac{\partial \mu_1^G}{\partial y_1}\right)_{T,P} dy_1 = 0 \tag{13.53}$$

Equation (13.53) relates differential changes in T, P, and y_1 along the phase boundary. We could have derived a similar equation for changes in T, P, and x_1, using the same approach but instead multiplying Eqns. (13.50) by y_1 and y_2 before adding them. In any case, Eqn. (13.53) is a reflection of the two degrees of freedom allowed by Gibbs' phase rule for a two-component, two-phase system. If we make changes in two variables, such as P and T, the change in the other variable is *completely specified*. Equation (13.53) is also intimately related to the Clapeyron relation, which we do indeed recover in the limit $x_1, y_1 \to 1$ with $\Delta S_1 \to \Delta s_{\mathrm{vap},1}$ and $\Delta V_1 \to \Delta v_{\mathrm{vap},1}$.

Equation (13.53) can be manipulated to show specific variances. For example, at constant pressure ($dP = 0$), we complete the derivative with respect to y_1 and find that the variation of the equilibrium temperature with composition obeys

$$\left(\frac{dT}{dy_1}\right)_{P,\,\text{phase boundary}} = \frac{\left[x_1 - x_2 \dfrac{y_1}{y_2}\right]\left(\dfrac{\partial \mu_1^G}{\partial y_1}\right)_{T,P}}{x_1 \, \Delta S_1 + x_2 \, \Delta S_2} \tag{13.54}$$

This equation has important implications for phase diagrams. Consider the azeotrope in Fig. 13.5. Because $x_1 = y_1$ at this point, Eqn. (13.54) shows that the slope of the phase boundary *must* be zero,

$$\left(\frac{dT}{dy_1}\right)_{P,\,\text{azeotrope}} = \frac{\left[x_1 - x_2 \dfrac{x_1}{x_2}\right]\left(\dfrac{\partial \mu_1^G}{\partial y_1}\right)_{T,P}}{x_1 \, \Delta S_1 + x_2 \, \Delta S_2} = 0 \tag{13.55}$$

Using (13.53), similar constraints could be shown to exist at constant temperature ($dT = 0$) or composition ($dy_1 = 0$). Thus, we conclude the following.

Thermodynamic consistency places constraints on the forms that mixture phase diagrams can take. The slopes of the phase boundaries in different state-space projections are tightly connected to compositions and partial molar quantities. These relationships are the multicomponent analogies of the pure-species Clapeyron relation.

13.8 Phase equilibrium in polymer solutions

Until now we have worked with the assumption that the ideal entropy of mixing can be written as $-Nk_B\Sigma_i x_i \ln x_i$. Such an approach works well when the molecular sizes of the components are comparable, but it fails for polymeric systems in which some of the species (e.g., polymers) are dramatically larger than others (e.g., solvent). Here we develop a simple model of a polymer–solvent system that includes a corrected mixing entropy term as well as mean-field energetic interactions, which is similar to the regular-solution model. This approach was initially developed by Paul Flory and Maurice Huggins in the 1940s, and hence is called *Flory–Huggins solution theory*, although our derivation will involve a simplified presentation that follows the approach of Hill.

To make progress, we utilize solution lattice models in which each polymer is a continuously connected chain of neighboring sites, as shown in Fig. 13.6. Let the number of polymers be N_1, each composed of M adjoined monomers. The total number of sites is V and all remaining, non-polymer sites are taken by solvent molecules such that their

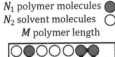

N_1 polymer molecules
N_2 solvent molecules
M polymer length

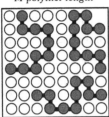

Figure 13.6. A lattice model of polymers in solvent can be used to understand their thermodynamic behavior. White solvent molecules occupy single sites, while dark polymers consist of multiple connected sites. In this particular diagram, the number of polymers is $N_1 = 3$, the number of solvent molecules is $N_2 = 34$, and the total number of lattice sites is $V = MN_1 + N_2$, where $M = 10$ is the polymer length.

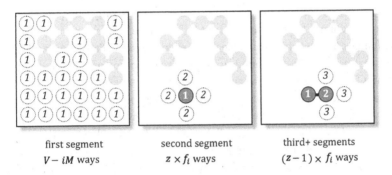

first segment	second segment	third+ segments
$V - iM$ ways	$z \times f_i$ ways	$(z-1) \times f_i$ ways

Figure 13.7. A consideration of the "growth" of the $(i + 1)$th polymer chain, in a mean field of the previous i chains, gives an approximate expression for the number of polymer configurations. The possibilities for the first and second segments laid down are distinct from each other and from those of the remaining $M - 2$ monomers. Upon placing each monomer, one considers neighboring free sites to be reduced by a fraction $f_i = (V - iM)/V$ due to a mean-field approximation of a uniform distribution of the previous polymers. In the diagram, the first polymer (light gray) has already been placed.

number is $N_2 = V - MN_1$. In other words, no vacant sites are present. We will also denote the number of nearest neighbors to a site as z, which is six in three dimensions.

Our first step is to compute the entropy of mixing for the polymer–solvent system by counting configurations, as illustrated in Fig. 13.7. We need to count just the polymers' configurations because, once these are in the lattice, there is only one way to fill all remaining sites with solvent. The easiest, albeit approximate, way to do this is through a sequential placement-and-growth method. That is, we place the first polymer in the lattice and count its configurations ω_1, then we place the second to obtain ω_2, and so on until we reach ω_{N_1}. Then the number of configurations is

$$\Omega = \frac{1}{N_1!} \prod_{i=1}^{N_1} \omega_i \tag{13.56}$$

where the factorial corrects for the indistinguishability of the polymers since they could have been placed in any order. The entropy directly follows

$$S = k_B \ln \Omega = k_B \ln \left(\prod_{i=1}^{N_1} \omega_i \right) - k_B \ln N_1! \tag{13.57}$$

To determine the ω terms, let us start with the first polymer and an empty lattice. There are clearly V places in which we can put the initial monomer. The second must be one of the z nearest neighbors to the first. The third, however, has only $(z - 1)$ choices, since the first monomer has already occupied one of the neighbors to the second. This also occurs for all of the remaining $M - 3$ monomers. By multiplying these choices, we arrive at an expression for ω_1,

$$\omega_1 = Vz(z - 1)^{M - 2} \tag{13.58}$$

Notice that we neglected the possibility that monomers early along the chain might also occupy neighbors to monomers towards the end of the chain, reducing the $(z - 1)$ possibilities. This would be the case if the polymer were collapsed or in a highly globular form that wrapped back around itself. There are approximations that one can use to account for such cases, but for simplicity we ignore them here.

Now consider placing and growing the second polymer. At this point we will *not* neglect the fact that some sites are already occupied by the first polymer because this will have a significant effect on solution behavior. For the second polymer's initial monomer, the number of available sites is reduced to $V - M$. Then, the second monomer ostensibly would have z locations as before, but that would be the case only if we ignored the first polymer. Instead, we will make a *mean-field* approximation by assuming that all previous monomers are distributed randomly in the lattice, neglecting polymer connectivity. This allows us to write the expected fraction of free sites neighboring monomer one of polymer two as $f_1 z$, where $f_1 = (V - M)/V$ gives the average volume fraction of monomer-free sites given that the first polymer is present. Similarly, we have $f_1(z - 1)$ choices for placing the remaining $M - 2$ monomers. So we find that the number of conformations for polymer two is

$$\omega_2 = (V - M)\left(\frac{V - M}{V}\right)^{M-1} z(z - 1)^{M-2} \tag{13.59}$$

Equations (13.58) and (13.59) are easily generalized. For the $(i + 1)$th polymer that we place and grow, we consider that the previous i polymers give a mean field of monomers with site volume fraction

$$f_i = \frac{V - iM}{V} \tag{13.60}$$

A general expression for ω_{i+1} then follows,

$$\omega_{i+1} = (V - iM)f^{M-1}z(z - 1)^{M-2}$$

$$= (V - iM)\left(\frac{V - iM}{V}\right)^{M-1} z(z - 1)^{M-2}$$

$$\approx (V - iM)^M \left(\frac{z - 1}{V}\right)^{M-1} \tag{13.61}$$

In the second line we simply approximated $z \approx z \dashv 1$. We are now able to write the entropy as

$$\frac{S}{k_B} = \ln \left(\prod_{i=0}^{N_1-1} \omega_{i+1} \right) - \ln N_1!$$

$$= M \ln \left(\prod_{i=0}^{N_1-1} (V - iM) \right) + (M-1)\ln \left(\prod_{i=0}^{N_1-1} \frac{z-1}{V} \right) - \ln N_1!$$

$$= M \ln \left(\prod_{i=0}^{N_1-1} (V - iM) \right) + (M-1)N_1 \ln \left(\frac{z-1}{V} \right) - \ln N_1! \qquad (13.62)$$

If we examine the first product, we see that it is given by $V(V-M)(V-2M) \times \cdots \times (N_2+1)$, which closely resembles a factorial. In fact, we can express it in such a way using a little mathematical manipulation,

$$\prod_{i=0}^{N_1-1} (V - iM) = M^{N_1} \left(\frac{V}{M} \right) \left(\frac{V}{M} - 1 \right) \left(\frac{V}{M} - 2 \right) \times \cdots \times \left(\frac{N_2}{M} + 1 \right)$$

$$= M^{N_1} \frac{(V/M)!}{(N_2/M)!} \qquad (13.63)$$

By substituting (13.63) into (13.62), applying Stirling's approximation to all factorials, and simplifying, we obtain

$$\frac{S}{k_B} = V \ln V - V - (N_1 \ln N_1 - N_1) - (N_2 \ln N_2 - N_2) + (M-1)N_1 \ln \left(\frac{z-1}{V} \right) \quad (13.64)$$

This is the entropy of the randomly dispersed polymer-solvent mixture. As we did for the regular solution, we need to compare this with the case prior to mixing in order to compute ΔS_{mix}. We will assume the latter situation involves two separate containers, one of pure but randomly configured polymers with MN_1 sites in total and one of pure solvent with N_2 sites. Since there is only a single configuration for the solvent container, its entropy is zero. The entropy of the polymer can be computed from (13.64) by letting $V \to MN_1$ and $N_2 \to 0$,

$$\frac{S_{\text{unmixed}}}{k_B} = MN_1 \ln(MN_1) - MN_1 - (N_1 \ln N_1 - N_1) + (M-1)N_1 \ln \left(\frac{z-1}{MN_1} \right) \quad (13.65)$$

To arrive at ΔS_{mix}, we subtract (13.65) from (13.64) and simplify,

$$\frac{\Delta S_{\text{mix}}}{k_B} = -N_1 \ln \left(\frac{MN_1}{V} \right) - N_2 \ln \left(\frac{N_2}{V} \right)$$

$$= -N_1 \ln \phi_1 - N_2 \ln \phi_2 \qquad \textbf{(13.66)}$$

This is our final expression for the mixing entropy. Note that we were able to express it in terms of the *volume fractions* of each species,

$$\phi_1 \equiv \frac{MN_1}{V} = \frac{MN_1}{MN_1 + N_2}, \qquad \phi_2 \equiv \frac{N_2}{V} = \frac{N_2}{MN_1 + N_2} \qquad (13.67)$$

Note that the volume fraction ϕ is distinct from the fugacity coefficient defined in Chapter 9, even though the variable is notated similarly. Importantly, the unequal size of the two species modifies the basic form of the ideal mixing entropy developed in Chapter 12. Clearly we recover the latter if we let $M = 1$. It is also easy to imagine that Eqn. (13.66) could be generalized to arbitrary solutions of differently sized molecular species.

The final component of the Flory–Huggins theory is the energy of mixing the polymer and solvent system. We assume that neighboring polymer monomers that are not bonded experience an interaction of w_{11}, neighboring solvent molecules an interaction of w_{22}, and neighboring solvent-to-polymer sites an interaction of w_{12}. Using the mean-field approximation, one can again derive an approximate expression for ΔE_{mix}, in the same manner as the approach used for the regular solution in Chapter 12. This task is left to the reader as a problem at the end of the chapter. The final result is

$$\Delta E_{mix} = k_B T V \chi_{12} \phi_1 \phi_2 \qquad (13.68)$$

As before, the chi parameter is defined by

$$\chi_{12} \equiv \frac{1}{k_B T} \left[(z-2) w_{12} - \frac{z-2}{2} w_{11} - \frac{z}{2} w_{22} \right] \qquad (13.69)$$

On putting the entropy and energy of mixing together, we can find a general expression for the Helmholtz free energy of mixing,

$$\Delta A_{mix} = \Delta E_{mix} - T \Delta S_{mix} = k_B T V \chi_{12} \phi_1 \phi_2 + k_B T N_1 \ln \phi_1 + k_B T N_2 \ln \phi_2 \quad (13.70)$$

It is convenient to express (13.70) in units of $k_B T$ and per the total number of lattice sites in order to make it intensive,

$$\frac{\Delta A_{mix}}{k_B T V} = \chi_{12} \phi_1 \phi_2 + \left(\frac{\phi_1}{M} \right) \ln \phi_1 + \phi_2 \ln \phi_2 \qquad (13.71)$$

The free energy of mixing gives rise to distinct phase behavior in this system. As with the regular solution, a negative value of χ_{12} indicates that favorable polymer–solvent interactions enhance the free energy of mixing, always leading to a single well-mixed phase. The more interesting case occurs when $\chi_{12} > 0$ and there is a competition between the energy and entropy of mixing. For large enough χ_{12}, the system can phase-separate into polymer-rich and solvent-rich phases. The Flory–Huggins model differs from the regular solution because the molecular size inequality leads to an asymmetry in the mixing entropy of Eqn. (13.66). A prominent effect is that the compositions of the two phases differ and, in particular, the solvent-rich phase can have extremely low amounts of polymer for large values of M.

The phase behavior of the Flory–Huggins model is illustrated in Fig. 13.8, in which $A_{mix}/(k_B T V)$ is shown as a function of ϕ_1 and at constant temperature and volume conditions. As M increases, the free energy curves become asymmetric, with their minima shifting towards higher values of ϕ_1. For sufficiently high values of χ_{12} and thus increasingly unfavorable energies of mixing, the curves develop regions of negative curvature. Such is the signature of phase separation and it can be shown that these

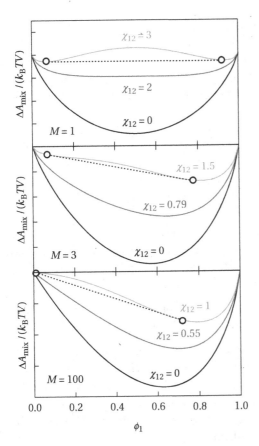

Figure 13.8. The phase behavior of the Flory–Huggins model can be understood by the dependence of the Helmholtz free energy of mixing on ϕ_1, here normalized by $k_B T$ and the total number of sites V. The top panel reflects the usual regular solution with $M = 1$ and equal-sized components. The middle and bottom panels show cases in which the polymer is small- to large-sized. For each M, three values of χ_{12} are illustrated: one below, one at, and one above the critical value for phase separation. The dotted common tangent lines show the free energy of the phase-separated state, with the points giving the equilibrium volume fractions of the solvent- and polymer-rich phases.

cases can lower the overall free energy by dropping down to a *common tangent line* (this is left as an end-of-chapter problem). As before, the end points of the common tangent give the compositions of the two coexisting phases, here expressed in terms of volume fractions. Interestingly, as M increases, the effect of the asymmetry in the free energy curve is to shift the composition of the solvent-rich phase to very low amounts of polymer. Indeed, such physical behavior is a hallmark of polymeric systems.

When systems contain molecules of very different sizes, the form of the mixing entropy changes from that of the ideal solution case. It involves the species volume fractions rather than mole fractions, and it has a significant effect on solution phase behavior.

13.9   Strong electrolyte solutions

Solutions containing mobile ions constitute another class of systems that deviate fundamentally from ideal solution behavior, even in the dilute case. The reason is that ions experience energetic interactions that are long-ranged, decaying with pair distance much slower than van der Waals forces, and these cannot ever be ignored. Here we consider a minimal model of their behavior, the so-called Debye–Hückel approach developed in the 1920s, which is valid in the very dilute ion limit for strong electrolytes that dissociate completely. For simplicity of exposition, let us treat the case in which a monovalent salt is dissolved and completely dissociated in a solvent, like NaCl in water. Thus the ions have $q_+ = +e$ and $q_- = -e$ valences, where e gives the fundamental unit of charge, and we take their bulk number densities as $\rho_{+,b} = \rho_{-,b} = \rho_b$. We will also assume that all ions have a hard-core diameter of a such that no two can come closer than a distance a apart.

Consider a particular "central" ion i in solution, say a cation, as shown in Fig. 13.9. Most of the surrounding molecules are solvent, but neighboring ions can interact through the Coulomb potential,

$$u(r_{ij}) = \frac{q_i q_j}{4\pi \epsilon r_{ij}} \tag{13.72}$$

where q_i and q_j are the charges of the ions (in our example, $\pm e$), r_{ij} is their scalar separation distance, and ϵ the permittivity or dielectric constant of the solvent medium. Sometimes a relative permittivity is used such that $\epsilon = \epsilon_r \epsilon_0$, where ϵ_0 is the permittivity of free space, a fundamental constant. In vacuum $\epsilon_r = 1$, whereas for water $\epsilon_r \approx 80$. The total electrostatic potential energy of the central ion is the sum of the contributions from all of the other co-ions and counterions,

$$U_i = \sum_{j \neq i} \frac{q_i q_j}{4\pi \epsilon r_{ij}} \tag{13.73}$$

While this expression may be intuitive, it will be helpful to write the energy of the ion in a different but equivalent manner through the spatially varying *electrostatic potential* Φ,

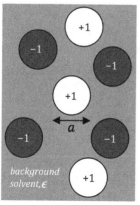

Figure 13.9. In an electrolyte solution, a central ion attracts a diffuse cloud enriched in counterions. Here, the solvent is modeled as a continuum with dielectric constant ϵ. For the purposes of illustration, monovalent ions of identical diameter a are shown. Note that the closest a counterion can approach a central ion is the distance a.

$$U_i = q_i\Phi(\mathbf{r}_i) \tag{13.74}$$

where Φ has units of energy per charge. By comparison with Eqn. (13.73), it is easy to see that the potential at a distance r due to an isolated point ion j must follow

$$\Phi = \frac{q_j}{4\pi\epsilon r} \tag{13.75}$$

so that Eqn. (13.73) is recovered for a collection of such ions. The advantage of the electrostatic-potential approach is that we can make a connection to the density of other ions in the system. *Poisson's equation*, one of the fundamental relations in electrostatics, quantifies this connection,

$$\nabla^2\Phi(\mathbf{r}) = -\frac{n_f(\mathbf{r})}{\epsilon} \tag{13.76}$$

Here, n_f is the *free charge density*, the total excess charge per volume at any given location in space. For the monovalent case, we have $n_f(\mathbf{r}) = e\rho_+(\mathbf{r}) - e\rho_-(\mathbf{r})$. More generally, the free charge density can be written as a sum over all ion species as $n_f(\mathbf{r}) = \sum_i q_i \rho_i(\mathbf{r})$, where i denotes each family of ions of charge q_i that have bulk density $\rho_{i,b}$ and local density $\rho_i(\mathbf{r})$.

Given a distribution of charge in space, Eqn. (13.76) is a second-order differential equation that can be solved for $\Phi(\mathbf{r})$. For a collection of point ions, the free charge density consists of delta-function charge distributions, $n_f(\mathbf{r}) = \sum_i q_i \delta[\mathbf{r} - \mathbf{r}_i]$, and the solution to the electrostatic potential in conjunction with (13.74) recovers Coulomb's law. However, (13.76) is more general because it allows for "smeared" ions or continuously varying charge densities that naturally arise from averaging over microstates. Indeed, this is the central aspect of this approach which will make it useful for our analysis.

To proceed, we focus our analysis on a central cation of radius a and examine its neighboring ions. Owing to Coulomb-type interactions, a region near the central ion will develop that will be enriched in anions and depleted in cations because the former are attracted while the latter are repelled. Let us call this region the central ion's "atmosphere" or "cloud" of counterions. If all ions have the same diameter a, none can come closer to the central one than a distance $r = a$, so $n_f(r < a) = 0$. On the other hand, we expect $n_f(r > a)$ to be negative owing to the anion cloud.

In very dilute systems, one can make several approximations to n_f in this region that permit a solution for Φ and hence give the favorable electrostatic potential energy of the central ion with its neighboring ion cloud. First, the free charge density and electrostatic potential will be isotropic around the central ion and have spherical symmetry. Equation (13.76) in spherical coordinates therefore reduces to

$$\nabla^2\Phi(r) = \frac{1}{r^2}\frac{d}{dr}\left(r^2\frac{d\Phi}{dr}\right) = -\frac{n_f(r)}{\epsilon} \tag{13.77}$$

The second and perhaps more significant approximation is that central-to-neighbor correlations are *pairwise* in nature. That is, the number of neighbors is sufficiently small that the detailed neighbor–neighbor interactions (i.e., anion–anion repulsions) are negligible. In this case, we can express the free charge density using so-called Boltzmann populations that are based directly on the interaction energies. For anions in the region $r > a$,

$$\frac{\rho_-(r)}{\rho_b} = \exp\left(-\frac{U}{k_B T}\right) = \exp\left(+\frac{e\Phi(r)}{k_B T}\right) \tag{13.78}$$

Equation (13.78) assumes that the density of anion neighbors, relative to the bulk, is enhanced by a *Boltzmann factor* given by the exponential of their negative interaction energy with the electrostatic potential. The RHS stems from Eqn. (13.74) with $q_- = -e$. We have not yet discussed Boltzmann factors, but it is sufficient to say that they give relative probabilities of microstates when a system is at constant temperature, rather than constant energy. Their formal derivation and analysis will be presented in detail in Chapter 16. In a similar manner, the density of cations around the central one for $r > a$ follows

$$\frac{\rho_+(r)}{\rho_b} = \exp\left(-\frac{e\Phi(r)}{k_B T}\right) \tag{13.79}$$

Considering Eqns. (13.78) and (13.79), one might expect that Φ will take on positive values such that the concentration of anions is increased and that of cations decreased near the central cation. To show this formally, we proceed with both equations to develop an expression for the free charge density in Poisson's equation in the region $r > a$,

$$\frac{1}{r^2}\frac{d}{dr}\left(r^2\frac{d\Phi(r)}{dr}\right) = -\frac{1}{\epsilon}[q_+\rho_+(r) + q_-\rho_-(r)]$$

$$= -\frac{e\rho_b}{\epsilon}\left[\exp\left(-\frac{e\Phi(r)}{k_B T}\right) - \exp\left(+\frac{e\Phi(r)}{k_B T}\right)\right] \tag{13.80}$$

Equation (13.80) gives the *Poisson–Boltzmann equation* for the monovalent case since it combines Poisson's equation with Boltzmann probabilities. It is a second-order, highly nonlinear differential equation in Φ that is difficult to solve in general. However, we have already made the assumption that the ions are dilute in applying Eqns. (13.78) and (13.79), so it is reasonable to further simplify on the basis that the electrostatic potential will therefore be quite small. Specifically, we can Taylor-expand the exponential terms to first order,

$$\exp\left(-\frac{e\Phi(r)}{k_B T}\right) \approx 1 - \frac{e\Phi(r)}{k_B T}, \qquad \exp\left(+\frac{e\Phi(r)}{k_B T}\right) \approx 1 + \frac{e\Phi(r)}{k_B T} \tag{13.81}$$

Substitution into Eqn. (13.80) then gives the *linearized Poisson–Boltzman equation*,

$$\frac{1}{r^2}\frac{d}{dr}\left(r^2\frac{d\Phi(r)}{dr}\right) = \frac{2e^2\rho_b}{\epsilon k_B T}\Phi(r)$$

$$= \kappa^2\Phi(r) \tag{13.82}$$

In the second line we lumped all of the constants together using a single term κ that has units of inverse distance. In fact, κ^{-1} is an important solution parameter called the *Debye length*. The advantage of this approach is that, if we had followed the Debye–Hückel derivation in the case of arbitrary ion valences, we would have obtained an identical equation but with a slightly modified expression for the Debye length. The general form is

$$\kappa^2 \equiv \frac{1}{\epsilon k_B T} \sum_i q_i^2 \rho_{i,b} \tag{13.83}$$

where again the summation proceeds not over individual ions but over ion species. Clearly the summation returns $2e^2 \rho_b$ for the monovalent case. More broadly, however, the Debye length has a direct connection to the so-called *ionic strength* of the solution,

$$I \equiv \frac{1}{2} \sum_i q_i^2 \rho_{i,b} \tag{13.84}$$

such that we could also write $\kappa^2 = 2I/(\epsilon k_B T)$.

The linearized Poisson–Boltzmann equation can now be solved to obtain a closed-form solution for the electrostatic potential. The first step is to re-express the Laplacian in spherical coordinates in a form suitable for solving the second-order, linear differential Eqn. (13.82). We note the identity

$$\frac{1}{r^2} \frac{d}{dr} \left(r^2 \frac{d\Phi}{dr} \right) = \frac{1}{r} \frac{d^2(r\Phi)}{dr^2} \tag{13.85}$$

which is easily verified by expanding the derivatives. Thus, we can rewrite (13.82) as

$$\frac{d^2\left(r\Phi(r)\right)}{dr^2} = \kappa^2 \left(r\Phi(r)\right) \tag{13.86}$$

Since $\kappa > 0$, it is straightforward to see that the solution is of the form

$$r\Phi(r) = c_1 \exp[\kappa r] + c_2 \exp[-\kappa r] \tag{13.87}$$

Alternatively,

$$\Phi(r) = c_1 \frac{\exp(\kappa r)}{r} + c_2 \frac{\exp(-\kappa r)}{r} \qquad \text{for } r > a \tag{13.88}$$

where c_1 and c_2 are constants that remain to be determined. Keep in mind that this expression for the electrostatic potential pertains only to the region $r > a$, where mobile ions are present and their populations can be approximated using Boltzmann probabilities. Inside the region $r < a$, no ions can penetrate and we must find a distinct form of the potential. There we can again apply the Poisson–Boltzmann equation but instead with a free charge density that is zero everywhere except at $r = 0$,

$$\frac{d^2\left(r\Phi(r)\right)}{dr^2} = 0 \tag{13.89}$$

With two simple integrations, the solution to this differential equation gives

$$\Phi(r) = c_3 + \frac{c_4}{r} \qquad \text{for } r < a \tag{13.90}$$

where c_3 and c_4 are two further constants of integration. We now have the form of $\Phi(r)$ both in and outside the central cation. However, we have four constants that must be determined by appropriate boundary conditions. One condition is that the electrostatic potential far from the central ion must vanish, which demands that c_1 is zero. A second condition is that the total fixed charge within the region $r < a$ where the central cation lies must be equal to $q_+ = +e$. This can be found by integrating the charge density there,

$$\int_0^a n_f(r)4\pi r^2 \, dr = +e \tag{13.91}$$

where $4\pi r^2$ accounts for the volume integral in spherical coordinates. If we use Poisson's relation in Eqn. (13.77) for n_f in terms of the potential Φ, we find

$$-4\pi\epsilon\left[r^2\frac{d\Phi}{dr}\right]_{r=a} = +e \tag{13.92}$$

On substituting the expression in Eqn. (13.90) for the inner potential, we obtain

$$4\pi\epsilon c_4 = +e \quad \rightarrow \quad c_4 = +\frac{e}{4\pi\epsilon} \tag{13.93}$$

which gives the constant c_4. It is interesting to note that the value of this constant, when used in Eqn. (13.90), gives a form that is reminiscent of Coulomb's law for the electrostatic potential very near a point charge, Eqn. (13.75), as we might expect.

With these considerations, we have only two remaining constants that must be determined, c_2 and c_3. These are finally specified by demanding that the electrostatic potential and its derivative remain continuous across the boundary $r = a$. Formally, these conditions are

$$\lim_{r\to a^-}\Phi(r) = \lim_{r\to a^+}\Phi(r), \qquad \lim_{r\to a^-}\frac{d\Phi(r)}{dr} = \lim_{r\to a^+}\frac{d\Phi(r)}{dr} \tag{13.94}$$

Using the outer and inner expressions for the potential from Eqns. (13.88) and (13.90) and the now-determined values for c_1 and c_4, the two boundary conditions become

$$c_3 + \frac{e}{4\pi\epsilon a} = c_2\frac{\exp(-\kappa a)}{a}, \qquad -\frac{e}{4\pi\epsilon a^2} = -c_2\frac{(1+\kappa a)\exp(-\kappa a)}{a^2} \tag{13.95}$$

The solutions for c_2 and c_3 immediately follow:

$$c_2 = \frac{e}{4\pi\epsilon}\frac{\exp(\kappa a)}{1+\kappa a}, \qquad c_3 = -\frac{e}{4\pi\epsilon a}\left(\frac{\kappa a}{1+\kappa a}\right) \tag{13.96}$$

So the full electrostatic potential is finally

$$\Phi(r) = \begin{cases} -\dfrac{e}{4\pi\epsilon a}\left(\dfrac{\kappa a}{1+\kappa a}\right) + \dfrac{e}{4\pi\epsilon r} & \text{for } r \leq a \\[4mm] \dfrac{e}{4\pi\epsilon r}\dfrac{\exp[-\kappa(r-a)]}{1+\kappa a} & \text{for } r \geq a \end{cases} \tag{13.97}$$

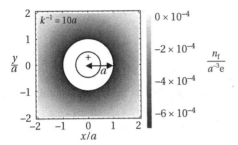

Figure 13.10. The free-charge density n_f (in units of $a^{-3}e$) is shown as a function of position surrounding a central cation in a monovalent system with $\kappa = 1/(10a)$. Here, a two-dimensional slice is illustrated at the $z = 0$ plane. Note that the free-charge density is uniformly negative due to the cloud of anions and it begins at the distance of their closest approach, a.

With the second equation in (13.97) we can now determine the free charge density surrounding the central ion using the Laplacian according to Poisson's equation, Eqn. (13.77). After some simplification,

$$n_f(r \geq a) = -\epsilon \, \nabla^2 \Phi(r)$$
$$= -\frac{e}{4\pi r} \frac{\kappa^2 \, \exp[-\kappa(r - a)]}{1 + \kappa a} \quad (13.98)$$

A two-dimensional slice of this density distribution is shown in Fig. 13.10. Clearly the free-charge density is negative, reflecting the cloud of anions that are attracted near the cation. In general, it can be shown that the total charge in this ion atmosphere is exactly equal and opposite to that of the central ion, with absolute magnitude e. (This task is left as an end-of-chapter problem.) Thus, when viewed at far distances, the central ion is completely *screened*. In other words, far-away ions positioned at large r have negligible interactions with the central ion because they see a region that essentially has net zero charge.

Such screening is an essential feature of electrolyte solutions and is qualitatively distinct from the bare Coulomb interaction that two lone opposite charges would experience in vacuum. Indeed, the form of Eqn. (13.97) shows that the electrostatic potential decays much faster than the pure Coulomb interaction in Eqn. (13.75). The main difference is the presence of the exponential decay in the former, which effectively converts the long-range $1/r$ interactions into shorter-range forces extending over a distance that scales as κ^{-1}. Of course, the present derivation considered only the monovalent case, but similar forms of the electrostatic potential and free charge density emerge for mixtures of ions of arbitrary valence.

We must be careful to keep in mind that the central approximation in the theory is the dilute nature of the electrolyte solution, which is the case for small κ. As a result Eqns. (13.97) and (13.98) are valid only when the Debye length κ^{-1} is relatively large and the screening is weak. Since the theory is only applicable in this regime, it seems perfectly reasonable to further assume that $\kappa^{-1} \gg a$ and take the limiting forms of Eqns. (13.97) and (13.98) when a becomes negligible. In the point-ion limit, these expressions become

$$\Phi(r > 0) = \frac{e \exp[-\kappa r]}{4\pi\epsilon r} \tag{13.99}$$

$$n_{\mathrm{f}}(r > 0) = -\frac{e\kappa^2 \exp[-\kappa r]}{4\pi r} \tag{13.100}$$

in which it is even clearer that the screening modifies the bare Coulomb interaction with an exponential decay with characteristic length scale κ^{-1}.

Now that we have determined the full electrostatic potential for a central ion, we can calculate the total electrostatic energy that it experiences due to its counterion cloud. Since we know the potential, the ion energy extends directly from (13.74) evaluated at the ion position, $r = 0$. However, there is a subtlety in that the potential described by Eqn. (13.99) also includes the contribution of the central ion itself. We must omit this because we would then add the interaction of the point ion with itself, which is infinite. This self-contribution is given by Eqn. (13.75), which we subtract from (13.99) to obtain the potential due to the counterion cloud alone,

$$\Phi_{\mathrm{cloud}}(r > 0) = \frac{e \exp[-\kappa r]}{4\pi\epsilon r} - \frac{e}{4\pi\epsilon r} \tag{13.101}$$

In the limit of small r, both terms in the RHS of (13.101) individually diverge, but their difference does not. We can see this by Taylor-expanding the exponential, $\exp[-\kappa r]/r - 1/r = -\kappa + \mathcal{O}(r)$. Thus at $r = 0$, we have

$$\Phi_{\mathrm{cloud}} = -\frac{e\kappa}{4\pi\epsilon} \tag{13.102}$$

Had we performed the derivation for arbitrary electrolytes, a more general form of this equation would involve the charge of the central ion q,

$$\Phi_{\mathrm{cloud}} = -\frac{q\kappa}{4\pi\epsilon} \tag{13.103}$$

For the remainder of this section, we will work with the general expression.

Our final task is to determine the total contribution of the electrostatic component to the system Gibbs free energy. Using the ideas developed in Chapter 12, we can write the total Gibbs free energy for a C-component system as an ideal part plus the contribution of the electrostatic energies,

$$G(T, P, N_1, \ldots, N_C) = \sum_i G_i(T, P, N_i) + Nk_{\mathrm{B}}T \sum_i x_i \ln x_i + \sum_i N_i \mu_{\mathrm{el}, i} \tag{13.104}$$

where the middle term on the RHS is the usual ideal mixing entropy and $\mu_{\mathrm{el}, i}$ gives the part of the chemical potential of species i that is due to electrostatic interactions. Note that this expression represents the electrostatic contribution to G using an integrated fundamental relation,

$$G_{\mathrm{el}} = \sum_i N_i \mu_{\mathrm{el}, i}$$

To determine this chemical potential, we must calculate the work required in order to add a single additional ion to the electrolyte system. Equivalently, we can think of this as the work required to convert a neutral particle into one that is fully charged. The differential work for such a process is given by

$$dW_{el, ion} = \Phi_{cloud} \, dq$$

$$= -\frac{\kappa}{4\pi\epsilon} q \, dq \tag{13.105}$$

where the subscript "el" indicates the electrostatic work. We integrate this expression from the uncharged to the fully charged state, obtaining

$$W_{el, ion} = -\frac{\kappa q^2}{8\pi\epsilon} \tag{13.106}$$

With these considerations, the chemical potential of species i is given by

$$\mu_i = \mu_{id, i} + \mu_{el, i} \qquad \text{with} \qquad \mu_{el, i} = -\frac{\kappa q_i^2}{8\pi\epsilon} \tag{13.107}$$

where $\mu_{id,i}$ is the ideal part. Referring back to Eqn. (13.104), we see that the electrostatic contribution to the Gibbs free energy is

$$G_{el} = \sum_i N_i \left(-\frac{\kappa q_i^2}{8\pi\epsilon} \right)$$

$$= -\frac{\kappa}{8\pi\epsilon} \sum_i N_i q_i^2$$

$$= -\frac{\kappa^3 V}{8\pi\epsilon} k_B T \tag{13.108}$$

In the last line, we used Eqn. (13.83) to rewrite the sum $\sum_i N_i q_i^2$ in terms of the Debye length. Together, Eqns. (13.104) and (13.108) give an expression for the total Gibbs free energy of the system; however, it is not yet in fundamental form because of the explicit presence of the volume V and the dependence of κ on the bulk ion densities. Even so, this result already illustrates two important physical features. First, G_{el} is always negative and shows that the presence of charges promotes mixing, which intuitively seems correct given that a central ion will attract a cloud of counterions. Second, the excess chemical potential in (13.107) implies a specific form of the activity coefficient. By comparison with Eqn. (13.36), we find that

$$\ln \gamma_i = -\frac{\kappa q_i^2}{8\pi\epsilon k_B T} \tag{13.109}$$

One can see that the activity coefficient scales as the square root of the salt concentration through the parameter κ. Indeed, experimental results for many salt solutions reveal this limiting behavior at low concentrations, supporting the validity of the Debye–Hückel model in that regime. Figure 13.11 illustrates the dependence of γ on the number density for a monovalent salt. Importantly, these results reinforce the failure of the ideal solution model alone to capture the behavior of electrolyte solutions at the dilute limit.

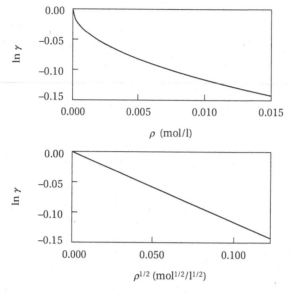

Figure 13.11. The logarithm of the activity coefficient in the Debye–Hückel model scales as the square root of the number density. Here the case for a monovalent salt is shown.

Namely, the electrostatic contributions to the chemical potential are nontrivial, reaching 5%–10% of $k_B T$ even at concentrations as low as 0.01 mol/L. Again, this behavior is due to the special long-ranged nature of electrostatic forces that, unlike van der Waals interactions, cannot be neglected even in very dilute systems.

> Electrostatic interactions can be long-ranged and exert non-negligible effects on solution thermodynamics, even in very dilute cases.

Problems

Conceptual and thought problems

13.1. Consider the binary mixture illustrated in Fig. 13.1 at the temperature labeled T_2. Sketch the behavior of this system in a P-x-y diagram in which P is the y-axis and the temperature is held constant at T_2.

13.2. Show that the relative volatilities of a mixture are constant and temperature-independent if (a) all of the pure species heats of vaporization are constant and equal and (b) the vapor and solution phases behave ideally.

13.3. Consider the Rachford–Rice approach to flash calculations. Show that the LHS of Eqn. (13.24) is a monotonic function of f, which facilitates the root-finding procedure. What kinds of solutions to f would be found if this approach were (incorrectly) applied in the cases $T < T_{bub}$ and $T > T_{dew}$?

13.4. The dissolution reaction for an electrolyte can be written in general terms as

$$M_{\nu_+} X_{\nu_-} \leftrightarrow \nu_+ M^{z_+} + \nu_- X^{z_-}$$

where the v terms give the stoichiometry of the ions upon dissociation and the z terms express their charge. For example, in the reaction $MgCl_2 \leftrightarrow Mg^{2+} + 2Cl^-$ one has $v_+ = 1$, $v_- = 2$, $z_+ = +2$, and $z_- = -1$. Clearly, for a neutral compound, one must have

$$v_+ z_+ + v_- z_- = 0$$

Let the chemical potentials of the ions M^{z+} and X^{z-} be denoted μ_+ and μ_-, respectively.

(a) Explain in physical terms why separately μ_+ and μ_- have no real physical significance and are immeasurable. Hint: consider the definition of the chemical potential.

(b) Prove that such a system, even though it has three species, is really described by only two chemical potentials. The first is that of the undissociated electrolyte given by μ_u. The second is the *mean chemical potential* of the ions, $\mu = v_+\mu_+ + v_-\mu_-$. Write the differential form of the Gibbs free energy for this case.

(c) Show that the activity coefficient of the ionic species can also be described by a mean value, $\gamma = (\gamma_+^{v_+}\gamma_-^{v_-})^{1/v}$, where $v = v_+ + v_-$.

Fundamentals problems

13.5. Derive expressions for the following derivatives along a phase-boundary for two-component liquid–vapor equilibrium. How do these behave at an azeotrope?

(a) $\left(\dfrac{dT}{dx_1}\right)_{P,\text{phase boundary}}$

(b) $\left(\dfrac{dP}{dy_1}\right)_{T,\text{phase boundary}}$ and $\left(\dfrac{dP}{dx_1}\right)_{T,\text{phase boundary}}$

(c) $\left(\dfrac{dP}{dT}\right)_{y_1,\text{phase boundary}}$ and $\left(\dfrac{dP}{dT}\right)_{x_1,\text{phase boundary}}$

13.6. A binary mixture is at liquid–vapor equilibrium and both phases behave ideally. Find expressions for the changes in liquid and vapor compositions with temperature at constant pressure, in terms of the compositions and the pure-phase vapor–liquid enthalpy differences Δh_1 and Δh_2. Namely, find

$$\left(\frac{dx_1}{dT}\right)_{P,\text{phase boundary}} = \dots \quad \text{and} \quad \left(\frac{dy_1}{dT}\right)_{P,\text{phase boundary}} = \dots$$

13.7. Consider a liquid mixture in equilibrium with its vapor. Show that the average or overall per-particle volume of the liquid, $v = V/N$, frequently obeys the following relation involving the liquid-phase mole fractions x_i and the partial pressures P_i in the vapor phase,

$$v\, dP = k_B T \sum_i x_i\, d\ln P_i \quad (\text{constant } T)$$

where $P = \sum_i P_i$ gives the total pressure.

13.8. Explicitly derive Eqn. (13.33) by filling in the missing steps.

13.9. Explicitly derive Eqn. (13.37) by filling in the missing steps.

13.10. Explicitly derive Eqns. (13.64) and (13.66) by filling in the missing steps.

13.11. Show that the polymer–solvent mixing entropy of Eqn. (13.66) might also be written in terms of mole fractions as

$$\frac{\Delta S_{\text{mix}}}{k_{\text{B}}} = -N_1 \ln\left(\frac{Mx_1}{1+(M-1)x_1}\right) - N_2 \ln\left(\frac{x_2}{M-(M-1)x_2}\right)$$

13.12. Consider a general C-component system of molecules of different sizes. Let component i be a polymer of length M_i (or a solvent if $M_i = 1$) and be present in the amount of N_i molecules.
   (a) Find general expressions for $\Delta S_{\text{mix}}/k_{\text{B}}$ and for $\Delta S_{\text{mix}}/(k_{\text{B}}V)$ involving the volume fractions of each species, ϕ_i.
   (b) Show that, if all of the components are of the same size, $M_1 = M_2 = \ldots = M_C$, one recovers the usual ideal solution entropy of mixing involving the species mole fractions.

13.13. Explicitly derive Eqn. (13.68) for ΔE_{mix} by making an analogy with the regular-solution approach described in Chapter 12.

13.14. Consider the Flory–Huggins model of a polymer–solvent system. Under conditions of phase separation at constant temperature and volume, the intensive Helmholtz free energy behaves like that shown in Figure 13.8. Consider conditions for which $\partial^2(\Delta A_{\text{mix}}/k_{\text{B}}TV)/\partial\phi_1^2 < 0$ such that the system phase-separates into a solvent-rich phase I with polymer fraction ϕ_1^{I} and a polymer-rich phase II with $\phi_1^{\text{II}} > \phi_1^{\text{I}}$. Under such conditions, the system reduces its free energy by moving to a common tangent line.
   (a) For any such solution in general, show that a line between two points in the $\Delta A_{\text{mix}}/(k_{\text{B}}TV)$ curves gives the corresponding (dimensionless, volume-normalized) free energy of a system with overall volume fraction ϕ_1 that is phase-separated into two phases with volume fractions specified at the line end points.
   (b) Show that when this line is tangent to the free energy curve at its end points, the chemical potentials in each phase are equal, $\mu_1^{\text{I}} = \mu_1^{\text{II}}$ and $\mu_2^{\text{I}} = \mu_2^{\text{II}}$.
   (c) Find M-dependent expressions for the critical volume fraction $\phi_{1,c}$ and exchange parameter $\chi_{12,c}$ above which no phase separation occurs.
   (d) The so-called theta temperature T_θ is the critical temperature for phase separation in the limit of very long polymers ($M \gg 1$). Find a scaling law for T_θ and its ultimate limit.

13.15. Consider a central point ion i surrounded by a number of point ions j. By comparing the Coulomb expression of (13.73) with (13.74), one can see that the electrostatic potential due to the j ions must be given by

$$\Phi(\mathbf{r}) = \sum_{j\neq i} \frac{q_j}{4\pi\epsilon|\mathbf{r}_j - \mathbf{r}|}$$

Show that, when used with Poisson's equation, (13.76), this is consistent with a delta-function distribution of ion charges in space,

$$n_f(\mathbf{r}) = \sum_{j \neq i} q_j \delta[\mathbf{r} - \mathbf{r}_j]$$

13.16. Using the Debye–Hückel theory, show that the total charge contained in the counterion atmosphere surrounding a central ion is equal and opposite to that of the ion, such that at long distances the central ion is completely "screened."

13.17. Consider the electrostatic Gibbs free energy G_{el} given by Eqn. (13.108). Find an expression for the electrostatic contribution to the system pressure, P_{el}. Subsequently, find an expression for the electrostatic Helmholtz free energy using $A_{el} = G_{el} - P_{el}V$. Hint: find the derivative $(\partial G_{el}/\partial V)_{T,N_i}$ and relate it to $(\partial P_{el}/\partial V)_{T,N_i}$ using standard thermodynamic relations. Then, integrate from the state of infinite volume where $P_{el} = 0$ to a state of finite V with finite P_{el}.

13.18. One can place a solution of ions in between two equally and oppositely charged plates to form a capacitor. In this case, cations will migrate towards the anode while anions will migrate towards the cathode. At equilibrium, there will be distinct distance-dependent concentrations of each species as a function of position with respect to the plates, $\rho_+(z)$ and $\rho_-(z)$, where $z = 0$ is the cathode and $z = L$ is the anode. That is, higher concentrations of cations will be present near the anode and higher concentrations of anions near the cathode. Here consider the case of a monovalent salt solution with $q_\pm = \pm e$.

(a) What must be constant across the distance z at equilibrium?

(b) Develop a mean-field lattice model of this system. Assume that the ions can occupy sites on a lattice with a total number of $V \geq N_- + N_+$ sites. Any remaining sites are occupied by solvent molecules, and only one molecule can occupy a site at any given time. The only energetic interactions present in the model are those due to a *mean electrostatic field* that accounts both for Coulombic interactions among the ions and for any *applied* electric fields. The form of the interaction is such that each cation experiences an energy $e\Phi$ and each anion an energy $-e\Phi$. Find an expression for the chemical potentials of each ion species. Express your answer in terms of the ion number densities $\rho_+ = N_+/V$ and $\rho_- = N_-/V$ and make use of Stirling's approximation.

(c) Find the two distributions $\rho_+(z)$ and $\rho_-(z)$ in terms of the distance-dependent electrostatic potential $\Phi(z)$. Assume that in the middle of the plates far from the electrodes, the potential is zero and the solution behaves as a bulk one such that $\rho_+ = \rho_- = \rho_b$.

In principle, with $\rho_+(z)$ and $\rho_-(z)$ from part (c), the field $\Phi(z)$ could then be found self-consistently by solving Poisson's equation with appropriate boundary conditions at the plates. This model was first developed by Bikerman [J. Bikerman, *Philos. Mag.*, Series 7, **33**, 384 (1942)].

13.19. A parallel-plate capacitor involves two planar interfaces separated by a dielectric medium. An electric potential Φ (with units of energy per charge) can be applied between the two plates, for example by connecting them to battery terminals. Owing to the potential, charges from within the medium migrate such that the surfaces develop equal and opposite net charges $+q$ and $-q$. The work associated with setting up this charge distribution is $\delta W = \Phi \, dq$ such that the fundamental equation for this system becomes

$$dE = T \, dS + \Phi \, dq + \mu \, dN$$

where μ gives the chemical potential for charges in the medium. Note that the capacitor is maintained in a fixed geometry.

(a) The differential capacitance of the system is defined by $C \equiv (\partial q / \partial \Phi)_{T,N}$. Show that this quantity must always be positive.

(b) At constant T, assuming a constant capacitance, show that the change in the chemical potential associated with charging the system is

$$\Delta \mu = -\frac{q^2}{2CN}$$

(c) A capacitor is connected to a battery that maintains a constant electric potential. It is also at fixed temperature. What quantity Z is minimized at equilibrium for these conditions? Express it in differential and integrated form.

(d) A capacitor is charged slowly and adiabatically, under nonisothermal conditions. Show that the effective capacitance, which is given by the rate of change $(\partial q / \partial \Phi)$ during this process, will be lower than the isothermal C as defined above.

Applied problems

13.20. At one bar, a mixture of mole fractions 0.20, 0.45, 0.25, and 0.10 of n-hexane, n-heptane, iso-heptane, and n-octane, respectively, is prepared. The saturation pressure of each species can be modeled by the equation $\log_{10} P^{\mathrm{vap}} = c_1 - c_2/(c_3 + T)$, where P^{vap} has units of mmHg, T has units of °C, and the constants c_1, c_2, and c_3 are given by Table 13.1.

Table 13.1 Values of the constants c_1, c_2, and c_3

Species	c_1	c_2	c_3
n-hexane	6.87024	1168.72	224.210
n-heptane	6.89385	1264.37	216.636
iso-heptane	6.87318	1236.03	219.545
n-octane	6.90940	1349.82	209.385

(a) Find the bubble-point temperature of this mixture and the corresponding vapor-phase compositions.

(b) Find the dew-point temperature and the corresponding liquid-phase compositions.

(c) At 92 °C, find both the liquid and vapor compositions and the fraction of the system that is vapor.

13.21. A mixture of ethanol and water forms an azeotrope at $x_1 = 0.894$ mole fraction ethanol, 78.2 °C, and 1 atm. The vapor pressures of ethanol and water at this temperature are 0.990 and 0.435 atm, respectively. It is proposed to describe the system by the two-parameter Margules model,

$$\frac{G_{ex}}{Nk_B T} = x_1 x_2 (ax_1 + bx_2)$$

Estimate values for the parameters a and b for this system.

13.22. Calculate the Debye length κ^{-1} for a 0.001 mol/l solution of NaCl in water at $T = 25$ °C. If the Debye–Hückel theory breaks down when κa exceeds, say, a value of 0.1, at what concentration will this occur? You may want to assume the ion radii are on the order of 1.5 Å.

FURTHER READING

K. Denbigh, *The Principles of Chemical Equilibrium*, 4th edn. New York: Cambridge University Press (1981).

K. Dill and S. Bromberg, *Molecular Driving Forces: Statistical Thermodynamics in Biology, Chemistry, Physics, and Nanoscience*, 2nd edn. New York: Garland Science (2010).

M. F. Doherty and M. F. Malone, *Conceptual Design of Distillation Systems*. Boston: McGraw-Hill (2001).

P. J. Flory, "Thermodynamics of high polymer solutions," *Journal of Chemical Physics* **9**, 660 (1941).

T. L. Hill, *An Introduction to Statistical Thermodynamics*. Reading, MA: Addison-Wesley (1960); New York: Dover (1986).

M. L. Huggins, "Solutions of long chain compounds," *Journal of Chemical Physics* **9**, 440 (1941).

D. A. McQuarrie, *Statistical Mechanics*. Sausalito, CA: University Science Books (2000).

A. Z. Panagiotopoulos, *Essential Thermodynamics*. Princeton, NJ: Drios Press (2011).

H. H. Rachford and J. D. Rice, "Procedure for use of electrical digital computers in calculating flash vaporization hydrocarbon equilibrium," *Journal of Petroleum Technology* **4**, 19 (1952).

J. M. Smith, H. V. Ness, and M. Abbott, *Introduction to Chemical Engineering Thermodynamics*, 7th edn. New York: McGraw-Hill (2005).

14 Solids

14.1 General properties of solids

The term "solids" denotes materials that generally have the following properties. From a microscopic perspective, the molecules in a solid are in a condensed, closely packed state, and they vibrate around a fixed equilibrium position. That is, molecules can be considered tethered near a specific location in space, since their diffusion is very slow relative to the time scales of observation. From a macroscopic point of view, solids have an *elastic modulus*. This means that the application of a stress to the material produces a strain as well as an opposing force that tends to return the solid to its original, unstrained state once the stress is removed. This contrasts with viscous behavior in which an applied stress results in continuous, permanent deformation, such as the flow of a liquid.

Generally speaking, there are two primary classes of solids. *Crystalline solids* are equilibrium states of matter in which the microscopic structure has a well-defined geometric pattern with long-range order: a crystalline lattice. In contrast to crystals, *amorphous solids* have no long-range order, meaning that they lack a lattice structure and regular positioning of the molecules. Glasses and many polymeric materials are amorphous. Frequently these systems are not at equilibrium, but evolve very slowly in time and are metastable with respect to a crystalline phase. They might be considered liquids of extremely high viscosity that are slowly en route to crystallization. However, typically the time scale to reach equilibrium is so long (perhaps longer than the age of the universe) that for all practical purposes the amorphous state appears solid and stable. Thus, in an empirical sense, often we can treat such systems as in quasi-equilibrium.

In earlier chapters, we sought expressions for the chemical potential μ for the systems examined because these provide a fundamental starting point for deriving other thermodynamic properties and for treating phase equilibrium. For gases and solutions, we developed these functions using microscopic models. While we treat a microscopic model of crystals later in this chapter, the resulting expression for the chemical potential will not be as simple as those determined thus far. Instead, for solids, we begin by making several broad generalizations that will enable us to study them without yet invoking a molecular picture. First, the temperature dependence of μ can be determined by integrating the fundamental relation

$$\left(\frac{\partial(\mu/T)}{\partial T}\right)_P = -\frac{h}{T^2} \tag{14.1}$$

which entails the molar enthalpy of the solid. Moreover, we have that $(\partial h/\partial T)_P = c_P$, so the enthalpy can be computed from the heat capacity. In the simplest approximation, the heat capacity is constant. Second, the pressure dependence of μ can be determined by integrating the relation

$$\left(\frac{\partial\mu}{\partial P}\right)_T = v \tag{14.2}$$

which entails the molar volume. For solids, this quantity is quite insensitive to temperature and pressure, and we can almost always integrate assuming constant v to write $\mu(T, P) \approx \mu(T, P_0) + (P - P_0)v$.

The thermodynamics of solids can be modeled in simple ways by considering the behavior of their chemical potential derivatives.

Moreover, for crystals the solid is frequently a pure state. That is, solutions of chemically distinct species often freeze into separate crystalline phases for each component. By separate, we mean crystals that are not mixed at the molecular level, though there can be "mixing" at a coarser level (μm to mm) due to the presence of multiple crystallites separated by grain boundaries.

The important implication of this last point is that solutions freezing into pure crystals eliminate the need for a crystal mixture model. From an applied perspective, this behavior also shows that freezing can be a powerful mechanism for separating solution components. Indeed, many industries requiring high-purity solids use crystallization from solution for this purpose, perhaps most notably in pharmaceutical manufacturing. Note that there are exceptions to this rule; some examples of non-pure crystals include hydrates and metallic alloys. Such cases complicate the analysis; however, often a first picture of solution freezing behavior can be obtained by assuming that the solid phases are pure. In any case, the considerations above are actually sufficient to solve a range of freezing problems, as we now explore.

14.2   Solid–liquid equilibrium in binary mixtures

Let us consider a mixture of two liquids that is slowly cooled at constant pressure until it begins to solidify. We will assume that species one acts as a solvent, while species two is rather dilute and behaves as a solute, permitting use of ideal solution models. This is essentially the freezing-point-depression problem that we considered in Chapter 12, with the exception that we will eventually consider the reverse case in which species one is dilute and species two is the solvent. The original case behaves such that, when the mixture is cooled below its melting temperature, pure solvent will crystallize out and attain equilibrium with the solution phase. At equilibrium between the solution and crystal, the chemical potentials of the solvent in each phase must be equal:

$$\mu^{*X}(T, P) = \mu^{*L}(T, P) + k_B T \ln x \rightarrow \Delta\mu = -k_B T \ln x \tag{14.3}$$

where implicitly all variables apply to the solvent and the superscripts "X" and "L" pertain to the crystal and liquid, respectively. Since $\Delta\mu = \mu^{*L} - \mu^{*X}$ is the difference in the pure-solvent per-particle Gibbs free energy between the two phases, we can use a derivative relationship to connect it to the corresponding enthalpy difference:

$$\left(\frac{\partial(\Delta\mu/T)}{\partial T}\right)_P = -\frac{\Delta h}{T^2} \tag{14.4}$$

Note that we define the delta quantities as pure liquid minus pure crystal, rather than the reverse. In Chapter 12, we treated the enthalpy difference as constant with temperature. Here we derive a slightly more general relationship that allows for changes in Δh with T. We assume constant heat capacities for both the liquid and crystal, and integrate to obtain the enthalpy of either phase relative to the pure-solvent melting temperature T_m,

$$\left(\frac{\partial h}{\partial T}\right)_P = c_P \quad \rightarrow \quad h = h_m + c_P(T - T_m) \tag{14.5}$$

By substituting an instance of (14.5) for each of the two phases into (14.4) and integrating at constant pressure, we obtain the equation

$$\int_{T_m}^T d(\Delta\mu/T) = -\int_{T_m}^T \frac{\Delta h_m + (T - T_m)\Delta c_P}{T^2} dT \tag{14.6}$$

Here, Δh_m is the enthalpy of melting or latent heat of fusion of pure solvent evaluated at T_m and Δc_P is the corresponding difference in heat capacities of the liquid and solid. To evaluate the integral, we use the fact that $\Delta\mu = 0$ at T_m because the liquid and crystal chemical potentials must be equal at the pure-component melting temperature. This gives

$$\frac{\Delta\mu}{T} = (\Delta h_m - T_m \Delta c_P)\left(\frac{1}{T} - \frac{1}{T_m}\right) - \Delta c_P \ln\left(\frac{T}{T_m}\right) \tag{14.7}$$

Finally, Eqn. (14.7) provides a suitable expression for $\Delta\mu$ in Eqn. (14.3), yielding

$$k_B T \ln x = -(\Delta h_m - T_m \Delta c_P)\left(1 - \frac{T}{T_m}\right) + T \Delta c_P \ln\left(\frac{T}{T_m}\right) \tag{14.8}$$

Alternatively,

$$x = \exp\left[-\left(\frac{\Delta h_m}{k_B T_m} - \frac{\Delta c_P}{k_B}\right)\left(\frac{T_m}{T} - 1\right) + \frac{\Delta c_P}{k_B} \ln\left(\frac{T}{T_m}\right)\right] \tag{14.9}$$

Equations (14.8) and (14.9) provide a way to determine a phase boundary for the equilibrium between a solvent–solute liquid mixture and a pure solvent crystal, namely the line given by $x(T)$. We can express this line in terms of the solute mole fraction,

$$x_{\text{solute}} = 1 - \exp\left[-\left(\frac{\Delta h_m}{k_B T_m} - \frac{\Delta c_P}{k_B}\right)\left(\frac{T_m}{T} - 1\right) + \frac{\Delta c_P}{k_B}\ln\left(\frac{T}{T_m}\right)\right] \tag{14.10}$$

Example 14.1 *Ethylene glycol is a compound frequently used in antifreeze formulations to lower the melting point of water. Determine an approximate phase diagram for dilute mixtures of ethylene glycol in water at atmospheric pressure. Assume ideal solution behavior and that the heat capacities of the liquid and solid phases are constant.*

When the solution is dilute in ethylene glycol, water is the solvent and thus we need its pure-phase properties, here evaluated at water's normal melting point (0 °C): $c_P^X = 37.0$ J/mol K, $c_P^L = 75.4$ J/mol K, $T_m = 273$ K, and $\Delta h_m = 6{,}020$ J/mol. Using Eqn. (14.10), we obtain an expression for the equilibrium mole fraction of ethylene glycol (EG),

$$x_{\text{EG}} = 1 - \exp\left[1.97\left(\frac{273}{T} - 1\right) + 4.62\ln\left(\frac{T}{273}\right)\right]$$

where T is in units of degrees Kelvin. This line is shown in the T-x diagram of Fig. 14.1.

Figure 14.1 shows that binary freezing of this sort happens over a range of temperatures. Continuing with the system in Example 14.1, let the initial ethylene glycol mole fraction be $x_{\text{EG},0} = 0.05$. At high temperatures, the mixture is entirely a liquid, but when it

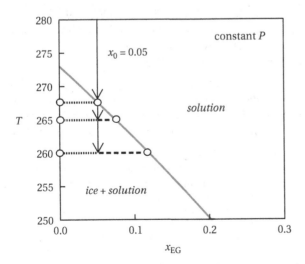

Figure 14.1. A phase diagram for dilute ethylene glycol in water based on ideal solution models illustrates phase separation for a solution initially with $x_{\text{EG},0} = 0.05$. Two phases are formed at low temperatures: an ice phase for which $x_{\text{EG}} = 0$ (left axis) and a solution phase with x_{EG} given by Eqn. (14.10) (gray line). When a solution of the two components is cooled into the two-phase region, the dotted and dashed tie lines give the relative total molar amounts of each phase.

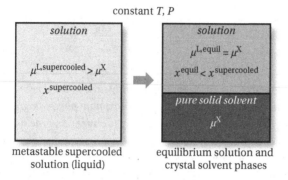

constant T, P

Figure 14.2. On the left, a supercooled liquid solution exists at conditions where the chemical potential of the solvent in the solution phase (μ^L) is greater than the chemical potential of the corresponding pure-solid solvent phase at the same temperature and pressure (μ^X). This creates a driving force for the formation of solid solvent. However, as solvent freezes out, the solution solvent mole fraction (x) and hence chemical potential decrease until the latter becomes equal to that of the solid phase. At this point equilibrium is reached, as shown on the right.

is cooled to about $T = 267$ K, an infinitely small amount of ice (pure solid solvent) forms. As the temperature is lowered further, increasing amounts of water freeze out, and thus the composition of the solution phase becomes enriched in ethylene glycol (solute). We can determine the relative amounts of each phase in the T–x diagram by using *tie lines*, in the same way as we did for liquid–vapor equilibrium in Chapter 13. Indeed, the number of moles of the solution phase is proportional to the difference $x_{EG} - 0$ (where zero indicates the solute fraction in ice) and that of the solid ice phase is proportional to $x_{EG} - x_{EG,0}$. This behavior differs significantly from the case of pure-solvent freezing, which occurs entirely at a single temperature.

A careful consideration of the chemical potentials shows why binary freezing happens in this manner. Imagine supercooling the solution to a temperature below the phase boundary such that initially it remains entirely in a metastable liquid state. At this point, the chemical potential of ice is lower than that of water in the solution. This creates a driving force for ice to form because doing so would lower the overall Gibbs free energy. As ice appears, however, the mole fraction of water in the solution phase decreases and thus so does the corresponding chemical potential. Yet, the chemical potential of the pure-ice phase remains constant, being fully determined by the temperature and pressure. As a result, ice only continues to freeze out of the solution until, due to the changing compositions, the solution chemical potential of water finally becomes equal to that of the solid phase. These ideas are illustrated in Fig. 14.2.

Example 14.2   *We can perform an equivalent analysis in which water is the dilute species while ethylene glycol acts as the solvent. Again, ideal solution models provide an adequate starting point in this opposite limit. Determine an approximate phase diagram for dilute mixtures of water in ethylene glycol in water at atmospheric pressure. Assume ideal solution behavior and that the heat capacities of the liquid and solid phases are constant.*

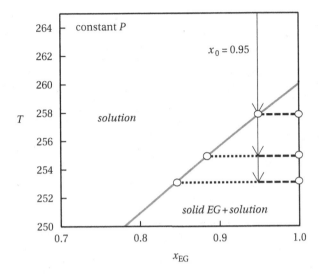

Figure 14.3. A phase diagram for dilute water in ethylene glycol illustrates phase separation for a solution initially with $x_{EG,0} = 0.95$. The two phases formed at low temperatures include a pure solid ethylene glycol phase (for which $x_{EG} = 1$) and a solution phase with x_{EG} given by Eqn. (14.9) (gray line). In the two-phase region, tie lines give the relative total molar amounts of each phase.

This case is the opposite of Example 14.1 in that ethylene glycol is now the solvent and we require its properties at its melting point, $c_P^X = 95$ J/mol K, $c_P^L = 140$ J/mol K, $T_m = 260$ K, and $\Delta h_m = 9,900$ J/mol. This time, we use Eqn. (14.9) because the ethylene glycol mole fraction corresponds to that of the solvent. With the numbers above, we then find

$$x_{EG} = \exp\left[-0.833\left(\frac{260}{T} - 1\right) + 5.41 \ln\left(\frac{T}{260}\right)\right]$$

where T is in units of degrees Kelvin, as before. This phase boundary is illustrated in Fig. 14.3.

Example 14.2 shows that we can calculate an entirely different phase boundary corresponding to the case in which the solvent is ethylene glycol and the solute is water. This kind of boundary is valid in the limit of ethylene glycol-rich systems near $x_{EG} = 1$, as shown in Fig. 14.3. Here the solution freezing behavior is qualitatively similar, but with one important distinction: the solid phase that forms consists of ethylene glycol, not water. That is, low temperatures involve phase equilibrium between a pure-solid ethylene glycol phase and a solution phase. As the temperature is decreased, increasing amounts of solid are formed, and the solution composition becomes depleted in ethylene glycol. Thus, an important feature of binary solid–liquid equilibrium is that the solid phase first formed depends on the initial composition of the mixture.

What is the behavior of this system over the full range of mole fractions? If we consider simultaneously the boundaries found in Examples 14.1 and 14.2, the overall phase

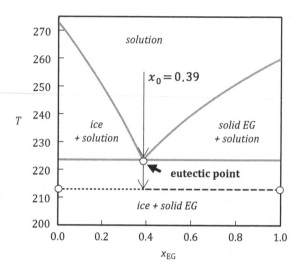

Figure 14.4. A phase diagram for mixtures of water and ethylene glycol shows that solutions initially with $x_{EG,0} = 0.39$ pass through a eutectic point when cooled. At lower temperatures only pure-solid phases are observed in coexistence, with the dotted and dashed tie lines giving their relative molar amounts.

diagram appears as in Fig. 14.4. The intersection of the two lines near 224 K and $x_{EG} = 0.39$ represents an important feature called the *eutectic point*. At temperatures below this, the boundary extending from the water-rich side suggests that the solution would need to become enriched in ethylene glycol in order to maintain equilibrium with the solid ice phase. At the same time, the boundary extending from $x_{EG} = 1$ suggests the opposite: that the solution would need to be enriched in water in order to coexist with solid ethylene glycol. As a result, there is no way the solution can adjust so as to continue to maintain equilibrium with both solid phases; its chemical potential will always be higher. Below the eutectic temperature, therefore, complete solidification occurs and the system consists of coexisting pure ice and pure solid ethylene glycol. These two phases may not demix at the bulk level, but rather can exist as small crystalline domains ("grains") that nonetheless still behave macroscopically.

The eutectic point delineates three composition regimes that have qualitatively distinct behaviors upon isobaric cooling. At lower compositions water freezes out into ice, whereas at higher ones ethylene glycol leaves the solution to form a solid. Both processes continue to produce increasing amounts of solid phase until the eutectic temperature is reached and the solution attains the eutectic composition. Alternatively, a solution beginning exactly at the eutectic mole fraction will remain a liquid all the way down to the eutectic temperature, at which point it will completely transform into coexisting solid phases. Thus the eutectic point gives the lowest possible temperature the mixture can reach as well as the composition that sustains the liquid phase (without any solid forming) over the widest temperature range. This fact can be exploited for applications; for example, menthol and camphor are both solids at room temperature but mixtures of them form a liquid above a eutectic point, which is useful in some pharmaceutical formulations.

Of course, Fig. 14.4 gives a highly approximate version of the actual phase behavior of ethylene glycol–water solutions. On the one hand, the boundaries drawn assume ideal solution behavior throughout the entire range of mole fractions. On the other, ethylene glycol and water form a hydrate phase – a distinct solid phase involving both components – at intermediate compositions that adds a small additional phase boundary near the eutectic point. These considerations aside, the ideal solution predictions do give a rather reasonable picture of the overall phase behavior of this system. The predicted eutectic point is at 224 K and $x_{EG} = 0.39$, while experiments show for metastable, hydrate-free systems that it occurs at 210 K and $x_{EG} = 0.34$ [J. B. Ott, J. R. Goates, and J. D. Lamb, *J. Chem. Thermodyn.* **4**, 123 (1972)]. In any case, the general ideas presented here apply to many binary liquid mixtures.

For ideal binary solutions that crystallize into pure solid phases, freezing occurs over a range of temperatures, with increasing amounts of a pure solid forming at lower temperatures. Moreover, the solid phase that is produced depends on the initial concentration of the mixture, and the temperature at which solidification begins is lower than the corresponding pure-component melting point (i.e., it is depressed). The **eutectic point** defines the lowest possible temperature at which the solution phase can exist. A system at the eutectic composition will transform directly from solution to coexisting solid phases at the eutectic temperature.

14.3 Solid–liquid equilibrium in multicomponent solutions

The considerations above show that, when a mixture is cooled, the species that first begins to solidify is concentration-dependent. It is not hard to imagine that the phenomenology becomes increasingly complicated as the number of components grows. Indeed, the solid–liquid phase behavior for solutions involving even a few components can be quite complex. One motivating question is as follows: which species will be the first to freeze out of a mixture of C different components? To proceed, we assume that only pure solids are formed, neglecting hydrates and other mixed solid phases.

We begin by considering the difference between the chemical potential of a component i in the solution phase and that of its pure solid state at the same temperature and pressure,

$$\delta\mu_i \equiv \mu_i^X(T,P) - \mu_i^L(T,P,\{x\}) \tag{14.11}$$

where $\mu_i^X(T,P)$ gives the chemical potential of solid i in pure form. Note that the chemical potential in the liquid depends in general on the entire solution composition. A general form may involve the use of an activity coefficient:

$$\delta\mu_i = \mu_i^X(T,P) - \mu_i^L(T,P) - k_B T \ln x_i - k_B T \ln \gamma_i(T,P,\{x\}) \tag{14.12}$$

Here the term $\mu_i^L(T, P)$ without a composition dependence indicates the chemical potential of pure liquid i. For the sake of simplicity, we will assume that the solution behaves ideally throughout the entire range of compositions, such that $\gamma_i = 1$ and

$$\delta\mu_i = \mu_i^X(T, P) - \mu_i^L(T, P) - k_B T \ln x_i \qquad (14.13)$$

The quantity $\delta\mu_i$ is useful because it indicates whether a component will remain in solution or whether it is favorable for part of it to crystallize out. By favorable, we mean a lowering of the overall system Gibbs free energy since the quantity $\delta\mu_i$ gives the change in G as one molecule of species i is transferred from solution to a phase of pure solid i. If $\delta\mu_i$ were positive then it would be unfavorable for component i to freeze, and it would remain completely in the solution phase. On the other hand, if $\delta\mu_i$ were equal to zero, the system would be at equilibrium with respect to transferring molecules between a solid phase of i and the solution. Therefore, we would expect to see i in both solution and solid form. Finally, if $\delta\mu_i$ were negative, then there would be a driving force for component i to form solid phase. At this point, the system would not be at equilibrium until enough i had frozen out and depleted its composition in the solution to establish a point where $\delta\mu_i = 0$.

From our discussion in the previous section, the simplest approximation for the difference in pure-phase chemical potentials involves the enthalpy of fusion. Taking Eqn. (14.7) with $\Delta c_P = 0$, we have

$$\mu_i^L(T, P) - \mu_i^X(T, P) \approx \Delta h_{m,i}\left(1 - \frac{T}{T_{m,i}}\right) \qquad (14.14)$$

Keep in mind that $T_{m,i}$ is the melting temperature and $\Delta h_{m,i}$ is the latent heat of fusion of pure i. Substitution into (14.13) gives

$$\delta\mu_i = -\Delta h_{m,i}\left(1 - \frac{T}{T_{m,i}}\right) + k_B T \ln x_i \qquad \mathbf{(14.15)}$$

Upon isobaric cooling, $\delta\mu_i$ changes sign and becomes negative for one component i when it becomes favorable to freeze. If all other components remain in solution, the temperature at which $\delta\mu_i = 0$ is given by

$$T = T_{m,i}\left(1 - \frac{k_B T_{m,i}}{\Delta h_{m,i}} \ln x_{i,0}\right)^{-1} \qquad (14.16)$$

where $x_{i,0}$ denotes the original composition of i in the solution prior to any solidification. Equation (14.16) predicts a different freezing temperature for each of the C components, and the first to freeze has the highest. It is similar to the expression for freezing-point depression developed in Chapter 12, although we will not simplify it using the assumption that $x_{i,0}$ is close to one because that might not be the general case here.

Imagine that component 1 is the first to freeze, and that this behavior occurs at a temperature denoted by T_1 found from Eqn. (14.16). At T_1 the values of $\delta\mu$ are positive for all other species and they remain in solution, and only an infinitesimally small amount of solid component 1 is formed. As the temperature is then lowered below T_1, the amount of solid phase increases as species 1 is removed from solution and, in turn, the compositions of the other components increase in the solution phase. The composition $x_1(T)$ in the solution changes so as to maintain crystal–solution equilibrium for component 1. This means that $\delta\mu_1 = 0$ as the temperature decreases, giving the relation

$$x_1(T) = \exp\left[-\frac{\Delta h_{m,1}}{k_B}\left(\frac{1}{T} - \frac{1}{T_{m,1}}\right)\right] \tag{14.17}$$

The freezing temperatures originally predicted for the other components by Eqn. (14.16) become meaningless at this point because the solution composition changes. Indeed, their concentrations are governed by

$$x_{i\neq1}(T < T_1) = x_{i,0}\frac{1 - x_1(T)}{1 - x_{1,0}} \tag{14.18}$$

This expression stems from a mole balance on all components exempting species 1:

$$(1 - x_{1,0})\,N_0 = (1 - x_1)\,N_S \tag{14.19}$$

where N_0 is the total number of molecules and N_S is the number in the solution phase. Equation (14.18) is easily recovered using $x_{i\neq1} = N_{i,0}/N_S = x_{i,0}N_0/N_S$ with (14.19).

As the temperature continues to lower, both the solution concentrations and the $\delta\mu$ values change for the unfrozen species. Their values are constrained by the relation

$$\delta\mu_{i\neq1} = -\Delta h_{m,i}\left(1 - \frac{T}{T_{m,i}}\right) + k_B T \ln\left[x_{i,0}\frac{1 - x_1(T)}{1 - x_{1,0}}\right] \tag{14.20}$$

where Eqn. (14.18) has been used to provide the now temperature-dependent concentration $x_{i\neq1}(T)$. Eventually, a second component, species 2, will reach the condition $\delta\mu_2 = 0$, and thus it will be favorable for it to begin forming a solid phase. Let the temperature at which this occurs be denoted T_2. As the temperature decreases further, both species 1 and species 2 continue to freeze, again concentrating the solution. The concentrations x_1 and x_2 both become governed by the conditions $\delta\mu_1 = 0$ and $\delta\mu_2 = 0$, which give two equations of the form in (14.17). The concentrations of the other components are given again by a mole balance:

$$x_{i>2}(T < T_2) = x_{i,0}\frac{1 - x_1(T) - x_2(T)}{1 - x_{1,0} - x_{2,0}} \tag{14.21}$$

Qualitatively speaking, similar behavior is found with further cooling. Namely, additional components solidify as they reach the condition $\delta\mu = 0$. Once a component begins to freeze, its $\delta\mu$ remains zero for all lower temperatures and thus its composition becomes governed by an equation like (14.17). The compositions of all species that are in solid–liquid equilibrium then determine the new composition of the solution phase through mole balances. Of course, these behaviors are predicated on ideal solution behavior and pure solid phases; clearly, more complicated behavior could occur if these assumptions were not valid.

Eventually, when only one component has yet to freeze, say component C, there will be a single temperature at which the solution vanishes entirely and only solid phases remain in coexistence. This is the point at which $\delta\mu_C$ finally reaches zero along with all of the other species. The corresponding temperature, T_C, is thus found by the simultaneous set of conditions

$$\delta\mu_i = 0 \quad \text{for all } i \tag{14.22}$$

Using Eqn. (14.15), this is equivalent to solving the $C + 1$ equations

$$T_C = T_{m,i} \left(1 - \frac{k_B T_{m,i}}{\Delta h_{m,i}} \ln x_i \right)^{-1} \quad \text{for} \quad i = 1, 2, \dots, C,$$

$$\sum_i x_i = 1 \tag{14.23}$$

The variables to be determined are the C compositions and the temperature T_C. Similar to the eutectic point discussed in the previous section, T_C gives the lowest possible temperature at which any solution can exist. For lower temperatures, $\delta\mu < 0$ for all species and therefore it is always favorable for solution-phase molecules to transfer to their respective solid phases. It is important to note here that T_C is distinct from a critical temperature. The C subscript simply indicates the index of the final species to freeze.

14.4   A microscopic view of perfect crystals

Crystals differ significantly from liquids and gases in one primary respect: they exhibit a well-defined pattern of microscopic ordering. That is, the positions of a crystal's atoms or molecules form a predictable, repeating lattice structure. The emergence of such order ultimately stems from both geometric and energetic considerations: repeated lattices often represent the most efficient ways to pack molecules in low-energy configurations at low temperatures or high pressures. In finite-temperature crystals, atoms vibrate around

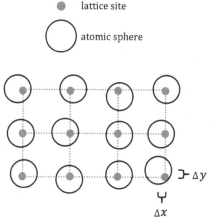

Figure 14.5. The atoms in a crystal vibrate around an equilibrium position corresponding to a lattice site, here shown as a simple square lattice. At any instant in time, an atom's position may be offset from its associated lattice site by an amount $\Delta \mathbf{r} = (\Delta x, \Delta y, \Delta z)$.

their lattice position but no one atom wanders substantially far from its "home" lattice site, at least on short time scales. Indeed, the diffusion of atoms in crystals tends to be orders of magnitude smaller than that of liquids. Figure 14.5 illustrates this scenario.

Here we derive a simple microscopic model of monatomic crystals starting from quantum considerations. As we will see, quantum mechanics is necessary to correctly describe crystal properties at low and sometimes even moderate temperatures. One of the main behaviors requiring a quantum description is that the heat capacity of crystals vanishes at absolute zero. To begin, we make a number of assumptions about the molecular picture.

The simplest model of a monatomic crystal has the following characteristics.

(1) N atoms each vibrate around a well-defined equilibrium lattice point in space. We will not assume a specific crystalline lattice, but merely demand that there is a one-to-one correspondence between atoms and lattice points.

(2) The atoms are all of the same type. Moreover, they are structureless and have spherically-symmetric interactions.

An atom remains in the vicinity of its lattice site because the interactions with its neighbors create a local *potential energy minimum*. In other words, the potential energy increases as an atom moves away from its equilibrium position. To make these ideas concrete, let the potential energy of the crystal be U_0 when each atom is at its lattice site, the positions of which are denoted by $\mathbf{r}_0^N$. We can consider the change in energy $\Delta U = U - U_0$ when atoms are displaced from their lattice positions by an amount $\Delta \mathbf{r}^N = \mathbf{r}^N - \mathbf{r}_0^N = (\Delta x_1, \Delta y_1, \ldots, \Delta z_N) = (x_1 - x_{1,0}, y_1 - y_{1,0}, \ldots, z_N - z_{N,0})$.

Let us imagine that a single atom is displaced in a single direction – say, atom 1 in the x direction – while all others remain fixed at their lattice sites. Figure 14.6 shows how the potential energy might change during this process, ultimately increasing sharply as the atom begins to "overlap" with its neighbors. In general this curve has a complex functionality because it represents a sum over many types of nonbonded molecular interactions from nearby atoms, including van der Waals forces, electrostatics, and

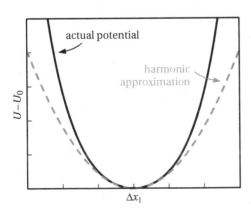

Figure 14.6. In general, an atom in a crystal sits near a local potential energy minimum. Perturbing one of its coordinates away from the equilibrium lattice position gives rise to an increase in energy, here illustrated with $\Delta x_1 = x_1 - x_{1,0}$. For small perturbations, the energy is well represented by a harmonic approximation.

close-range repulsive forces. However, at small enough perturbations one can Taylor-expand the potential energy about the minimum to second order:

$$\Delta U(x_1) = \Delta U(x_{1,0}) + \Delta x_1 \frac{\partial \Delta U}{\partial x_1}(x_{1,0}) + \frac{\Delta x_1^2}{2} \frac{\partial^2 \Delta U}{\partial x_1^2}(x_{1,0}) + \cdots$$

$$\approx \frac{\Delta x_1^2}{2} \frac{\partial^2 \Delta U}{\partial x_1^2}(x_{1,0}) = \frac{\Delta x_1^2}{2} \frac{\partial^2 U}{\partial x_1^2}(x_{1,0}) \tag{14.24}$$

In the second line, the zeroth- and first-order terms vanish because the expansion is performed around an energy minimum with $\Delta U = 0$. We can also equate derivatives of ΔU with those of U itself since U_0 is a constant. The final functionality of Eqn. (14.24) is quadratic in distance and reminiscent of springs in mechanics. A system with this kind of quadratic behavior of the energy has the general name *harmonic oscillator.*

We continue the analogy with a spring by recognizing that the effective force constant implied by (14.24) is given by the $\partial^2 U / \partial x_1^2$ term. A spring's force constant dictates the frequency with which it vibrates, v. The latter quantity is given by

$$v = \frac{1}{2\pi} \sqrt{\frac{1}{m} \frac{\partial^2 U}{\partial x_1^2}(x_{1,0})} \tag{14.25}$$

Here the frequency is related to the mass of the atom m and the curvature at the bottom of the potential energy well in which it resides. The form of the frequency simply extends from standard mechanics of harmonic springs. Importantly, we can rewrite Eqn. (14.24) using v in place of the potential energy curvature:

$$\Delta U(x_1) = \frac{\Delta x_1^2}{2} m(2\pi v)^2 \tag{14.26}$$

In general, the frequency depends on the crystal lattice structure and the particular intermolecular energetic interactions specific to the species at hand. It is also dependent on the number and proximity of nearby atoms, and thus is affected by density.

14.5   The Einstein model of perfect crystals

Einstein developed the first successful model of crystals, and his general approach was the following. While above we examined the change in potential energy for a perturbation in one direction of one atom, in general we could change all three coordinates of each of the N atoms. With each perturbation away from the lattice positions, the potential energy would increase. In essence, Einstein's approximation was that each such perturbation is independent and equivalent.

The **Einstein model of crystals** assumes the following.

(1) Each direction of each atom gives an independent harmonic oscillator, such that they number $3N$.
(2) The frequencies of the harmonic oscillators are the same, being equal to v.

According to these assumptions, the equation that describes potential energy changes in the system is given by

$$\Delta U = 2\pi^2 mv^2 (\Delta x_1^2 + \Delta y_1^2 + \cdots + \Delta z_N^2) \tag{14.27}$$

Said differently, a crystal's energy as a function of its atomic coordinates might be approximated as a set of $3N$ independent harmonic oscillators.

The next step in our analysis is to compute the energy eigenspectrum from the potential in Eqn. (14.27) using quantum mechanics. Without loss of generality, we will assume $U_0 = 0$. Schrödinger's equation for a one-dimensional oscillator reads

$$-\frac{\hbar}{2m} \left(\frac{\partial^2 \psi}{\partial x^2} \right) + (2\pi^2 mv^2)x^2 \psi = \epsilon \psi \tag{14.28}$$

where ϵ gives the energy eigenvalue of the oscillator. Though we do not derive them explicitly, the solutions are readily found by solving this second-order differential equation using standard techniques,

$$\epsilon = \left(n + \frac{1}{2} \right) hv, \qquad n = 0, 1, 2, 3, \ldots \tag{14.29}$$

The variable n is the *quantum number* and specifies the state of the oscillator. Equation (14.29) shows that the allowable energies form a ladder of equally spaced values that each differ by hv. Since there are $3N$ independent harmonic oscillators according to (14.27), we obtain one such energy spectrum for each of them. Thus we have $3N$ energies ϵ_i specified by $3N$ quantum numbers n_i.

To develop the thermodynamic properties of this model, we must determine a form for the density of states $\Omega(E, V, N)$. Given a total energy E, this function counts the number of ways that we can pick the $3N$ n_i values such that

$$E = \sum_i \epsilon_i = \sum_i \left(n_i + \frac{1}{2} \right) hv \tag{14.30}$$

Another way of writing (14.30) is

$$\sum_i n_i = n_t, \qquad \text{where} \qquad n_t \equiv \left(\frac{E}{hv} - \frac{3N}{2} \right) \tag{14.31}$$

In other words, specifying the total energy is equivalent to specifying the sum of all quantum numbers. We can therefore write for the crystal that

$$\Omega(E, V, N) = \text{ways to assign integers to all } n_i \text{ such that} \sum_i n_i = n_t \tag{14.32}$$

A simple change of perspective helps to evaluate the density of states. We can think of n_t as the number of energy quanta that we parcel out and assign to the $3N$ oscillators. These quanta are indistinguishable in the sense that it does not matter in what order we make assignments, and more than one quantum can be associated with a given oscillator since any n_i can be greater than one. The problem becomes one of assigning n_t indistinguishable objects to $3N$ bins in which more than one object can occupy each bin. We considered an analogous problem in Example 2.2 for placing particles on lattice sites. The number of combinations is

$$\Omega\,(E, V, N) = \frac{(n_{\mathrm{t}} + 3N - 1)!}{(3N - 1)!n_{\mathrm{t}}!} \qquad (14.33)$$

Taking the logarithm, and simplifying using Stirling's approximation and $3N - 1 \approx 3N$ for large N, we find the entropy to be

$$S(E, V, N) = -k_{\mathrm{B}} \left[3N \ln \left(\frac{3N}{n_{\mathrm{t}} + 3N} \right) + n_{\mathrm{t}} \ln \left(\frac{n_{\mathrm{t}}}{n_{\mathrm{t}} + 3N} \right) \right] \qquad (14.34)$$

Note that the energy dependence is through the variable n_v, which is given by (14.31). Indeed, Eqn. (14.34) is in fundamental form, and in principle we can derive from it all of the thermodynamic properties of the crystal. Let us focus specifically on the temperature–energy relationship, which we obtain through a derivative,

$$\frac{1}{T} = \left(\frac{\partial S}{\partial E} \right)_{V,N}$$

$$= \left(\frac{\partial S}{\partial n_{\mathrm{t}}} \right)_{V,N} \left(\frac{\partial n_{\mathrm{t}}}{\partial E} \right)_{V,N}$$

$$= -\frac{k_{\mathrm{B}}}{h v} \ln \left(\frac{n_{\mathrm{t}}}{n_{\mathrm{t}} + 3N} \right) \qquad (14.35)$$

After substituting for n_{t} and rearranging, we find

$$E(T, V, N) = 3Nh v \left(\frac{1}{2} + \frac{1}{e^{h v/k_{\mathrm{B}} T} - 1} \right) \qquad (\mathbf{14.36})$$

Notice that, as the temperature reaches zero, we have all of the following behaviors:

$$T \to 0, \quad E \to \frac{3Nh v}{2}, \quad n_{\mathrm{t}} \to 0, \quad S \to 0, \quad \Omega \to 1 \qquad (14.37)$$

These behaviors are quantum in nature. At absolute zero the crystal has a single accessible microstate. Interestingly, it sits at a ground-state energy that is finite, even though the potential in Eqn. (14.27) has a minimum at zero. The "residual" energy can be viewed as a manifestation of quantum uncertainty since the state $\Delta U = 0$ would imply that all particles are positioned exactly at their lattice sites. Moreover, the entropy is in absolute terms and vanishes at $T = 0$ for the perfect crystal. A classical treatment would not be able to identify an absolute value of S and would actually predict it to diverge towards negative infinity at absolute zero. All of these points have important implications for the third law of thermodynamics, as discussed in Chapter 15.

Another distinctive feature of this crystal model is the behavior of the heat capacity, which follows from the temperature derivative of (14.36),

$$c_V^{\mathrm{Einstein}} = 3k_{\mathrm{B}} \left(\frac{h v}{k_{\mathrm{B}} T} \right)^2 \frac{e^{h v/k_{\mathrm{B}} T}}{(e^{h v/k_{\mathrm{B}} T} - 1)^2} \qquad (14.38)$$

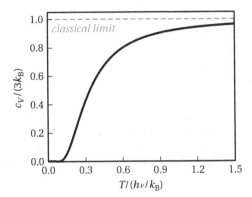

Figure 14.7. The heat capacity of the Einstein model of a perfect crystal decreases towards zero at absolute zero. At higher temperatures, it approaches the classical limit of $c_V = 3k_B$. Note that the heat-capacity and temperature axes are both expressed in dimensionless terms.

The combination of variables $h\nu/k_B$, which has units of temperature, appears frequently in this expression. It becomes convenient to introduce an "Einstein temperature" Θ_E to simplify the presentation,

$$\frac{c_V^{\text{Einstein}}}{3k_B} = \left(\frac{\Theta_E}{T}\right)^2 \frac{e^{\Theta_E/T}}{(e^{\Theta_E/T} - 1)^2} \quad \text{with} \quad \Theta_E \equiv \frac{h\nu}{k_B} \quad (14.39)$$

Figure 14.7 depicts the temperature dependence of c_V. At high temperatures the heat capacity approaches $3k_B$. This is the value that we would have calculated for *every* temperature if we had not used quantum mechanics to compute the entropy, but instead had taken a classical approach. (To see this, you can take the limit of the expression above as $h \to 0$. An end-of-chapter problem also develops the classical model explicitly.) In fact, one of the early successes of statistical mechanics and quantum theory was to predict that the heat capacity of a crystal should decrease to a value of zero at absolute zero. In the early twentieth century, this had been observed experimentally but could not be explained using purely classical arguments.

> The Einstein model of crystals predicts a vanishing heat capacity as the temperature approaches absolute zero. This is quantum behavior and contrasts with classical models, which always give a heat capacity of $3k_B$ for monatomic crystals.

For a point of reference, the values of Θ_E for some typical metallic crystals are 2,240 K for diamond, 415 K for iron, 343 K for copper, and 225 K for silver. Thus, for many of these substances, quantum effects are important at room temperature. Copper, for example, has a room temperature value of $T/\Theta_E = 0.86$, and the heat capacity is roughly 90% of the classical value. Therefore, classical models of crystals can fail even at moderate conditions.

What about other thermodynamic properties of the Einstein crystal? A Legendre transformation of Eqn. (14.34) ultimately leads to a Helmholtz free energy of the form

$$A(T, V, N) = -3Nk_\text{B}T \ln \left(\frac{e^{-\Theta_\text{E}/2T}}{1 - e^{-\Theta_\text{E}/T}} \right) \tag{14.40}$$

The pressure could be derived by taking the volume derivative of this expression. However, that would require knowledge of the dependence of v on density, for which we do not yet have a relationship. In general, v will change with volume because the shape of an atom's potential energy minimum will as well.

We can also address the chemical potential. Formally, we would need the dependence of v on density to take the N derivative of A. However, for crystals the pressure–volume term PV appearing in the Gibbs free energy is typically small compared with the Helmholtz free energy because V is small and changes little with pressure. Therefore a useful approximation is $\mu = G/N = A/N + Pv \approx A/N$.

14.6 The Debye model of perfect crystals

The Einstein model postulates that perturbations of atoms away from their lattice positions independently increase the system potential energy. This is a rather harsh assumption, and one can be more rigorous in the analysis. This approach was originally pursued by Peter Debye, a Dutch physical chemist, in an effort to achieve better agreement between the model and experimental heat-capacity measurements. It is termed the Einstein–Debye model or simply the Debye model of crystals.

A key aspect of this approach is to consider the energy minimum as a function not of a single degree of freedom as in Fig. 14.6, but of all atomic displacements at once. If there were only two displacements, the picture would resemble Fig. 14.8. In reality there are $3N$ atomic positions and we must conceptualize – even if it is impossible to directly

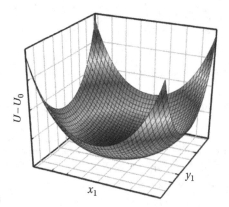

Figure 14.8. The potential energy landscape of a harmonic crystal has a minimum when all atoms sit at their lattice site positions. While only two displacement axes are shown above (x_1, y_1), in general this landscape exists in a $(3N + 1)$-dimensional space in which there exists an axis for each particle positional coordinate.

visualize – an *energy landscape* in which there are $3N$ axes corresponding to different displacements and a final additional axis measuring the system potential energy. In other words, the energy landscape is a $3N$-dimensional surface that has a minimum where all particle coordinates have the values of their lattice site positions.

If we consider simultaneous perturbations in the atomic coordinates, the Taylor expansion for the change in potential energy becomes more complicated and involves cross second derivatives between pairs of positions,

$$\Delta U = \frac{\Delta x_1^2}{2}\frac{\partial^2 U}{\partial x_1^2} + \frac{\Delta x_1\,\Delta y_1}{2}\frac{\partial^2 U}{\partial x_1\,\partial y_1} + \cdots + \frac{\Delta z_N\,\Delta y_N}{2}\frac{\partial^2 U}{\partial z_N\,\partial y_N} + \frac{\Delta z_N^2}{2}\frac{\partial^2 U}{\partial z_N^2} \tag{14.41}$$

In contrast to Eqn. (14.27), here we have a harmonic term for each combination of pairwise perturbations among the $3N$ coordinates, giving a total of $9N^2$ terms on the RHS. The expansion can be represented using a matrix formalism,

$$\Delta U = \frac{1}{2}\begin{bmatrix}\Delta x_1\\\Delta y_1\\\Delta z_1\\\Delta x_2\\\vdots\\\Delta z_N\end{bmatrix}^{\mathrm{T}}\begin{bmatrix}\dfrac{\partial^2 U}{\partial x_1^2} & \dfrac{\partial^2 U}{\partial x_1\,\partial y_1} & \cdots & \dfrac{\partial^2 U}{\partial x_1\,\partial z_N}\\[2mm]\dfrac{\partial^2 U}{\partial y_1\,\partial x_1} & \dfrac{\partial^2 U}{\partial y_1\,\partial y_1} & & \dfrac{\partial^2 U}{\partial y_1\,\partial z_N}\\[2mm]\vdots & & \ddots & \vdots\\[2mm]\dfrac{\partial^2 U}{\partial z_N\,\partial x_1} & \dfrac{\partial^2 U}{\partial z_N\,\partial y_1} & \cdots & \dfrac{\partial^2 U}{\partial z_N\,\partial z_N}\end{bmatrix}\begin{bmatrix}\Delta x_1\\\Delta y_1\\\Delta z_1\\\Delta x_2\\\vdots\\\Delta z_N\end{bmatrix} \tag{14.42}$$

Equation (14.42) can be expressed compactly by defining a vector of atom displacements $\mathbf{q}$ and the so-called *Hessian* matrix of all second partial derivatives $\mathbf{H}$,

$$\Delta U = \frac{1}{2}\mathbf{q}^{\mathrm{T}}\mathbf{H}\mathbf{q} \tag{14.43}$$

An important property of $\mathbf{H}$ is that it is a symmetric matrix since the order in which the second derivatives are taken does not matter. As a result we can express it as $\mathbf{H} = \mathbf{P}^{\mathrm{T}}\mathbf{D}\mathbf{P}$, where $\mathbf{D}$ is a diagonal matrix of the eigenvalues of $\mathbf{H}$ and $\mathbf{P}$ is a matrix of the corresponding eigenvectors. Some matrix algebra allows us to rewrite (14.43) as

$$\Delta U = \frac{1}{2}\mathbf{q}^{\mathrm{T}}\mathbf{P}^{\mathrm{T}}\mathbf{D}\mathbf{P}\mathbf{q}$$

$$= \frac{1}{2}(\mathbf{P}\mathbf{q})^{\mathrm{T}}\mathbf{D}(\mathbf{P}\mathbf{q}) \tag{14.44}$$

Equation (14.44) has a very important physical interpretation: it suggests that the relevant coordinates are not the original particle positions, but rather their projections along the eigenvectors contained in $\mathbf{P}$, that is, the set of coordinates implied by $\mathbf{P}\mathbf{q}$. We can trade the original $3N$ atomic position axes in the $(3N + 1)$-dimensional energy landscape for a new set of axes that extend along the eigenvectors of $\mathbf{H}$. The direction of each axis corresponds to a linear combination of atomic positions, and represents a so-called *normal mode* for atomic displacements in the system. Some modes involve quite localized vibrations of individual atoms, while others entail collective rearrangements of many atoms. Because **D** is diagonal, the harmonic potential energy in (14.44) simplifies considerably,

$$\Delta U = \frac{1}{2} \sum_{i=1}^{3N} \lambda_i \, \Delta s_i^2 \tag{14.45}$$

Here, λ_i is the ith eigenvalue and Δs_i is the change in position of the atoms along the ith eigenvector, equivalently along the ith axis generated from **P**. It represents a linear combination of positions of all of the atom positions, $\Delta s_i = a_i \, \Delta x_1 + b_i \, \Delta y_1 + c_i \, \Delta z_1 + d_i \, \Delta x_1 + \cdots$, where the coefficients are given by the values in the eigenvector i. Moreover, because the Hessian matrix can be diagonalized in this way, the system can move independently along each of the eigenvectors, and they form an orthogonal set. In turn, we can re-express the potential energy as a sum that involves $3N$ independent harmonic oscillators with $3N$ associated frequencies,

$$\Delta U = 2\pi^2 m \sum_{i=1}^{3N} v_i^2 \, \Delta s_i^2 \tag{14.46}$$

where, by comparison with (14.45), $v_i^2 = \lambda_i / (4\pi^2 m)$. The eigenvalues λ_i and hence frequencies v_i depend on the lattice structure, the specific intermolecular interactions, and the density.

If all of the frequencies v_i were identical, Eqn. (14.46) would recover the Einstein model. In general, however, they are not. By analyzing the behavior of standing waves in a continuum solid, Debye suggested an approximate distribution for the $3N$ values of v_i,

$$\wp(v) = \frac{3v^2}{v_m^3} \quad \text{for} \quad v < v_m \tag{14.47}$$

where $\wp(v)dv$ gives the fraction of oscillators having a frequency in the range $v \pm dv/2$. Here, v_m is a cutoff frequency that is a lattice-, density-, and material-specific constant. The model assumes that there are no vibrations in the crystal that occur with frequencies above v_m. Physically this seems reasonable because the fastest vibrations and hence highest possible frequencies should occur at the level of individual atomic vibrations. With this distribution, the observed thermodynamic properties are a sum of Einstein-like oscillators with $3N$ different frequencies distributed according to $\wp(v)$.

The **Debye model of crystals** assumes that a distribution of frequencies characterizes the $3N$ harmonic oscillators of a monatomic crystal. This distribution applies to the eigenmodes of atomic vibrations that stem from diagonalization of the Hessian matrix of second derivatives.

The distribution in (14.47) provides a simple way to compute thermodynamic properties: because the oscillators are independent, we simply need to perform a weighted average of the Einstein model over the spectrum of different frequencies. For example, the heat capacity is given by

$$c_V^{\text{Debye}}(T, V) = \int_0^{\nu_m} c_V^{\text{Einstein}}(T, V)\wp(\nu)d\nu$$

$$= 3k_B \left(\frac{h\nu}{k_B T}\right)^2 \frac{e^{h\nu/(k_B T)}}{(e^{h\nu/(k_B T)} - 1)^2}\left(\frac{3\nu^2}{\nu_m^3}\right)d\nu \qquad (14.48)$$

With some simplification, this becomes

$$c_V^{\text{Debye}}(T, V) = 9k_B \left(\frac{T}{\Theta_D}\right)^3 \int_0^{\Theta_D/T} \frac{x^4 e^x}{(e^x - 1)^2}\,dx, \quad \text{where} \quad \Theta_D \equiv \frac{h\nu_m}{k_B} \qquad (14.49)$$

Here we defined the Debye temperature Θ_D in a manner analogous to that of the Einstein approach. However, these two characteristic temperatures differ for a given material: Θ_E entails an averaged vibrational frequency while Θ_D involves the maximum frequency. In any case, both models give the same high-temperature classical limit of $c_V = 3k_B$. Indeed, the limit $T \to \infty$ of Eqn. (14.49) shows this.

The main feature of the Debye model that makes it superior to the Einstein one is the manner in which it describes the heat capacity at very low temperatures near absolute zero. As $T \to 0$ the integral in (14.49) approaches a limiting value of $12\pi^4/45$. Therefore the heat capacity at low temperatures behaves as

$$c_V^{\text{Debye}}(T, V) = \frac{12k_B\pi^4}{5}\left(\frac{T}{T_D}\right)^3 \quad \text{(low } T) \qquad (14.50)$$

Detailed and precise experiments at low temperature have shown that a wide range of perfect crystals obey the relation $c_V \propto T^3$ near absolute zero. The Einstein model misses this behavior, instead giving $c_V \propto T^{-2}\exp(-\Theta_E/T)$ as $T \to 0$. The Debye model was an important theoretical achievement because it rationalized the universality of this behavior and gave a quantitative origin in terms of quantum mechanics.

The Debye model of crystals correctly predicts the experimentally observed T^3 dependence of the heat capacity in crystals at low temperatures, and demonstrates the origins of this behavior in terms of quantum mechanics.

Problems

Conceptual and thought problems

14.1. Dry ice is a solid form of carbon dioxide that at standard conditions sublimates to a gas. A system is prepared in which dry ice is in equilibrium with a gas phase of CO_2 and N_2. Assume the gas phase to behave ideally. Indicate whether each of the following is true or false, or there is not enough information to tell. Explain.
  (a) Increasing P at constant T will increase the mole fraction of CO_2 in the gas phase.
  (b) Increasing T at constant P will increase the partial pressure of CO_2 in the gas phase.
  (c) Adding more N_2 at constant T and P will decrease the mole fraction of CO_2 in the gas phase.

14.2. For binary solid liquid equibrium, show rigorously that the "freezing out" of the solvent from a metastable supercooled liquid phase results in an overall decrease in the Gibbs free energy. Remember that $G/N \neq \mu$ for a multicomponent system.

14.3. Using ideal solution models for a binary system, prove that the free energy of the solution phase will always be greater than that of the state with two pure solid phases below the eutectic temperature.

14.4. Figures 14.1 and 14.4 show that a solution of water with a dilute amount of ethylene glycol will freeze over a range of temperatures, beginning at one below 273 K and then completely solidifying at the eutectic temperature. In between, as the temperature is lowered, more solid ice is formed and the solution becomes richer in ethylene glycol. However, in the case that the solution is extremely dilute, $x_{EG} \approx 0$, these diagrams still show that solid–liquid equilibrium occurs over a range of temperatures. How is this possible in light of the fact that pure liquid water freezes at the single temperature 273 K, as Gibbs' phase rule would require? Reconcile the apparent discrepancy.

14.5. Equation (14.23) shows that the lowest possible temperature at which a multi-component solution can exist is *independent* of its original composition $\{x_{i,0}\}$. How is this possible when the issue of which species solidify first is highly dependent on composition? Provide a clear rationale.

Fundamentals problems

14.6. Prove that the lengths of the tie lines in Figs. 14.1, 14.3, and 14.4 are proportional to the molar amounts of each phase. For example in Fig. 14.1, show that the ratio of the numbers of moles in the ice and solution phases is given by $(x_{EG} - x_{EG,0})/(x_{EG,0} - 0)$. This is the so-called lever-arm rule.

14.7. Show that a one-dimensional particle of mass m and potential $U = (k/2)(x - x_0)^2$ vibrates with a frequency of $\sqrt{k/m}/(2\pi)$.

14.8. In general, the eutectic temperature in a binary mixture depends on the pressure. A line describing a family of eutectic points can be constructed in the P, T plane. Consider the slope of this line, $(dP/dT)_{\text{eutectic}}$.

(a) Find a general expression for this slope in terms of the per-particle entropies (s_i^X) and volumes (v_i^X) in the pure crystal phases, and the partial molar entropies $(\overline{S}_i)$, volumes $(\overline{V}_i)$, and mole fractions $(x_1$ and $x_2)$ in the solution phase.

(b) Find an expression in the case that the solution phase is ideal. Show that the result can be expressed in terms of the pure-phase enthalpies and volumes of melting, $\Delta h_{\text{m},i}$ and $\Delta v_{\text{m},i}$, in addition to the mole fractions.

14.9. Consider the microscopic model of monatomic crystals discussed in the text. Let the positions of all of the atoms at any instant be described by $\mathbf{r}^N$, a vector with $3N$ components. Similarly, let the positions of all of the atoms at their energy-minimizing lattice positions be $\mathbf{r}_0^N$. We might approximate the potential energy about the minimum in a harmonic manner, as $U(\mathbf{r}^N) \approx U(\mathbf{r}_0^N) + a|\mathbf{r}^N - \mathbf{r}_0^N|^2$, where a is a constant. Assume the potential energy at the minimum is zero, $U(\mathbf{r}_0^N) = 0$. Here, you will show that the value of the intensive constant-volume heat capacity in the classical approximation is always $3k_B$, independently of T.

(a) For the moment, neglect kinetic energies and let $E = U(\mathbf{r}^N)$. This will allow you to compute excess properties beyond the ideal gas case. Show that the number of classical microstates is proportional to $E^{3N/2}$. Hint: envision a $3N$-dimensional space in which the vector $\mathbf{r}^N - \mathbf{r}_0^N$ can be plotted; a surface of constant $|\mathbf{r}^N - \mathbf{r}_0^N| = R$ gives a hypersphere of radius R.

(b) Show that the configurational or excess part of c_V (neglecting kinetic energies) is given by $c_V^{\text{ex}} = (3/2)k_B$. The total heat capacity is given by $c_V = c_V^{\text{ig}} + c_V^{\text{ex}}$, where $c_V^{\text{ig}} = (3/2)k_B$. Therefore, the classical heat capacity is $3k_B$ for a monatomic crystal.

14.10. Prove that c_V for the Einstein and Debye models of crystals approaches $3k_B$ as $T \to \infty$.

14.11. Explicitly derive the Helmholtz free energy for the Einstein model of crystals starting with Eqns. (14.34) and (14.36).

14.12. Using the Einstein model, show that the vapor pressure above a low-temperature monatomic crystal is expected to follow the form

$$\ln P^{\text{vap}} = c_1 + c_2 \ln T + \frac{c_3}{T}$$

Here, c_1, c_2 and c_3 are constants; find expressions for them in terms of molecular properties.

14.13. A purely classical model of a crystal that is based on a harmonic oscillator shows that its per-particle Helmholtz free energy can be written as $a = -3k_BT \ln(T/\Theta)$, where Θ is analogous to an Einstein temperature and relates to the curvature of the potential energy minimum. Consider coexistence between two crystalline phases 1 and 2 that have distinct functionalities for their classical Einstein temperatures: $\Theta_1 = c_1\rho_1$ and $\Theta_2 = c_2\rho_2^2$, where the c terms are constants and the ρ terms are number densities. Find expressions for the equilibrium coexistence pressure P and latent entropy of melting Δs along this solid–solid phase boundary. Note that your answers should depend only on temperature since there is one degree of freedom as per Gibbs' phase rule.

Applied problems

14.14. One of water's most familiar and unusual properties is its negative volume of melting: the density of ice is 917 kg/m^3 and that of liquid water is 1,000 kg/m^3 at the normal melting point. Some other useful properties of water for this problem are $c_{P,\text{ice}} \approx 38$ J/K mol, $c_{P,\text{liq.}} \approx 75$ J/K mol, $\Delta h_m \approx 6$ kJ/mol at 0°C, and $\mathcal{M}_{\text{H}_2\text{O}} \approx 18$ g/mol.

(a) It has been claimed that ice skating is possible because the pressure of the blade on the ice lowers its freezing point, induces melting, and provides a slippery liquid layer. What is the expected freezing temperature for a 65-kg person on skates that have a total area in contact with the ice of 30 cm^2 (blades 0.5 cm wide and 30 cm long)? Is it likely the ice will actually melt?

(b) A small amount of solute is added to liquid water at 0°C and atmospheric pressure at mole fraction $x_{\text{solute}} = 0.05$. The mixture is then cooled at constant pressure until 10% of the original water is present as ice. How much heat must be removed per mole of water, Q/n, to accomplish this task? Assume that the solute contributes negligible heat capacity to the solution.

14.15. 100 grams of NaCl are added to 1 L of liquid water at standard conditions.

(a) Calculate the temperature at which water will freeze at 1 atm.

(b) How will the salt affect the pressure at which the solution will freeze at 0°C?

(c) Repeat part (a) in the case that the salt is MgCl$_2$.

14.16. Consider the triple point of water, which occurs around 273 K and 610 Pa. If a small amount of salt is added, both the temperature and the pressure of this point will change. Find the new triple point of water for a system with mole fraction 0.02 of NaCl. You may assume that water vapor behaves as an ideal gas, that the enthalpies of fusion and vaporization (Δh_{fus} and Δh_{vap}) are roughly constant, that no salt is present in the solid or gas phases, and that the liquid is an ideal solution. Note that $\Delta h_{\text{fus}} \approx 6$ kJ/mol, $\Delta h_{\text{vap}} \approx 45.0$ kJ/mol, and $\Delta v_m \approx -0.02$ m^3/mol. Hint: find two equations with two unknowns that are the small changes ΔT and ΔP of the triple-point conditions, valid in the limit $x_{\text{salt}} \ll 1$.

14.17. Consider what happens when a bottle of beer and a bottle of vodka are put into a freezer at atmospheric pressure. For the following questions, assume ideal solutions, constant enthalpies of vaporization, and that the beer and vodka are 4% and 40% alcohol by mass, respectively. The data in Table 14.1 may be helpful.

Table 14.1 Data on water and ethanol

Parameter	Pure water	Pure ethanol
T_m (K)	273	159
Δh_{fus} (kJ/mol)	6.01	4.96
c_{liquid} (g/ml)	1	0.789
$c_{crystal}$ (g/ml)	0.93	–
Molecular weight (g/mol)	18.0	46.1

(a) Find the temperatures at which each bottle will first begin to freeze.

(b) Find the temperatures at which each bottle will completely freeze, with no liquid present at lower temperatures. What is the composition of the liquid solution at temperatures slightly above this freezing temperature? (Note: in reality, this temperature is highly hypothetical since it neglects the nonidealities of each liquid.)

(c) For each bottle, find the expected fraction that is solid ice (mole basis) for a typical household freezer set to $-18\,°C$. How much more concentrated is the solution in ethanol than at room temperature?

(d) An alternative way to freeze the solutions is to change the pressure at $0\,°C$. Upon changes in pressure, assume constant molar densities and that the water freezes first. Must the pressure be increased or decreased in order to induce freezing in these solutions at $0\,°C$, and why is that the case? At $0\,°C$ for the beer, what pressure would have to be reached in order for freezing to begin, and why is this result unrealistic?

14.18. If interfacial properties contribute to the bulk thermodynamic behavior of a system, the fundamental equation must be modified to include surface effects:

$$dG = -S\,dT + V\,dP + \mu\,dN + \gamma\,d\mathcal{A}$$

where $\mathcal{A}$ is the surface area of the system and γ is its surface tension with the surrounding environment. Consider a nanoparticle made of iron oxide (Fe_2O_3). In bulk, iron oxide can exist in two different crystalline forms, hematite (H) and maghemite (M). Both of these have roughly the same density ρ. The transition between these two phases occurs at temperature T_1 with a molar latent heat $\Delta h = h_M - h_H > 0$. In iron oxide nanoparticles, however, surface energetics are important and can modify the relative stabilities of the two phases since $\Delta\gamma = \gamma_M - \gamma_H < 0$. For a nanoparticle of radius R at constant N and P conditions, show that the temperature at which it will spontaneously transform from hematite to maghemite is given approximately by

$$T_2 = T_1\left(1 + \frac{3\,\Delta\gamma}{\rho R\,\Delta h}\right)$$

14.19. Consider a hypothetical stagnant and deep ocean that is at constant temperature $T = -5\,°C$. In principle the ocean would possess a thick layer of ice on its surface, but, due to the hydrostatic pressure, would have liquid water at low enough depths. Recall that the pressure head at a depth h in a liquid is given by $\rho g h$, where ρ is the mass density and g is the gravitational constant. Assume ideal solution behavior and that the densities of liquid water and ice are constant at 1.0 and 0.93 g/ml, respectively.

(a) Treating it as pure water, at what depth will there be liquid–ice coexistence in the ocean?

(b) Instead, sea water contains a variety of ions (due to salts) that total roughly 0.02 on a mole fraction basis. What is the depth in this case? Assume that the liquid always has a mole fraction 0.02 of ions, regardless of the amount of ice present.

(c) What aspects of this model make it unrealistic for actual oceans?

14.20. Pure cholesterol (molecular weight 387) is a solid at room temperature and nearly insoluble in water. Its solubility in ethanol (molecular weight 46) is more significant, being 2.2% by mass at $20\,°C$ and 3.5% at $40\,°C$. Estimate the heat per mole of cholesterol associated with dissolving it in ethanol. That is, find the enthalpy of dissolution. Hint: consider equilibrium between solid cholesterol and a cholesterol–ethanol solution. Then, find a relationship between the temperature dependence of the solution concentrations and cholesterol's pure and partial molar enthalpies. The overall solution enthalpy can be found from an integrated relation.

FURTHER READING

K. Denbigh, *The Principles of Chemical Equilibrium*, 4th edn. New York: Cambridge University Press (1981).

T. L. Hill, *An Introduction to Statistical Thermodynamics*. Reading, MA: Addison-Wesley (1960); New York: Dover (1986).

L. D. Landau and E. M. Lifshitz, *Statistical Physics*, 3rd edn. Oxford: Butterworth-Heinemann (1980).

D. A. McQuarrie, *Quantum Chemistry*. Mill Valley, CA: University Science Books (1983).

D. A. McQuarrie, *Statistical Mechanics*. Sausalito, CA: University Science Books (2000).

15 The third law

In Chapters 4 and 5 we introduced the molecular basis for the first two laws of thermodynamics. In particular, we developed a physical interpretation for the entropy in terms of microstates. However, we have yet to address two subtle questions regarding this relationship. *Is there any such thing as an absolute value of the entropy?* That is, does it make sense to identify an exact numerical value of S for a particular system and a particular state point, as opposed to the more modest calculation of a change in entropy between two state points? Furthermore, *what is the behavior of the entropy and other thermodynamic functions as the temperature approaches absolute zero?*

As we shall see in this chapter, both of these questions cannot be addressed using the first two laws alone. Instead, we must introduce a third law of thermodynamics that provides a context for understanding absolute entropies and absolute zero. The third law is not conceptually as straightforward as the others, first because it is not needed in many practical calculations far away from absolute zero, and second and more importantly, because it can be presented in several quite different ways. In particular, we will describe a number of distinct formulations of the third law and attempt to provide some molecular interpretation for their rationale. Unlike the first and second laws, however, there are many subtleties associated with these formulations that in some cases remain actively discussed. We will not attempt to synthesize and reconcile the details here, but present only the main ideas. The interested reader is referred to the references at the end of the chapter for further information.

15.1 Absolute entropies and absolute zero

On what basis can we speak of an absolute entropy? Boltzmann's microscopic definition is a natural place to begin, $S = k_B \ln \Omega$. As we have seen, the interpretation of Ω is that it counts the number of microscopic states of a system. However, the behavior of Ω depends on whether or not we take a classical or quantum viewpoint of the world. In the former, we cannot count microstates exactly; there are infinitely many since the positions and velocities of atoms can change continuously. As a result, classical systems cannot have an absolute entropy. Still, it is worth remembering that, while we cannot count absolute numbers of configurations, we can compare relative numbers and hence obtain entropy differences. For example, if a volume V containing a single particle is expanded to $2V$, the number of microstates doubles such that $\Delta S = k_B \ln(2V/V) = k_B \ln 2$.

On the other hand, a quantum description is distinct. Quantum mechanics by way of Schrödinger's equation leads to a discrete set of states available to a system. For example, a particle in volume $V = L^3$ can exist in states whose energy is given by $\epsilon = \left(h^2/8mL^2 \right) \left(n_x^2 + n_y^2 + n_z^2 \right)$, where n_x, n_y, and n_z are all positive integers. The quantum numbers make it possible to count microstates exactly, and therefore an absolute entropy seems possible. Indeed, the world is quantum in nature and therefore the idea of absolute entropies is real and significant.

Are these considerations alone sufficient to define entropies on absolute terms? Consider the possibility that we had defined the entropy using a modified Boltzmann equation,

$$S = k_B \ln \Omega + S_{ref} \tag{15.1}$$

in which S_{ref} is a universal, material-independent constant. Would this re-defining of the entropy imply any *physical* changes to the equilibrium behavior of a system, as given by the mathematical relationships surrounding the entropy function and the second law? Indeed, not a single physical property or behavior would be affected. There is no physical constraint preventing a definition for the entropy that includes a nonzero value for S_{ref}. Recall that our original justification for the entropy was the rule of equal a priori probabilities, which demands that the macrostate with the most microstates dominates equilibrium behavior. This means that having two systems sharing a conserved quantity $X = E$, V, or N maximizes the following at equilibrium:

$$\Omega_{tot} = \Omega_1(X_1)\Omega_2(X_2) \quad (X_1 + X_2 \text{ constant}) \tag{15.2}$$

The corresponding condition for maximization is

$$\frac{\partial \ln \Omega_1}{\partial X_1} = \frac{\partial \ln \Omega_2}{\partial X_2} \tag{15.3}$$

This result provided the justification for introducing an entropy function, $S = k_B \ln \Omega$, whose derivatives give the properties of the two systems that are equal at equilibrium (T, P, and μ). However, we could equally well have used a function $S = k_B \ln \Omega + S_{ref}$ and the maximization condition would have been the same. Indeed, the state functions T, P, and μ all extend from derivatives of S such that S_{ref} would vanish in the relationship between E, V, N and these properties for any substance. Therefore, at a fundamental level, the following applies.

We are able to have an absolute entropy because we take the constant S_{ref} to be equal to zero by convention. That is, we define the entropy using the expression $S = k_B \ln \Omega$.

The implication of this standard is that the entropy reaches zero when $\Omega = 1$. That is, the entropy vanishes only when the system is at conditions at which only a single microstate is available. Typically $\Omega = 1$ occurs at a *ground state of a system*, a single microstate with a lowest possible energy. We have seen one microscopic model that shows this behavior: the picture of monatomic crystals developed in

Chapter 14. As $T \to 0$ the energy of this crystal reaches a minimum and the number of microstates approaches unity. We might expect that all perfect monatomic crystals approach the same value of entropy at absolute zero, and we take this value to be zero by the convention per the definition of the entropy. Since quantum mechanics demands that Ω must be a positive integer, this approach shows that all substances have $S \geq 0$.

However, the notion of "convention" is essential here since there are subtleties associated even with this seemingly straightforward example. The main issue is that there exist real degrees of freedom that can give rise to degeneracies and hence multiple microstates, even for a perfect monatomic, elemental crystal in its ground state. For example, the atoms in the crystal can have different nuclear states owing to distinct nuclear spins. Or, there may exist even further differences at the deeply subatomic level that differentiate atoms of the same element, some of which might not be detectable or may even yet remain to be discovered. Such features give rise to very subtle but nonetheless extant variations in the atoms, leading to different "subspecies" of them. In turn, degeneracies associated with the number of ways of arranging these subspecies in the crystalline lattice can emerge.

As a result, we must specify the resolution at which we define the zero entropy of a perfect crystal. These considerations lead to one of several formulations of the third law of thermodynamics.

A first formulation of the **third law of thermodynamics** states that the entropy of a perfect monatomic (elemental) crystal is taken to be zero at absolute zero, at which $\Omega = 1$. Here, Ω appearing in Boltzmann's equation omits any degeneracies due to nuclear spins or other subatomic degrees of freedom. For all substances and conditions, therefore, the entropy is always greater than or equal to zero.

This statement paraphrases ideas originally put forth by Planck and later qualified by Lewis and Randall in the early twentieth century. It is a formulation of the third law that relies on both a molecular and quantum picture of the world. One can see that it is rather specialized in that it speaks to a quite limited class of substances and states.

Indeed, does the zero-entropy state apply generally to all kinds of crystals at absolute zero, including molecular ones? It seems plausible for any given crystal to have a unique ground state of lowest energy, leading to $\Omega \to 1$ and $S \to 0$ as $T \to 0$. On the other hand, some crystalline materials may retain degeneracies at absolute zero for at least two reasons. First, atoms may be distinct isotopes of the same element, again leading to multiple subspecies that can be arranged in numerous ways on the crystalline lattice. Second, non-monatomic molecular crystals may permit each molecule to adopt multiple orientations at its lattice site. Figure 15.1 shows two examples involving solid phases of carbon monoxide and water, in which there are multiple ways to orient the molecules while still maintaining the overall lattice structure. Both of these isotopic and orientational degeneracies give rise to a residual entropy at absolute zero.

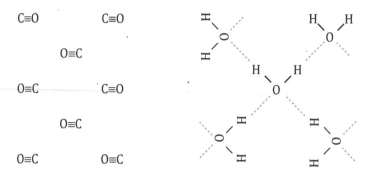

Figure 15.1. Carbon monoxide (left) and water (right) can form crystalline lattices that have residual entropy at absolute zero due to orientational degeneracies. In the former, CO molecules can point in one of two directions. In the latter, a central water molecule has multiple ways of forming hydrogen bonds (donor versus acceptor) with respect to its four tetrahedrally oriented neighbors.

For non-monatomic, non-isotopic perfect crystals, the ground-state degeneracy can be greater than one due to possible isotopic and orientational rearrangements of the molecules. These systems can have a ground-state entropy S_0 as $T \to 0$ that is finite and positive, also called a **residual entropy**.

Let us consider the orientational degeneracies illustrated in Fig. 15.1 in more detail. In solid carbon monoxide, each molecule can adopt one of two orientations that are flipped versions of each other and the residual entropy is close to $S_0 = Nk_B \ln(2)$. In the case of ice, each water molecule hydrogen bonds to four nearest neighbors, but there are multiple ways to decide which neighbors provide the donor (hydrogen) versus acceptor (oxygen) bonding sites. The residual entropy due to the multiplicity of hydrogen-bonding networks in a perfect ice crystalline lattice is about $S_0 = Nk_B \ln(3/2)$.

Of course, orientational perturbations within a lattice structure are likely to result in energetic differences such that a single, completely ordered orientational configuration is indeed a unique ground state of the system. Why then is there a residual entropy, since one might expect that the lattice itself will become orientationally ordered as $T \to 0$ as the system is pushed towards this lowest energy state? In practice, such an ordering process might not be possible from a kinetic point of view. The energy differences among distinct orientational arrangements may be too small to favor a unique (lowest-energy) one when the crystal forms at finite temperature. At the same time, the kinetic barriers to rearranging into a new orientational state become increasingly large as the system is cooled. Thus, the orientational disorder "freezes in" along with the formation of the crystal and does not change all the way down to absolute zero.

A similar argument for the "freezing in" of lattice disorder or randomness exists for crystals involving multiple isotopes. In either case, it is important to note that the behavior is system-dependent. Some substances, such as helium, will spontaneously phase-separate into different crystal phases at low temperatures, one of each isotope. In these cases, the kinetic barrier is not so great as to prevent the formation of isotopically pure states. We must therefore view residual entropy in general as an operational necessity to address actual kinetic constraints present in experimental practice.

15.2   Finite entropies and heat capacities at absolute zero

The considerations above show that, as $T \to 0$, the entropy reaches a value that is zero or positive for any substance. This behavior emerges because nature obeys quantum mechanics so that microstates can be counted exactly. An important feature of these results is that the entropy attains a finite value at absolute zero and does not diverge. In classical models, on the other hand, the entropy is always predicted to trend towards negative infinity as the temperature approaches absolute zero.

The idea that the entropy reaches a finite value was originally suggested by Einstein during the early developments of the third law and quantum theory. While this statement alone does not provide a complete version of that law, it does have important consequences for the behavior of the heat capacities near absolute zero. Let us expand the constant-volume heat capacity of a pure substance around $T = 0$,

$$C_V \approx c_0 + c_1 T + c_2 T^2 + \cdots \tag{15.4}$$

where the constants of the expansion c_0, c_1, c_2, and so on in general depend on the molar volume and size of the system. Now consider the change in entropy at constant volume when the temperature is varied from T_1 to $T_2 = T_1 + \Delta T$, where ΔT is a small enough temperature increment that the system does not pass through a phase transition. The corresponding change in entropy is

$$S(T_2) - S(T_1) = \int_{T_1}^{T_2} \left(\frac{\partial S}{\partial T} \right)_{V,N} dT = \int_{T_1}^{T_2} \frac{C_V}{T} dT$$

$$= c_0 \ln \left(\frac{T_2}{T_1} \right) + c_1 (T_2 - T_1) + \frac{c_2}{2} \left(T_2^2 - T_1^2 \right) + \cdots \tag{15.5}$$

We can examine the specific case in which T_1 approaches absolute zero and T_2 is some small but finite temperature,

$$\lim_{T_1 \to 0} [S(T_2) - S(T_1)] = \left[\lim_{T_1 \to 0} c_0 \ln \left(\frac{T_2}{T_1} \right) \right] + c_1 T_2 + \frac{c_2}{2} T_2^2 + \cdots \tag{15.6}$$

Because the entropy is finite at absolute zero, the LHS of this equation is also finite. The RHS, however, can remain finite only if the constant c_0 is strictly zero. Therefore, we must conclude that this is indeed the general case, and as a result the heat capacity must vanish at absolute zero for any substance per (15.4),

$$C_V \to 0 \text{ as } T \to 0 \tag{15.7}$$

We could have constructed a similar argument in the constant-pressure case, giving

$$C_P \to 0 \text{ as } T \to 0 \qquad (15.8)$$

The end result is that both heat capacities tend towards zero at absolute zero simply because the entropy must remain finite there. We saw a specific instance of this behavior in Chapter 14 in models of ideal crystals, but the phenomenon is quite general.

15.3   Entropy differences at absolute zero

The previous discussion shows that a perfect crystal reaches a minimum value of the entropy at absolute zero. If the crystal lacks a residual entropy, the value of this entropy is taken to be zero. For other systems, we can also say in general that a minimum value of the entropy is reached at absolute zero and that it is greater than or equal to that of a perfect monatomic crystal.

Given these results, it has been postulated that the difference in entropy between two stable equilibrium states of a substance approaches zero at absolute zero. This is another manifestation of the third law of thermodynamics, and the idea originates from Nernst around the year 1906 and is sometimes called the Nernst or Nernst–Simon theorem. A concise but nuanced version was later put forth by Fowler and Guggenheim, which is paraphrased here.

A second formulation of the **third law of thermodynamics** states that for any isothermal process involving only condensed phases in equilibrium, or for meta-stable phases in local equilibrium that remain metastable during the process, $\Delta S \to 0$ as $T \to 0$.

The crux of this statement is that changes in the entropy of a system approach zero for processes that occur near absolute zero. Such a process could correspond to a change in pressure, volume, magnetization, electric field, surface area, or any other macroscopic constraint associated with a work term in the fundamental equation. A key idea is that the phases must exist at least in local thermodynamic equilibrium, at a local free energy minimum, which excludes out-of-equilibrium states such as glasses.

The qualification regarding metastable phases is an important one. It means that during the process, any degrees of freedom that are "frozen in," such as isotopic mixing or crystalline orientational degeneracies, must remain that way. Similarly, this restriction allows a process that involves a transition between two phases that are both metastable with respect to a third phase, but remain in local equilibrium with respect to each other. In essence, the class of processes relevant to this statement of the third law cannot involve irreversible transitions from a state in which an internal constraint of some kind exists to one in which the constraint is absent. An example of such an irreversibility would be a crystal involving a molecule that, upon compression or expansion, decomposes into smaller molecules.

The justification for this postulate is that any stable state at absolute zero has a minimum value of the entropy. If two such states are perfect monatomic crystals, then their entropies both approach zero and thus $\Delta S = 0$ is automatically satisfied for transitions between them at $T = 0$. If there are residual orientational entropies but these entropies are the same for each state, then $\Delta S = 0$ as well. However, it is difficult to prove this postulate on more general grounds. That being said, detailed measurements on a large number of systems do support the generality of the Nernst theorem.

The implications of these ideas are quite significant. First, consider a small change in pressure in an isothermal process as one approaches absolute zero. The differential change in entropy is given by $dS = (\partial S/\partial P)_T \, dP$. According to the theorem, this change must be zero. Therefore,

$$\left(\frac{\partial S}{\partial P}\right)_T \to 0 \text{ as } T \to 0 \tag{15.9}$$

By a Maxwell relation, this implies

$$-\left(\frac{\partial S}{\partial P}\right)_T = \left(\frac{\partial V}{\partial T}\right)_P = V\alpha_P \to 0 \text{ as } T \to 0 \tag{15.10}$$

which shows that the thermal expansion coefficient vanishes at absolute zero.

Second, a phase transition generally involves a molar entropy change Δs. A process that is performed near absolute zero that induces a phase transition, for example by a change in pressure or volume, will result in a latent entropy that approaches $\Delta s \to 0$. The associated phase boundary must then have a zero slope in the P-T diagram according to the Clapeyron equation,

$$\left(\frac{dP}{dT}\right)_{\text{phase boundary}} = \frac{\Delta s}{\Delta v} \to 0 \text{ as } T \to 0 \tag{15.11}$$

Equation (15.11) applies to phases that are in (at least local) equilibrium with respect to each other, such as many crystal–crystal transitions. The helium isotope ^{4}He is a particularly interesting example because at low temperatures it can exist either as a liquid (a quantum superfluid) or as a crystal. A schematic phase diagram is shown in Fig. 15.2. Many experimental measurements have indeed suggested that the slope of the boundary between the superfluid and crystal approaches zero near absolute zero.

In total, we have seen that the Nernst theorem and the finiteness of the entropy place rather significant constraints on the behavior of thermodynamic functions as the temperature nears absolute zero.

As a system approaches absolute zero, the response functions c_V, c_P, and α_P all approach zero and any phase boundary between stable states attains a zero slope, $(dP/dT) = 0$.

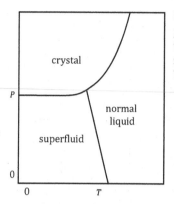

Figure 15.2. The isotope ^{4}He can exist in three different phases at low temperatures, here shown schematically for the ranges $0 < T < 3$ K and $0 < P < 5$ MPa. Near absolute zero, experiments show that the slope of the phase boundary between the superfluid and the crystal approaches zero.

15.4   Attainability of absolute zero

So far we have considered the third law in relation to the existence of molecules. Such was also the case when we initially examined the first and second laws. However, we eventually found that those laws could also be formulated at a purely macroscopic level: we could think of the first law in terms of heat and work rather than constant-energy Newtonian mechanics, and we could define the entropy in terms of reversible heat transfers $dS = \delta Q_{rev}/T$ rather than microstates. It therefore may come as no surprise that the third law also has a purely macroscopic formulation.

> A third formulation of the **third law of thermodynamics** states that the temperature absolute zero is unattainable in any process with a finite number of steps and in finite time.

This form of the law was also proposed by Nernst as an alternative to his earlier theorem and is arguably one of its most general presentations. Indeed, no reference to the molecular world is required and this statement can be directly tested by experiments; countless experimental efforts support the unattainability of absolute zero. Some view this presentation of the third law as the most fundamental, requiring the least qualifications.

There is a close connection between the unattainability statement and the molecular quantum picture. We first need to specify what we mean by attainability. In order to reach absolute zero, ultimately we must perform some kind of isentropic process such as a reversible adiabatic expansion that accomplishes cooling. Or, the process could be more complicated and consist of a combination of isentropic and isothermal steps, so long as the last step is isentropic. The final step can never be isothermal because that would presume the availability of a heat reservoir already

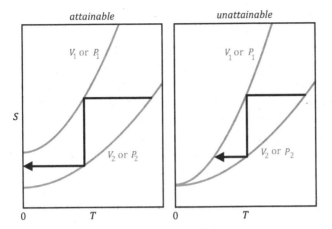

Figure 15.3. An S–T diagram illustrates the connection between the attainability of absolute zero and the Nernst theorem. On the left, two isochores (or isobars) converge at different values of the entropy at absolute zero. A process (black arrow) that starts at high temperature and consists of an isentropic expansion, an isothermal compression, and a second isentropic expansion ultimately is able to reach absolute zero. On the right, the isochores converge at the same value of the entropy at absolute zero, which is consistent with the Nernst theorem. In this case, an infinite number of process steps alternating between isentropic and isobaric elements would be necessary in order to reach absolute zero.

at absolute zero. Moreover, attainability requires that the beginning and end states of the isentropic steps be bounded; for example, we cannot have expansion to an infinite volume.

Figure 15.3 uses S–T diagrams to illustrate how such processes behave in two different scenarios. In the left one, all isochores of the system approach distinct values of the entropy at absolute zero, which violates the Nernst theorem. This allows one to construct a process involving a finite series of isentropic and isothermal steps that directly reaches absolute zero and results in attainability. In the right diagram, all isochores instead approach the same entropy at absolute zero. It is clear that a process would then require an infinite number of alternating isentropic and isothermal steps in order to reach $T = 0$. No isentropic process can be performed over a bounded range of volumes that leads to absolute zero. In other words, the Nernst theorem and the unattainability statement are intimately linked. Here we considered only volume as the control parameter during the isentropic steps, but this variable could equally be the pressure or any other term appearing in a process involving reversible work for the system.

Another way of looking at this problem is the following: any arbitrary process of cooling can be decomposed into a series of reversible adiabatic and isothermal steps. That means a process can be discussed in terms of hopping between different isentrops and isotherms in the phase diagram. At absolute zero, however, the isentrop ($dS = 0$) is identical to the isotherm ($dT = 0$) as a result of the Nernst postulate. Hence, there is no way to reach absolute zero by hopping between these two kinds of iso-lines a finite number of times.

A final related derivation involves an expansion of the entropy near $T = 0$. Let the leading-order behavior of the heat capacity at low temperatures be described by $C_V = c(V)T^n$, where $c(V)$ is simply a volume-dependent constant and $n > 0$ is an exponent characterizing the temperature dependence. For example, in the Debye crystal model $n = 3$. Note that n cannot be strictly equal to zero in order for C_V to vanish at absolute zero as required for a finite entropy. We have

$$\left(\frac{\partial S}{\partial T}\right)_V = c(V)T^{n-1} \tag{15.12}$$

We can integrate the heat capacity from absolute zero to an arbitrary small temperature T to find a general form for the entropy,

$$S(T,V) = n^{-1}c(V)T^n + S_0(V) \tag{15.13}$$

where $S_0(V)$ is the ground-state entropy. Now presume that we perform a reversible, adiabatic expansion from (T_1, V_1) to (T_2, V_2). Since the process is isentropic, we have

$$0 = S(T_2, V_2) - S(T_1, V_1)$$
$$= n^{-1}\left[c(V_2)T_2^n - c(V_1)T_1^n\right] + S_0(V_2) - S_0(V_1) \tag{15.14}$$

Let us consider the possibility that a final temperature of absolute zero can be reached, with $T_2 = 0$. In this case Eqn. (15.14) becomes

$$n^{-1}c(V_1)T_1^n = S_0(V_2) - S_0(V_1) \tag{15.15}$$

If the Nernst theorem is correct, then the RHS of this expression is zero and the only temperature that T_1 can possibly be is absolute zero itself. Therefore, an isentropic process that accomplishes cooling from a finite temperature to absolute zero is impossible in the case that S_0 is independent of V. On the other hand, if the Nernst theorem is not valid, then Eqn. (15.15) could provide for a finite initial T_1.

This leaves us with an unsettling conundrum. How can we calculate absolute entropies if there is no way to reach the ground state of a system where we can definitively identify the entropy as zero? In general, we determine absolute entropies by fitting experimental data to theoretical forms at low temperatures and then extrapolating down to $T = 0$. Crystals are particularly convenient for this analysis because both experiments and the Debye model give strong support to the following functionality for their heat capacity:

$$C_V(T \to 0) = cT^3 \tag{15.16}$$

where c is a constant. With this model, c can be determined by fitting results from low- but finite-temperature heat-capacity experiments to this form. Subsequently, the entropy at absolute zero can be related to that at a finite temperature T_1 by

$$S(T_1) - S(0) = \int_0^{T_1} \frac{C_V}{T} \, dT$$

$$= \int_0^{T_1} cT^2 \, dT$$

$$= \frac{1}{3} cT_1^3 \qquad\qquad (15.17)$$

If the crystal is monatomic or has no orientational degeneracies, then the constant $S(0)$ can be taken as zero by the convention discussed above, and the absolute entropy at T_1 can be calculated exactly.

Problems

Conceptual and thought problems

15.1. Without the third law, how could the Helmholtz free energy $A(T, V, N)$ of a system change at high temperatures, preserving the same physical behavior?

15.2. It is often stated that at absolute zero all molecular motion ceases. Explain why this is not entirely correct and needs qualification. Hint: is it possible to measure motion to arbitrary precision?

15.3. Show that the equation of state for any substance should have the following leading-order behavior near $T = 0$:

$$P(T,\rho) = P_0(\rho) + b(\rho)T^2$$

where $P_0(\rho)$ is the pressure at absolute zero and $b(\rho)$ is a density-dependent constant.

15.4. Consider the two-state model discussed in earlier chapters, in which each particle can be in an "up" position with energy b or a "down" position with energy 0. Under what conditions will the average energy of the system be greater than $Nb/2$? In what direction will energy move if such a system is brought into contact with a large body at a finite, positive temperature? Considering this, is the system hotter or colder than the body?

15.5. The Ising model of ferromagnets consists of a cubic lattice where each site is occupied by a spin. Spins can either be up or down, and when nearest-neighbor spins are in the same orientation, they experience a net attractive interaction of $-\epsilon$. Consider a three-dimensional lattice with $N = L^3$ spins.
   (a) What are the ground states of this system, what is their energy, and how many are there?
   (b) What is the entropy of the system at the ground-state energy? Show that even though there are multiple ground states, the entropy is negligible in the thermodynamic limit, $N \to \infty$.

15.6. The equation of state for a certain gas is well described by $P = k_BT/(v - b)$ over a range of conditions near ambient. Can this equation be correct as $T \to 0$? Justify your answer mathematically.

15.7. Prove that C_V and C_P become equal as a system approaches absolute zero. Hint: prove that $(C_P - C_V)/T$ goes to zero as $T \to 0$.

15.8. For a single-component system, indicate whether each quantity is positive, negative, or zero, or there is not enough information to tell.

(a) $\left(\dfrac{\partial T}{\partial S}\right)_{P,\mu}$

(b) $\left(\dfrac{\partial S}{\partial T}\right)_{V,\mu}$

(c) $\left(\dfrac{\partial h}{\partial P}\right)_T$ as $T \to 0$

(d) $\left(\dfrac{\partial^2 h}{\partial T^2}\right)_P$

(e) $\dfrac{1}{\mu}\left(\dfrac{\partial A}{\partial \mu}\right)_{T,V}$

(f) $\left(\dfrac{\partial P}{\partial T}\right)_v$ as $T \to 0$

(g) $\left(\dfrac{\partial P}{\partial T}\right)_h$ as $T \to 0$

(h) $\left(\dfrac{\partial v}{\partial P}\right)_T$ as $T \to 0$

(i) $\left(\dfrac{\partial \mu}{\partial T}\right)_P$

(j) $\overline{V}_i$

15.9. If Eqn. (15.1) were adopted with $S_{ref} > 0$, would the first law be affected in any way? Specifically, would the definition of reversible heat Q_{rev} be modified?

15.10. Can normal gases exist at absolute zero, even in the simultaneous limit $P \to 0$? Provide a proof.

15.11. The residual entropy of carbon monoxide at absolute zero has been found experimentally to be around 4.6 J/mol K. Explain why this is less than the estimate $k_B \ln 2$ one might expect on the basis of Fig. 15.1.

15.12. Molten liquid silica can crystallize to form quartz, which has a chiral crystal structure even though its molecular building blocks are achiral. As a result, the perfect crystalline solid corresponds to one of two enantiomeric structures, with entirely L or R chirality. Explain why the entropy for this system effectively becomes zero at absolute zero, even though there are two possible ground states.

Fundamentals problems

15.13. Prove that, for any classical model, the entropy diverges like $S \to -\infty$ as $T \to 0$. Hint: consider the ground-state (global) potential energy minimum as a function of the $3N$ particle coordinates.

15.14. A crystal has a constant-pressure heat capacity at low temperatures that scales as $c_P \sim T^3$. How will its volume scale with T at constant P?

15.15. Ice has been found to have a residual entropy of about 3.5 J/mol K.

(a) Show that this can be approximated by $S_0 = Nk_B \ln(3/2)$. Consider a central water molecule and its four hydrogen-bonded nearest neighbors in the tetrahedral lattice structure of ice. Of the central water's four hydrogen bonds, two involve its own hydrogens and are donors. How many ways are there to pick which bonds are the donors versus the acceptors? Next, consider whether a given arrangement of central hydrogen donors is likely to be compatible with the neighboring molecules' donor and acceptors. What is the probability that the two donor hydrogens of the central molecule will be pointing towards acceptor (rather than donor) sites on the two neighboring molecules?

(b) Outline a method by which the residual entropy of water could be calculated experimentally. Hint: the absolute entropy is known from quantum mechanics for an ideal gas state.

15.16. A Carnot refrigerator can be used to cool a system to lower temperatures. It requires a net input of work in order to move heat from a cold reservoir at T_C to a hot one at T_H, effectively cooling the former. Let us consider the use of a Carnot refrigerator in cooling a system (the cold reservoir) towards absolute zero.

(a) A Carnot refrigerator achieves maximum possible efficiency and least required work according to the second law. Show that the applied differential work δW required to remove an amount of heat δQ_C from the cold reservoir is given by the following, where δW is positive and δQ_C negative:

$$\delta W = \frac{T_H - T_C}{T_C}(-\delta Q_C)$$

(b) The system (cold reservoir) is cooled at constant volume and has a temperature-dependent heat capacity $C_V(T)$. Imagine that the system begins at $T_C = T_0$ and is taken all the way to $T_C = 0$. Find an integral expression for the total work required in order to accomplish this task.

(c) Classical statistical mechanics predicts $C_V = 3Nk_B$ at low temperatures. Show that the work required in this case becomes infinite. In other words, show that the second law implies the unattainability of absolute zero if the world behaves classically and thus no separate third law is needed.

(d) Quantum statistical mechanics predicts $C_V = cNT^3$ at low temperatures, where c is a density-dependent constant. Show that the work required in this case becomes finite. In other words, show that a separate law regarding the unattainability of absolute zero is needed if the world behaves quantum-mechanically.

15.17. What are the implications of the third law for the thermodynamic properties of systems with the following kinds of macroscopic variables, field parameters, and reversible work contributions?

(a) A surface of area $\mathcal{A}$ and surface tension γ, with $\delta W = \gamma \, d\mathcal{A}$.

(b) A band of length L and tension τ, with $\delta W = \tau \, dL$.

(c) A magnet of magnetization M and magnetic field $\mathcal{H}$, with $\delta W = \mathcal{H} \, dM$.

(d) An object of charge q in an electrostatic potential Φ, with $\delta W = \Phi \, dq$.

FURTHER READING

H. Callen, *Thermodynamics and an Introduction to Thermostatistics*, 3rd edn. New York: Wiley (1985).

K. Denbigh, *The Principles of Chemical Equilibrium*, 4th edn. New York: Cambridge University Press (1981).

R. H. Fowler and E. A. Guggenheim, *Statistical Thermodynamics*. Cambridge: Cambridge University Press (1939).

A. Y. Klimenko, "Teaching the third law of thermodynamics," *The Open Thermodynamics Journal* **6**, 1 (2012).

B. G. Kyle, "The third law of thermodynamics," *Chemical Engineering Education* **28**, 176 (1994).

G. N. Lewis and M. Randall, *Thermodynamics and the Free Energy of Chemical Substances*. New York: McGraw-Hill (1923).

E. M. Loebl, "The third law of thermodynamics, the unattainability of absolute zero, and quantum mechanics," *Journal of Chemical Engineering Education* **37**, 361 (1960).

16 The canonical partition function

16.1 A review of basic statistical-mechanical concepts

Earlier chapters introduced microscopic bases for the laws of thermodynamics and then proceeded to generalize them to macroscopic principles that involved concepts like heat and work. We now revisit the molecular world in order to fully develop the framework of statistical mechanics, which provides the microscopic underpinnings of macroscopic thermodynamic properties.

Let us recall the most important conceptualizations discussed so far. Most prominent is the link between microstates and entropy embodied in Boltzmann's equation, $S = k_B \ln \Omega$. A microstate is simply one "configuration" of the system, and the density of states Ω counts how many are compatible with a given set of macroscopic descriptors like E, V, and N. In a classical system of N atoms, a microstate is characterized by a list of the $3N$ positions $\mathbf{r}^N$ and $3N$ momenta $\mathbf{p}^N$ for a total of $6N$ pieces of information. In a quantum system, on the other hand, microstates are characterized by quantum numbers. In general, we can calculate the total, potential, and kinetic energies for any given microstate. In the classical case, the potential energy function depends only on the positions $U(\mathbf{r}^N)$ and the kinetic energy function depends on the momenta $K(\mathbf{p}^N)$.

The principle of equal a priori probabilities leads directly to the second law and motivates the form of Boltzmann's equation. It states that an isolated system at equilibrium visits each microstate consistent with the macrostate with equal probability. That is, the system spends an equal amount of time in each of the $\Omega(E, V, N)$ microstates. If two systems 1 and 2 can share energy, volume, and/or particles, the number of microstates for the combined system (which is isolated) is then given by an expression like

$$\Omega(X_T) = \sum_{X_1} \Omega_1(X_1)\Omega_2(X_T - X_1) \approx \Omega_1\left(X_1^*\right)\Omega_2(X_T - X_1^*)$$

$$\text{where } \frac{\partial \ln \Omega_1\left(X_1^*\right)}{\partial X_1} = \frac{\partial \ln \Omega_2\left(X_2^*\right)}{\partial X_2}$$

(16.1)

Here, X denotes any conserved quantity like E, V, or N such that $X_T = X_1 + X_2$ is constant. Generally speaking, we found that the sum above is extremely peaked around some particular value X_1^* in macroscopic systems; by maximizing the summed quantity we found that X_1^* is given by the condition in the last line of Eqn. (16.1). The implication is that the number of microstates at $X_1 = X_1^*$ is overwhelmingly large and the system spends nearly all of its time in this state; therefore, we almost always observe the precise equilibrium value X_1^* because *fluctuations* away from it are small and rare. For example, when the exchanged quantity is energy, the maximization condition requires that $T_1 = T_2$ and we expect the systems to approach specific values of E_1 and E_2 at equilibrium upon thermal energy exchange, even though their actual energies can fluctuate with time by tiny, microscale amounts.

These ideas motivate the macroscopic notion that the entropy of an isolated system reaches a maximum at equilibrium with respect to any internal degree of freedom. Earlier chapters in this book explored the implications of this idea both in general and for specific kinds of systems. In particular, we found the conditions for equilibrium when systems were not isolated but coupled to various kinds of baths, and these scenarios naturally suggested a number of alternative thermodynamic potentials such as the Helmholtz and Gibbs free energies. Most of our analyses until this point have constituted a classical or macroscopic approach to thermodynamics. Indeed, all three laws of thermodynamics can be taken as postulates that do not require any reference to the molecular world.

In the next few chapters, we now revisit the idea of entropy maximization, coupling with baths, and thermodynamic potentials from a *microscopic* perspective that is the essence of statistical mechanics. From this point of view we are interested in characterizing not only the macroscopic behavior of a system, but also the detailed probabilities with which each microstate is visited. In other words, the difference between macroscopic and microscopic perspectives is as follows.

A **macroscopic** perspective concerns the relevant thermodynamic potential at the conditions of interest and its derivatives, which interrelate macroscopic quantities like E, V, N, T, P, and μ.

On the other hand, a **microscopic** perspective addresses the probability $\wp_m$ associated with every microstate m. The collection of these probabilities characterizes a particular thermodynamic **ensemble** for a system. These probabilities can be used to determine averages that give rise to the macroscopic observables, such as the average energy $\langle E \rangle = \sum_m \wp_m E_m$.

16.2   Microscopic equilibrium in isolated systems

Above we reviewed much of the properties of isolated systems for which E, V, and N are constant and cannot change. At a macroscopic level, the relevant thermodynamic functions are $S(E, V, N)$ or alternatively $E(S, V, N)$. At a microscopic level, the probability of each microstate can be written compactly as

$$\wp_m = \frac{\delta_{E_m, E}}{\Omega(E, V, N)} \tag{16.2}$$

where E_m is the total energy of a microstate m and the Kronecker delta function is such that $\delta_{x,y} = 1$ if $x = y$ and $\delta_{x,y} = 0$ otherwise. A particular set of microstates and their probabilities constitutes an ensemble, and we have a special name for the ensemble described by Eqn. (16.2).

> The **microcanonical ensemble** corresponds to an isolated system at equilibrium with constant E, V, and N. In this ensemble each microstate that has the same energy as the macroscopic energy E appears with equal probability; other microstates are not visited.

As shown in Chapter 4, a convenient way to express the density of states involves enumerating all microstates at a given V and N, and filtering for those with the same energy as the macroscopic one:

$$\Omega(E, V, N) = \sum_{\text{all microstates } n \text{ at } V, N} \delta_{E_n, E} \tag{16.3}$$

Another interpretation is the following: if we demand that each configuration has a probability proportional to $\delta_{E_n, E}$, then we see that the density of states is the normalizing factor in the proportionality constant to ensure that the probabilities sum to unity, $\sum_n \wp_n = 1$. Such probability-normalizing factors have a special role in statistical mechanics.

> Quantities that normalize the microstate probabilities so that they sum to one are called **partition functions** in statistical mechanics, and they involve sums over all microstates. The density of states $\Omega(E, V, N)$ is therefore the **microcanonical partition function**.

We will see shortly that logarithms of partition functions are always related to thermodynamic potentials as functions of their natural variables. In this case, the logarithm of the microcanonical partition function gives S/k_B.

16.3   Microscopic equilibrium at constant temperature

We now examine the microscopic behavior of a system that is not isolated but at constant temperature. That is, the system is connected to a very large bath with which it can exchange energy. In this case, the composite of the system and a bath constitutes an isolated supersystem that can be considered in terms of a microcanonical ensemble. However, we are interested only in the probabilities of microstates in the system, not those in the bath, so we must consider the situation carefully.

Let E be the energy of the system, E_B that of the bath, and $E_T = E + E_B$ the fixed total. The index m denotes a microstate of the system, while l refers to a microstate of the bath. It is important to recognize that a microstate for the supersystem is given by the combination of a particular system microstate m and bath microstate l, denoted lm. If the system and bath were to have fixed energies E and E_B, the number of supersystem microstates would simply count all possible combinations of compatible l and m values, giving $\Omega(E)\Omega_B(E_B)$. Here we omit V and N terms since these are fixed. In reality, the system can exchange energy with the bath, so we must consider all possible combinations of E and E_B as would be suggested by Eqn. (16.1). The total number of supersystem microstates can thus be written as

$$\Omega_T(E_T) = \sum_E \Omega(E)\Omega_B(E_T - E) \tag{16.4}$$

The sum is performed over all values for the system energy, in general from negative to positive infinity, although one or both of the densities of states may be zero for some energies. (Consider, for example, that $\Omega = 0$ when the energy is less than the system ground-state energy.) The bath energy of course is always the difference with the total energy, which is constant. The probability of a composite microstate lm according to (16.2) is then

$$\wp_{lm} = \frac{\delta_{E_{lm}, E_T}}{\Omega_T(E_T)} \tag{16.5}$$

E_m is the energy of microstate m of the system and E_{Bl} that of microstate l in the bath, with $E_{lm} = E_m + E_{Bl}$ that of the combined microstate.

At this point, we have an expression for the supersystem microstate probabilities, but we really only care about those of the system. That is, we want to know the probability $\wp_m$ of system microstate m regardless of the bath microstate l with which it is paired. To find this, we simply need to sum $\wp_{lm}$ over all bath microstates l,

$$\wp_m = \sum_{\text{all } l} \wp_{lm} \tag{16.6}$$

Because the probability $\wp_{lm}$ is zero for any state for which $E_m + E_{Bl} \neq E_T$, we need to perform the sum only over microstates l that have energy $E_{Bl} = E_T - E_m$. There are exactly $\Omega_B(E_T - E_m)$ of these states and $\wp_{lm} = 1/\Omega_T$ for each of them. The sum in (16.6) is thus

$$\wp_m = \frac{\Omega_B(E_T - E_m)}{\Omega_T(E_T)} \tag{16.7}$$

Note that $\wp_m$ depends on E_m, which can fluctuate because the system is in contact with the bath. This contrasts with the isolated case in which the system energy is fixed.

Our next approach is to simplify Eqn. (16.7) in light of the fact that the ideal bath is infinitely large. Since the denominator is the same for every m, we can take a simpler perspective for now by simply absorbing it into a constant of proportionality,

$$\wp_m \propto \Omega_B(E_T - E_m) \tag{16.8}$$

To account for the size of the bath, we Taylor-expand its entropy to provide a suitable form for Ω_B through Boltzmann's relation. We do not expand Ω_B directly because we expect the entropy to be of order N_B due to extensivity:

$$k_B \ln \Omega_B(E_T - E_m) = S_B(E_T - E_m)$$

$$= S_B(E_T) - E_m \left(\frac{\partial S_B(E_T)}{\partial E_B} \right) + \frac{E_m^2}{2} \left(\frac{\partial^2 S_B(E_T)}{\partial E_B^2} \right) + \cdots \quad (16.9)$$

Now, we replace the entropy of the bath with an intensive version scaled by the size of the bath, $S_B(E_B) = N_B s_B(E_B/N_B)$:

$$S_B(E_T - E_m) = N_B s_B(E_T/N_B) - E_m \left(\frac{\partial s_B(E_T/N_B)}{\partial e_B} \right) + \frac{E_m^2}{2N_B} \left(\frac{\partial^2 s_B(E_T/N_B)}{\partial e_B^2} \right) + \cdots$$

$$(16.10)$$

This transformation allows the size of the system to remain constant while the bath grows infinitely large. In this limit we have

$$\lim_{N_B \to \infty} \frac{E_T}{N_B} = \lim_{N_B \to \infty} \frac{E_m + N_B e_B}{N_B} = e_B \quad (16.11)$$

and thus

$$\lim_{N_B \to \infty} S_B(E_T - E_m) = N_B s_B(e_B) - E_m \left(\frac{\partial s_B(e_B)}{\partial e_B} \right) \quad (16.12)$$

The higher-order terms in the Taylor expansion drop out completely since they involve N_B raised to increasingly negative powers. Importantly, the entropy derivative for the bath gives its inverse temperature, $1/T_B$. Here we will simply denote the temperature by T, dropping the subscript, since it is through the bath that the temperature is ultimately defined for the system.

By simplifying the bath entropy in this manner, we can finally obtain a limiting expression for Ω_B that allows us to simplify the microstate probabilities in Eqn. (16.8),

$$\wp_m \propto \exp\left[\frac{N_B s_B(e_B)}{k_B} - \frac{E_m}{k_B T} \right] \quad (16.13)$$

Note that the first term in the exponential is independent of the microstate m, so we can absorb it into the constant of proportionality,

$$\wp_m \propto \exp\left(-\frac{E_m}{k_B T} \right) \quad \textbf{(16.14)}$$

This important result shows that the probability of a particular microstate *of the system* has an exponential dependence on its energy. Indeed, it is a rather remarkable relationship for several reasons. For one, the relative probability has nothing to do with the bath beyond the temperature it establishes. That is, no other property of the bath affects the distribution of energies in the system. Ultimately, this is because the bath is so large.

Moreover, the energy of the system when coupled to a bath is no longer fixed as it was in the isolated case. Instead, it can fluctuate as different microstates are visited according to $\wp_m$. Microstates with higher energies are visited with exponentially less frequency than those with low energies, and the degree of exponential damping is controlled by the inverse temperature. At low temperatures, the microstates with the lowest energies have the highest relative probabilities. In principle, as $T \to 0$, the probability of the lowest-energy microstate (or microstates) dominates and all other probabilities become infinitely small. At high temperatures, the differences in probabilities for microstates with different energies begin to disappear. In principle, as $T \to \infty$, all microstates have the same probability regardless of energy.

The exponential in (16.14) giving the relative probability of microstates at constant temperature is called the *Boltzmann factor*, and it appears frequently in statistical mechanics. The combination of variables $1/k_B T$ occurs so frequently that it is common to define a new variable called the *thermodynamic beta* that has units of inverse energy,

$$\beta \equiv \frac{1}{k_B T} \tag{16.15}$$

When this is used, the Boltzmann expression becomes

$$\wp_m \propto \exp(-\beta E_m) \tag{16.16}$$

Recall that a particular set of microstates and their probabilities constitutes an ensemble. The probabilities embodied by Eqn. (16.16) constitute a particular ensemble in the constant temperature case.

> The **canonical ensemble** corresponds to a system at fixed V and N that is held at constant temperature T through coupling to an ideal heat bath. In this ensemble, the probability of each microstate m is proportional to $\exp(-\beta E_m)$.

It may seem unusual that we have not yet discussed the constant of proportionality in the canonical probabilities. We absorbed into it many factors that involved properties of the bath. Perhaps surprisingly, the absolute values of the probabilities do not depend on the details of the bath at all. This is because they must be normalized; their sum over all system microstates must equal one. We can thus immediately write that the probabilities must have the exact form

$$\wp_m = \frac{e^{-\beta E_m}}{\sum_n e^{-\beta E_n}} \tag{16.17}$$

where the sum in the denominator is performed over all microstates in the system. It is trivial to see that (16.17) directly leads to $\sum_m \wp_m = 1$.

In the discussion of the microcanonical ensemble, we commented that the normalizing factor for microstate probabilities is called a partition function in statistical mechanics. In the canonical ensemble, we therefore define a *canonical partition function*:

$$Q(T, V, N) \equiv \sum_{\text{all } n \text{ at } V, N} e^{-\beta E_n} \tag{16.18}$$

Here, Q is a function of V and N because these specify the list of microstates used in the sum; it is also a function of T through β. Using Q, we can easily express the microstate probabilities by

$$\wp_m = \frac{e^{-\beta E_m}}{Q(T, V, N)} \tag{16.19}$$

The reason why we identify the probability-normalization constant Q with a unique variable and name is because partition functions relate to thermodynamic potentials. We can begin to see this by computing the average energy in the canonical ensemble,

$$\langle E \rangle = \sum_n \wp_n E_n$$

$$= \frac{1}{Q} \sum_n E_n e^{-\beta E_n} \tag{16.20}$$

Notice, however, that the expression within the sum looks similar to Q except for a factor involving the microstate energy E_n. A clever manipulation allows us to express the average energy in terms of a derivative of Q,

$$\langle E \rangle = \frac{1}{Q} \sum_n -\frac{\partial}{\partial \beta} e^{-\beta E_n}$$

$$= \frac{1}{Q} \left(-\frac{\partial Q}{\partial \beta} \right)$$

$$= -\frac{\partial \ln Q}{\partial \beta} \tag{16.21}$$

On substituting $\beta = 1/k_B T$ and carrying out some careful rearrangement, we arrive at the following particularly informative equation:

$$\frac{\langle E \rangle}{T^2} = \frac{\partial (k_B \ln Q)}{\partial T} \tag{16.22}$$

Compare Eqn. (16.22) with the classical derivative relation found in Chapter 7,

$$\frac{E}{T^2} = -\frac{\partial (A/T)}{\partial T} \tag{16.23}$$

At a macroscopic level, the average energy $\langle E \rangle$ is identical to the single, classical value of E observed for a system in thermal equilibrium with a heat bath, neglecting all of the tiny fluctuations. The similarity of Eqns. (16.22) and (16.23) therefore suggests that

$$A(T, V, N) = -k_B T \ln Q(T, V, N) \tag{16.24}$$

This is indeed the case, and it can be shown with more rigorous arguments as outlined in the texts by Hill and McQuarrie. The important result is that the canonical partition function Q is related to the Helmholtz free energy. Intuitively, one might anticipate this result on the basis that A is the relevant thermodynamic potential at constant T, V, N conditions. Nonetheless, the connection has significant implications, the most important of which is as follows.

If we can determine the canonical partition function Q, then we can obtain the Helmholtz free energy A in fundamental form. From that, we can then obtain all of the thermodynamic properties of the system.

As we have seen many times, one approach to evaluating the thermodynamics of a microscopic model involves computing the entropy from the density of states, $S = k_B \ln \Omega$. This gives the entropy in fundamental form. Here we now see that there is an alternative way through the canonical ensemble. Namely, we can use Q to recover $A = -k_B T \ln Q$ in fundamental form. Either way, we can obtain the entire macroscopic thermodynamic properties of the model.

Switching from one partition function to another, such as going from Ω to Q, is a kind of microscopic analogy of a Legendre transform.

Example 16.1 *Consider the two-state model described in earlier chapters. Compute the microcanonical and canonical partition functions, and use both to find the energy E as a function of T.*

We have already found the density of states for this model in Example 3.1. It gave the entropy

$$S = Nk_B \left[-\left(\frac{E}{Nb}\right) \ln\left(\frac{E}{Nb}\right) - \left(1 - \frac{E}{Nb}\right) \ln\left(1 - \frac{E}{Nb}\right) \right]$$

The temperature is given by the E derivative,

$$\frac{1}{T} = \left(\frac{\partial S}{\partial E}\right)_N = \frac{k_B}{b} \ln\left[\frac{Nb}{E} - 1\right]$$

Upon inverting this relationship, the final result for the energy is

$$E = Nb \frac{e^{-\beta b}}{1 + e^{-\beta b}}$$

Now, we perform the same computation in the canonical ensemble, giving

$$Q = \sum_n e^{-\beta E_n} = \sum_n e^{-\beta \sum_i \epsilon_i}$$

where ϵ_i gives the energy of particle i. The sum over n is a sum over all microstates, i.e., a sum over all possible combinations of particles in up or down positions.

It is instructive to consider the case of two particles and only four microstates in total. The partition function would be

$$Q = e^{-\beta 0 - \beta 0} + e^{-\beta 0 - \beta b} + e^{-\beta b - \beta 0} + e^{-\beta b - \beta b}$$

where the zeros are explicitly shown to illustrate the different microstates. We could rewrite this as

$$Q = \sum_{\epsilon_1 \in \{0, b\}} \sum_{\epsilon_2 \in \{0, b\}} e^{-\beta \epsilon_1 - \beta \epsilon_2}$$

In general, for N particles, the sum becomes

$$Q = \sum_{\epsilon_1 \in \{0, b\}} \sum_{\epsilon_2 \in \{0, b\}} \cdots \sum_{\epsilon_N \in \{0, b\}} e^{-\beta \sum_i \epsilon_i}$$

Each sum applies to a single particle and includes only two terms: one for the up state ($\epsilon_i = b$) and one for the down state ($\epsilon_i = 0$). Notice that the particles are non-interacting such that $E = \epsilon_1 + \epsilon_2 + \cdots + \epsilon_N$. That means the sum can be rewritten as

$$Q = \sum_{\epsilon_1 \in \{0, b\}} \sum_{\epsilon_2 \in \{0, b\}} \cdots \sum_{\epsilon_N \in \{0, b\}} e^{-\beta \epsilon_1} \times e^{-\beta \epsilon_2} \times \cdots \times e^{-\beta \epsilon_N}$$

Whenever a nested sum involves a product of individual terms that are separable with respect to each of the sum indices, it can be factored,

$$Q = \left(\sum_{\epsilon_1 \in \{0, b\}} e^{-\beta \epsilon_1} \right) \left(\sum_{\epsilon_2 \in \{0, b\}} e^{-\beta \epsilon_2} \right) \times \cdots \times \left(\sum_{\epsilon_N \in \{0, b\}} e^{-\beta \epsilon_N} \right)$$

Each individual summation here gives the same result. Therefore, we can simplify to

$$Q = \left(\sum_{\epsilon \in \{0, b\}} e^{-\beta \epsilon} \right)^N = (1 + e^{-\beta b})^N$$

The Helmholtz free energy then follows from Eqn. (16.24),

$$A = -k_B T \ln Q = -N k_B T \ln \left(1 + e^{-\beta b} \right)$$

As we would expect, it is extensive and scales linearly with N. From A, we can find E using

$$E = -T^2 \frac{\partial (A/T)}{\partial T} = N k_B T^2 \left(\frac{d\beta}{dT} \right) \frac{d}{d\beta} \ln(1 + e^{-\beta b})$$

$$= N k_B T^2 \left(-\frac{1}{k_B T^2} \right) \frac{-b e^{-\beta b}}{1 + e^{-\beta b}} = N b \frac{e^{-\beta b}}{1 + e^{-\beta b}}$$

We find that we recover the same expression for the average energy as in the microcanonical case.

Though it may not seem so from Example 16.1, computing the properties of model systems in the canonical ensemble is often much easier than in the microcanonical one. In the latter, we must count microstates that have a given total energy in order to compute the density of states. This energy constraint usually necessitates a deep

Table 16.1 A comparison of the statistical mechanical properties of the microcanonical and canonical ensembles

Property	Microcanonical ensemble	Canonical ensemble
Constant conditions	E, V, N	T, V, N
Microstate probabilities	$\wp_m = \delta_{E_m, E}/\Omega(E, V, N)$	$\wp_m = e^{-\beta E_m}/Q(T, V, N)$
Partition function	$\Omega(E, V, N) = \displaystyle\sum_{\text{all } n \text{ at } V, N} \delta_{E_n, E}$	$Q(T, V, N) = \displaystyle\sum_{\text{all } n \text{ at } V, N} e^{-\beta E_n}$
Thermodynamic potential	$S = k_B \ln \Omega(E, V, N)$	$A = -k_B T \ln Q(T, V, N)$

consideration of combinatorics that can become quite complex for nontrivial molecular interactions. On the other hand, in the canonical ensemble, every microstate enters into the partition sum but each contributes a different probability weight as given by the Boltzmann factor. As a result, often we can avoid the complicated issues of combinatorics if we can evaluate sums over microstates.

Table 16.1 summarizes the basic differences between the microcanonical and canonical ensembles. In what was presented thus far, we considered only a single component system. However, these relations readily extend to multicomponent cases in which there exists more than one particle number. Indeed, one can replace the N appearing in Table 16.1 with $\{N\}$ or $\mathbf{N} = (N_1, N_2, \ldots)$ and retain complete generality.

16.4   Microstates and degrees of freedom

In order to compute the canonical partition function, one needs to perform a sum over all microstates of the system. Let us consider this set of microstates in more detail. In simple and coarse-grained models like the two-state system and the lattice gas, the complete set of microstates is often discrete and can be enumerated exactly. It is for this reason that such simple models appear so frequently in statistical mechanics. In systems described by quantum mechanics, a set of discrete states also emerges; the states are indexed by a set of quantum numbers, although their determination can be more complicated owing to the need to solve Schrödinger's equation. In classical mechanics, microstates are defined by continuous position and momenta variables and thus there are infinitely many of them. Therefore, the partition sums of Eqns. (16.3) and (16.18) become integrals, an approach that we will discuss in Chapter 18. Regardless of the nature of the molecular model, however, we can make a general statement about the microstate ensembles.

A microstate is defined by specific values for all of the **microscopic degrees of freedom** in the system. A sum over all microstates is a sum over all possible combinations of values of these degrees of freedom. For degrees of freedom that are continuous, this sum actually involves integrals.

Think of the set of degrees of freedom as all of the numbers that you would need to write down to completely specify the microscopic configuration of a system, such that, if you sent that description to a friend, they would be able to reconstruct the configuration exactly.

With these considerations, we can rewrite partition sums over microstates instead as sums over values for the degrees of freedom,

$$
\Omega(E, V, N) = \sum_{\mathcal{X}_1}\sum_{\mathcal{X}_2}\cdots\sum_{\mathcal{X}_N} \delta_{E(\mathcal{X}_1, \mathcal{X}_2, \ldots, \mathcal{X}_N), E}
$$

$$
Q(T, V, N) = \sum_{\mathcal{X}_1}\sum_{\mathcal{X}_2}\cdots\sum_{\mathcal{X}_N} e^{-\beta E(\mathcal{X}_1, \mathcal{X}_2, \ldots, \mathcal{X}_N)}
$$

(16.25)

Here each $\mathcal{X}$ variable indicates one degree of freedom in the system, such as the x position of particle 1 or the y momentum of particle 5 in a classical model, the third quantum number in a quantum model, or the up versus down state of a particle in a two-state model. There are many microscopic degrees of freedom $\mathcal{N}$; they total $6N$ for the positions and momenta in a classical system. Each sum implies a looping through all possible values for that degree of freedom, such as all possible values of a given quantum number. It is perhaps best to think of each sum as enumerating the elements of a long list of options for the given $\mathcal{X}$. In general, the energy of a microstate is a function of all of the degrees of freedom, which we write as $E_m = E(\mathcal{X}_1, \mathcal{X}_2, \ldots, \mathcal{X}_N)$. In other words, the microstate index is given by a list of specific values for all the $\mathcal{X}$ terms, that is, $m = \{\mathcal{X}_1, \mathcal{X}_2, \ldots, \mathcal{X}_N\}$.

Equation (16.25) is a very general way to express the partition sums that often leads to important simplifications. Namely, what happens if we can write the energy as a sum of two different terms depending on *mutually exclusive* subsets of the degrees of freedom? We might be able express the total energy as a linear sum of terms,

$$
E(\mathcal{X}_1, \mathcal{X}_2, \ldots, \mathcal{X}_N) = E_1(\mathcal{X}_1, \ldots, \mathcal{X}_k) + E_2(\mathcal{X}_{k+1}, \ldots, \mathcal{X}_N)
$$

(16.26)

where the first energy term E_1 depends only on degrees of freedom 1 through k and the second term E_2 on the remaining $k + 1$ through $\mathcal{N}$. The ordering of the degrees of freedom is immaterial at this point; it is merely important that the energy can be subdivided in this way. An obvious example is that of classical systems, for which we can write $E = K + U$; here, the kinetic energy depends only on the momentum degrees of freedom, while the potential energy only on the positional ones.

When a decomposition of the energy as per (16.26) can be performed, we say that the two subsets of degrees of freedom are *non-interacting* since they independently affect the energy. This simplifies the expression for the canonical partition function because the sums can be factored in a way analogous to Example 16.1,

$$
Q(T, V, N) = \sum_{\mathcal{X}_1}\sum_{\mathcal{X}_2}\cdots\sum_{\mathcal{X}_N} e^{-\beta E_1(\mathcal{X}_1, \ldots, \mathcal{X}_k)} e^{-\beta E_2(\mathcal{X}_{k+1}, \ldots, \mathcal{X}_N)}
$$

$$
= \left[\sum_{\mathcal{X}_1}\cdots\sum_{\mathcal{X}_k} e^{-\beta E_1(\mathcal{X}_1, \ldots, \mathcal{X}_k)}\right]\left[\sum_{\mathcal{X}_{k+1}}\cdots\sum_{\mathcal{X}_N} e^{-\beta E_2(\mathcal{X}_{k+1}, \ldots, \mathcal{X}_N)}\right]
$$

$$
= Q_1(T, V, N) Q_2(T, V, N)
$$

(16.27)

In the last line, we introduced two new partition function sums that act on subsets of the degrees of freedom. This derivation shows that, any time the energy function can be split into multiple linear terms like this, the overall partition function can be factored.

> For energy functions that are linearly separable in the system degrees of freedom – equivalently, that have non-interacting, mutually exclusive subsets of degrees of freedom – the canonical partition function can be written as a product of canonical partition functions for the individual components of the energy function.

Owing to the constraint of total energy, however, a simple factorization such as this is not generally possible in the microcanonical ensemble.

Example 16.2 *The famous Ising model describes a ferromagnet consisting of a lattice of N atomic spins as illustrated in Fig. 16.1. Each spin i can point in one of two directions: up with $\sigma_i = +1$ or down with $\sigma_i = -1$. Nearest-neighbor spins contribute an energy $-J\sigma_i\sigma_j$ that favors spins pointing in the same direction and penalizes oppositely aligned ones. In addition, a magnetic field of strength $\mathcal{H}$ can interact with individual spins to give energetic contributions of the form $\mathcal{H}\sigma_i$. $\mathcal{H}$ can be positive or negative, depending on the direction of the field. Find the partition function, the free energy, and the average energy as a function of temperature for the case in one dimension and with $\mathcal{H} = 0$.*

From the problem statement, we see that the energy is formally given by

$$E = -J \sum_{i,j \text{ neighbors}} \sigma_i\sigma_j - \mathcal{H}\sum_i \sigma_i$$

But, for this example, $\mathcal{H} = 0$, and we can neglect the final term. In one dimension, particles have only two neighbors. Using Eqn. (16.25), we write the partition function as

$$Q = \sum_{\sigma_1 \in \{-1, +1\}} \sum_{\sigma_2 \in \{-1, +1\}} \cdots \sum_{\sigma_N \in \{-1, +1\}} e^{\beta J(\sigma_1\sigma_2 + \sigma_2\sigma_3 + \cdots + \sigma_{N-1}\sigma_N)}$$

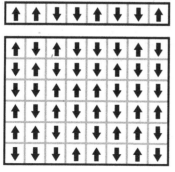

Figure 16.1. The Ising model describes the properties of materials possessing a collection of magnetic spins. Each lattice site contains a spin that either can point up or down. Both one- (top) and two-dimensional (bottom) versions are depicted here.

It will help to make a variable transformation for σ_2 through σ_N in order to evaluate these sums. We define

$$t_i = \sigma_{i-1}\sigma_i \quad \rightarrow \quad \sigma_i = \frac{t_i}{\sigma_{i-1}}$$

Because any $\sigma = \pm 1$, we also have that any $t = \pm 1$. We can then rewrite the partition function as

$$Q = \sum_{\sigma_1 \in \{-1,+1\}} \sum_{t_2 \in \{-1,+1\}} \cdots \sum_{t_N \in \{-1,+1\}} e^{\beta J(t_2 + t_3 + \cdots + t_N)}$$

The energy is separable in the new degrees of freedom as per Eqn. (16.27),

$$Q = \sum_{\sigma_1 \in \{-1,+1\}} \left(\sum_{t \in \{-1,+1\}} e^{\beta J t} \right)^{N-1} = 2(e^{-\beta J} + e^{\beta J})^{N-1}$$

The factor of two comes from the sum over the first spin (σ_1). The Helmholtz free energy is given by Eqn. (16.24) and becomes

$$A = -k_B T(N-1)\ln(e^{-\beta J} + e^{\beta J}) - k_B T \ln 2 \approx -N k_B T \ln(e^{-\beta J} + e^{\beta J})$$

where the second line gives the limiting large-N behavior. The energy stems from (16.21),

$$\langle E \rangle = -\left(\frac{\partial \ln Q}{\partial \beta}\right) = -(N-1)J\frac{e^{\beta J} - e^{-\beta J}}{e^{\beta J} + e^{-\beta J}} \approx -NJ \tanh(\beta J)$$

As $T \rightarrow \infty$, we have that $\langle E \rangle \rightarrow 0$, as one might expect. In that case, it is equally likely that neighboring particles will be aligned and that they will not, so the average pair interaction is zero. In contrast as $T \rightarrow 0$, we have that the limit of the tanh function becomes unity and $\langle E \rangle \rightarrow -NJ$. This indicates that all spins are aligned in the same direction.

It is interesting to note that the exact solution to the Ising model described in the example above is extremely complicated in two dimensions and was accomplished in 1944 by Lars Onsager, who later won a Nobel Prize for his contributions to the theory of magnetism. The difficulty of the two-dimensional case stems from the need to treat correlations among multiple neighbors, whereas in one dimension each spin can be "referenced" with respect to a single previous neighbor as illustrated through the variable transforms above. No exact solutions are available in three dimensions, so approximate methods or computer simulations must be used. In any case, the Ising model provides a foundation for understanding many different kinds of phase transitions in statistical mechanics, and its behavior is equivalent upon certain variable transforms to that of the lattice-gas model.

16.5   The canonical partition function for independent molecules

In Example 16.1 we found a particular simplification of the partition sum when particles lack interactions with each other. In fact, one can generalize this situation to arbitrary systems using the ideas presented in the previous section. Consider a case in which the energy of a system can be expressed as

$$E(\mathcal{X}_1, \mathcal{X}_2, \ldots, \mathcal{X}_N) = \sum_i \epsilon_i(\mathcal{X}_i) \tag{16.28}$$

where ϵ_i is the part of the energy that depends on the degrees of freedom $\mathcal{X}_i$ associated with particle i. By using a single $\mathcal{X}_i$ for each particle, we are not assuming that the particle has only a single degree of freedom but rather envision that each $\mathcal{X}_i$ can be a vector of values. For example, classical monatomic particles would have six degrees of freedom stored in each $\mathcal{X}_i$, namely their positions and momenta.

In any case, when Eqn. (16.28) describes a system, this means that particle i's degrees of freedom do not affect the energies of any other particles. In turn, the particles are non-interacting or *independent*. If two particles experienced a pairwise interaction such as an attraction or repulsion, this expression could not be applied because the energy would require terms involving two particles' degrees of freedom.

For independent particles, the canonical partition function simplifies considerably because it is separable in each of the particle degrees of freedom. We begin by writing the single exponential in (16.25) as a product of exponentials, which is possible by application of (16.28),

$$Q = \sum_{\mathcal{X}_1} \sum_{\mathcal{X}_2} \cdots \sum_{\mathcal{X}_N} \left(\prod_i e^{-\beta \epsilon_i(\mathcal{X}_i)} \right) \tag{16.29}$$

The interpretation of the sums is similar to what we found before: they simply enumerate all possible values for the degrees of freedom of a given particle. The key step is that we can re-factor the sums and products just as we did in the previous section:

$$Q = \left(\sum_{\mathcal{X}_1} e^{-\beta \epsilon_1(\mathcal{X}_1)} \right) \left(\sum_{\mathcal{X}_2} e^{-\beta \epsilon_2(\mathcal{X}_2)} \right) \times \cdots \times \left(\sum_{\mathcal{X}_N} e^{-\beta \epsilon_N(\mathcal{X}_N)} \right) = \prod_i \left(\sum_{\mathcal{X}_i} e^{-\beta \epsilon_i(\mathcal{X}_i)} \right) \tag{16.30}$$

The last equation shows that the total partition function can be written as a product of single-particle partition functions:

$$Q = \prod_i q_i, \quad \text{where} \quad q_i \equiv \sum_{\mathcal{X}_i} e^{-\beta \epsilon_i(\mathcal{X}_i)} \tag{16.31}$$

If all of the particles are the same, then all of the q_i will be the same, since they will have the same degrees of freedom and energy levels. This means that the total partition function can be written as

$$Q = q^N \quad \text{(independent distinguishable particles)} \tag{16.32}$$

Here we have qualified that this expression applies to *distinguishable particles*. Those are cases for which we would be able to identify the index number of each particle if we were shown a snapshot of the system (e.g., what the values of i are). In contrast, we cannot perform such an identification for *indistinguishable particles*. Quantum mechanics dictates that a configuration of indistinguishable particles does not change when one switches the identities of two particles by swapping their degrees of freedom. In other words, there is only one microstate for a particular configuration regardless of any swaps one can perform among the particle identities.

For indistinguishable particles, the canonical partition function derived above over-counts the number of microstates by the number $N!$ of possible swaps among particle identities. (Technically, this factor is correct only as long as particles are in distinct states, which is always true for classical particles and approximately correct for high-temperature quantum systems.) On correcting for the overcounting, we find

$$Q = \frac{q^N}{N!} \qquad \text{(independent indistinguishable particles)} \qquad \textbf{(16.33)}$$

This expression readily generalizes to multicomponent systems. If there are two types of particles, A and B, for example, we can write

$$Q = \frac{q_A^{N_A} q_B^{N_B}}{N_A! N_B!} \qquad (16.34)$$

since A particles are distinguishable from B ones but not from each other, etc. These factors highlight an important concept.

> For independent molecules, the canonical partition function Q can be written as a product of individual, single-particle partition functions q_i. If the particles are indistinguishable, we must introduce correction factors of $1/N!$ for each distinct species.

Example 16.3 *Consider a lattice-based ideal gas similar to that presented in Chapter 3. Each gas molecule can occupy one of V/v lattice sites, where v is the volume of a lattice "pixel." Here, we will also assume that each molecule i can adopt one of two energy states due to internal isomerization, with $\epsilon_i = 0$ or b. Find the Helmholtz free energy.*

In this case, the particles remain independent of each other but are indistinguishable such that Eqn. (16.33) applies. The degrees of freedom of a single molecule are two-fold: the specific lattice site that it occupies and the particular isomer state. Only the latter degree of freedom affects the particle energy. According to (16.31), we then have

$$q = \sum_{\substack{\text{lattice} \\ \text{sites}}} \sum_{\epsilon \in \{0, b\}} e^{-\beta \epsilon} = \left(\frac{V}{v}\right)(1 + e^{-\beta b})$$

The Helmholtz free energy is thus

$$A = -k_B T \ln \left(\frac{q^N}{N!} \right) \approx -k_B T \ln \left(\frac{q^N e^N}{N^N} \right)$$

$$= -N k_B T \ln \left[\left(\frac{Ve}{Nv} \right) (1 + e^{-\beta b}) \right]$$

where the second line employs Stirling's approximation. Notice that A is extensive; however, it cannot be defined in absolute terms for a classical model because the pixel volume v is not defined in a classical sense.

The microstate probabilities for independent particles are also separable. Consider the distinguishable case. We begin with Eqn. (16.19) and substitute Eqn. (16.28),

$$\wp(\mathcal{X}_1, \mathcal{X}_2, \ldots, \mathcal{X}_N) = \frac{\prod_i e^{-\beta \epsilon_i(\mathcal{X}_i)}}{\sum_{\mathcal{X}_1'} \sum_{\mathcal{X}_2'} \cdots \sum_{\mathcal{X}_N'} \left(\prod_i e^{-\beta \epsilon_i(\mathcal{X}_i')} \right)} \tag{16.35}$$

Here, the primes on the $\mathcal{X}$ terms in the denominator are used to sum the canonical partition function, while the unprimed versions designate the particular microstate $m = \{\mathcal{X}_1, \mathcal{X}_2, \ldots, \mathcal{X}_N\}$ whose probability is of interest. As before, the denominator can be factored by switching the order of the sums and product, giving

$$\wp(\mathcal{X}_1, \mathcal{X}_2, \ldots, \mathcal{X}_N) = \frac{e^{-\beta \epsilon_1(\mathcal{X}_1)}}{\sum_{\mathcal{X}_1'} e^{-\beta \epsilon_1(\mathcal{X}_1')}} \frac{e^{-\beta \epsilon_2(\mathcal{X}_2)}}{\sum_{\mathcal{X}_2'} e^{-\beta \epsilon_2(\mathcal{X}_2')}} \times \cdots \times \frac{e^{-\beta \epsilon_N(\mathcal{X}_N)}}{\sum_{\mathcal{X}_N'} e^{-\beta \epsilon_N(\mathcal{X}_N')}}$$

$$= \wp(\mathcal{X}_1)\wp(\mathcal{X}_2) \times \cdots \times \wp(\mathcal{X}_N) \tag{16.36}$$

The last line shows that the probability that a particle adopts specific values of its degrees of freedom is independent of the other particles: the total probability for a given microstate m is simply the product of the individual probabilities for the degrees of freedom, $\wp(\mathcal{X}_1)$, $\wp(\mathcal{X}_2)$, etc. That is, the individual particles obey probability functions that are reduced in dimension, with the general form

$$\wp(\mathcal{X}) = \frac{e^{-\beta \epsilon(\mathcal{X})}}{\sum_{\mathcal{X}'} e^{-\beta \epsilon(\mathcal{X}')}} \tag{16.37}$$

It is as if we can consider each particle in isolation from the others, and simply compute partition functions like $Q(T, V, N = 1)$. For example, the distribution of the position of particle 1 in the system would be uninfluenced by the locations of all other particles. Of course, this would be expected on physical grounds if the energy could be expressed as per Eqn. (16.28) and there were no cross-interaction terms.

Example 16.4 *Consider the two-state model solved in Example 16.1. What is the probability that particle 1 is in the down position? What is the probability that all N particles are in the down position?*

Since the two-state model involves an energy function like Eqn. (16.28), we can treat the particles as being independent. The probability that an individual particle is in the down position is then given by Eqn. (16.37),

$$\wp(\mathcal{X}_1 = \text{down}) = \frac{e^{-\beta 0}}{e^{-\beta 0} + e^{-\beta b}} = \frac{1}{1 + e^{-\beta b}}$$

Notice that this probability is the same for any particle, since each particle has the same two states and energy levels. The probability that *all* of the particles will be in the down position is given by the complete microstate probability as factored in Eqn. (16.36). This simply introduces N factors identical to those above,

$$\wp(\mathcal{X}_1 = \text{down}, \mathcal{X}_2 = \text{down}, \dots, \mathcal{X}_N = \text{down}) = \left(\frac{1}{1 + e^{-\beta b}} \right)^N$$

Problems

Conceptual and thought problems

16.1. Consider a simple model of dimerization. Two molecules are present on a cubic three-dimensional lattice with number of lattice sites V. When molecules sit next to each other, there is a net favorable energy of $-\epsilon$. The system is maintained at a constant temperature T. Neglect edge effects in the lattice, i.e., assume periodic boundary conditions.
   (a) How many microstates are there in total, and how many have the two particles as neighbors?
   (b) What is the canonical partition function?
   (c) What is the absolute probability of a single microstate that has the two particles as neighbors?
   (d) What is the total probability of seeing a dimer, in any microstate? Express your answer using the concentration of particles, $c = 2/V$, instead of V.

16.2. Why might you expect the canonical partition function to scale exponentially with the system size, thus always ensuring that the Helmholtz free energy is extensive?

16.3. In a quantum sense, independent particles have distinct values for their degrees of freedom provided that they have different quantum numbers. With this in mind, explain why Eqn. (16.33) is generally valid only in the limit of high energies and temperatures. Hint: what happens if two independent particles have the same quantum numbers? Alternatively, consider the $T \to 0$ case in which each particle is in its ground state. Has any overcounting occurred? What should q and Q be in this case?

16.4. Consider two particles, each of which can exist in n distinct states. The energy of a particle i in its state $k_i = 1, 2, \ldots, n$ is given by bk_i, where b is a constant with units of energy.

(a) If the particles are independent, $E = bk_1 + bk_2$. Show explicitly that the probability for the state of particle 1 is independent of the probability for the state of particle 2. Namely, show that $\wp(k_1, k_2) = \wp(k_1)\wp(k_2)$.

(b) Consider the case in which the particles are not independent, and the energy is actually given by $E = bk_1 + bk_2 + ck_1k_2$, where c is a constant with units of energy that measures the interaction between the two particles. Prove that $\wp(k_1, k_2) \neq \wp(k_1)\wp(k_2)$.

16.5. Consider the two-state model analyzed in Example 16.4.

(a) Does the two-state model consist of distinguishable or indistinguishable particles?

(b) What is the probability that particles 1 and 2 will both be in the down position?

(c) What is the probability that *exactly* two particles of the N will both be in the down position, regardless of their index.

16.6. When we say "independent particles" we really mean "very weakly interacting particles" instead of "fully non-interacting particles." Why?

Fundamentals problems

16.7. Derive Eqn. (16.7) on more rigorous mathematical grounds. Use the property of the Kronecker delta that $\delta_{x,y} = \delta_{x-z,y-z}$ and let $z = E_m$ to perform the sum explicitly.

16.8. For any ensemble the entropy can be computed by evaluating

$$S/k_B = -\sum_n \wp_n \ln \wp_n$$

where the sum is performed over all microstates. Prove rigorously that this expression gives the entropy both for the microcanonical case and for the canonical case.

16.9. The average energy can be written as

$$\langle E \rangle = \sum_n \wp_n E_n$$

(a) Find the total differential of $\langle E \rangle$. The term dE_n can be associated with the reversible work required to change the energy of microstate n. Find an expression for the reversible net average heat transfer in terms of the quantities $\wp_n$ and E_n.

(b) If the entropy is written as

$$S/k_B = -\sum_n \wp_n \ln \wp_n$$

show that then the macroscopic definition $dS = \delta Q_{rev}/T$ must be true in the canonical ensemble. Hint: take the total differential of the microscopic entropy expression.

16.10. Use Eqn. (16.22) to show for independent particles that

$$\langle E \rangle = \sum_i \langle \epsilon_i \rangle$$

where $\langle \epsilon_i \rangle$ gives the average energy of particle i.

16.11. Prove in the canonical ensemble that, as $T \to 0$, the microstate probability $\wp_m$ approaches a constant for any ground state m with lowest energy E_0 but is otherwise zero for $E_m > E_0$. What is the constant?

16.12. This chapter examines the two-state system, but consider instead the infinite-state system consisting of N non-interacting particles. Each particle i can be in one of an infinite number of states designated by an integer, $n_i = 0, 1, 2, \ldots$. The energy of particle i is given by $\epsilon_i = \epsilon n_i$, where ϵ is a constant. Note: you may need the series sum

$$\sum_{i=0}^{\infty} r^i = \frac{1}{1-r}$$

(a) If the particles are distinguishable, compute $Q(T, N)$ and $A(T, N)$ for this system.
(b) Repeat part (a) for the indistinguishable case.
(c) In cases (a) and (b), what is the probability that a particle will be in its ground state?
(d) What is the magnitude of the fluctuations in energy,

$$\sigma_E^2 = \langle (E - E)^2 \rangle = \langle E^2 \rangle - \langle E \rangle^2$$

Note that

$$E^2 = \sum_n \wp_n E_n^2$$

(e) How do the fluctuations behave as $T \to 0$? Is this intuitive?

16.13. Using the canonical ensemble, derive the Helmholtz free energy for the ideal monatomic gas starting from quantum mechanics. Compare your result with that found in Chapter 9, for which we used a microcanonical approach. Note that ideal gas molecules are independent and indistinguishable. You may want to use the following approximation:

$$\sum_{n=1}^{\infty} e^{-cn^2} \approx \int_0^{\infty} e^{-cn^2} \, dn = \sqrt{\frac{\pi}{4c}}$$

which is valid for small values of the constant c. When will this approximation break down in the case of the ideal gas?

16.14. Consider the one-dimensional Ising model described in Example 16.2.

(a) What is the average $\langle \sigma_i \sigma_{i+1} \rangle$ as a function of temperature? Hint: what is the relation of this quantity to $\langle E \rangle$?

(b) As $T \to 0$, how many ways are there for the energy to approach its ground-state value?

(c) Find an expression for the heat capacity. What is its behavior both at $T \to 0$ and at $T \to \infty$? Explain.

(d) Find a temperature at which the heat capacity is a maximum. What is the physical basis for this temperature?

16.15. Using Example 16.3, find expressions for c_V, P, and μ in terms of the state parameters T, V, and N. Which of these depend on the magnitude of the pixel discretization v? Explain on physical grounds why this is the case.

16.16. A very common technique for evaluating statistical-mechanical models is *perturbation theory*. In this approach, one writes the total energy function as the sum of one part E_0 that is solvable (or reasonably approximated) plus a small second contribution E_1 that is more difficult to evaluate. That is, the energy of a microstate n is given by

$$E_n = E_{0,n} + E_{1,n}$$

Show that the Helmholtz free energy in this case can be approximated as

$$A \approx A_0 - k_B T \ln(1 - \beta \langle E_1 \rangle_0) \approx A_0 + \langle E_1 \rangle_0$$

where

$$A_0 = -k_B T \ln Q_0 = -k_B T \ln \left(\sum_n e^{-\beta E_{0,n}} \right)$$

and

$$\langle E_1 \rangle_0 = \frac{1}{Q_0} \sum_n E_{1,n} e^{-\beta E_{0,n}}$$

Applied problems

16.17. Liquid–crystalline systems form phases of a variety of types. One transition involves moving from an isotropic phase to the so-called nematic one in which

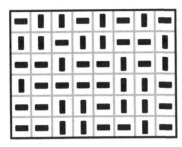

Figure 16.2. A simple model of a two-dimensional liquid-crystalline system involves particles that can adopt one of two orientations.

particles become orientationally aligned. A highly simplified two-dimensional lattice model of this behavior might be developed as shown in Fig. 16.2. There are N particles, each confined to a lattice site, that can have one of two orientations. When two neighboring particles are in opposite orientations, they incur an energetic penalty of ϵ, a positive constant.

(a) Develop a mean-field model for this system by finding the Helmholtz free energy $A = E - TS$ as a function of the fraction f of particles in the vertical orientation. Use the microcanonical expression involving the density of states to find the orientational entropy and use an average "field" of orientations to find a mean energy. This is the so-called Bragg–Williams approach.

(b) Find an equation that can be solved to give the equilibrium value of f. How many solutions are there at low T and what happens to them as $T \to 0$?

(c) What is the temperature above which a phase transition no longer occurs?

16.18. Repeat the liquid–crystalline problem of Fig. 16.2 in the three-dimensional case. Hint: you will need one more variable like f since there are three possible orientations along the x-, y-, and z-axes.

16.19. Consider again the liquid–crystalline problem of Fig. 16.2 in the two-dimensional case. Rather than developing a mean-field model using microcanonical ideas (the so-called Bragg–Williams treatment), use the canonical ensemble in what is often termed a self-consistency approach. Begin by assuming that particles are independent of each other and exist in a mean field of orientations described by f.

(a) Find an expression for the single-particle partition function q in terms of f.

(b) Find an expression for the fraction of time a particle spends in the vertical direction, $f = \wp_{vertical}$. Hint: you need to consider just a single particle, its partition function, and probabilities in order to do this.

(c) Show that your result in part (b) gives a self-consistent equation that can be solved to give a value of f as a function of temperature. Show that this equation is the same as the one you found in the alternate approach taken in the problem above.

16.20. Consider in general the Ising model described in Example 16.2.

(a) What is the ground state of the system at $T = 0$ in the cases $\mathcal{H} > 0$, $\mathcal{H} < 0$, and $\mathcal{H} = 0$ exactly?

(b) Take the case that $\mathcal{H} = 0$ and the system is in three dimensions. Develop a mean-field model in which a central spin is surrounded by six neighbors of average spin $m = \langle \sigma \rangle$. Show that the canonical partition function can be written as $Q = q^N$, and find the single-spin partition function q.

(c) The value m used in part (b) must satisfy a mean-field self-consistency criterion. Namely, the average spin predicted by the Boltzmann probabilities must be consistent with that used in the Boltzmann factors themselves. Show that the average spin can be expressed as $m = \wp_{up} - \wp_{down}$. Then, find these two probabilities in the single-spin case using the Boltzmann factors and q. Show that this leads to a consistency equation that can be solved to find m.

(d) How many solutions are there for m? At what temperature does the number of solutions change?

(e) Repeat this mean-field analysis in part (c) for the case of finite $\mathcal{H}$.

16.21. Proteins are biopolymers that have the following rather remarkable behavior: when placed in aqueous solution a given protein (i.e., a particular sequence of amino-acid monomers) will spontaneously acquire a unique three-dimensional native configuration. This behavior is termed "folding," and any particular protein sequence always folds to the same structure, i.e., folding is a thermodynamic process that drives towards an equilibrium state. Proteins can be unfolded by heating above a folding temperature T_f.

Consider the following highly simplified model of protein folding. A protein consists of a sequence of N monomers with $N \gg 1$. Each monomer can independently adopt one of $(g + 1)$ discrete conformations (e.g., rotations around backbone chemical bonds). Of these conformations for a monomer, one of them corresponds to the native, folded conformation for that residue and incurs a favorable energy $-\epsilon$; the remaining conformations have zero energy. The total number of conformations for the entire protein is $(g + 1)^N$.

(a) For a given total energy E, write an expression for the density of states of this system $\Omega(E)$ in terms of N, g, and ϵ. Let the variable x equal the fraction of monomers that are in their native conformation.

(b) Find $x(T)$ and find the temperature at which the protein is half folded. This might be considered the folding temperature T_f.

(c) Instead of this microscopic analysis, consider a macroscopic picture in which protein folding is considered to be a first-order phase transition, whereby a protein changes discontinuously from an unfolded state with lower Gibbs free energy at $T > T_f$ to a folded state with lower Gibbs free energy at $T < T_f$. At the temperature $T = T_f$, the Gibbs free energies of the two possible "phases" are equal, and they coexist with each other (i.e., the protein can be either folded or unfolded). It can be shown that the folding fraction at any temperature is given by a ratio involving the Gibbs free energies of the folded and unfolded phases,

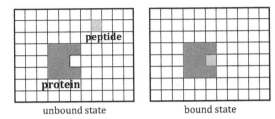

unbound state   bound state

Figure 16.3. A lattice model of peptide association with a protein involves the former finding a specific binding site to which it can associate with high affinity. Here, the peptide is represented by a single lattice site, as is the protein's binding cavity.

$$x(T) = \frac{e^{-G_{folded}(T)/k_B T}}{e^{-G_{unfolded}(T)/k_B T} + e^{-G_{folded}(T)/k_B T}} = \left(1 + e^{\Delta G_{fold}(T)/k_B T}\right)^{-1}$$

where $\Delta G_{fold} \equiv G_{folded} - G_{unfolded}$. Assume that the enthalpy of folding ΔH_{fold} is constant. Develop an expression for $\Delta G_{fold}(T)$, and hence for $x(T)$, in terms of ΔH_{fold} and T_f using macroscopic arguments. Assume constant pressure throughout.

(d) If $\Delta H_{fold} = -N\epsilon$, how do your results for $x(T)$ in parts (b) and (c) compare? Which approach will have a sharper transition?

16.22. Short peptides, polymeric molecules consisting typically of five to ten amino-acid monomers or *residues*, play important roles in cells as signaling agents and hormones. Peptides can operate by binding to a site on a larger protein, which can inhibit protein function or cause a conformational change that modulates protein activity. In solution, peptides tend to be fairly "floppy" molecules that can experience different conformational states. When bound to proteins, on the other hand, they typically adopt a single conformation.

Consider the following model of a single peptide binding to a protein, as shown in Fig. 16.3. All of space (volume V) is divided into pixels that are each the size of the peptide, v_{pep}. A protein is at a fixed location in this space; its volume is v_{prot} and it occupies a number of pixels $v_{prot}/v_{pep} > 1$ (i.e., the protein is larger than the peptide). In the bound state, the peptide occupies a single pixel within the protein, and it has a single conformational state. In solution, the peptide can occupy any one of the other pixels in the volume, and each of its N_{res} residues can adopt any one among m conformations independently of the other residues. The only energy of the system is an energy $-\epsilon$ experienced by the peptide when bound to the protein. Assume that the volume of the protein is very small, such that $V - v_{prot} \approx V$.

(a) What are the degrees of freedom in this system? How many microstates are there in total? Compute the canonical partition function Q for this system.

(b) The binding constant is defined as $K = $ [bound]/[unbound], where the brackets indicate concentration or, equivalently here, probability. Find the binding constant for the system. What changes in T and V will maximize K?

(c) For most of the 20 naturally occurring amino acids, one can approximate $m \approx 3$ (i.e., each residue can be in an extended, an alpha-helical, or a "loop" conformation). However, for the amino acid proline, $m \approx 1$ due to its particular chemical architecture, which makes it highly rigid. In nature, many peptide binders are proline-rich. In light of your expressions for (b), explain why this is so.

16.23. A gas is in equilibrium with a solid surface onto which its molecules can adsorb. The surface exposes M adsorption sites, each of which can be either free or occupied by a single gas molecule. Adsorbed molecules do not interact with each other, but there is a favorable energetic decrease of amount $-\epsilon$ each time a site is occupied.

(a) Find the canonical partition function for N molecules adsorbed on M sites at a temperature T.

(b) Find an expression for the chemical potential of an adsorbed molecule, μ_{ads}.

(c) At equilibrium at constant T, show that the dependence of the fraction of occupied sites, $x = N/M$, is given by the so-called *Langmuir adsorption isotherm*,

$$x = \frac{cP}{1 + cP}$$

where c is a constant. How does c vary with temperature?

FURTHER READING

D. Chandler, *Introduction to Modern Statistical Mechanics*. New York: Oxford University Press (1987).

K. Dill and S. Bromberg, *Molecular Driving Forces: Statistical Thermodynamics in Biology, Chemistry, Physics, and Nanoscience*, 2nd edn. New York: Garland Science (2010).

T. L. Hill, *An Introduction to Statistical Thermodynamics*. Reading, MA: Addison-Wesley (1960); New York: Dover (1986).

T. L. Hill, *Statistical Mechanics: Principles and Selected Applications*. New York: McGraw-Hill (1956); New York: Dover (1987).

E. A. Jackson, *Equilibrium Statistical Mechanics*. Mineola, NY: Dover (1968).

L. D. Landau and E. M. Lifshitz, *Statistical Physics*, 3rd edn. Oxford: Butterworth-Heinemann (1980).

D. A. McQuarrie, *Statistical Mechanics*. Sausalito, CA: University Science Books (2000).

17 Fluctuations

17.1 Distributions in the canonical ensemble

In the last chapter, we found that the canonical ensemble gives the probabilities of microstates at constant temperature, volume, and number of particles,

$$\wp_m = \frac{e^{-\beta E_m}}{Q(T, V, N)} \tag{17.1}$$

where m is the index of a microstate and Q is the canonical partition function,

$$Q(T, V, N) \equiv \sum_{\text{all microstates } n \text{ at } V, N} e^{-\beta E_n} \tag{17.2}$$

We saw that the latter is related to the Helmholtz free energy through $A = -k_B T \ln Q$. It became apparent that the canonical approach provided a second route, in addition to the density of states, to determine the behavior of microscopic models. That is, provided that we can evaluate (17.2), even approximately, we obtain a fundamental thermodynamic potential in terms of its natural variables. Hence, we can then apply all of the mathematics of classical thermodynamics to determine the *complete macroscopic* thermodynamic behavior of the system. Everything is accessible once we have a fundamental potential – state variables, response functions, phase behavior, work and heat limits in processes – anything that involves a thermodynamic variable.

In this chapter, we move beyond that perspective to show that statistical mechanics actually provides much more information than is accessible from the point of view of classical thermodynamics and hence is useful for many tasks beyond the determination of thermodynamic potentials. Indeed, Eqn. (17.1) illustrates that statistical mechanics provides the microstate probabilities in complete detail. This enables one to determine all kinds of *averages* for a system at constant temperature. Let us consider that there exists some property X that can be associated with a microstate. For example, X could be

- the total energy
- the kinetic or potential energy
- one component of the potential energy, such as the electrostatic part
- the distance between two atoms, such as the end-to-end distance of a polymer molecule

- the distance from a molecule to an attractive surface
- the degree of rotation around a particular chemical bond
- the number of nearest neighbors to a particular molecule
- the density of a fluid near an interface and the forces exerted on it
- the number of bonds formed, such as hydrogen bonds or chemical bonds

In the examples described above, it is clear that, if one knew the detailed molecular configuration of the system at some point in time, the property could be evaluated exactly. Thus, a general idea is that we can define a property X that takes on a particular value for each microstate n, denoted by X_n. For example, the potential energy U of microstate n would be U_n.

With these formalities, it is possible to construct *ensemble averages* that give the observed value of the property X when the distribution of microstates is taken into account. In any ensemble, we can write the expectation value of X as

$$\langle X \rangle = \sum_{\text{all } n \text{ at } V, N} X_n \wp_n \tag{17.3}$$

Of course, the exact form of $\wp_n$ depends on the particular ensemble at hand. At constant T, V, and N it follows Eqn. (17.1), and we already saw in Chapter 16 that the average for $X = E$ could be expressed as a derivative of the partition function Q. On the other hand, Eqn. (17.3) shows a quite general mathematical relation that may, or might not, be expressible as a partition function derivative, depending on the nature of X.

Averages often correspond to the single, observable value of some macroscopic property, such as the macroscopic energy. However, undetectable microscopic fluctuations of very small magnitudes may occur. For example, the energy of a system in contact with a heat bath experiences tiny variations, even though classical thermodynamics would predict a unique value at equilibrium for a given temperature. An advantage of statistical mechanics is that these fluctuations can be exactly quantified. For example, the variance of any property X can be computed using

$$\sigma_X^2 = \langle (X - \langle X \rangle)^2 \rangle$$
$$= \langle X^2 \rangle - 2\langle X \rangle \langle X \rangle + \langle X \rangle^2$$
$$= \langle X^2 \rangle - \langle X \rangle^2$$
$$= \sum_{\text{all } n \text{ at } V, N} X_n^2 \wp_n - \left(\sum_{\text{all } n \text{ at } V, N} X_n \wp_n \right)^2 \tag{17.4}$$

Indeed, the entire probability distribution of X can be determined. One simply needs to sum the probabilities of all microstates having a particular value of X. As we have seen before, finding microstates that obey certain constraints requires a delta function,

$$\wp(X) = \sum_n \wp_n \delta_{X_n, X} \tag{17.5}$$

Here, the Kronecker delta ensures that the probability $\wp(X)$ is the total probability due to only those microstates with $X_n = X$.

17.2   The canonical distribution of energies

Let us make the ideas of the previous section more specific by calculating the distribution of total energy in the canonical ensemble,

$$\wp(E) = \sum_n \wp_n \delta_{E_n, E} \tag{17.6}$$

With Eqn. (17.1) for $\wp_n$, we find

$$\wp(E) = \sum_n \frac{e^{-\beta E_n}}{Q(T, V, N)} \delta_{E_n, E} \tag{17.7}$$

Since the only nonzero terms in this sum are those for which $E_n = E$, we can replace E_n in the Boltzmann factor with E. This lets us pull all terms but the delta function outside of the sum, since they are independent of n,

$$\wp(E) = \frac{e^{-\beta E}}{Q(T, V, N)} \sum_n \delta_{E_n, E} \tag{17.8}$$

The final summation is simply a definition for the density of states, so we arrive at the rather instructive result

$$\wp(E) = \frac{\Omega(E, V, N) e^{-\beta E}}{Q(T, V, N)} \tag{17.9}$$

Equation (17.9) gives the distribution of energy E in the system when it is coupled to a heat bath at temperature T. Notice that it appears similar to the microstate probability, except for the presence of a prefactor Ω that accounts for degeneracies of microstates with the same energy. This should be intuitive: the probability that the system resides at a given energy level should be proportional to both the number of microstates at that level and the corresponding Boltzmann factor. The partition function Q in the denominator does not depend on energy and thus again acts as a normalizing factor for the distribution. Indeed, the sum $\sum_E \wp(E)$ equals one, as we show shortly.

Just as we are able to derive a distribution of energies from a distribution of micro-states, we can express the canonical partition function as a summation over *energy levels* rather than microstates as in Eqn. (17.2). First take note of the identity

$$\sum_E \delta_{E_n, E} = 1 \tag{17.10}$$

where E_n is the energy of an arbitrary microstate n and the sum is taken over all energy levels. Equation (17.10) is simply a statement of the fact that, given a microstate value E_n, there is one E in the energy spectrum that equals it. Insertion into Eqn. (17.2) gives

$$Q(T, V, N) = \sum_{\text{all } n \text{ at } V, N} \left[e^{-\beta E_n} \sum_E \delta_{E_n, E} \right]$$

$$= \sum_{\text{all } n \text{ at } V, N} \sum_E e^{-\beta E_n} \delta_{E_n, E} \tag{17.11}$$

We use the same trick as before, namely the fact that $E_n = E$ for the nonzero terms allows us to re-express the Boltzmann factor and switch the summation order,

$$Q(T, V, N) = \sum_{\text{all } n \text{ at } V, N} \sum_E e^{-\beta E} \delta_{E_n, E}$$

$$= \sum_E e^{-\beta E} \left[\sum_{\text{all } n \text{ at } V, N} \delta_{E_n, E} \right] \tag{17.12}$$

The term in brackets now becomes the density of states, providing another but equivalent way to form the canonical partition function that involves a summation over energy levels rather than microstates,

$$Q(T, V, N) = \sum_E \Omega(E, V, N) e^{-\beta E} \tag{17.13}$$

This transformation should also be intuitive: it simply lumps together microstates with identical energies and hence identical Boltzmann factors, and uses the density of states to account for their degeneracies. Though the idea is simple, Eqn. (17.13) has quite important implications. It shows that the partition function in the microcanonical ensemble Ω can be transformed into the canonical partition function Q through a summation involving the Boltzmann factor. In fact, this summation resembles a discrete version of a mathematical operation called the *Laplace transform*, here performed in the energy domain. Indeed, for continuous energy spectra, the transform is not discrete and becomes an actual Laplace transform; those points will be discussed with respect to classical systems in Chapter 18. We find such transformations to be a general feature of partition functions.

One partition function can be converted to another using a procedure similar to a **Laplace transform**. The transformation is with respect to an extensive variable that fluctuates by coupling to a bath. This is a microscopic analogy of Legendre transforms for macroscopic thermodynamic potentials.

If we replace the density of states in Eqn. (17.13) with the entropy, the distribution $\wp(E)$ takes a particularly informative form:

$$\wp(E) \propto \exp\left[\frac{S(E, V, N)}{k_B} - \frac{E}{k_B T} \right] \tag{17.14}$$

where we have omitted the term involving Q since it does not affect the form of the distribution but only its normalization. Equation (17.14) shows that there are two competing effects determining the probabilities of energy levels. S is a concave, generally increasing function of E, but the second term decreases linearly with E. The net effect of these features is a maximum at an intermediate energy. The maximum is dramatically peaked because the competing terms exist within the exponential, and both S and E are extensive quantities that scale as N. For macroscopic systems, the peak in the distribution is so sharp that $\wp(E)$ appears as essentially one for only a single energy, and zero for all others.

At what energy E^* does the distribution $\wp(E)$ have its maximum? If the energy levels are densely packed or if the energy spectrum is continuous, we can find this by differentiation of the term inside the exponential of Eqn. (17.14),

$$0 = \frac{\partial}{\partial E}\left[\frac{S(E, V, N)}{k_B} - \frac{E}{k_B T}\right] \tag{17.15}$$

On simplifying, we find that the condition for E^* is

$$\left(\frac{\partial S(E^*, V, N)}{\partial E}\right)_{V,N} = \frac{1}{T} \tag{17.16}$$

This is not all that surprising. The peak of the distribution corresponds to the energy at which the E derivative of the entropy equals the inverse temperature of the bath. By examining only the most probable energy, we recover the familiar condition that we would expect from macroscopic thermodynamics.

In light of the peaked nature of $\wp(E)$, we can gain additional insight into the canonical partition function. If we approximate the canonical ensemble as one in which only the most probable energy E^* is visited, we might estimate Q through Eqn. (17.13) as

$$Q(T, V, N) \approx \Omega(E^*, V, N)e^{-\beta E^*} \tag{17.17}$$

Essentially this approach recognizes that one term in the summation is so much greater than all others that we can approximate the sum with its maximum value. We already applied this *maximum-term method* much earlier in Chapter 4 when we addressed energy exchange and thermal equilibrium between two systems. The approach in Eqn. (17.17) is similar except for the fact that we have replaced one of the systems with a bath. Keep in mind that the maximum-term approximation succeeds for macroscopic systems because they consist of so many particles.

Equation (17.17) is particularly informative for the connection between the partition functions and their corresponding thermodynamic potentials. Taking its logarithm gives

$$-A(T, V, N)/T = S(E^*, V, N) - E^*/T \tag{17.18}$$

This result is particularly striking because it is *exactly* a Legendre transform of the entropy. Written in another way for clarity,

$$\mathcal{L}_E[S] = S - E\left(\frac{\partial S}{\partial E}\right) \tag{17.19}$$

Clearly the value of E^* in Eqn. (17.18) is consistent with the Legendre transform since it is determined by the condition of Eqn. (17.16). We also see that (17.18) satisfies the integrated relationship $A = E - TS$, as one would expect. The reader is encouraged to convince themselves of these facts in detail by revisiting the ideas of Chapter 7. In any case, the maximum-term method applied at a microscopic level leads to profound insight into macroscopic thermodynamic relations.

The **maximum-term method** gives the most likely state of the system, ignoring small fluctuations away from it. A maximum-term approximation of the microscopic partition function is equivalent to performing a Legendre transform of a macroscopic thermodynamic potential. In the thermodynamic limit, that is, for very large systems, the maximum-term method becomes a near-exact approximation because the fluctuations become so small.

For nanoscopic systems, say on the order of 10^4–10^6 molecules or less, the distribution in energies is not so sharply peaked. This means that the maximum-term method fails and macroscopic thermodynamics and Legendre transforms are not exactly the most appropriate approach to treating these systems. Instead, the full canonical partition function as per Eqn. (17.13) must be used. Indeed, many systems of current interest in science and nanotechnology exist at the nanoscale. The power of statistical mechanics is the ability to treat such systems with rigor, without macroscopic simplifications.

Statistical mechanics also renders more nuanced the idea of different constant conditions. In classical thermodynamics, we have full access to all properties of a system if we use a thermodynamic potential of its natural variables, regardless of the actual constant conditions. For example, $A(T, V, N)$ and $S(E, V, N)$ ultimately contain the same, complete set of macroscopic thermodynamic information, even though the former arises due to constant-temperature conditions whereas the latter emerges in isolated systems. Statistical mechanics, however, shows that there are important but subtle differences between these two conditions and the properties they predict. In the microcanonical ensemble, there are rigorously no fluctuations in energy. On the other hand, the energy fluctuates in the canonical ensemble even if only by very small amounts, while it is the temperature that becomes rigorously constant due to coupling with an infinite heat bath. It turns out that the physical behavior of small systems can actually be distinct depending on what is held constant, and statistical mechanics provides the correct framework for this kind of analysis. In light of these observations, one can make the following general statement about ensembles.

For macroscopic systems, it makes no difference which ensemble we use to compute bulk properties. We could use either $S = k_B \ln \Omega$ or $A = -k_B T \ln Q$ and still extract all of the same thermodynamic properties using fundamental relations, Legendre transforms, etc. At the microscopic level, however, there are differences between the ensembles in terms of the microscopic properties and the fluctuations the system experiences.

Example 17.1 *Find expressions for the canonical partition function and Helmholtz free energy of the two-state model in three different ways: (1) by summation over microstates, (2) by summation over energy levels, and (3) using the maximum-term approximation.*

Approach (1). In Example 16.1 we computed the partition function by summing over microstates. Because the particles are independent, we found that we could factor this sum as a product of individual particle sums:

$$Q = \sum_{\text{microstates } n} e^{-\beta \sum_i \epsilon_i} = \sum_{\epsilon_1 \in \{0, b\}} \cdots \sum_{\epsilon_N \in \{0, b\}} e^{-\beta \epsilon_1} \times \cdots \times e^{-\beta \epsilon_N}$$

$$= \left(\sum_{\epsilon \in \{0, b\}} e^{-\beta \epsilon} \right)^N = (1 + e^{-\beta b})^N$$

In turn the Helmholtz free energy follows,

$$A = -k_B T \ln Q = -Nk_B T \ln(1 + e^{-\beta b})$$

Approach (2). We use Eqn. (17.13), which involves a Laplace-like transform of the density of states. Let M stand for the number of particles that are in the up position. The energy is then $E = Mb$ such that $M = E/b$. The sum over energy levels is

$$Q = \sum_{E=0}^{Nb} \frac{N!}{M!(N-M)!} e^{-\beta E}$$

We have explicitly included the limits on the sum: the total energy of the system can vary between $E = 0$ (all particles in the down state) and $E = Nb$ (all up). Since the energy levels are directly related to the number of up particles, we can switch to a sum over M:

$$Q = \sum_{M=0}^{N} \frac{N!}{M!(N-M)!} e^{-\beta b M}$$

This expression has the form of the famous binomial theorem in mathematics:

$$\sum_{k=0}^{n} \frac{n!}{k!(n-k)!} x^k = (1+x)^n$$

In this example, we have $x = e^{-\beta b}$, $k = M$, and $n = N$. Therefore the partition function is

$$Q = (1 + e^{-\beta b})^N$$

which is identical to the result found in approach (1).

Approach (3). We begin by constructing the entropy using Stirling's approximation. We did this in Example 3.1 and the result was

$$S = Nk_B \left[-\left(\frac{E}{Nb} \right) \ln \left(\frac{E}{Nb} \right) - \left(1 - \frac{E}{Nb} \right) \ln \left(1 - \frac{E}{Nb} \right) \right]$$

After some mathematical manipulation, the energy derivative simplifies to

$$\left(\frac{\partial S}{\partial E} \right)_N = -\frac{k_B}{b} \ln \left(\frac{E}{Nb - E} \right)$$

We set this equal to $1/T$ as per Eqn. (17.16) to find E^*. It is given by

$$E^* = Nb \frac{e^{-\beta b}}{1 + e^{-\beta b}}$$

This is the relationship found already in Example 16.1. We now substitute into the relationship

$$A = E^* - TS(E^*, N)$$

to find the Helmholtz free energy. After a fair amount of mathematical manipulation, the result is the same as before, namely

$$A = -Nk_{\mathrm{B}}T \ln(1 + e^{-\beta b})$$

Of course, approach (3) required the use of Stirling's approximation, while neither approach (1) nor approach (2) made any such simplifications. It is for this reason that the maximum-term method gives, perhaps fortuitously, an exactly identical result. On the other hand, the former two approaches may be regarded as more rigorous for small systems. Indeed, the energy spectrum in this model is discrete and, technically, the maximum term should not be found through differentiation with respect to the non-continuous E.

17.3   Magnitude of energy fluctuations

We have said that fluctuations in energy are very small in macroscopic systems at constant temperature. Let us quantify the magnitude of those fluctuations by calculating the energy variance,

$$\sigma_E^2 = \overline{E^2} - \overline{E}^2$$
$$= \sum_E E^2 \wp(E) - \left(\sum_E E \wp(E) \right)^2 \tag{17.20}$$

Because we have the full probability distribution in energy from Eqn. (17.13), we can easily compute each of these terms. However, before we proceed directly with the formula for $\wp(E)$, we can make some simplifying observations. Chapter 16 showed that the average energy can be formed from a β derivative of Q. This can be generalized to an average of E^ν, where ν is an arbitrary exponent, in the following way:

$$\langle E^\nu \rangle = \sum_E E^\nu \wp(E)$$

$$= \frac{1}{Q} \sum_E E^\nu \Omega(E) e^{-\beta E}$$

$$= \frac{(-1)^\nu}{Q} \frac{\partial^\nu}{\partial \beta^\nu} \sum_E \Omega(E) e^{-\beta E}$$

$$= \frac{(-1)^\nu}{Q} \frac{\partial^\nu Q}{\partial \beta^\nu} \tag{17.21}$$

Here we omitted the constant independent variables that did not involve E for simplicity. With these considerations, the mean-squared energy fluctuations from Eqn. (17.20) are

$$\sigma_E^2 = \frac{1}{Q}\frac{\partial^2 Q}{\partial \beta^2} - \frac{1}{Q^2}\left(\frac{\partial Q}{\partial \beta}\right)^2$$

$$= \frac{\partial^2 \ln Q}{\partial \beta^2} \tag{17.22}$$

The last line stems from a simple product rule for derivatives

$$\frac{\partial^2 \ln Q}{\partial \beta^2} = \frac{\partial}{\partial \beta}\frac{1}{Q}\frac{\partial Q}{\partial \beta} = -\frac{1}{Q^2}\frac{\partial Q}{\partial \beta}\frac{\partial Q}{\partial \beta} + \frac{1}{Q}\frac{\partial^2 Q}{\partial \beta^2} \tag{17.23}$$

Returning to σ_E^2 in (17.22), we can simplify this to

$$\sigma_E^2 = \frac{\partial}{\partial \beta}\frac{\partial \ln Q}{\partial \beta} = -\frac{\partial \langle E \rangle}{\partial \beta} \tag{17.24}$$

where Eqn. (17.21) with $v = 1$ was used to introduce $\langle E \rangle$. At this point, it will be more informative to switch from β to $T = (k_B \beta)^{-1}$,

$$\sigma_E^2 = k_B T^2 \left(\frac{\partial \langle E \rangle}{\partial T}\right)_{V,N} \tag{17.25}$$

We have now explicitly indicated the constant conditions for this ensemble that were suppressed throughout the derivation. Equation (17.25) reveals an important connection to measurable quantities. Namely, because $\langle E \rangle$ gives the observed macroscopic energy, we can link the fluctuations to the heat capacity,

$$\sigma_E^2 = C_V k_B T^2 \tag{17.26}$$

This result is quite striking. Equation (17.26) shows that the magnitude of the energy fluctuations, which are far too small to detect macroscopically, is actually related to a quantity that can be measured with great ease, namely the constant-volume heat capacity. This relationship is valid for *any* system, since no approximations are involved in its derivation. Macroscopic thermodynamics says nothing about this connection; a statistical-mechanical perspective is required here.

How can something undetectable be related to a measurable quantity? We need to probe deeper by considering the scaling of the terms involved with system size. The extensive heat capacity C_V and equivalently the derivative $\partial E / \partial T$ in Eqn. (17.25) both grow linearly with the size of the system. We need to compare a typical energy fluctuation with the average energy, which is also extensive. A measure of the former is the energy standard deviation, σ_E. We thus construct

$$\frac{\sqrt{\sigma_E^2}}{E} = \sqrt{\frac{C_V k_B T^2}{E}} \propto \frac{\sqrt{N}}{N} \propto N^{-1/2} \tag{17.27}$$

In other words, the *relative* magnitude of energy fluctuations decreases as the inverse square root of the system size. For macroscopic systems, with $N = O(10^{23})$, the typical relative fluctuations are extremely small, of the order 10^{-11}! Therefore, from a macroscopic perspective we hardly ever detect these fluctuations in energy, even if they are connected to the heat capacity.

At constant temperature, the heat capacity measures the mean-squared fluctuations in energy. Relative to the average, these fluctuations scale as $N^{-1/2}$, so macroscopic systems appear to have constant, well-defined energies.

Are there cases in which fluctuations become observable at a macroscopic scale? Of course, for small systems, as discussed before, these fluctuations can be significant and are relevant to many biological and nanotechnological settings. Another case where fluctuations become important is at a critical point. There, C_V diverges and fluctuations become so large that they are noticeable at a macroscopic scale. For the liquid–vapor critical point, these fluctuations scatter light and result in the phenomenon called *critical opalescence*.

Example 17.2   *Find an expression for the heat capacity and the energy fluctuations of the two-state model. What is the behavior of the heat capacity as $T \to 0$? On the other hand at large temperatures, for what value of N do the fluctuations become approximately one part in a million?*

The average energy of the two-state model was solved in Example 16.1:

$$\langle E \rangle = Nb \frac{e^{-\beta b}}{1 + e^{-\beta b}}$$

To determine σ_E^2, it is convenient to use Eqn. (17.24),

$$\sigma_E^2 = -\frac{\partial \langle E \rangle}{\partial \beta} = Nb^2 \frac{e^{-\beta b}}{(1 + e^{-\beta b})^2}$$

The heat capacity then stems from (17.26),

$$C_V = k_B \beta^2 \sigma_E^2 = Nb^2 k_B \frac{\beta^2 e^{-\beta b}}{(1 + e^{-\beta b})^2}$$

The limit of this expression as $T \to 0$ and $\beta \to \infty$ is

$$\lim_{\beta \to \infty} C_V \to Nb^2 k_B (\beta^2 e^{-\beta b}) \to 0$$

Ultimately the limit must be handled by l'Hôpital's rule, but the final result is that the exponentials win out and the heat capacity vanishes. This is expected since the system should reside in its single ground state at that point. An interesting side note is that the low-temperature dependence of C_V for the two-state model has the same functional form as that of the Einstein model of crystals.

The fractional energy fluctuations are

$$\frac{\sigma_E}{\langle E \rangle} = \frac{e^{\beta b/2}}{\sqrt{N}}$$

At high temperatures, $\beta \to 0$ and the exponential approaches unity. Thus, if the LHS of this expression is 10^{-6}, then N is of the order of 10^{12}.

Problems

Conceptual and thought problems

17.1. Use Eqn. (17.6) to explicitly prove that the sum over energy probabilities in the canonical ensemble is normalized, $\sum_E \wp(E) = 1$. More generally, prove that the sum over any property distribution given by (17.5) must be unity.

17.2. There is no strict requirement that S be an increasing function of E, although it should remain concave. Indeed, in systems with degrees of freedom that can become *saturated* or that have a finite maximum possible energy, the density of states will necessarily decrease with energy at some point (since $\Omega = 0$ beyond this maximum energy).
   (a) Sketch $S(E)$ for a system with a minimum and maximum possible energy.
   (b) Show that such systems have regions of energies corresponding to negative temperatures.
   (c) Show that, when coupled to a realistic heat bath at positive temperature, the system can never reach energies near the maximum possible one. Consider Eqn. (17.14).
   (d) Construct an argument for why classical systems can never have negative temperature.

17.3. Consider a metastable system at constant T, V, N conditions that lies at a local free energy minimum. For example, this may be a liquid that is supercooled with respect to a crystalline phase or superheated with respect to a vapor phase.
   (a) In the two cases described above, what is the final state of the system when it reaches global stability?
   (b) The Helmholtz free energy should decrease when the system transforms to a more stable state. Explain why this must be the case according to the canonical partition function. You may want to consider the nature of the sums in (17.2) and (17.13). How does metastability play out in terms of microstates?

17.4. Show that, if $S(E)$ is convex in some region of the energy spectrum ($\partial^2 S/\partial E^2 > 0$), then one expects two maxima in $\wp(E)$. To what physical phenomenon does this correspond?

17.5. As shown in Example 17.2 and in Chapter 14, both the two-state model and the Einstein model of crystals show a heat capacity that scales proportionally to $T^{-2}\exp(-\text{constant}/T)$ when $T \to 0$. Explain why this is so.

17.6. For a system of independent molecules, the energy can be written as $E = \sum_i \epsilon_i$, where ϵ_i gives the energy of a single molecule i.

(a) In the canonical ensemble, compare the magnitude of the fluctuations in energy of a single molecule, σ_ϵ, with the mean energy $\langle \epsilon \rangle$ and with the fluctuations in the entire system, σ_E.

(b) Repeat the previous part for the microcanonical ensemble.

17.7. The canonical ensemble can be characterized by the form for its distribution of energies, $\wp(E)$. Although in the text we derived this through explicit coupling to a bath, it can actually be found on the basis of much simpler arguments involving the probabilities of independent systems. Consider two systems 1 and 2 that are each maintained at the same constant temperature T. Let their energies at any one point in time be denoted by E_1 and E_2, respectively. At equilibrium, the canonical energy distribution is given by a universal function $f(E)$ such that

$$\wp(E_1) = c_1 f(E_1) \quad \text{and} \quad \wp(E_2) = c_2 f(E_2)$$

where c_1 and c_2 are constants. The entire supersystem consisting of both 1 and 2 can also be analyzed and, since it too is in contact with a heat bath, must have a distribution

$$\wp(E_T) = c_T f(E_T)$$

where $E_T = E_1 + E_2$ is the supersystem energy and c_T is a constant. However, because systems 1 and 2 are large and independent, the probabilities are related by $\wp(E_T) = \wp(E_1)\wp(E_2)$ and thus the function $f(E)$ must satisfy

$$c_T f(E_1 + E_2) = c_1 c_2 f(E_1) f(E_2)$$

(a) What determines the constants c_T, c_1, and c_2?

(b) Show that the relation above requires that f has an exponential form,

$$f(E_1) = e^{-\beta E_1}, \quad f(E_2) = e^{-\beta E_2}, \quad f(E_T) = e^{-\beta E_T}$$

in which β is a constant that is the same for systems 1 and 2. In other words, show that the independence of the energy probabilities in the canonical ensemble demands that they take a form involving the Boltzmann factor. Hint: begin by examining the partial derivatives of the constraint relation above.

Fundamentals problems

17.8. From a macroscopic perspective, how does one obtain C_V from the fundamental potential $A(T, V, N)$? Using this and the relation $A = -k_B T \ln Q$, find an expression for C_V in terms of Q. Show that the result is identical to that found in Eqn. (17.26).

17.9. In the infinite-state model, particles are independent and each experiences an energy of $\epsilon_i = \epsilon n_i$, where ϵ is a constant and $n_i \geq 0$ an integer. Find the distribution of states for a particle, $\wp(n)$.

17.10. Find the fluctuations in the energy σ_E^2 for
(a) the infinite-state model and
(b) the one-dimensional Ising model with $\mathcal{H} = 0$.

17.11. Assume that the canonical-ensemble energy distribution $\wp(E)$ can be approximated as a continuous Gaussian or normal distribution, with average energy $\langle E \rangle$ and standard deviation σ_E.
(a) Write an expression for the distribution in terms of $\langle E \rangle$ and C_V. Simplify the expression, as much as possible, for the case of a monatomic ideal gas.
(b) For the ideal gas, show that the probability of the system visiting an energy $E = (1 - x)\langle E \rangle$ is proportional to $\exp[-3x^2 N/4]$, where x is a small value for small fluctuations. What does this mean for the probability of an energy fluctuation that is 99.99999% of the average energy, relative to the probability of the mean energy, in a mole of a macroscopic ideal gas? Do you think such a fluctuation is detectable?

17.12. Consider E to be a continuous variable. Prove that the distribution $\wp(E)$ implied by Eqn. (17.9) is essentially Gaussian for macroscopic, single-phase systems. Hint: you may want to Taylor-expand the entropy around the peak value E^*.

17.13. Suppose that the energy of a system depends linearly on an intensive macroscopic parameter y as
$$E_n = yX_n + \text{(additional terms)}$$
where X_n is a characteristic of a microstate n. For example, y might correspond to the strength of an applied electric, magnetic, or gravitational field and X to the corresponding dipoles or elevations of particles in the microstates.
(a) Show that $\langle X \rangle = -k_B T \, (\partial \ln Q / \partial y)$.
(b) Show that $\sigma_X^2 = (k_B T)^2 (\partial^2 \ln Q / \partial y^2)$.
(c) Find $(\partial \langle X \rangle / \partial y)$. Show that this is always negative and is consistent with thermodynamic stability criteria.

17.14. Show that a general expression for the dependence of the Helmholtz free energy on a parameter in the microscopic energy function ξ is
$$\left(\frac{\partial A}{\partial \xi} \right)_{T,V,N} = \left\langle \frac{\partial E}{\partial \xi} \right\rangle = \sum_n \wp_n \frac{\partial E_n}{\partial \xi}$$
where the brackets indicate canonical averaging.

17.15. The mean-field approximation neglects correlated fluctuations between a molecule and its neighbors. Often it suggests that the energy function can be written as
$$E = E(\mathcal{X}_1, \mathcal{X}_2, \ldots) \approx \sum_i \epsilon_i(\psi)$$
where $\mathcal{X}_i$ gives the degrees of freedom of particle i and the rightmost term expresses the energy as a sum of single-particle energies ϵ_i that depend on some

effective field ψ. For example, ψ may be related to the average density of neighbors as in the lattice gas, or to the average composition as in the regular solution. Find an expression for the partition function of such a system. Show that the free energy of the system can be written as

$$A \approx N \langle \epsilon \rangle - TS_0$$

where $\langle \epsilon \rangle$ is an average particle energy, and $S_0 = k_B \ln \Omega_0$ with Ω_0 being the number of ways of organizing the particles regardless of the energy of each configuration. Note that this result was used in Chapters 10 and 12 for the lattice gas and regular solution.

Applied problems

17.16. DNA consists of two *complementary* strands of nucleotide base sequences: each adenine (A) in one strand is paired with a thymine (T) in the other, and each cytosine (C) with a guanine (G). A–T pairing involves two complementary hydrogen bonds and C–G involves three. Mismatches, however, can occur between A–C and G–T pairs, which results in there being one hydrogen bond fewer than the maximum possible number. During DNA replication, numerous protein enzymes and machines unpair the two complementary strands and make independent copies of each of these two templates. These machines involve sophisticated error-correcting mechanisms to ensure that mismatches do not occur during the synthesis of the new strands, which would result in mutations to the genetic code.

What would happen instead if DNA replication obeyed equilibrium Boltzmann populations? Consider a single nucleotide along the template DNA sequence. Assume that there are two possibilities during replication: a correctly matched nucleotide and an incorrectly matched one. The difference in energy between the two states is roughly 2 kcal/mol due to the hydrogen bond. Using Boltzmann populations for these two states, what is the probability of a mismatch occurring at $T = 300$ K? Compare your answer with the error rate achieved by the DNA replication machinery, which is roughly 1 in 10^7.

FURTHER READING

D. Chandler, *Introduction to Modern Statistical Mechanics*, New York: Oxford University Press (1987).

K. Denbigh, *The Principles of Chemical Equilibrium*, 4th edn. New York: Cambridge University Press (1981).

K. Dill and S. Bromberg, *Molecular Driving Forces: Statistical Thermodynamics in Biology, Chemistry, Physics, and Nanoscience*, 2nd edn. New York: Garland Science (2010).

T. L. Hill, *An Introduction to Statistical Thermodynamics*. Reading, MA: Addison-Wesley (1960); New York: Dover (1986).

T. L. Hill, *Statistical Mechanics: Principles and Selected Applications*. New York: McGraw-Hill (1956); New York: Dover (1987).

E. A. Jackson, *Equilibrium Statistical Mechanics*. Mineola, NY: Dover (1968).

L. D. Landau and E. M. Lifshitz, *Statistical Physics*, 3rd edn. Oxford: Butterworth-Heinemann (1980).

D. A. McQuarrie, *Statistical Mechanics*. Sausalito, CA: University Science Books (2000).

18 Statistical mechanics of classical systems

18.1 The classical canonical partition function

When quantum effects are not significant, we can approximate the behavior of a system using a classical representation. Here we derive the form of the partition function in such cases. For simplicity, consider a single-component system composed of monatomic molecules. We begin with the discrete version of the partition function introduced in Chapter 16,

$$Q(T, V, N) = \sum_{\text{all } n \text{ at } V, N} e^{-\beta E_n}$$

$$= \sum_{\mathcal{X}_1} \sum_{\mathcal{X}_2} \cdots \sum_{\mathcal{X}_N} e^{-\beta E(\mathcal{X}_1, \mathcal{X}_2, \dots, \mathcal{X}_N)} \tag{18.1}$$

where n is an index running over all microstates of the system and the sums in the second line indicate exhaustive enumeration of all possible values for each microscopic degree of freedom $\mathcal{X}_i$. For a classical system of N particles in three dimensions, each microstate involves $\mathcal{N} = 6N$ degrees of freedom: the three positions and three momenta for each particle. These variables are of course continuous, so we rewrite the summations in Eqn. (18.1) as integrals,

$$Q(T, V, N) = c \int \int \cdots \int e^{-\beta H(\mathbf{p}^N, \mathbf{r}^N)} \, dp_{1,x} \, dp_{1,y} \cdots dp_{N,z} \, dx_1 \, dy_2 \cdots dz_N$$

$$= c \int e^{-\beta H(\mathbf{p}^N, \mathbf{r}^N)} \, d\mathbf{p}^N \, d\mathbf{r}^N \tag{18.2}$$

where c is a normalization constant that will be discussed shortly. The quantities $\mathbf{r}^N$ and $\mathbf{p}^N$ are the vectors of all positions and momenta. Equation (18.2) shows that the microstate energy is given by the classical Hamiltonian and that the partition function involves a total of $6N$ integrals. This $6N$-dimensional integral can become quite complicated for macroscopic systems in which N is very large. The second line introduces a notational convention that we will adopt moving forward, including the use of a single integral sign and the representation of the $6N$ differentials by $d\mathbf{p}^N \, d\mathbf{r}^N$. The reader can always infer the dimensionality of the integrals from the number of integrand variables.

The limits of the integrals in Eqn. (18.2) require some consideration. Each of the momentum variables has limits $-\infty < p < \infty$ since there are no physical constraints on them. Thus, these degrees of freedom and the associated integrals are *unbounded*. In contrast, the allowed values for the position variables depend on the volume of system and are *bounded*. For a cubic box of side length L, one has $0 < x < L$, $0 < y < L$, and $0 < z < L$. It is therefore through the limits of the position integrals that the partition function acquires a dependence on V.

Of course, the system need not be cubic, and this would seem to raise an important conundrum. Owing to the position integral limits, will the shape of the volume V have an effect on the properties of a system? Certainly for bulk, homogeneous materials we do not expect this to be the case – water at standard conditions behaves identically regardless of the particular glass in which it is contained. Indeed, for large systems in which interfaces do not direct physical behavior, the system shape will have negligible effect on the partition function in Eqn. (18.2) for a given volume. A rough argument is the following: an arbitrary macroscopic system can be effectively decomposed into smaller yet still macroscopic subsystems between which boundary interactions are negligible. The total free energy can be written as the sum of subsystem free energies and, in turn, the total partition function can be expressed as a product of subsystem ones. Since the subdivision is arbitrary, it stands that the partition function should not depend on shape. Of course, this would not be the case for systems of very small scale or in which interfacial interactions are important.

We now return to the issue of the constant c that may have seemed arbitrarily introduced in Eqn. (18.2). Its purpose actually stems from the dimensionality of Q. Namely, the $6N$ integrals give rise to a partition function that is not dimensionless since they produce units of [momentum]^{3N} × [distance]^{3N} owing to the integrand variables. This means that one can never calculate absolute values of purely classical partition function integrals since the results depend on the units used. Instead, we can only compute ratios of partition functions, leading to differences in free energies. This result mirrors our considerations of the third law of thermodynamics in Chapter 15, where we found that a classical description is inadequate for providing an absolute entropy. The general rule is as follows.

> In a purely classical description composed of continuous degrees of freedom in the form of atomic positions and momenta, we cannot compute absolute values of partition function integrals. Instead, we can only compute ratios of these integrals at different states.

In reality we know that the world is not classical but behaves quantum-mechanically. In the high-temperature limit, however, quantum systems behave like classical ones because the energy spectrum becomes dense and appears continuous. Therefore, we introduce the correction factor c into the classical canonical partition function to make it agree with the quantum behavior in the high-temperature limit. The form of this correction finally gives

$$Q(T, V, N) = \frac{1}{h^{3N} N!} \int e^{-\beta H(\mathbf{p}^N, \mathbf{r}^N)} \, d\mathbf{p}^N \, d\mathbf{r}^N \tag{18.3}$$

Here h is Planck's constant and $N!$ is the quantum correction for indistinguishable particles. Notice that the factor of h provides a natural metric to make the canonical partition function dimensionless since it has units of [momentum] $\times$ [distance]. In fairness, Eqn. (18.3) is introduced here in a quite cursory manner, but the interested reader is referred to advanced statistical-mechanical texts for a more rigorous analysis, such as the text by Hill. In any case, the quantum-corrected partition function allows one to use the classical approximation at high temperatures to compute *absolute* partition functions and hence absolute free energies.

It becomes possible to simplify the classical partition function further, since we know that the total energy is the sum of the potential and kinetic ones,

$$H(\mathbf{p}^N, \mathbf{r}^N) = K(\mathbf{p}^N) + U(\mathbf{r}^N) \tag{18.4}$$

This enables us to write

$$Q(T, V, N) = \frac{1}{h^{3N} N!} \left[\int e^{-\beta K(\mathbf{p}^N)} \, d\mathbf{p}^N \right] \left[\int e^{-\beta U(\mathbf{r}^N)} \, d\mathbf{r}^N \right] \tag{18.5}$$

Here we can separate the kinetic and potential energy contributions because the two terms involve mutually exclusive degrees of freedom, namely the momenta and positions. Therefore, the exponentials are factored as $\exp(-\beta K - \beta U) = \exp(-\beta K) \exp(-\beta U)$, and the integrals can be separated using the general mathematical property

$$\iint f(x) g(y) \, dx \, dy = \left[\int f(x) dx \right] \left[\int g(y) dy \right] \tag{18.6}$$

With the separation above, the kinetic integrals can be evaluated analytically using

$$K(\mathbf{p}^N) = \frac{1}{2m} \sum_i (p_{i,x}^2 + p_{i,y}^2 + p_{i,z}^2) \tag{18.7}$$

Substituting into the integral expression gives

$$\int e^{-\beta K(\mathbf{p}^N)} \, d\mathbf{p}^N = \int e^{-(\beta/2m) \sum_i \left(p_{i,x}^2 + p_{i,y}^2 + p_{i,z}^2 \right)} \, d\mathbf{p}^N$$

$$= \left[\int e^{-(\beta/2m)p_{1,x}^2} \, dp_{1,x} \right] \times \cdots \times \left[\int e^{-(\beta/2m)p_{N,z}^2} \, dp_{N,z} \right]$$

$$= \left[\int_{-\infty}^{\infty} e^{-(\beta/2m)p^2} \, dp \right]^{3N} \tag{18.8}$$

The last line in Eqn. (18.8) arises from the fact that the $3N$ integrals are identical for particles of the same mass. The final result is a Gaussian integral that is easily evaluated,

$$\int e^{-\beta K(\mathbf{p}^N)} \, d\mathbf{p}^N = \left(\frac{2\pi m}{\beta} \right)^{3N/2} \tag{18.9}$$

Finally, the total canonical partition function becomes

$$Q(T, V, N) = \frac{1}{N!} \left(\frac{2\pi m}{h^2 \beta} \right)^{3N/2} \int e^{-\beta U(\mathbf{r}^N)} \, d\mathbf{r}^N \tag{18.10}$$

The combination of the constants in front of the integral is a familiar grouping: it is related to the thermal de Broglie wavelength $\Lambda(T) = (h^2/2\pi m k_B T)^{1/2}$. On simplifying, we arrive at our final expression for the canonical partition function,

$$Q(T, V, N) = \frac{Z(T, V, N)}{\Lambda(T)^{3N} N!} \quad \text{where } Z \equiv \int e^{-\beta U(\mathbf{r}^N)} \, d\mathbf{r}^N \tag{18.11}$$

Here we introduced a new variable $Z(T, V, N)$ that is called the *configurational partition function*. Z depends only on the positional part of the degrees of freedom and the potential energy function. All particles with the same mass will have identical contributions from their kinetic degrees of freedom (the momenta), but may have distinct modes of interacting that will result in different contributions from the configurational part.

> The kinetic and potential energy contributions to the canonical partition function are separable. The kinetic contribution can be evaluated analytically and, for simple spherical particles, it gives the thermal de Broglie wavelength. The potential energy contribution gives rise to a **configurational partition function**.

We find that the task of evaluating the properties of a classical system actually amounts to the problem of determining the configurational partition function Z. In general this is a complicated task since the multidimensional integral over the $3N$ positions is nontrivial to evaluate in the general case, owing to the complex functionality of typical potential energy functions. A significant body of statistical mechanics has developed sophisticated expansion and perturbation techniques that can be used to approximate the configurational integral.

In the special case of identical non-interacting classical particles, in which the potential energy is separable into individual particle contributions, one can write

$$Q(T, V, N) = \frac{q^N}{N!} \quad \text{where} \quad q \equiv \frac{1}{\Lambda(T)^3} \int e^{-\beta u(\mathbf{r})} \, d\mathbf{r} \tag{18.12}$$

Here, q is the single-particle partition function and u gives the single-particle potential energy. This potential energy is generally called a "field" since it is one-body in nature. It could be a physical field, such as an electric, magnetic, or gravitational one. Alternatively, $u(\mathbf{r})$ could emerge from a mean-field treatment of particle interactions.

Equation (18.11) is easily generalized to molecular and multicomponent systems. In such cases, one simply incurs a distinct factorial term $N!$ for each particle/atom type and the potential-energy function accounts both for intermolecular and for intramolecular interactions,

$$Q(T, V, \{N\}) = \frac{Z(T, V, \{N\})}{\prod_i \Lambda_i(T)^{3N_i} N_i!}, \quad \text{where} \quad Z \equiv \int e^{-\beta U(\mathbf{r}^N)} \, d\mathbf{r}^N \tag{18.13}$$

Example 18.1 *Show that the quantum-corrected classical partition function gives the correct free energy for an ideal gas.*

For an ideal gas, $U(\mathbf{r}^N) = 0$. Therefore,

$$Z(T, V, N) = \int e^0 \, d\mathbf{r}^N = V^N$$

The total canonical partition function is

$$\begin{aligned} Q(T, V, N) &= \frac{V^N}{\Lambda(T)^{3N} N!} \\ &= \left(\frac{V}{N}\right)^N \frac{e^N}{\Lambda(T)^{3N}} \end{aligned}$$

The second line makes use of Stirling's approximation. With $A = -k_B T \ln Q$, we find

$$A = -N k_B T \ln \left(\frac{V}{\Lambda(T)^3 N}\right) - N k_B T$$

This is the same result as what we found in Chapter 9, in which we computed $S = k_B \ln \Omega$ using quantum mechanics and subsequently performed a Legendre transform to find A.

18.2 Microstate probabilities for continuous degrees of freedom

The canonical ensemble provides microstate probabilities at a given set of constant conditions T, V, and N, but the treatment of these probabilities in classical systems is special because there are infinitely many microstates. Let us consider one particular microstate m that is specified by a set of values for each classical degree of freedom, equivalently, the $6N$-dimensional vector $(\mathbf{p}_m^N, \mathbf{r}_m^N)$. The probability that the system will visit this microstate is

$$\wp(\mathbf{p}_m^N, \mathbf{r}_m^N) = \frac{e^{-\beta H(\mathbf{p}_m^N, \mathbf{r}_m^N)}}{\int e^{-\beta H(\mathbf{p}^N, \mathbf{r}^N)} \, d\mathbf{p}^N \, d\mathbf{r}^N} \tag{18.14}$$

There is a subtle difference between this probability and what we obtained for discrete systems in Chapter 16. Here we must talk about differential probabilities or probability densities. Specifically, the quantity $\wp(\mathbf{p}_m^N, \mathbf{r}_m^N) d\mathbf{p}^N \, d\mathbf{r}^N$ is proportional to the probability that the system is in a microstate between $p_{1,x} - dp_{1,x}/2$ and $p_{1,x} + dp_{1,x}/2$, $x_1 - dx_1/2$ and $x_1 + dx_1/2$, and so on and so forth. In short, it gives the probability that the system lies within a differential element $d\mathbf{p}^N d\mathbf{r}^N$ centered around the phase-space point $(\mathbf{p}_m^N, \mathbf{r}_m^N)$. General properties of probabilities of this sort are as follows.

$\wp(x, y)$ for continuous variables x and y is defined such that $\wp(x, y)dx\,dy$ gives the **differential probability** that x and y lie within an infinitesimally small region dx centered around x and dy centered around y. Therefore, $\wp(x, y)$ has dimensions of $[x]^{-1}\,[y]^{-1}$ and $\wp(x, y)dx\,dy$ is dimensionless. In this sense, $\wp(x, y)$ is called a **probability density** since it gives probability per units of x and y.

Just as the sum of the microstate probabilities in a discrete system must equal one, the integral of the differential probabilities in a classical system is also normalized,

$$\int \wp(\mathbf{p}^N, \mathbf{r}^N)d\mathbf{p}^N\,d\mathbf{r}^N = 1 \tag{18.15}$$

Here the limits of the integrals are the same as those mentioned above for Eqn. (18.2).

As for the canonical partition function, we find that we can separate the probability for a given microstate m into kinetic and potential energy contributions. The approach begins by factoring the exponential in the Boltzmann factor,

$$\wp(\mathbf{p}_m^N, \mathbf{r}_m^N) = \frac{e^{-\beta K(\mathbf{p}_m^N)} e^{-\beta U(\mathbf{r}_m^N)}}{\displaystyle\int e^{-\beta K(\mathbf{p}^N)} e^{-\beta U(\mathbf{r}^N)}\,d\mathbf{p}^N\,d\mathbf{r}^N} \tag{18.16}$$

The integrals can then be factored as well,

$$\wp(\mathbf{p}_m^N, \mathbf{r}_m^N) = \left[\frac{e^{-\beta K(\mathbf{p}_m^N)}}{\displaystyle\int e^{-\beta K(\mathbf{p}^N)}\,d\mathbf{p}^N}\right]\left[\frac{e^{-\beta U(\mathbf{r}_m^N)}}{\displaystyle\int e^{-\beta U(\mathbf{r}^N)}\,d\mathbf{r}^N}\right]$$

$$= \wp(\mathbf{p}_m^N)\wp(\mathbf{r}_m^N) \tag{18.17}$$

Each of the terms in brackets is itself a probability dependent on the Boltzmann factor, with a normalization integral in the denominator. The implication is that we can treat a classical canonical system as being composed of two subsystems: one entails the kinetic and the other the configurational degrees of freedom. Importantly, Eqn. (18.17) suggests that the joint probability of one set of momenta and one set of positions is given by the product of the two separate probabilities. As a consequence, the particle momenta are entirely independent of the positions.

In the canonical ensemble, the configurational and kinetic contributions to the partition function are entirely separable. The total partition function Q can be written as a product of partition functions for these two kinds of degrees of freedom. Moreover, the probability distribution of atomic configurations is independent of the distribution of momenta. The former depends only on the potential energy function, and the latter only on the kinetic energy function.

If we are interested only in the distribution of the momenta, we can use the analytical integrals presented earlier for the kinetic energy to find that

$$\wp(\mathbf{p}^N) = \frac{e^{-\beta K(\mathbf{p}^N)}}{h^{3N}\Lambda(T)^{-3N}} \tag{18.18}$$

This expression gives the probability with which a particular set of momenta for all atoms will be attained by the system at any one random instant in time. On the other hand, if we are interested in the distribution of the atomic configurations, $\wp(\mathbf{r}^N)$, the considerations above show that

$$\wp(\mathbf{r}^N) = \frac{e^{-\beta U(\mathbf{r}^N)}}{Z(T,V,N)} \tag{18.19}$$

Equation (18.19) provides the joint probability that the particles will adopt the positions $\mathbf{r}^N$. Notice that no kinetic energy terms remain in Eqn. (18.19) and no potential energy ones in Eqn. (18.18). In other words, there is a complete separation of the probability distributions of momenta and of spatial configurations; these degrees of freedom do not "interact with" or influence each other. The likelihood of the system visiting a particular configuration of atomic positions is completely independent of the velocity of each atom in that configuration. For example, a particle sitting right next to a container wall would have exactly the same distribution of momenta as if it were in the bulk. Table 18.1 summarizes the two kinds of degrees of freedom.

Importantly, the configurational distribution allows one to calculate many different system properties that depend on the positions of the atoms. Consider a property $X(\mathbf{r}^N)$ in which we have made this dependence explicit. Its average is given by a weighted sum over configurational microstates using Eqn. (18.19),

$$\langle X \rangle = \int \wp(\mathbf{r}^N) X(\mathbf{r}^N) d\mathbf{r}^N \tag{18.20}$$

Similarly, fluctuations in X can be estimated from its variance in the ensemble,

$$\sigma_X^2 = \langle X^2 \rangle - \langle X \rangle^2$$

$$= \left[\int \wp(\mathbf{r}^N) X(\mathbf{r}^N)^2 \, d\mathbf{r}^N \right] - \left[\int \wp(\mathbf{r}^N) X(\mathbf{r}^N) d\mathbf{r}^N \right]^2 \tag{18.21}$$

Table 18.1 Kinetic and configurational degrees of freedom make separable contributions to the classical canonical partition function

Property	Kinetic	Configurational
Contribution to total partition function	$\int e^{-\beta K(\mathbf{p}^N)} \, d\mathbf{p}^N$	$\int e^{-\beta U(\mathbf{r}^N)} \, d\mathbf{r}^N$
Probability distribution	$\wp(\mathbf{p}_m^N) = \dfrac{e^{-\beta K(\mathbf{p}_m^N)}}{\int e^{-\beta K(\mathbf{p}^N)} \, d\mathbf{p}^N}$	$\wp(\mathbf{r}_m^N) = \dfrac{e^{-\beta U(\mathbf{r}_m^N)}}{\int e^{-\beta U(\mathbf{r}^N)} \, d\mathbf{r}^N}$

Finally, the entire probability distribution of X can be found if one inserts a *Dirac delta function* into the configurational integral,

$$\wp(X') = \int \wp(\mathbf{r}^N)\delta[X(\mathbf{r}^N) - X']d\mathbf{r}^N \tag{18.22}$$

The Dirac delta function is a sort of continuous analogue of the Kronecker one, and in Eqn. (18.22) it filters for microstates that have a given value of $X = X'$, summing their probabilities. Loosely speaking, it has the following mathematical properties,

$$\delta[x] = \begin{cases} \infty & \text{if } x = 0 \\ 0 & \text{otherwise} \end{cases}$$

and

$$\int_{-\infty}^{\infty} \delta[x]dx = 1 \tag{18.23}$$

Equations (18.20)–(18.22) should appear similar to the discrete-state versions presented in Chapter 17. The distinction in this case of course is that the summations over microstates become integrals over the classical position variables. While we have considered only position-dependent properties, Eqns. (18.20)–(18.22) are readily generalized to the full microstate probability distribution $\wp(\mathbf{p}^N, \mathbf{r}^N)$ for properties that might depend on both sets of degrees of freedom, $X(\mathbf{p}^N, \mathbf{r}^N)$. Similarly, one can formulate averages in the kinetic degrees of freedom alone using $\wp(\mathbf{p}^N)$ for properties $X(\mathbf{p}^N)$.

It is interesting to note that the configurational distribution for independent particles can be factored in a similar manner to what was done in Eqn. (18.17). In this special case, the particles have only one-body potential energy terms $u(\mathbf{r})$, lacking mutual interactions. One can then write

$$\wp(\mathbf{r}^N) = \wp(\mathbf{r}_1)\wp(\mathbf{r}_2) \times \cdots \times \wp(\mathbf{r}_N) \tag{18.24}$$

where each individual distribution is given by

$$\wp(\mathbf{r}_m) = \frac{e^{-\beta u(\mathbf{r}_m)}}{\int e^{-\beta u(\mathbf{r})}\, d\mathbf{r}} \tag{18.25}$$

As before, the microstate label m distinguishes the distribution variable from that in the integrand in the denominator. This mirrors the discussion about independent degrees of freedom in Chapter 16 as well as the result in Eqn. (18.12).

Example 18.2   *A single classical particle exists in a one-dimensional world and sits in a harmonic energy well, for which $U = \alpha x^2$, where α is a constant with units of [energy]/ [distance]^2. The particle is in an infinite volume and its position x can vary continuously* *among any real values. Compute $\langle x \rangle$, $\langle x^2 \rangle$, and the constant-volume heat capacity as a function of T.*

We begin by calculating the single-particle partition function in the one-dimensional case,

$$q = \frac{1}{\Lambda(T)} \int_{-\infty}^{\infty} e^{-\alpha\beta x^2} \, dx = \frac{1}{\Lambda(T)} \left(\frac{\pi}{\alpha\beta}\right)^{1/2} = \frac{\pi}{\beta h} \left(\frac{2m}{\alpha}\right)^{1/2}$$

The microstate distribution is Gaussian,

$$\wp(x) = \frac{e^{-\alpha\beta x^2}}{\displaystyle\int_{-\infty}^{\infty} e^{-\alpha\beta x^2} \, dx} = \left(\frac{\alpha\beta}{\pi}\right)^{1/2} e^{-\alpha\beta x^2}$$

The averages are thus

$$\langle x \rangle = \int_{-\infty}^{\infty} x \wp(x) \, dx = 0$$

and

$$\langle x^2 \rangle = \int_{-\infty}^{\infty} x^2 \wp(x) \, dx = \left(\frac{\alpha\beta}{\pi}\right)^{1/2} \int_{-\infty}^{\infty} x^2 e^{-\alpha\beta x^2} \, dx = \frac{1}{2\alpha\beta}$$

The first moment of course vanishes because of the symmetry of the particle in x. We see that the second moment is linear in temperature. Because $\langle U \rangle = \alpha\langle x^2 \rangle$, the potential energy also depends linearly on temperature.

To find the heat capacity, we can use the relationship developed in Chapter 17 to find the average energy,

$$\langle E \rangle = -\frac{\partial \ln q}{\partial \beta} = k_B T$$

which clearly shows that the heat capacity is constant and equal to $C_V = k_B$.

Example 18.3 *Consider the particle in Example 18.2 to exist in a two-dimensional world, with $U = \alpha(x^2 + y^2) = \alpha r^2$. Show that the joint probability distribution is factorable as $\wp(x,y) = \wp(x)\wp(y)$ and find $\wp(x)$, $\wp(y)$, and $\langle r \rangle$.*

We can begin with Eqn. (18.25) and then factor the exponentials,

$$\wp(x,y) = \frac{e^{-\alpha\beta(x^2+y^2)}}{\displaystyle\iint e^{-\alpha\beta(x^2+y^2)} \, dx \, dy} = \frac{e^{-\alpha\beta x^2} e^{-\alpha\beta y^2}}{\displaystyle\iint e^{-\alpha\beta x^2} e^{-\alpha\beta y^2} \, dx \, dy}$$

Because the integrals are now factorable, so is the probability distribution,

$$\wp(x, y) = \left[\frac{e^{-\alpha\beta x^2}}{\int e^{-\alpha\beta x^2}\, dx} \right] \cdot \left[\frac{e^{-\alpha\beta y^2}}{\int e^{-\alpha\beta y^2}\, dy} \right]$$

$$= \left[\left(\frac{\alpha\beta}{\pi} \right)^{1/2} e^{-\alpha\beta x^2} \right] \left[\left(\frac{\alpha\beta}{\pi} \right)^{1/2} e^{-\alpha\beta y^2} \right]$$

$$= \wp(x)\wp(y)$$

In the second line we evaluated the Gaussian integrals in the denominators. The final line shows that the individual probability distributions for x and y are separable, identical, and each equal to that found in Example 18.2.

We now find the average distance from the origin, $\langle r \rangle$. Start with Eqn. (18.20),

$$\langle r \rangle = \int\int \wp(x, y) \sqrt{x^2 + y^2}\, dx\, dy$$

$$= \left(\frac{\alpha\beta}{\pi} \right) \int\int e^{-\alpha\beta(x^2 + y^2)} \sqrt{x^2 + y^2}\, dx\, dy$$

At this point, a transformation to polar coordinates r and ϕ helps,

$$\langle r \rangle = \left(\frac{\alpha\beta}{\pi} \right) \int_0^{2\pi} \int_0^\infty e^{-\alpha\beta r^2} r^2\, dr\, d\phi$$

$$= 2\alpha\beta \int_0^\infty e^{-\alpha\beta r^2} r^2\, dr$$

$$= \frac{1}{2} \left(\frac{\pi}{\alpha\beta} \right)^{1/2}$$

As we might expect, when T grows larger, the average distance from the origin increases.

Example 18.4 *A classical diatomic ideal gas molecule can be described as two bonded atoms tethered by a harmonic spring, with $u_{bond} = \alpha(r - r_0)^2$, where r_0 is the equilibrium bond distance and α is a force constant as in the previous example. Find an expression for the heat capacity of such a molecule, assuming that the atoms are of the same type (e.g., O_2 or N_2).*

Since ideal gas molecules do not interact, we can use a single-molecule partition function,

$$q = \frac{1}{\Lambda(T)^6 2!} \int\int e^{-\alpha\beta(|\mathbf{r}_1 - \mathbf{r}_2| - r_0)^2}\, d\mathbf{r}_1\, d\mathbf{r}_2$$

If the volume of the system were comparable to the distance r_0 this would be a difficult integral to solve. However, since the gas is ideal we expect the volume to be large. This means effectively that we can place the first atom at a location in space and then consider it to be the origin for the placement of the second atom; the latter will occupy only distances very close to the first and thus for all practical purposes never approach a volume boundary. Thus we can write the integral as,

$$q = \frac{1}{2\Lambda(T)^6} \left(\int d\mathbf{r}_1 \right) \left(\int e^{-\alpha\beta(|\mathbf{r}_2|-r_0)^2} d\mathbf{r}_2 \right)$$

The first three-dimensional integral clearly gives the volume. The second is easily evaluated by a transformation to spherical coordinates r_2, θ, and ϕ, and a variable substitution $r_2 = r + r_0$,

$$q = \frac{V}{2\Lambda(T)^6} \int\int\int e^{-\alpha\beta r^2} (r + r_0)^2 \sin\theta \, dr \, d\theta \, d\phi$$

Presuming that α is large because bonds are stiff degrees of freedom, we can approximate the Gaussian integral using limits on r from $-\infty$ to ∞, even though the true lower bound of r is $-r_0$. We can do this because the Boltzmann factor will become small rapidly as r deviates from 0, and so those contributions to the integral become negligible. The integral evaluates to

$$q = \frac{V}{2\Lambda(T)^6} \left[2\left(\frac{\pi}{\alpha\beta}\right)^{3/2} + 4r_0^2 \left(\frac{\pi^3}{\alpha\beta}\right)^{1/2} \right]$$

$$\approx \frac{V}{2\Lambda(T)^6} \left[4r_0^2 \left(\frac{\pi^3}{\alpha\beta}\right)^{1/2} \right]$$

We neglected the first term in the brackets because it is a factor of $1/\alpha$ smaller than the second, and α is presumed to be large. Finally, we substitute the expression for the de Broglie wavelength to obtain a partition function,

$$q = \frac{2Vr_0^2\pi^{3/2}}{\alpha^{1/2}\beta^{7/2}} \left(\frac{2\pi m}{h^2}\right)^3$$

The average energy follows:

$$\langle E \rangle = -\frac{\partial\ln q}{\partial\beta} = \frac{7}{2}k_BT$$

This shows that $c_V = (7/2)k_B$ for this fully classical, diatomic ideal gas. Indeed, this is $2k_B$ greater than the monatomic case. Physically, the increase is due to the bond stretching and rotation of the diatomic molecule. In reality, neither of these degrees of freedom is generally well described by a classical treatment for gases at standard conditions. Rather, a quantum-mechanical approach is usually needed, which gives rise to a discrete set of energy levels in both cases. When these are considered in detail, the heat capacity of the diatomic gas is found to vary with T, from about $(3/2)k_B$ at low temperatures to the fully classical value of $(7/2)k_B$ at high ones. Even at room temperature this fully classical value is often not yet reached, and many diatomic gases have $c_V \approx (5/2)k_B$.

18.3   The Maxwell–Boltzmann distribution

The distribution of momenta is general and analytically tractable, so we analyze it in greater detail. Specifically, we find the famous Maxwell–Boltzmann distribution of particles' velocities at a given temperature. Starting with Eqn. (18.18), let us focus on the first particle and the x component of its momentum, $p_{1,x}$. We want to find the one-dimensional distribution $\wp(p_{1,x})$, as opposed to the joint $3N$-dimensional distribution $\wp(\mathbf{p}^N)$. We can think of this as finding the probability that we will see a particular value of $p_{1,x}$ regardless of the values of momenta of all the other components and particles. This implies that we enumerate all of the microstates in which $p_{1,x}$ takes on a particular value and sum up their probabilities. Here, however, the degrees of freedom are continuous and the sum corresponds to an integral. These ideas stem from a general feature of such multivariate probability distributions.

> Given a continuous **joint probability density** $\wp(x, y, z)$, the **marginal probability distribution** is given by integrating the former over the variables not of interest. For example, $\wp(x) = \int\int \wp(x, y, z) dy\, dz$ describes the probabilities of different values of x, regardless of the values of y and z. As expected, the dimensions of $\wp(x)$ are $[x]^{-1}$, since the dimensions of $\wp(x, y, z)$ are $[x]^{-1} [y]^{-1} [z]^{-1}$ and the integral adds dimensions of $[y]\, [z]$.

With these general considerations, the distribution of $p_{1,x}$ follows from

$$\wp(p_{1,x}) = \int \wp(\mathbf{p}^N) dp_{1,y}\, dp_{1,z}\, d\mathbf{p}_2 \cdots d\mathbf{p}_N \tag{18.26}$$

In other words, we integrate the kinetic probability distribution over all of the other $3N - 1$ degrees of freedom, not including $p_{1,x}$. This integral is analytic,

$$\wp(p_{1,x}) = \frac{\Lambda(T)^{3N}}{h^{3N}} \int e^{-(\beta/2m) \sum_i \left(p_{i,x}^2 + p_{i,y}^2 + p_{i,z}^2\right)} dp_{1,y}\, dp_{1,z}\, d\mathbf{p}_2 \cdots d\mathbf{p}_N$$

$$= \frac{\Lambda(T)^{3N}}{h^{3N}} e^{-(\beta/2m)p_{1,x}^2} \left[\int e^{-\beta/2mp^2} dp\right]^{3N-1} \tag{18.27}$$

In the second line, we factor out the part of the exponential that is due to $p_{1,x}$, and the remaining $3N - 1$ integrals are identical. By evaluating the Gaussian integrals as before, we obtain

$$\wp(p_{1,x}) = \frac{\Lambda(T)^{3N}}{h^{3N}} e^{-(\beta/2m)p_{1,x}^2} \frac{h^{3N-1}}{\Lambda(T)^{3N-1}}$$

$$= \frac{\Lambda(T)}{h} e^{-(\beta/2m)p_{1,x}^2} \tag{18.28}$$

It would have been equally valid to determine this result by recognizing that each momentum degree of freedom is linearly independent in the kinetic energy function, so we are able to factor their probabilities as described in Chapter 16 and as illustrated for independent particles in Eqn. (18.24). Namely,

$$\wp(\mathbf{p}^N) = \wp(p_{1,x})\wp(p_{1,y}) \times \cdots \times \wp(p_{N,z}) \qquad (18.29)$$

Since there is no difference between the particles or the momentum components by symmetry, we can say that in general the probability any particle will adopt a particular momentum at temperature T is given by

$$\wp(p_x) = (2\pi m k_B T)^{-1/2} \exp\left(-\frac{p_x^2}{2m k_B T}\right) \qquad (18.30)$$

This is the *Maxwell–Boltzmann* distribution for one component of a particle's momentum. Equivalent expressions are found for p_y and p_z. Of course, $\wp(p_x)$ is a probability density and the quantity $\wp(p_x)dp_x$ gives the differential probability that a particle has one of its components of momentum in the range $p_x - dp_x/2$ to $p_x + dp_x/2$. As in the canonical partition function integrals, a momentum component p can vary between negative and positive infinity, and thus the integral of $\wp(p_x)$ over that range evaluates to one.

The Maxwell–Boltzmann distribution makes several significant statements about the universality of kinetic degrees of freedom at equilibrium. First, the form of the distribution is Gaussian, as shown in Fig. 18.1; its peak lies at $p_x = 0$ and the peak probability increases at lower temperatures. Second, the distribution is symmetric and there is no net momentum in either direction. In other words, there can never be any *flow* within the system; transport and hydrodynamic processes are inherently nonequilibrium because they require an asymmetric momentum distribution. Third, and most importantly, the form of the Maxwell–Boltzmann distribution applies to any system, since it makes no reference to the potential energy function. It also applies at any density because the final distribution is independent of the system volume.

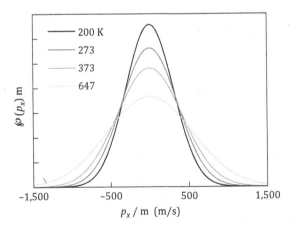

Figure 18.1. The Maxwell–Boltzmann distribution predicts the probabilities for molecular momenta at a given temperature. Here, results are illustrated for water molecules at a low temperature, the normal melting point, the normal boiling point, and the critical temperature. The axes have been normalized by the mass of a water molecule for the purposes of illustrating the distribution of velocities.

The canonical distribution of atomic momenta given by the **Maxwell–Boltzmann** expression is *universal*, depending only on the mass of the individual molecules and the temperature of the system. It is independent of the potential energy function and of the density.

There are two interpretations of this distribution. One is that $\wp(p_x)$ gives the frequencies for a given particle's x component of momentum when viewed over long periods of time. Alternatively, $\wp(p_x)$ gives the distribution of p_x over a large number of particles at one instant in time. These two interpretations are equivalent at equilibrium.

The average kinetic energy can be evaluated using the momentum distribution. The $3N$ momenta are independent, so we can write

$$\langle K \rangle = \frac{1}{2m} \left(\langle p_{1,x}^2 \rangle + \langle p_{1,y}^2 \rangle + \cdots + \langle p_{N,z}^2 \rangle \right)$$

$$= \frac{3N}{2m} \langle p_x^2 \rangle$$

$$= \frac{3N}{2m} \int_{-\infty}^{\infty} p_x^2 \wp(p_x) \, dp_x \tag{18.31}$$

The second line uses the fact that all of the averages must be the same because there is no difference between particles or components. Substituting (18.30) and evaluating the integral leads to

$$\langle K \rangle = \frac{3}{2} N k_B T \tag{18.32}$$

In plain words, the average kinetic energy is proportional to the number of particles and to the temperature. Equation (18.32) is quite general, even for multicomponent systems, so long as they can be described by atoms or particles rather than rigid bodies (for which rotational kinetic energies would become relevant).

At constant temperature, the average kinetic energy is only a function of the number of particles and the temperature, regardless of the density or the nature of the atoms (their mass and interactions).

In principle, the average kinetic energy could be measured from a large number of snapshots of a system taken randomly in time. Equation (18.32) would then allow one to find an estimate of the temperature. In turn, we say that the average kinetic energy is an *estimator* of the temperature. However, as we shall see shortly, it is only at the level of an average that this is true.

Let us now consider the so-called *speed* of a particle v, defined as the magnitude of the velocity vector,

$$v = \frac{1}{m} \sqrt{p_x^2 + p_y^2 + p_z^2} \tag{18.33}$$

Unlike a component of the momentum or velocity vectors, the speed is always a positive number. To find the distribution of the speed, Eqn. (18.33) shows that we need to consider the joint distributions of the individual component momenta. Because the momenta are independent in the kinetic energy, we can express the joint distribution for one particle as a product of the individual x, y, and z component distributions, as in Eqn. (18.29),

$$\wp(p_x, p_y, p_z) = \wp(p_x)\wp(p_y)\wp(p_z)$$

$$= (2\pi m k_B T)^{-3/2} \exp\left(-\frac{p_x^2 + p_y^2 + p_z^2}{2m k_B T}\right)$$

$$= (2\pi m k_B T)^{-3/2} \exp\left(-\frac{|p|^2}{2m k_B T}\right) \tag{18.34}$$

We want to remove the orientational contribution to this distribution, leaving only the absolute value of the momentum. Let us convert (18.34) into spherical coordinates with the substitutions $p_x = |p|\sin\theta \cos\phi$, $p_y = |p|\sin\theta \sin\phi$, and $p_z = |p|\cos\theta$:

$$\wp(|p|, \theta, \phi) = (2\pi m k_B T)^{-3/2} \exp\left(-\frac{|p|^2}{2m k_B T}\right) p^2 \sin\theta \tag{18.35}$$

Here, the probability distribution acquired a term $p^2 \sin\theta$ because the differential probabilities $\wp(p_x, p_y, p_z)dp_x\, dp_y\, dp_z = \wp(|p|, \theta, \phi)d|p|d\theta\, d\phi$ must be equivalent. (For example, both probability distributions must integrate to unity, and this term would naturally appear in performing the variable transform to spherical coordinates.) Integrating over the angular variables θ and ϕ gives

$$\wp(|p|) = 4\pi(2\pi m k_B T)^{-3/2} p^2 \exp\left(-\frac{|p|^2}{2m k_B T}\right) \tag{18.36}$$

Finally, we make the substitution $|p| = mv$. This also introduces a factor of m due to the equivalence of differential probabilities $\wp(|p|)d|p| = \wp(v)dv$. As a result, we arrive at the distribution of speeds:

$$\wp(v) = 4\pi\left(\frac{m}{2\pi k_B T}\right)^{3/2} v^2 \exp\left(-\frac{mv^2}{2k_B T}\right) \tag{18.37}$$

Figure 18.2 illustrates the distribution of molecular speeds for water at different temperatures. As expected, the distribution broadens to larger velocities as the temperature increases. This can be quantified by computing the average,

$$\langle v \rangle = \int v\wp(v)dv = \left(\frac{8k_B T}{\pi m}\right)^{1/2} \tag{18.38}$$

We see that the average velocity we would expect from a particle depends monotonically on the temperature. This relationship, however, applies only to the *average*. In reality, the Maxwell–Boltzmann distribution shows that particles must be *distributed* in velocity at a given temperature, and with a very specific form for that distribution. This does not mean that some particles have lower temperature than others, since it is the entire

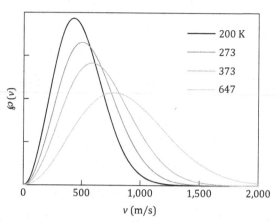

Figure 18.2. The Maxwell–Boltzmann distribution predicts the probabilities for molecular speeds, that is, absolute velocities. As in Figure 18.1, water is shown as an example.

system that is coupled to the bath. Thus, while it is tempting to think of the temperature in terms of molecular velocities, one must realize that this does not define the temperature and the connection exists only at the ensemble level.

> The velocity of a molecule is not a measure of its temperature. Instead, the ensemble average velocity of all molecules, or of a single molecule over long periods of time, provides an estimator of the temperature. The temperature is not an intrinsic property of a molecule, but an aggregate statistical property of the system ensemble and the bath. The notion of a temperature, therefore, is inherently a property of very large systems of molecules.

18.4   The pressure in the canonical ensemble

In the previous section we uncovered universal relationships among the average kinetic energy, molecular speeds, and the macroscopic temperature. Naturally, one might seek a similar interpretation of the macroscopic pressure: can one relate a microscopic average to this quantity? Since the Helmholtz free energy is given by $A = -k_B T \ln Q$ and because the pressure stems from $P = -(\partial A / \partial V)_{T,N}$, we begin by expressing the latter in terms of a derivative of the canonical partition function,

$$P = \frac{k_B T}{Q} \left(\frac{\partial Q}{\partial V} \right)_{T,N} \tag{18.39}$$

Using Eqn. (18.11), we find

$$P = \frac{k_B T}{Z} \left(\frac{\partial Z}{\partial V} \right)_{T,N}$$

$$= \frac{k_B T}{Z} \frac{\partial}{\partial V} \int e^{-\beta U(\mathbf{r}^N)} \, d\mathbf{r}^N \tag{18.40}$$

and therefore the pressure depends only on the configurational part of the partition function. The derivative of the integral above requires some careful thought because the variable limits depend on the system volume. That is, the integral is evaluated over all positions available in a volume V. For a cubic box, each particle's x component can vary between 0 and $L = V^{1/3}$ and similarly for the y and z components. We need to take the derivative of an integral with respect to a variable in the limit.

The simplest approach to this problem is to remove the $V^{1/3}$ dependence from the integral limits themselves through a change of variables. We non-dimensionalize each particle coordinate by the size of the box, introducing new variables $s_{1,x} = V^{-1/3}x_1$, $s_{1,y} = V^{-1/3}y_1$, and so on. After this substitution, each integral runs between the values 0 and 1 for a cubic box, or between two otherwise V-independent limits in the general case of an arbitrarily shaped volume. We have

$$P = \frac{k_B T}{Z} \frac{\partial}{\partial V} \int e^{-\beta U(V^{1/3}\mathbf{s}^N)} V^N \, d\mathbf{s}^N \qquad (18.41)$$

where $d\mathbf{s}^N = ds_{1,x} \, ds_{1,y} \ldots ds_{N,z}$, as is consistent with our usual notation. The term V^N arises from the $3N$ changes of variables, each of which gave a factor of $V^{1/3}$. We have also written the potential energy function using the new non-dimensionalized coordinates,

$$U(\mathbf{r}_1, \mathbf{r}_2, \ldots, \mathbf{r}_N) = U(L\mathbf{s}_1, L\mathbf{s}_2, \ldots, L\mathbf{s}_N) = U(V^{1/3}\mathbf{s}^N) \qquad (18.42)$$

It is important to remember that the potential energy depends on the real positions, not the scaled ones. The notation $V^{1/3}\mathbf{s}^N$ implies that the prefactor $V^{1/3}$ individually scales each component of $\mathbf{s}^N$.

To finally evaluate the derivative, we use the chain rule because there are two terms involving V, one within the potential energy and one involving the factor V^N:

$$P = \frac{k_B T}{Z} \int \left[e^{-\beta U(V^{1/3}\mathbf{s}^N)} N V^{N-1} - \beta \left(\frac{\partial U}{\partial V} \right)_{\mathbf{s}^N} e^{-\beta U(V^{1/3}\mathbf{s}^N)} V^N \right] d\mathbf{s}^N \qquad (18.43)$$

By factoring out the common terms and moving back from the non-dimensionalized coordinates to the usual ones, we obtain

$$P = \frac{k_B T}{Z} \int \left[\frac{N}{V} - \beta \left(\frac{\partial U}{\partial V} \right)_{\mathbf{s}^N} \right] e^{-\beta U(\mathbf{r}^N)} \, d\mathbf{r}^N$$

$$= \frac{N k_B T}{V} \frac{1}{Z} \int e^{-\beta U(\mathbf{r}^N)} \, d\mathbf{r}^N - \frac{1}{Z} \int \left(\frac{\partial U}{\partial V} \right)_{\mathbf{s}^N} e^{-\beta U(\mathbf{r}^N)} \, d\mathbf{r}^N$$

$$= \frac{N k_B T}{V} - \frac{1}{Z} \int \left(\frac{\partial U}{\partial V} \right)_{\mathbf{s}^N} e^{-\beta U(\mathbf{r}^N)} \, d\mathbf{r}^N \qquad (18.44)$$

Notice that the first integral in the second line is equivalent to Z, and thus cancels out with the $1/Z$ factor. The second term requires us to compute the derivative of the potential energy function with respect to volume at fixed scaled positions $\mathbf{s}^N$. The approach is somewhat involved,

$$\left(\frac{\partial U(\mathbf{r}^N)}{\partial V}\right)_{\mathbf{s}^N} = \left(\frac{\partial U(V^{1/3}s_{1,x}, \ldots, V^{1/3}s_{N,z})}{\partial V}\right)_{\mathbf{s}^N}$$

$$= \frac{\partial U}{\partial x_1}\frac{d(V^{1/3}s_{1,x})}{dV} + \cdots + \frac{\partial U}{\partial z_N}\frac{d(V^{1/3}s_{N,z})}{dV}$$

$$= (-f_{1,x})\left(\frac{1}{3}V^{-2/3}s_{1,x}\right) + \cdots + (-f_{N,z})\left(\frac{1}{3}V^{-2/3}s_{N,z}\right)$$

$$= (-f_{1,x})\left(\frac{1}{3V}x_1\right) + \cdots + (-f_{N,z})\left(\frac{1}{3V}z_N\right)$$

$$= -\frac{1}{3V}\sum_i \mathbf{f}_i \cdot \mathbf{r}_i \tag{18.45}$$

In the third line, we introduced the forces on each atom i, given by the derivative of the potential energy with respect to position, $\mathbf{f}_i = -\partial U/\partial \mathbf{r}_i$. We then switched back to the dimensional coordinates in the fourth line. The final result shows that the derivative of interest involves a sum of dot products of particles' forces and positions, and is thus configuration-dependent. Substituting (18.45) into Eqn. (18.44) gives

$$P = \frac{Nk_{\mathrm{B}}T}{V} + \frac{1}{3VZ}\int\left(\sum_i \mathbf{f}_i \cdot \mathbf{r}_i\right)e^{-\beta U(\mathbf{r}^N)}\,d\mathbf{r}^N \tag{18.46}$$

Notice, however, that the integral can be rewritten as a sum over the configurational probabilities by bringing the factor of Z inside and using Eqn. (18.19),

$$P = \frac{Nk_{\mathrm{B}}T}{V} + \frac{1}{3V}\int\left(\sum_i \mathbf{f}_i \cdot \mathbf{r}_i\right)\wp(\mathbf{r}^N)\,d\mathbf{r}^N \tag{18.47}$$

The significance of the integral is that it represents a canonical average per Eqn. (18.20). We can finally express the pressure in the following compact manner,

$$P = \frac{Nk_{\mathrm{B}}T}{V} + \frac{1}{3V}\left\langle\sum_i \mathbf{f}_i \cdot \mathbf{r}_i\right\rangle \tag{18.48}$$

The average sum of forces times positions is called the *internal virial* or the *pressure virial*, and is often denoted by the symbol W,

$$W \equiv \sum_i \mathbf{f}_i \cdot \mathbf{r}_i \tag{18.49}$$

which is not to be confused with the work appearing in the first law. The expression in (18.48) shows that the pressure relates to an average over forces and positions of particles. If there are no potential energies and hence no forces, the virial is zero. In this case, $W = 0$ and the pressure simply reduces to the ideal gas one.

In the canonical ensemble, the pressure involves an ideal gas term plus an additional contribution that stems from the **internal virial** or **pressure virial**. The latter derives from interaction forces acting among and on the particles.

18.5 The classical microcanonical partition function

For a classical system, the microcanonical partition function requires special treatment because a delta function is involved. In discrete systems, for example, we found

$$\Omega(E, V, N) = \sum_{\mathcal{X}_1}\sum_{\mathcal{X}_2}\cdots\sum_{\mathcal{X}_N} \delta_{E(\mathcal{X}_1,\mathcal{X}_2,\ldots,\mathcal{X}_N),E} \tag{18.50}$$

This equation implies an enumeration of all possible combinations of values for particle degrees of freedom and a counting of each time a microstate has the same energy as the macroscopic energy. For classical systems, we change the sums to integrals, but we must also change the Kronecker to a Dirac delta function,

$$\Omega(E, V, N) = \frac{1}{h^{3N}N!}\int\int\cdots\int \delta\left[H(\mathbf{p}^N, \mathbf{r}^N) - E\right]d\mathbf{p}^N\, d\mathbf{r}^N \tag{18.51}$$

The limits on the integrals are the same as those discussed for the canonical version, and we have already added the quantum correction. In simple terms, the Dirac delta function is infinitely peaked around the point where its argument is zero. In the microcanonical partition function, it "selects out" those configurations which have the same energy as the specified energy. It can be shown that evaluating the multidimensional integral with the delta function is the same as finding the $6N$-dimensional area in phase space where the energy adopts a specified value. It is useful to note that the Dirac delta function has the inverse units of its argument.

The microcanonical partition function is not separable into kinetic and potential energy components in the same way as the canonical one is. It is also not separable in a simple way for independent molecules. Moreover, due to difficulties in working with the Dirac delta function, it is almost always much easier to compute Q rather than Ω for classical systems. However, it is relatively easy to show that the relationship between the two bears great resemblance to the case for discrete systems, involving a Laplace-like transform over the energy,

$$Q(T, V, N) = \int_{-\infty}^{\infty} \Omega(E, V, N)e^{-\beta E}\, dE \tag{18.52}$$

Example 18.5 *Show that the Sackur–Tetrode equation for the entropy and hence density of states of an ideal monatomic gas recovers the correct canonical partition function when Eqn. (18.52) is used.*

From Chapter 9, the Sackur–Tetrode entropy can be written as

$$\frac{S}{k_B} = N\ln\left[\left(\frac{E}{N}\right)^{3/2}\left(\frac{V}{N}\right)\right] + cN, \quad \text{where} \quad c = \frac{5}{2} + \frac{3}{2}\ln\left(\frac{4\pi m}{3h^2}\right)$$

The density of states therefore has the form

$$\Omega(E, V, N) = \left(\frac{E}{N}\right)^{3N/2}\left(\frac{V}{N}\right)^N e^{5N/2}\left(\frac{4\pi m}{3h^2}\right)^{3N/2}$$

for $E > 0$ only. Substitution into Eqn. (18.52) gives

$$Q(T, V, N) = \frac{V^N e^{5N/2}}{N^{5N/2}} \left(\frac{4\pi m}{3h^2}\right)^{3N/2} \int_0^\infty E^{3N/2} \exp(-\beta E) dE$$

Let us make the variable transform $x = \beta E$. The integral becomes a relatively simple one that evaluates to a factorial,

$$Q(T, V, N) = \frac{V^N e^{5N/2}}{N^{5N/2}} \left(\frac{4\pi m}{3h^2}\right)^{3N/2} \frac{1}{\beta^{3N/2+1}} \int_0^\infty x^{3N/2} \exp(-x) dx$$

$$= \frac{V^N e^{5N/2}}{N^{5N/2}} \left(\frac{4\pi m k_B T}{3h^2}\right)^{3N/2} \left(\frac{3N}{2}\right)!$$

In the second line, we made the approximation $\beta^{3N/2+1} \approx \beta^{3N/2}$ since an extra factor of β will become negligible in the thermodynamic limit. (Consider for example that all other terms in the partition function scale exponentially with N.) Finally, we express the factorial using Stirling's approximation and group terms,

$$Q(T, V, N) = V^N \frac{e^{5N/2} e^{-3N/2} N^{3N/2}}{N^{5N/2}} \left(\frac{3}{2}\right)^{3N/2} \left(\frac{4\pi m k_B T}{3h^2}\right)^{3N/2}$$

$$= \frac{V^N e^N}{N^N} \left(\frac{2\pi m k_B T}{h^2}\right)^{3N/2}$$

$$= \left(\frac{V}{N}\right)^N \frac{e^N}{\Lambda(T)^{3N}}$$

This is identical to the result that we found in Example 18.1.

Problems

You may find the following integral relations useful in the subsequent problems:

$$\int_0^\infty e^{-cx^2} dx = \frac{\pi^{1/2}}{2c^{1/2}}, \qquad \int_0^\infty x e^{-cx^2} dx = \frac{1}{2c}$$

$$\int_0^\infty x^2 e^{-cx^2} dx = \frac{\pi^{1/2}}{4c^{3/2}}, \qquad \int_0^\infty x^4 e^{-cx^2} dx = \frac{3\pi^{1/2}}{8c^{5/2}}$$

Conceptual and thought problems

18.1. Show formally that uncertainty in the constant c appearing in the classical partition function of Eqn. (18.2) is equivalent to the inability to define an absolute entropy.

18.2. Which thermodynamic properties does indistinguishability and the factor $N!$ in Eqn. (18.3) affect?

18.3. How would the heat capacity of the classical diatomic ideal gas in Example 18.4 change if there were no bonding energy present? In other words, what is the contribution of u_{bond} to c_V relative to the case that the "molecule" consists of two free particles? Does this make physical sense?

18.4. Consider water at its normal boiling point. How is it possible that a water molecule in the dense liquid phase has the same average velocity as that in a gas phase, as the Maxwell–Boltzmann distribution would predict?

18.5. How is it that the average molecular speed in Eqn. (18.38) does not predict the correct average kinetic energy in Eqn. (18.32)? That is, explain why

$$Nm\langle v \rangle^2/2 \neq (3/2)Nk_\mathrm{B}T$$

18.6. Consider two equally sized ions with opposite charges, $+1$ and -1. One ion is fixed at the origin in three-dimensional space. The ions interact in accord with Coulomb's law, $u(r) = q_1q_2/(4\pi\epsilon_0 r)$, where r is the distance between them, but due to steric overlap r cannot be less than a value of σ (the diameter of the ions). The system is at a constant temperature T.
   (a) Write an expression proportional to the probability that the free ion has coordinates (x, y, z).
   (b) Write an expression proportional to the probability that the ions are separated by a distance r.

18.7. A nonideal gas is initially at equilibrium when it is suddenly and rapidly compressed isothermally to ten times its original pressure. The system then returns to equilibrium. Indicate whether each of the following increases, decreases, or stays the same, or there is not enough information to tell.
   (a) The average kinetic energy.
   (b) The average potential energy.
   (c) The Gibbs free energy.

18.8. Pure monatomic A and B are held at constant temperature and pressure in two compartments separated by a wall. The wall is removed and A and B mix, forming a regular solution with $\chi < 0$. Indicate whether the following quantities increase, decrease, or stay the same, or there is not enough information to specify the answer without knowing more about the system.
   (a) The average kinetic energy.
   (b) The average potential energy.
   (c) The system volume.

18.9. Indicate whether or not each of the following is always true regarding the distribution of molecular speeds in a single-component, monatomic system.
   (a) The mean speed is the same as the most probable speed.

(b) If the distribution is not Maxwell–Boltzmann and the system is coupled to a heat bath, the system is not at equilibrium.

(c) The mean speed does not change upon adiabatic, quasi-static compression.

18.10. Does Eqn. (18.18) violate Galilean invariance? What happens to the microstate probabilities and the partition function if the velocity vector of every atom is shifted by an identical amount? How are the forms of these functions affected?

Fundamentals problems

18.11. Consider the classical canonical partition function. Show that, for any system at any state, the total free energy and entropy can be written as the sum of an ideal gas part and an excess part,

$$A(T, V, N) = A_{IG}(T, V, N) + A_{ex}(T, V, N)$$

and

$$S(T, V, N) = S_{IG}(T, V, N) + S_{ex}(T, V, N)$$

Derive general expressions for these two excess quantities in terms of the configurational partition function.

18.12. Explicitly derive Eqns. (18.24) and (18.25) for the case of independent particles with

$$U(\mathbf{r}^N) = \sum_i u(\mathbf{r}_i)$$

where $u(\mathbf{r})$ is a one-body "field-like" potential.

18.13. Prove that the distribution of an arbitrary property $\wp(X)$ as given by Eqn. (18.22) is normalized. That is, show that its integral is unity.

18.14. Imagine that the particle in Example 18.3 exists in a three-dimensional world. Its potential energy is given by $U = a(x^2 + y^2 + z^2) = ar^2$, where r is the radial distance from the origin.

(a) Compute the single particle partition function, $q(T)$.

(b) Compute $\langle x \rangle$ and $\langle x^2 \rangle$ as functions of T. How do these averages compare with those of y and z?

(c) Compute $\langle r \rangle$ and $\langle r^2 \rangle$ as functions of T.

(d) Compute the heat capacity.

18.15. What is the heat capacity of a particle that sits in an n-dimensional world with a potential energy analogous to that in Example 18.3, that is, $U = ar^2$?

18.16. A one-dimensional harmonic oscillator has a potential energy given by $U = m(2\pi v)^2 x^2 / 2$, where v is its frequency. The corresponding energy eigenvalues that are found using quantum mechanics are $\epsilon = (n + 1/2)hv$, where $n = 0, 1, 2, 3, \ldots$.

(a) Find the quantum partition function q and heat capacity c_V for this system.

(b) What are the limiting values of the heat capacity for $T \to 0$ and $T \to \infty$? At what value of $k_B T/(h\nu)$ does the heat capacity reach 90% of its classical $(T \to \infty)$ value?

(c) In diatomic nitrogen, the bond between the two atoms is well described by a harmonic oscillator with $h\nu/k_B \approx 3{,}300$ K. Calculate the contributions of the bond vibrations to nitrogen's heat capacity at room temperature. Do they behave classically or quantum-mechanically?

18.17. Using the Maxwell–Boltzmann distribution, find an expression for the fluctuations in the kinetic energy, $\sigma_K^2 = \langle K^2 \rangle - \langle K \rangle^2$.

18.18. Consider the Maxwell–Boltzmann distribution of absolute speeds, $\wp(v)$.

(a) Explicitly derive the expression for the average speed $\langle v \rangle$ of Eqn. (18.38).

(b) Find an expression for the most probable speed v in the Maxwell–Boltzmann distribution. How does it compare with the mean $\langle v \rangle$? Does this make sense?

(c) Find an expression for the variance in the speed.

18.19. Prove that Eqn. (18.52) is indeed correct and gives rise to Eqn. (18.3).

18.20. Consider a monatomic ideal gas in a cubic volume V that is subject to a gravitational field. The potential energy experienced by any one particle is equal to mgz, where m is the particle mass, g is the gravitational constant, and z is the height of the particle.

(a) Compute the single-molecule classical canonical partition function, q. Use this result to compute the full classical partition function Q for N molecules.

(b) For a single particle, find the probability distribution for its elevation, $\wp(z)$. Using this result, find the density distribution, $\rho(z)$, for all the particles in the volume, where ρ is in units of molecules per volume.

18.21. A cubic volume V contains N ideal gas particles at temperature T. One wall of the volume is a surface onto which particles can adsorb. A particle is considered adsorbed onto the surface when it is within a distance l from it, wherein it experiences an attractive potential energy $-\epsilon$. As a working simplification, consider all particles to be point masses that are independent of each other. What is the temperature at which 50% of the molecules are adsorbed, on average? One might consider this to be a "desorption temperature."

18.22. Consider a system of N structureless particles of mass m in a cubic volume $V = L^3$. At the boundary interface along the $x = 0$ side of the cube is an attractive wall that draws the particles towards it. The energy of interaction of one particle with the wall is given by $u = \epsilon x$, where x is the x coordinate of the particle and ϵ is a positive constant.

(a) Consider the particles to be non-interacting. Find the per-particle excess entropy in the canonical ensemble. Note that $S_{ex} = S - S_{ig}$, where S_{ig} is that for the ideal gas.

(b) Instead, consider the particles to be weakly repulsive towards one another. Rather than explicitly accounting for the detailed interactions of each pair of molecules, assume that you can describe the potential energy of a particle using a mean-field approximation:

$$u = \epsilon x + \alpha \rho(x)$$

where $\rho(x)$ is the average number density (molecules per volume) as a function of the location in the x direction and α is a positive constant characterizing the interparticle interactions. Here, you are simply assuming that a repulsive energy increases linearly with the number of nearby particles. Find a self-consistent integral equation whose solution determines $\rho(x)$. Hint: find the probability distribution in space of any given particle, which should relate to the mean density.

(c) Another, but ultimately equivalent, method of finding the mean-field density profile is to functionally minimize the free energy of a single particle with respect to $\rho(x)$. Show this and indicate what constraint you would need to impose on the form of $\rho(x)$ during the minimization.

18.23. Consider the virial contribution to the pressure for a system whose molecules interact exclusively through pairwise forces, that is, $U = \sum_{i<j} u(r_{ij})$, where $u(r)$ is a pair potential.

(a) Show that the virial is given by

$$W = -\sum_{i<j} \frac{du(r_{ij})}{dr} r_{ij}$$

Notice that this summation is one involving $N(N-1)/2$ pair terms, rather than the N terms in Eqn. (18.49).

(b) Find an expression for the virial of a fluid described by the Lennard-Jones potential.

(c) For what pair interaction will the virial be equal to the potential, $W = U$?

18.24. "Hard-spheres" (Fig. 18.3) are among the simplest models of solids, liquids, and granular systems, and have a long history in statistical mechanics. They are molecules of diameter σ that interact with the pairwise potential energy function:

$$u(r_{ij}) = \begin{cases} \infty & r_{ij} < \sigma \\ 0 & r_{ij} \geq \sigma \end{cases}$$

In other words, hard spheres experience only an excluded-volume interaction. Note that the total potential energy for a given configuration is given by

$$U = \sum_{i<j} u(|\mathbf{r}_i - \mathbf{r}_j|)$$

where i and j are particle indices. Prove that the excess quantities A_{ex}/T and S_{ex} for hard spheres depend only on the density, independent of temperature. Here, excess indicates the total minus the ideal gas contribution.

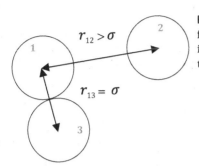

Figure 18.3. Hard spheres are model systems of many fluid and solid states. They have only excluded-volume interactions in that no two particles can come closer than a distance σ apart.

18.25. The term "soft spheres" is used to describe a system of classical particles interacting through a pairwise inverse-power repulsive potential of the form

$$u(r) = cr^{-n}$$

where c is a constant coefficient and n describes the "hardness" of the repulsion: as n grows the potential increases more rapidly at close distances. This model can approach the hard-sphere limit for $n \to \infty$, provided that c also scales exponentially with n. The utility of soft spheres as a model system stems from the fact that their non-dimensionalized excess thermodynamic properties depend only on a single variable that is a combination of volume and temperature, $\phi = V^{n/3}T$. Note that the fundamental equation for excess properties in a single-component system at constant T, V, and N is

$$dA_{ex} = -S_{ex}\, dT - P_{ex}\, dV + \mu_{ex}\, dN$$

where each quantity with subscript "ex" is relative to the ideal gas, e.g., $P_{ex} = P - k_B T/v$.

(a) Show that the non-dimensional excess Helmholtz free energy, $A_{ex}/k_B T$, can be uniquely expressed as a function of ϕ for soft spheres.

(b) The result in part (a) can be summarized as $A_{ex}/k_B T = f(\phi)$. Show then that the soft sphere excess entropy, S_{ex}, also only depends on ϕ.

(c) The exponent of a given soft-sphere system can be found by examining the relationship between T and V along a line of constant S_{ex}, for which ϕ is also constant per the result in part (b). Show that the exponent is given by

$$\frac{n}{3} = \frac{V}{C_{V,ex}}\left(\frac{\partial P_{ex}}{\partial T}\right)_V$$

where the excess heat capacity is $C_{V,ex} \equiv (\partial U/\partial T)_V$.

(d) Starting with the expression for the exponent given in part (c), show that the exponent can also be related to the cross-fluctuations in the potential energy and virial in the canonical ensemble as

$$n = \frac{\langle W \rangle \langle U \rangle - \langle WU \rangle}{\langle U^2 \rangle - \langle U \rangle^2}$$

$$= \frac{-\langle (W - \langle W \rangle)(U - \langle U \rangle)\rangle}{\langle (U - \langle U \rangle)^2 \rangle}$$

$$= \frac{-\langle \Delta W\, \Delta U \rangle}{\langle \Delta U^2 \rangle}$$

where $W \equiv \sum_i \mathbf{f}_i \cdot \mathbf{r}_i$, $\Delta W = W - \langle W \rangle$, and $\Delta U = U - \langle U \rangle$.

18.26. A general virial theorem states that the following canonical average must hold for any microscopic degree of freedom $\mathcal{X}$ that contributes to the system Hamiltonian:

$$\left\langle \mathcal{X} \frac{\partial H}{\partial \mathcal{X}} \right\rangle = k_B T$$

(a) Show that, if $\mathcal{X}$ is a momentum variable, this expression implies that the kinetic energy of a system will be equal to $3Nk_B T/2$.

(b) Show that, if $\mathcal{X}$ is a position variable, this expression implies Eqn. (18.48) for the pressure. Hint: you will need to consider that there are external or boundary forces that keep a system within its volume.

(c) Derive the virial theorem in the canonical ensemble starting with the canonical average and $\wp \propto \exp(-\beta H)$. You will need the following identity:

$$\mathcal{X} \frac{\partial}{\partial \mathcal{X}} e^{-\beta H} = \frac{\partial}{\partial \mathcal{X}} (\mathcal{X} e^{-\beta H}) - e^{-\beta H}$$

18.27. Show that the microcanonical partition function for a monatomic classical system can be written as

$$\Omega(E, V, N) = \frac{1}{h^{3N} N!} \int_{-\infty}^{\infty} \Omega_{\text{kinetic}}(E - \epsilon, V, N) \Omega_{\text{config}}(\epsilon, V, N) d\epsilon$$

where the convolution integral involves separate density of states functions for the kinetic and configurational degrees of freedom,

$$\Omega_{\text{kinetic}}(E, V, N) = \int \delta[K(\mathbf{p}^N) - E] d\mathbf{p}^N$$

$$\Omega_{\text{config}}(E, V, N) = \int \delta[U(\mathbf{r}^N) - E] d\mathbf{r}^N$$

The kinetic contribution can be evaluated analytically by transforming the integral above into hyperspherical coordinates so that the delta function can be treated. Show for large N that the result is

$$\Omega_{\text{kinetic}}(E, V, N) = (2\pi m)^{3N/2} E^{3N/2} \Gamma(3N/2)^{-1}$$

where Γ is the gamma function. With these results, you can show that another way to express the classical microcanonical partition function is

$$\Omega(E, V, N) = \frac{1}{N! \Gamma(3N/2)} \left(\frac{2\pi m}{h^2} \right)^{3N/2} \int [E - U(\mathbf{r}^N)]^{3N/2} \theta[E - U(\mathbf{r}^N)] d\mathbf{r}^N$$

where θ is the Heaviside step function.

Applied problems

18.28. In this problem, you will derive the famous van der Waals equation of state. This fluid model makes the following approximations to evaluate the configurational integral: (1) a particle is not influenced by other particles in a detailed way, but instead sees an average potential energy field due to a random distribution of the other particles in the volume; (2) particles do not interact in a detailed way, but through a mean field and can be considered independent of each other; and (3) particles cannot come closer to each other than a center-to-center distance of σ, the diameter of one particle.

(a) From a central van der Waals atom, show that the average number of particles that lie within a differential region dr a distance r away from it is given by $4\pi \rho r^2\, dr$.

(b) Consider that the central particle interacts in a pairwise fashion with others through a van der Waals potential energy function, $u(r) = -Cr^{-6}$, where C is a constant and r is the distance separating the atom centers, for $r > \sigma$. Compute the average potential energy ϵ experienced by the central particle interacting with a random distribution of other particles around it. In turn, show that the total potential energy of the system is given by

$$E = -\frac{4\pi C\rho N}{6\sigma^3}$$

(c) Consider the excluded volume per particle, the average amount of space a particle prevents others from occupying. If one particle is present, a volume of $4\pi\sigma^3/3$ is excluded from the locations where a second particle could be placed. Thus one might approximate on the basis of this simple two-particle consideration that the average excluded volume per particle is $b = 2\pi\sigma^3/3$. With this approximation, show that the canonical partition function for the system is

$$Q = \frac{(V - Nb)^N e^{-\beta N\epsilon/2}}{\Lambda(T)^{3N} N!}$$

(d) Show that the equation of state for this system is the following, with $a \equiv 2\pi C/(3\sigma^3)$ and v equal to the volume per particle:

$$P = \frac{k_B T}{v - b} - \frac{a}{v^2}$$

(e) What would happen if a different attractive potential had been chosen? It turns out that not all energy functions will work with the assumption of a random particle distribution. Particles that interact through pairwise energy functions of the form $u(r) = Cr^{-n}$, where $n \le 3$, cannot be treated as in part (b) and require special consideration. Why is this so?

18.29. The Lindemann criterion is an empirical hypothesis that states that a crystal will melt when the root-mean-squared fluctuation in the distance of atoms from their lattice sites reaches a certain fraction f of the lattice spacing, a. Consider a (classical) monatomic crystal on a cubic lattice.

(a) Assume that each particle is independent of its neighbors, and sits in a harmonic potential energy well around the lattice site, $u(\Delta x, \Delta y, \Delta z) = \alpha(\Delta x^2 + \Delta y^2 + \Delta z^2)$, where α is a constant. Find the probability $\wp(r)$ that the particle is a distance r away from its lattice site. Assume that the quantity $\alpha\beta a^2$ is very large such that $\exp(-\alpha\beta a^2) \approx \exp(-\infty)$.

(b) Find the average $\langle r^2 \rangle$. In turn, show that the melting temperature according to the Lindemann criterion is given by $T_m = 2\alpha a^2 f^2/(3k_B)$.

(c) If the atoms act through a repulsive pairwise potential of the form $u(r) = Cr^{-n}$, it can be shown that the pressure dependence of α must be $\alpha(P) = bP^m$, where b and m are constants. Show that such a case implies that the entropy of melting is given by

$$\Delta s = \frac{1}{m} \frac{P\,\Delta v}{T_m(P)}$$

where Δv is the volume of melting.

18.30. Consider a classical ideal gas of point dipoles in three dimensions. The potential energy of a dipole of magnitude p in an external electric field of strength $\mathcal{E}$ is given by $U(\theta) = -p\mathcal{E}\cos\theta$, where θ is the angle the dipole makes with respect to the field direction (assume for simplicity that it is along the z-axis). In general, each dipole has five configurational degrees of freedom: three positional (x, y, and z) and two rotational (θ and ϕ).

(a) A system is placed under conditions at which T, V, N, and $\mathcal{E}$ are constant. Find the single-particle partition function q and the heat capacity c_V of this gas. Hint: a clever variable transform will allow you to evaluate the integral.

(b) Find the probability distribution $\wp(\theta)$ for any given molecule.

(c) Starting from $\mathcal{E} = 0$, a small electric field is applied slowly and adiabatically at constant volume conditions. Outline a way that you could find an expression $\Delta T = \dots$ for the temperature change of the gas given the single-particle partition function $q(T, V, \mathcal{E})$ and the heat capacity $c_V(T, V, \mathcal{E})$, both of which were solved in part (a).

18.31. Supercapacitors are of great interest as high-density energy-storage devices. A basic design is shown in Fig. 18.4. Two fixed parallel plates carry equal and opposite net surface charges $+Q$ and $-Q$ and are separated by a net-neutral sea of mobile ions. Assume that the ions are monatomic, monovalent, all of equal mass, and can move within a confined volume $V = \mathcal{A}L$ (see Fig. 18.4). The total potential energy of the system is given by

$$U = \frac{Q^2 L}{2\epsilon\mathcal{A}} - \frac{Qp(\mathbf{r}^{2N})}{\epsilon\mathcal{A}} + U_{\text{ion}-\text{ion}}(\mathbf{r}^{2N})$$

where ϵ is the dielectric constant, $p = \sum_i q_i x_i$ is the instantaneous net mobile ion dipole, q_i is the charge of ion i, and x_i is its x position. In this expression, the three terms account for plate–plate, plate–ion, and ion–ion interactions, respectively.

A voltmeter connected between the two plates reads a corresponding electric-potential drop ψ with units of energy per charge. The work associated with setting up the charges on the two plates, and hence the energy stored, is given by $\delta W = \psi\, dQ$.

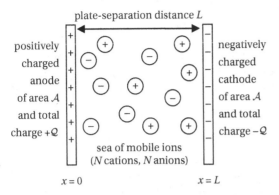

plate-separation distance L

positively charged anode of area A and total charge $+Q$

negatively charged cathode of area A and total charge $-Q$

sea of mobile ions
(N cations, N anions)

$x = 0$

$x = L$

Figure 18.4. A simple capacitor involves a sea of mobile ions of equal and opposite charges confined between two oppositely charged plates.

The differential capacitance of the system is $C \equiv (\partial Q / \partial \psi)_{T, V, N}$. Show that the inverse of C is given by the following, where σ_p^2 gives the fluctuations in the net mobile ion dipole p:

$$C^{-1} = \frac{L}{\epsilon A} \left(1 - \frac{\sigma_p^2}{\epsilon k_B T A L} \right)$$

18.32. N molecules of mass m of a dilute, monatomic gas exist in a vertically oriented piston. The gas is maintained at constant temperature T through coupling to a heat bath, and it is further subject to a strong gravitational field g. The piston is then compressed such that the height of the container is changed from L_1 to $L_2 < L_1$.

(a) Find an expression that gives the minimum work per molecule of gas for this process.

(b) Find an expression for the per-particle energy change Δe for this process.

18.33. The *Widom insertion* technique is frequently used to compute the chemical potential of a system in molecular simulations. It relies on the following theoretical result, shown here for single-component systems at constant T, V, N conditions:

$$\mu = \mu_{ig} - k_B T \ln \langle e^{-\beta \Delta U} \rangle$$

where μ_{ig} is the chemical potential of an ideal gas at the same temperature and density, and ΔU is the change in potential energy upon inserting one additional particle at a completely random location. The angle brackets denote canonical averaging.

(a) Derive this result. You may want to begin with

$$\mu = (\partial A / \partial N)_{T, V} \approx A(T, V, N + 1) - A(T, V, N)$$

which is valid for large N. You will also want to make the two configurational integrals have the same number of integrand variables, perhaps by introducing a factor of

$$\frac{1}{V}\int d\mathbf{r} = 1$$

(b) The same result can be applied to a dilute species dissolved in a solvent to provide a molecular derivation of Henry's law. For example, it can be shown that the solubility of a gas in a liquid solvent follows the relation

$$x_i = \frac{y_i P}{K}, \text{ where } K = \frac{k_B T \rho_{\text{solvent}}}{\langle e^{-\beta \Delta U} \rangle}$$

Derive this result. What will be the nature and sign of the term $\beta \, \Delta U$ in the exponential for small concentrations of low-solubility dissolved gas? What about for high-solubility cases? Hint: begin by considering a gas phase in equilibrium with a liquid.

FURTHER READING

D. Chandler, *Introduction to Modern Statistical Mechanics*, New York: Oxford University Press (1987).

K. Dill and S. Bromberg, *Molecular Driving Forces: Statistical Thermodynamics in Biology, Chemistry, Physics, and Nanoscience*, 2nd edn. New York: Garland Science (2010).

T. L. Hill, *An Introduction to Statistical Thermodynamics*. Reading, MA: Addison-Wesley (1960); New York: Dover (1986).

T. L. Hill, *Statistical Mechanics: Principles and Selected Applications*. New York: McGraw-Hill (1956); New York: Dover (1987).

E. A. Jackson, *Equilibrium Statistical Mechanics*. Mineola, NY: Dover (1968).

D. A. McQuarrie, *Statistical Mechanics*. Sausalito, CA: University Science Books (2000).

19 Other ensembles

19.1 The isothermal–isobaric ensemble

In Chapters 16–18, we derived the microstate probabilities for constant-temperature systems using heat baths. The same approach can be used to understand the behavior of systems at other constant conditions. We begin by finding the specific form of these probabilities at both constant temperature and pressure, when the system is coupled to an infinitely large energy and volume bath. However, our approach will emphasize general features of partition functions and ensembles that are applicable to arbitrary constant conditions and baths.

Conceptually, the isothermal–isobaric derivation proceeds in a similar fashion as the approach to the canonical ensemble. In this case, however, the boundaries of the system fluctuate as it exchanges volume with the bath. We consider the system plus bath to constitute an isolated microcanonical supersystem of constant total energy and volume. The probability of a microstate m in the system is given by the number of times a combined microstate ml appears, where l is any microstate of the bath. Each combined microstate appears with equal frequency according to the law of equal a priori probabilities. Thus the likelihood that the system is in m regardless of the state of the bath is proportional to the number of bath microstates compatible with m's energy and volume,

$$\wp_m = \sum_l \wp_{ml}$$
$$\propto \Omega_B(E_B, V_B)$$
$$\propto \Omega_B(E_T - E_m, V_T - V_m)$$
$$\propto \exp\left[\frac{S_B(E_T - E_m, V_T - V_m)}{k_B}\right] \tag{19.1}$$

where V_T and E_T are the total constant volume and energy of the supersystem. Note here that a microstate m in the system is characterized both by a specific configuration (and hence energy) and by a value of the system volume. We now Taylor-expand the bath entropy in both E_B and V_B and take the limit in which the bath is infinitely large,

$$\wp_m \propto \exp\left[\frac{S_B(E_B, V_B)}{k_B} - \frac{E_m}{k_B}\frac{\partial S_B(E_B, V_B)}{\partial E_B} - \frac{V_m}{k_B}\frac{\partial S_B(E_B, V_B)}{\partial V_B} + \text{terms of order} \leq 1/N_B\right]$$
$$\tag{19.2}$$

We can absorb the first term in the exponential into the constant of proportionality since it is independent of the microstate m. We can also simplify the bath entropy derivatives by recognizing their relationship to the bath temperature and pressure, which we simply denote by T and P without subscripts. Our final expression for the probability of a microstate in the system is

$$
\begin{aligned}
\wp_m &\propto \exp\left[-\frac{E_m}{k_{\mathrm{B}}T} - \frac{PV_m}{k_{\mathrm{B}}T}\right] \\
&\propto \exp\left[-\beta E_m - \beta PV_m\right]
\end{aligned}
\tag{19.3}
$$

Relative to the canonical ensemble, we see that we obtain an additional term in the Boltzmann factor involving the pressure P (specified by the bath) and the volume associated with microstate m. The term PV_m might be viewed as an additional work term that contributes to the total energy of microstate m. Because the probabilities of the system microstates must sum to one, the constant of proportionality can be determined exactly,

$$
\wp_m = \frac{e^{-\beta E_m - \beta PV_m}}{\Delta(T, P, N)}
\tag{19.4}
$$

$$
\Delta(T, P, N) \equiv \sum_V \sum_{\text{all } n \text{ at } V, N} e^{-\beta E_n - \beta PV}
\tag{19.5}
$$

Here, we introduce the *isothermal–isobaric partition* function Δ that serves as a normalizing factor for the microstate probabilities. Notice that Δ involves two sums, the first of which spans all possible volumes of the system, from zero to infinity. The second sum is nested within the first and enumerates all microstates at a given value of V and N. As a result, Δ is a function only of T, P, and N. Intuitively, it should no longer be a function of V because now the system volume can fluctuate by exchange with the bath.

The particular form of Δ in Eqn. (19.5) allows us to rewrite it in terms of other partition functions. We can factor out the pressure term,

$$
\Delta(T, P, N) = \sum_V e^{-\beta PV} \sum_{\text{all } n \text{ at } V, N} e^{-\beta E_n}
\tag{19.6}
$$

The second sum is the canonical partition function, such that

$$
\Delta(T, P, N) = \sum_V e^{-\beta PV} Q(T, V, N)
\tag{19.7}
$$

From Chapter 17, we also know that Q can be expressed in terms of an energy sum involving the microcanonical partition function, Ω. This leads to

$$
\Delta(T, P, N) = \sum_V \sum_E e^{-\beta E - \beta PV} \Omega(E, V, N)
\tag{19.8}
$$

Equation (19.8) shows that we can find the isothermal–isobaric partition function from the density of states using a sum over variables that fluctuate by exchange with a bath. This turns out to be a rather general principle.

The microcanonical partition function Ω can be transformed into other partition functions by summing over the macroscopic variables that are allowed to fluctuate through coupling to a bath. The weight of each term in the sum is given by a Boltzmann factor that includes the fluctuating variables. Each fluctuating variable X in the Boltzmann exponential is multiplied by $-1/k_B$ times its conjugate in the entropy function $(\partial S/\partial X)$.

This principle clearly succeeds in describing (19.8): it suggests that Δ would involve a sum over both energies and volumes, and that the Boltzmann exponential would contain $-E(\partial S/\partial E) - V(\partial S/\partial V) = -E/T - PV/T$ divided by k_B. The reason the principle works is actually quite simple. When one couples a system to a bath, one always obtains $\wp_m \propto \Omega_B$, and the Taylor expansion of the bath entropy naturally leads to these kinds of variable combinations.

The microstate distribution in Eqn. (19.4) gives the probability that the system will have a volume V and a particular configuration of particles with total energy E. If we are interested in the macroscopic distribution of energy and volume, we must sum the probabilities in (19.4) over all microstates that are consistent with particular values of E and V. There happen to be $\Omega(E, V, N)$ of them:

$$\wp(E, V) = \frac{\Omega(E, V, N)e^{-\beta E - \beta PV}}{\Delta(T, P, N)} \qquad (19.9)$$

This leads to another general principle.

In any ensemble, the joint probability distribution for the macroscopic parameters that fluctuate is related to the microscopic Boltzmann factor multiplied by the density of states Ω.

If we are interested in the volume probability distribution alone, $\wp(V)$, we can sum over the energies in the joint distribution,

$$\wp(V) = \sum_E \wp(E, V)$$

$$= \frac{e^{-\beta PV}}{\Delta(T, P, N)} \sum_E \Omega(E, V, N)e^{-\beta E}$$

$$= \frac{e^{-\beta PV} Q(T, V, N)}{\Delta(T, P, N)} \qquad (19.10)$$

In the last line, we were able to show that $\wp(V)$ is related to the canonical partition function Q weighted by a Boltzmann factor. Importantly, just as in the canonical ensemble, the macroscopic distributions of energy and volume are sharply peaked for large systems. Thus, even though the volume and energy can fluctuate in the isothermal–isobaric ensemble, these fluctuations are too small to observe in macroscopic systems.

One might anticipate on the basis that Δ is a function only of T, P, and N that it is related to the macroscopic thermodynamic potential $G(T, P, N)$. This is indeed the case, and the maximum-term approximation makes this clear. $\wp(E, V)$ of Eqn. (19.9) is so sharply peaked that we can approximate the partition sum of (19.8) with a single value of the summand,

$$\Delta(T, P, N) = \sum_E \sum_V e^{-\beta E - \beta PV} \Omega(E, V, N)$$

$$= \sum_E \sum_V \exp[-\beta E - \beta PV + S(E, V, N)/k_B]$$

$$\approx \exp[-\beta E^* - \beta PV^* + S(E^*, V^*, N)/k_B] \tag{19.11}$$

where E^* and V^* are the values corresponding to the maximum term in the sum, which is determined by setting the E and V derivatives in the exponential equal to zero,

$$\left(\frac{\partial S(E^*, V^*, N)}{\partial E}\right)_{V,N} = \frac{1}{T}, \quad \left(\frac{\partial S(E^*, V^*, N)}{\partial V}\right)_{E,N} = \frac{P}{T} \tag{19.12}$$

We can rewrite the maximum-term approximation as

$$k_B \ln \Delta(T, P, N) = -E^*/T - PV^*/T + S(E^*, V^*, N) \tag{19.13}$$

With the conditions for E^* and V^* above, we note that the RHS gives the Legendre transform of the entropy with respect to both energy and volume. On rearranging, we find

$$-k_B T \ln \Delta(T, P, N) = E^* + PV^* - TS \tag{19.14}$$

The RHS of (19.14) shows that Δ relates to the macroscopic Gibbs free energy, at least when fluctuations are neglected. The relationship is actually more general, and we can establish that indeed

$$G(T, P, N) = -k_B T \ln \Delta(T, P, N) \tag{19.15}$$

Considering the results we have seen so far for the microcanonical, canonical, and isothermal–isobaric ensembles, we might anticipate another general principle in statistical mechanics.

> The logarithm of the partition function for any ensemble, times k_B, gives a macroscopic thermodynamic potential that is a Legendre transform of the entropy. The transform is performed with respect to the fluctuating macroscopic variables in the ensemble, i.e., those that are coupled to a bath.

For classical systems, we write the isothermal–isobaric partition function by converting the sums to integrals in Eqn. (19.5). We also introduce the quantum correction factor as we did in Chapter 18,

$$\Delta(T, P, N) = \int_0^\infty e^{-\beta PV} \left\{ \frac{1}{h^{3N} N!} \int e^{-\beta H(\mathbf{p}^N, \mathbf{r}^N)} \, d\mathbf{p}^N \, d\mathbf{r}^N \right\} dV \tag{19.16}$$

The outer integral is performed over V and thus determines the limits of the $3N$ inner integrals over the position variables. We can also express Eqn. (19.16) in terms of the classical configurational integral,

$$\Delta(T, P, N) = \int_0^\infty e^{-\beta PV} Q(T, V, N)dV$$

$$= \frac{1}{\Lambda(T)^{3N} N!} \int_0^\infty e^{-\beta PV} Z(T, V, N)dV \qquad (19.17)$$

In the second line, Q could be separated into its kinetic and configurational parts. Equation (19.17) indicates that, if we can compute the configurational integral Z and its V dependence, we can transform it to the isothermal–isobaric partition function, and hence to the Gibbs free energy.

In fact, there is a subtlety with the classical partition function in (19.17): it has dimensions of volume due to the units of the integrand variable. There are several ways to treat this issue, which generally introduce a correction prefactor that has units of inverse volume, but we will not address this here. For macroscopic systems, these prefactors are small relative to the other terms and make a vanishing contribution to the Gibbs free energy in the limit that the number of particles grows large. Small systems, however, do require careful treatment of these factors.

Example 19.1   *Find an expression for the average volume $\langle V \rangle$ in the isothermal–isobaric ensemble.*

Equation (19.10) gives a convenient starting point,

$$\langle V \rangle = \sum_V V \wp(V) = \frac{1}{\Delta} \sum_V V e^{-\beta PV} Q(T, V, N)$$

where the independent variables of Δ have been suppressed for clarity. The sum on the RHS can be expressed as a pressure derivative,

$$\langle V \rangle = \frac{-k_B T}{\Delta} \frac{\partial}{\partial P} \sum_V e^{-\beta PV} Q(T, V, N)$$

$$= -\frac{k_B T}{\Delta} \left(\frac{\partial \Delta}{\partial P} \right)_{T, N}$$

$$= -k_B T \left(\frac{\partial \ln \Delta}{\partial P} \right)_{T, N}$$

With Eqn. (19.15) we immediately recognize this derivative as $(\partial G/\partial P)_{T,N}$, which we would have expected on macroscopic thermodynamic grounds.

Example 19.2   *Find the isothermal–isobaric partition function for the ideal monatomic gas.*

We begin with the result from Chapter 18 that $Z(T, V, N) = V^N$ for an ideal gas. Insertion into Eqn. (19.17) then gives

$$\Delta(T, P, N) = \frac{1}{\Lambda(T)^{3N} N!} \int_0^\infty e^{-\beta P V} V^N \, dV$$

Let us substitute $x = \beta P V$,

$$\Delta(T, P, N) = \frac{(\beta P)^{-(N+1)}}{\Lambda(T)^{3N} N!} \int_0^\infty e^{-x} x^N \, dx$$

The integral evaluates to $N!$, and we can also make the approximation $N + 1 \approx N$ for the exponent on βP. We find

$$\Delta(T, P, N) = \frac{(\beta P)^{-(N+1)}}{\Lambda(T)^{3N} N!} N!$$

$$\approx [\beta P \Lambda(T)^3]^{-N}$$

We now construct the Gibbs free energy per particle, which is equivalent to the chemical potential for the single-component system,

$$\mu(T, P) = G/N$$

$$= -\frac{k_B T}{N} \ln \Delta(T, P, N)$$

$$= k_B T \ln \left(\frac{\Lambda(T)^3}{k_B T} \right) + k_B T \ln P$$

$$= \mu^\circ(T) + k_B T \ln P$$

This is precisely the expression for the ideal gas chemical potential that we determined in Chapter 9.

19.2   The grand canonical ensemble

The *grand canonical ensemble* results when a system at fixed volume is coupled to a bath with which it can exchange particles and energy. It is "grand" in the sense that it encompasses fluctuations in particle number; that is, it is an open system. As a result of its coupling to the bath, this ensemble is at constant temperature and chemical potential.

The derivation of ensemble probabilities proceeds almost identically to the procedure for the isothermal–isobaric ensemble, only replacing the fluctuating V with N,

$$\wp_m \propto \exp\left[\frac{S_{\mathrm{B}}(E_{\mathrm{T}} - E_m, N_{\mathrm{T}} - N_m)}{k_{\mathrm{B}}}\right] \tag{19.18}$$

In order to Taylor-expand the entropy, we assume that we can write it as a continuous function of the number of particles, even though we know that this is a discrete variable in reality. Since the bath is infinitely large, this assumption is reasonable:

$$\wp_m \propto \exp\left[\frac{S_{\mathrm{B}}(E_{\mathrm{B}}, N_{\mathrm{B}})}{k_{\mathrm{B}}} - \frac{E_m}{k_{\mathrm{B}}}\frac{\partial S_{\mathrm{B}}(E_{\mathrm{B}}, N_{\mathrm{B}})}{\partial E_{\mathrm{B}}} - \frac{N_m}{k_{\mathrm{B}}}\frac{\partial S_{\mathrm{B}}(E_{\mathrm{B}}, N_{\mathrm{B}})}{\partial N_{\mathrm{B}}} + \text{terms of order} \leq 1/V_{\mathrm{B}}\right] \tag{19.19}$$

Here we let the bath become ideal by allowing its volume to grow to infinity since its particle number is not fixed due to the exchange. On substituting for the entropy derivatives and introducing the normalizing factor, we arrive at the following microstate probabilities:

$$\wp_m = \frac{e^{-\beta E_m + \beta \mu N_m}}{\Xi(T, V, \mu)} \tag{19.20}$$

where

$$\Xi(T, V, \mu) \equiv \sum_{N}\sum_{\text{all } n \text{ at } V, N} e^{-\beta E_n + \beta \mu N} \tag{19.21}$$

$\Xi(T, V, \mu)$ is the grand canonical partition function, and it involves two nested sums. The first proceeds over numbers of particles in the system, spanning from zero to infinity. The second is nested within the first, and it is a sum over all microstates for a given V and N. Ξ is a function only of T, V, and μ, since the bath specifies the temperature and the chemical potential.

Just like the isothermal–isobaric partition function, we can express Ξ in several other ways through Laplace-like transforms of other partition functions:

$$\Xi(T, V, \mu) = \sum_{N} e^{\beta \mu N} Q(T, V, N)$$

$$= \sum_{N}\sum_{E} e^{-\beta E + \beta \mu N} \Omega(E, V, N) \tag{19.22}$$

It is often useful to introduce a quantity $\lambda \equiv \exp[\beta \mu]$. This is sometimes called the *absolute activity* or *absolute fugacity* since the total chemical potential is given by $k_{\mathrm{B}}T \ln \lambda$, which bears a resemblance to chemical-potential expressions for nonideal gases and solutions. With this new variable, the grand partition function becomes

$$\Xi(T, V, \mu) = \sum_{N} \lambda^N Q(T, V, N), \quad \text{where} \quad \lambda \equiv \exp[\beta \mu] \tag{19.23}$$

An explicit representation of the sum shows that it involves increasing particle numbers, including the case $N = 0$,

$$\Xi(T, V, \mu) = 1 + \lambda Q(T, V, 1) + \lambda^2 Q(T, V, 2) + \cdots \tag{19.24}$$

The joint distribution of energies and particle numbers follows the microstate probabilities in Eqn. (19.20),

$$\wp(E, N) = \frac{\Omega(E, V, N)e^{-\beta E + \beta \mu N}}{\Xi(T, V, \mu)} = \frac{\Omega(E, V, N)e^{-\beta E}\lambda^N}{\Xi(T, V, \mu)} \tag{19.25}$$

The particle distribution alone is found by summing (19.25) over energies,

$$\wp(N) = \frac{e^{\beta \mu N} Q(T, V, N)}{\Xi(T, V, \mu)} = \frac{\lambda^N Q(T, V, N)}{\Xi(T, V, \mu)} \tag{19.26}$$

For classical systems, the particle number is naturally discrete, so we can simply write the total grand canonical partition function as

$$\Xi(T, V, \mu) = \sum_N \frac{\lambda^N Z(T, V, N)}{\Lambda(T)^{3N} N!} \tag{19.27}$$

To which thermodynamic potential does $\Xi(T, V, \mu)$ relate? Recall the general principle from our previous discussion: the relevant thermodynamic potential corresponds to a Legendre transform of the entropy with respect to the fluctuating quantities. Here both the energy and the number of particles fluctuate, and we construct a potential Y such that

$$Y = \mathcal{L}_{E,N}[S(E, V, N)]$$

$$= S - E\left(\frac{\partial S}{\partial E}\right) - N\left(\frac{\partial S}{\partial N}\right)$$

$$= S - \frac{E}{T} + \frac{\mu N}{T} \tag{19.28}$$

On substituting the thermodynamic identity $E = TS - PV + \mu N$, we find that this Legendre transform gives

$$Y = PV/T \tag{19.29}$$

The potential Y is not a named thermodynamic potential that we have considered before, but it is conceptually no different. Since the Legendre transform swapped E for T and N for μ, Y has independent variables such that the complete function is $Y(T, V, \mu)$. This, of course, would be expected because a system coupled to a particle and energy bath would have these particular constant conditions. On the basis of our earlier discussion, we therefore find

$$Y(T, V, \mu) = k_B \ln \Xi(T, V, \mu) \tag{19.30}$$

On substituting the integrated form for Y, we arrive at

$$PV = k_B T \ln \Xi(T, V, \mu) \tag{19.31}$$

The grand canonical partition function enables us to compute the pressure as a function of T, V, and μ. Moreover, because the thermodynamic potential Y is fundamental, we can extract all other macroscopic thermodynamic properties from its derivatives, just as is the case for the partition functions in other ensembles.

The grand canonical ensemble and partition function are very important constructs in statistical mechanics. The reason lies in the fact that the sum over particle numbers in Ξ enables one to express the thermodynamic properties as a series of increasingly complicated configurational integrals,

$$\Xi(T, V, \mu) = 1 + \frac{\lambda Z(T, V, 1)}{\Lambda(T)^3} + \frac{\lambda^2 Z(T, V, 2)}{\Lambda(T)^6 2!} + \frac{\lambda^3 Z(T, V, 3)}{\Lambda(T)^9 3!} + \cdots \qquad (19.32)$$

The dimensionality of a configurational integral $Z(T, V, N)$ is $3N$, which in general makes it very difficult to evaluate the integrals analytically for large particle numbers. With the expansion above, it is possible to compute configurational integrals for small numbers of particles and then drop higher-order terms. Such approximations typically work well for gases and dilute solutions. The grand canonical approach is also the basis of many other statistical-mechanical approximations and theories, including so-called field-based ones; the interested reader is referred to specialized texts such as those listed in the references.

Example 19.3 *Find the grand canonical partition function for the ideal monatomic gas.*
The configurational partition function for an ideal gas is $Z(T, V, N) = V^N$. Substitution into (19.27) gives

$$\Xi(T, V, \mu) = \sum_{N=0}^{\infty} \frac{\lambda^N V^N}{\Lambda(T)^{3N} N!} = \sum_{N=0}^{\infty} \left(\frac{\lambda V}{\Lambda(T)^3} \right)^N \frac{1}{N!}$$

We can use the Taylor expansion of the exponential term to compute the sum. Namely, we have

$$e^x = \sum_{n=0}^{\infty} \frac{x^n}{n!}$$

In this case $x = \lambda V / \Lambda^3$, so the grand partition function evaluates to

$$\Xi(T, V, \mu) = \exp\left[\frac{\lambda V}{\Lambda(T)^3} \right]$$

Using Eqn. (19.31), we see that

$$PV = k_B T \ln \Xi(T, V, \mu) = k_B T \frac{\lambda V}{\Lambda(T)^3}$$

Alternatively,

$$P = \frac{k_B T}{\Lambda(T)^3} e^{\beta\mu} = e^{\beta\mu - \beta\mu^\circ(T)}$$

This is a simple inversion (to solve for P) of the function $\mu(T, P)$ for the monatomic ideal gas that we found in Chapter 9.

19.3   Generalities and the Gibbs entropy formula

There exist many different kinds of ensembles other than the four presented thus far. In multicomponent systems, for example, it is possible to have fluctuations in one, some, or all of the particle numbers. There may be other macroscopic quantities that can fluctuate by exchange with a bath beyond E, V, and N. Examples include an area, length, charge, or any other extensive quantity relevant to a reversible work term in the fundamental equation. Clearly the combination of different baths allows many possible ensembles.

For all of the ensembles studied thus far, we can generalize the form of the partition function. Consider a system with macroscopic quantities X_1 and X_2 that can fluctuate by coupling to a bath. Let us designate y_1 and y_2 as the conjugate variables to these quantities that are found by differentiation of the entropy. For example, they would be $1/T$, P/T, and $-\mu/T$ for E, V, and N, respectively. The y terms are the constant conditions that the bath establishes. With this notation in mind, we can write any generic partition function Υ with fluctuating variables X_1 and X_2 as

$$\Upsilon(y_1, y_2, X_3) = \sum_{X_1} \sum_{X_2} \exp\left[-\frac{y_1 X_1 + y_2 X_2}{k_B}\right] \Omega(X_1, X_2, X_3) \qquad (19.33)$$

where Ω is the microcanonical partition function. If we allowed only one fluctuating quantity, there would be only one sum and one term in the exponential. Similarly, if we had more fluctuating quantities (more X terms), there would be additional sums and exponential terms. Using the same notation, we can also write the probability of any one microstate,

$$\wp_m = \frac{1}{\Upsilon(y_1, y_2, X_3)} \exp\left[-\frac{y_1 X_{1,m} + y_2 X_{2,m}}{k_B}\right] \qquad (19.34)$$

where the values of the X terms in the exponential depend on the particular microstate m. At a macroscopic level, the thermodynamic potential to which $\wp$ relates is given by a Legendre transform of the entropy,

$$k_B \ln \Upsilon(y_1, y_2, X_3) = S - y_1 X_1 - y_2 X_2 \qquad (19.35)$$

It turns out that in any ensemble we can represent the entropy by the famous Gibbs formula,

$$S = -k_B \sum_m \wp_m \ln \wp_m \qquad \mathbf{(19.36)}$$

where the sum is over all microstates of the system. Here we will prove that using the general notation above. We start by substituting the microstate probability from (19.34),

$$S = -\sum_m \wp_m[-y_1 X_{1,m} - y_2 X_{2,m} - k_B \ln \Upsilon(y_1, y_2, X_3)] \qquad (19.37)$$

Note that the y terms are constant since they are properties established by the bath. Therefore, we can rewrite the entropy as

$$S = y_1\langle X_1\rangle + y_2\langle X_2\rangle + k_B \ln \Upsilon(y_1, y_2, X_3)\tag{19.38}$$

Here we could factor Υ out of the sum since it is constant as well. On noting that the average values of the X terms are just the macroscopic values, and using the expression in Eqn. (19.35) for $\ln \Upsilon$, we find that

$$\begin{aligned}
S &= y_1\langle X_1\rangle + y_2\langle X_2\rangle + S - y_1\langle X_1\rangle - y_2\langle X_2\rangle\\
&= S\tag{19.39}
\end{aligned}$$

which verifies Eqn. (19.36).

> For any statistical-mechanical ensemble, the entropy is given by the **Gibbs entropy formula,**
>
> $$S = -k_B \sum_m \wp_m \ln \wp_m$$
>
> where the sum is over all microstates of the system.

Interestingly, the Gibbs version of the entropy has deep connections to information theory, a broad area that is concerned with the quantification of information content and the signal requirements for communicating it. In fact, a nearly identical formula due to Claude Shannon in 1948 is used to quantify the uncertainty in the information contained in a message. In retrospect, the connection is not too surprising. The Boltzmann version of the entropy, $S = k_B \ln \Omega$, is in some sense a quantification of "information loss." That is, the density of states tells us how much we do not know about the microscopic state of an isolated system: the larger its value, the greater the number of possible microscopic states available and hence the more significant our naïveté from a macroscopic perspective.

Many texts take the Gibbs entropy formula as a basic starting point for developing ensemble theory and thermodynamics itself. This book began with the Boltzmann formulation since it offers a more mechanistic view of the microscopic world that does not require an understanding of information theory, but the interested reader is encouraged to explore the profound connections between the two approaches on his or her own.

Problems

Conceptual and thought problems

19.1. For a single-component system, is there such a thing as a partition function corresponding to a system that can exchange energy, volume, and particles with a bath? What is the value of this partition function, and what is the nature of this ensemble?

19.2. Consider a two-component system of particles of types A and B. The system is held at constant T and P. In addition, the system is coupled to an infinite reservoir of A, which can diffuse into or out of the system (e.g., via a membrane), but the amount of B is fixed. Write down expressions for the following quantities:

(a) the partition function for this ensemble, in terms of the microcanonical one $\Omega(E, V, N_A, N_B)$;

(b) the thermodynamic potential to which (a) relates; and

(c) the microstate probability distribution, $\wp_m$.

19.3. A rigid container contains components A, B, and C held at constant T, but allows only the latter two species to exchange with an infinite reservoir of the components through a membrane. In this reservoir, a catalyst that enables the reaction $B \leftrightarrow C$ is present. Find an expression for the partition function for this ensemble, in terms of the microcanonical one, and the thermodynamic potential to which it relates.

19.4. For what sized cubic volume will the density fluctuations in an ideal gas amount to 10% of the average density? Consider the two cases $P = 1\,atm$ and $P = 0.001\,atm$ at $T = 0\,°C$. The latter conditions are typical of those in the upper atmosphere, where density fluctuations occur on length scales comparable to the wavelength of light, scattering it. Because the scattering has a larger effect on light with short wavelengths, the sky often appears blue.

19.5. In the so-called *semi-grand canonical ensemble*, the total number of molecules is fixed but molecules can freely change type such that the composition can vary. This is particularly useful in molecular simulation studies. Consider a semi-grand system of components A and B in which the two species can freely interconvert.

(a) The system is at constant T, V, and $N = N_A + N_B$ conditions, although N_A and N_B can vary as one particle type transforms into another. There is one additional combination of variables that must be specified as constant in this ensemble. What is it?

(b) Find an expression for the partition function for this ensemble in terms of the microcanonical one $\Omega(E, V, N_A, N_B)$, and give the thermodynamic potential to which it relates.

Fundamentals problems

19.6. Explicitly derive the limiting forms of Eqns. (19.2) and (19.19) for idealized, infinite-sized baths.

19.7. Find an expression for the volume distribution $\wp(V)$ in the isothermal–isobaric ensemble for the ideal gas. What is the probability for observing a volume fluctuation 0.0001% away from the mean for a mole of gas at standard conditions?

19.8. Find expressions for the average enthalpy $\langle H \rangle$, enthalpy fluctuations σ_H^2, and constant-pressure heat capacity C_P in terms of the isothermal–isobaric partition function.

19.9. Consider a single-component system held at constant T and P in the isothermal-isobaric ensemble.

(a) Show how the compressibility κ_T relates to the average volume $\langle V \rangle$ and the fluctuations in volume, $\sigma_v^2 = \langle V^2 \rangle - \langle V \rangle^2$.

(b) For a single-phase system, how do the fluctuations in volume scale with the system size N?

(c) State conditions are chosen so as to place the system on its line of liquid-vapor phase coexistence. How do the fluctuations in volume scale with the system size in this case? What does this imply for the behavior of the compressibility in the limit of infinite system size?

19.10. Complete a detailed derivation of the equations governing the grand canonical ensemble.

(a) Explicitly derive Eqns. (19.22) and (19.26).

(b) Make a case for Eqn. (19.31) using the maximum-term method.

19.11. Consider the grand canonical ensemble.

(a) Show that the average number of particles is given by

$$\langle N \rangle = \lambda \left(\frac{\partial \ln \Xi}{\partial \lambda} \right)_{T, V}$$

(b) Show that the fluctuations in particle number are given by

$$\sigma_N^2 = k_B T \left(\frac{\partial \langle N \rangle}{\partial \mu} \right)_{T, V}$$

and assess their magnitude relative to the average number.

(c) Show that independent particles have $\langle N \rangle = \sigma_N^2$.

19.12. Prove that the potential $PV = k_B T \ln \Xi(T, V, \mu)$ in the grand canonical ensemble is indeed fundamental. That is, show how the quantities E, P, N, and S can be found from it.

19.13. Both in the isobaric–isothermal ensemble and in the grand canonical ensemble, two quantities fluctuate. Therefore, it is possible to define a covariance that measures the degree of correlation between the fluctuating quantities:

$$\text{cov}(X, Y) = \langle (X - \langle X \rangle)(Y - \langle Y \rangle) \rangle$$

(a) Show that $\text{cov}(X, Y) = \langle XY \rangle - \langle X \rangle \langle Y \rangle$.

(b) In the isothermal–isobaric ensemble, find an expression for $\text{cov}(E, V)$ in terms of thermodynamic quantities and response functions. Hint: it will be easier to find $\text{cov}(H, V)$ and note the connection to $\text{cov}(E, V)$ using $E = H - PV$.

19.14. Let X and Y be extensive quantities that fluctuate in a given ensemble. Show that the fluctuations measured by $\langle X^2 \rangle - \langle X \rangle^2$ are always proportional to a thermodynamic quantity that must be positive by stability, which is consistent with the fact that a variance is always non-negative. On the other hand, show that the covariance $\langle XY \rangle - \langle X \rangle \langle Y \rangle$ is not related to a stability criterion and may be positive or negative.

19.15. As discussed in Chapter 16, two independent degrees of freedom have separable probability distributions, $\wp(\mathcal{X}_1, \mathcal{X}_2) = \wp(\mathcal{X}_1)\wp(\mathcal{X}_2)$, if they make distinct, additive contributions to the microstate energy. In this case, show that the Gibbs entropy formula leads to additive entropies per

$$\frac{S}{k_B} = -\sum_{\mathcal{X}_1}\sum_{\mathcal{X}_2} \wp(\mathcal{X}_1, \mathcal{X}_2)\ln \wp(\mathcal{X}_1, \mathcal{X}_2) = \frac{S_1}{k_B} + \frac{S_2}{k_B}$$

where

$$\frac{S_1}{k_B} = -\sum_{\mathcal{X}_1} \wp(\mathcal{X}_1)\ln \wp(\mathcal{X}_1), \qquad \frac{S_2}{k_B} = -\sum_{\mathcal{X}_2} \wp(\mathcal{X}_2)\ln \wp(\mathcal{X}_2)$$

19.16. Consider a one-dimensional elastic band as a model of a linear polymer. The end-to-end distance of the polymer is L. A force τ can be applied to stretch the polymer and extend its length.

(a) What is the fundamental equation ($dS = ?$) for this system? Consider a statistical-mechanical ensemble in which the temperature and force are held constant. Write a general expression for the partition function Υ in this ensemble, in terms of the density of states Ω. Be sure to include all the appropriate independent variables in your expression.

(b) For the ensemble in (a), show that the average length is given by

$$\langle L \rangle = \frac{1}{\beta} \left(\frac{\partial \ln \Upsilon}{\partial \tau} \right)_T$$

(c) For the ensemble in (a), show that the mean-squared fluctuation in length is given by

$$\sigma_L^2 = \frac{1}{\beta^2} \left(\frac{\partial^2 \ln \Upsilon}{\partial \tau^2} \right)_T$$

19.17. Find expressions for the average magnetization $\langle M \rangle$ and its fluctuations σ_M^2 for the general Ising model of nonzero field $\mathcal{H}$ introduced in Chapter 16.

Applied problems

19.18. A common way to write the equation of state for a nonideal gas is the so-called virial expansion:

$$\frac{P}{k_B T} = \rho + B_2(T)\rho^2 + B_3(T)\rho^3 + \cdots$$

where $\rho = N/V$ and the temperature-dependent coefficients B_i are called the ith virial coefficients; they depend only on temperature and the nature of the interactions between the specific gas molecules involved. In the ideal case $B_i = 0$ for all $i \geq 2$ and the ideal gas is recovered. Here, you will find expressions for the coefficients.

(a) Starting from the grand canonical partition function and using the Taylor expansion $\ln x \approx (x - 1) - (x - 1)^2/2 + \cdots$ for $\ln \Xi$, show that the following relationship holds:

$$\frac{PV}{k_B T} = Q_1 \lambda + \frac{2Q_2 - Q_1^2}{2}\lambda^2 + \cdots \text{ (higher-order terms in } \lambda)$$

Here, $Q_1 \equiv Q(T, V, N = 1)$ and $Q_2 \equiv Q(T, V, N = 2)$.

(b) Using a grand canonical expression for N in terms of a derivative of $\ln \Xi$, show that the density is given by

$$\rho = \frac{Q_1}{V}\lambda + \frac{2Q_2 - Q_1^2}{V}\lambda^2 + \cdots$$

(c) Substitute this expression for the density into the virial expansion. By comparing terms of similar order λ with the result from part (a), show that the second virial coefficient is given by

$$B_2(T) = \frac{Q_1^2 - 2Q_2}{2Q_1^2/V} = \frac{Z_1^2 - Z_2}{2Z_1^2/V}$$

(d) Show that the second virial coefficient for hard spheres is independent of temperature and given by $B_2 = 2\pi\sigma^3/3$. Recall that hard spheres have a pair potential of the form $u(r_{ij} < \sigma) = \infty$ and $u(r_{ij} \geq \sigma) = 0$.

19.19. A cubic container of volume $V = L^3$ is filled with an ideal monatomic gas. Molecules can adsorb on one of its six surfaces, resulting in an attractive energy $-\epsilon$ with $\epsilon > 0$. The absorption region is defined as a zone that extends a constant distance b from the interface. Assume that the molecules remain non-interacting in the adsorbate phase.

(a) The system is compressed isotropically, reversibly, and isothermally at T from V_1 to $V_2 < V_1$. Find an expression for the work required per molecule. Is it less or greater than the case for which the surface had been non-adsorbing?

(b) Derive an expression for the fluctuations in the number of adsorbed molecules, $\sigma_{N_{ads}}^2$, given the bulk gas conditions T and P.

19.20. Liquid water exhibits a temperature of maximum density (TMD) when cooling or heating at constant pressure. At atmospheric pressure, this temperature is around $4\,°C$. In general, however, one can construct a line in the T, P plane that gives the TMD for any pressure.

(a) Consider the properties of the TMD line. Along it, what must be true about the thermal expansion coefficient? Prove also that $(\partial P/\partial T)_v = 0$ along this line. Be careful to address all of the terms in your proof.

(b) Show that the temperature derivative of the average volume is given by the following isothermal–isobaric ensemble average:

$$\left(\frac{\partial V}{\partial T}\right)_P = \frac{\langle HV\rangle - \langle H\rangle\langle V\rangle}{k_B T^2}$$

$$= \frac{\langle (H - \langle H\rangle)(V - \langle V\rangle)\rangle}{k_B T^2}$$

$$= \frac{\langle \delta H\,\delta V\rangle}{k_B T^2}$$

where δH and δV are fluctuations away from the mean enthalpy and volume. Thus, the thermal expansion coefficient α_P measures the correlation between enthalpy and volume fluctuations. Below the TMD, low-enthalpy fluctuations in the system have higher volume. Explain how this is consistent with hydrogen-bonding interactions and the tetrahedral network of water.

19.21. Consider the parallel association of two identical, square macroscopic surfaces with side length L. Classical ideal gas molecules can diffuse in between the surfaces and adsorb on either of the two interior walls. When the molecules are within a distance b of the surface they experience an attractive energy of $-\epsilon$, where $\epsilon > 0$ is a constant. Let the surface–surface separation distance be h. The system is at constant temperature T.

(a) Let the gas molecules have a thermal de Broglie wavelength $\Lambda(T)$, and let their number between the surfaces be constant and given by N. Write expressions for the single-molecule and total partition functions q and Q. Assume that $h > 2b$.

(b) Find an expression for the force required to hold the surfaces a given distance h apart, using your result from part (a).

(c) In actuality, the number of molecules between the surfaces can vary as they enter or exit the interstitial space due to coupling to a bulk ideal gas phase. If the surfaces are held fixed, what constant conditions describe the inter-surface phase? Find the partition function in the relevant ensemble. It may be useful to know that

$$e^x = \sum_{n=0}^{\infty} \frac{x^n}{n!}$$

(d) Considering that the molecules are in equilibrium with a bulk ideal gas phase at constant T and P, show that the average number of molecules $\langle N\rangle$ in the interstitial region is given by an expression of the form

$$\langle N\rangle = \frac{PV}{k_B T}\left[C\left(\frac{b}{h}\right) + 1\right]$$

where V is the volume of the interstitial region and C is a constant dependent on temperature only. Find C.

19.22. Consider a drop of liquid water suspended in a near-vacuum in a zero-gravity environment. As it nears equilibrium, the droplet shape approaches that of a sphere. Let γ give the water–vacuum surface tension, and let A be the surface area of the drop. Neglect evaporation and assume that the drop is roughly maintained at constant temperature through occasional interactions with molecules in the near-vacuum.

(a) What thermodynamic potential X is minimized as the drop comes to equilibrium? Give its integrated and differential forms.

(b) Prove that the so-called area compressibility modulus, K, must be positive:

$$K \equiv A\left(\frac{\partial \gamma}{\partial A}\right)_{T,P,N}$$

(c) Find an expression for the partition function Y for this system in terms of the density of states $\Omega(E, V, N, A)$, and show its relation to the potential in part (a).

(d) Show how the fluctuations in the area relate to K and that their relative magnitude grows as $N^{-1/2}$.

FURTHER READING

D. Chandler, *Introduction to Modern Statistical Mechanics*, New York: Oxford University Press (1987).

K. Dill and S. Bromberg, *Molecular Driving Forces: Statistical Thermodynamics in Biology, Chemistry, Physics, and Nanoscience*, 2nd edn. New York: Garland Science (2010).

T. L. Hill, *An Introduction to Statistical Thermodynamics*. Reading, MA: Addison-Wesley (1960); New York: Dover (1986).

T. L. Hill, *Statistical Mechanics: Principles and Selected Applications*. New York: McGraw-Hill (1956); New York: Dover (1987).

E. A. Jackson, *Equilibrium Statistical Mechanics*. Mineola, NY: Dover (1968).

D. A. McQuarrie, *Statistical Mechanics*. Sausalito, CA: University Science Books (2000).

20 Reaction equilibrium

20.1 A review of basic reaction concepts

Generally, reactions may be categorized into several types. *Chemical* reactions involve bond breaking and formation, such as the famous water-gas shift, $CO + H_2O \leftrightarrow CO_2 + H_2$. *Physical* reactions involve the association and dissociation of molecules through non-covalent interactions; examples include the hybridization of DNA strands into a helix, the binding of a ligand to a protein, the formation of a micelle from individual surfactants, and the adsorption of a molecule from a bulk to a surface phase. In any case, what defines a reaction is a specific, stoichiometrically constrained change in the balance of species present, regardless of what one considers to be a distinct species. Here, we study such cases at a very general level by examining the equilibrium behavior of reacting systems at both the macroscopic and microscopic levels. In other words, we will consider the limiting, long-time behavior of a system with one or more reactions.

It is worthwhile to recall the basic perspective on reversible reactions that stems from introductory chemistry. Namely, a reversible reaction can be described by an equilibrium constant K_{eq}. Consider, for example, the reaction

$$A + 2B \overset{K_{eq}}{\leftrightarrow} C \tag{20.1}$$

The equilibrium constant of such reactions is often related to the concentrations. A frequent notation is that $[i]$ gives the concentration of species i in moles per volume (molarity), but we could just as well describe concentration in terms of molecules per volume as we have done throughout this text, using ρ_i. The latter remains our primary convention in this chapter, although it is simple to note that the two kinds of concentrations differ only by a factor of Avogadro's number $\mathcal{N}_A$. For the reaction in (20.1), the equilibrium constant then follows:

$$K_{eq} = \frac{[C]}{[A][B]^2} \quad \text{or} \quad K_{eq} = \frac{\rho_C}{\rho_A \rho_B^2} \tag{20.2}$$

Here, the general rule is that reactant concentrations are multiplied in the denominator, products in the numerator, and the powers on each are determined by the stoichiometric coefficients. Notice that we might consider the case of irreversible reactions as a special subset of reversible ones in which the equilibrium constant is very large.

While these basic "rules" for equilibrium constants might seem all too familiar, we should be suspicious enough to ask why they should be this way. Why are the concentrations raised to their stoichiometric coefficients? How is it possible that one can even write down an equilibrium constant in the manner of (20.2)? To what thermodynamic properties does K_{eq} relate, and how can it be determined from molecular interactions? These are all questions that we will address in this chapter. In particular, we will see that the basic form of Eqn. (20.2) actually relies on several approximations and is not the most general way to address reaction equilibrium from a thermodynamic perspective. Hence, we begin from very basic principles.

20.2 Reaction equilibrium at the macroscopic level

Let us characterize a reacting system using the classical, macroscopic thermodynamic-potential analysis. Take the specific reaction above as an example, involving components A, B, and C. We consider its behavior at constant temperature and pressure. The relevant thermodynamic potential is the Gibbs free energy, and it is minimized at equilibrium. If the reaction in Eqn. (20.1) does not occur, such that we can independently control the respective amounts of each species, the Gibbs free energy is then given by

$$G\left(T, P, N_A, N_B, N_C\right) \tag{20.3}$$

where the independent variables are indicated explicitly. However, if the reaction is allowed to proceed, this perspective changes. We can no longer independently control the variables N_A, N_B, and N_C; they can change as the system approaches equilibrium. The changes in the molecule numbers are subject to a conservation rule that stems from the stoichiometric coefficients,

$$dN_A = -dN_C$$
$$dN_B = -2 \, dN_C \tag{20.4}$$

These equations imply that for every one A molecule and two B molecules lost due to a reaction event, a single C molecule is gained. There is a system internal degree of freedom that can vary due to the reaction and that minimizes G as it comes to equilibrium. We could think of that degree of freedom as the number of C molecules that react relative to the initial conditions, but it would be just as valid to use either the number of reacted A molecules or the number of reacted B molecules instead; given any of these changes, the others are found by the relations in Eqn. (20.4). More generally, there is one reaction and hence one *extent of reaction*.

If the reaction proceeds uninhibited from some initial specified composition of the three components, the Gibbs free energy will be minimized at long times. The change in G with the amounts of the species at constant T and P is

$$dG = \left(\frac{\partial G}{\partial N_A}\right)_{T, P, N_B, N_C} dN_A + \left(\frac{\partial G}{\partial N_B}\right)_{T, P, N_A, N_C} dN_B + \left(\frac{\partial G}{\partial N_C}\right)_{T, P, N_A, N_B} dN_C$$

$$= \mu_A \, dN_A + \mu_B \, dN_B + \mu_C \, dN_C \tag{20.5}$$

With the stoichiometric considerations in Eqn. (20.4), we can write

$$dG = (-\mu_A - 2\mu_B + \mu_C)dN_C \tag{20.6}$$

When the Gibbs free energy is minimized, $dG = 0$. According to (20.6), this means that the reaction reaches equilibrium when the chemical potentials satisfy the condition

$$\mu_A + 2\mu_B = \mu_C \tag{20.7}$$

If Eqn. (20.7) is not satisfied, the Gibbs free energy is not minimized. The system reaches equilibrium only when the state conditions are such that (20.7) holds. We find the following generality.

> The thermodynamic quantity that predicts reaction equilibrium is the chemical potential.

In fact this is the first case in which we have seen an equality of chemical potentials involving multiple types of species. In situations of non-reactive "chemical equilibrium" of the type discussed in Chapter 4, we found that chemical potentials were equal for particles of the same type that could be exchanged between two systems or phases. Here we see that reactions are required in order for equalities like Eqn. (20.7) that involve chemical potentials for *distinct* species to emerge. One can think of the latter case in similar "exchange" terms, only with the exchange being among different components.

Notice that the equilibrium condition for the chemical potentials reduces the number of macroscopic degrees of freedom. The state of a non-reacting three-component, single-phase system depends on four intensive variables according to Gibbs' phase rule, such as T, P, x_A, and x_B. However, if we know that there is a single reaction involving A, B, and C, we will obtain an equation like (20.7) that relates these state variables because the chemical potentials are of the general form $\mu_i(T, P, x_A, x_B)$. In turn, we will be able to specify only three state parameters, such as T, P, and x_A, because x_B and $x_C = 1 - x_A - x_B$ would then automatically adjust so as to satisfy Eqn. (20.7). A general principle is as follows.

> For every independent reversible reaction in a system, the number of macroscopic degrees of freedom is reduced by one.

These considerations addressed a specific example, but we can treat arbitrary cases using the following notation for a given reaction:

$$\sum_i \nu_i M_i = 0 \tag{20.8}$$

Here, ν_i is a stoichiometric coefficient and M_i is the chemical symbol for component i. The ν_i are defined such that they take on negative values for reactants (the LHS of the reaction) and positive values for products (the RHS). Our previous example would be

$$-A - 2B + C = 0 \tag{20.9}$$

with $\nu_A = -1$, $\nu_B = -2$, and $\nu_C = 1$.

For one instance of the reaction from left to right – for one chemical event – the change in the number of molecules of each component is v_i. If we let the variable ξ indicate the number of such events, then a differential change in ξ leads to corresponding changes in the amount of each component,

$$dN_i = v_i \, d\xi \tag{20.10}$$

The variable ξ is called the *extent of reaction*. In this case, the total change in the Gibbs free energy for an arbitrary reaction is

$$dG = \sum_i \left(\frac{\partial G}{\partial N_i}\right)_{T,P,N_{j\neq i}} dN_i$$

$$= d\xi \sum_i v_i \mu_i \tag{20.11}$$

The second line made use of Eqn. (20.10). With $dG = 0$, we find that the general expression for reaction equilibrium is

$$\sum_i v_i \mu_i = 0 \tag{20.12}$$

Of course, if multiple reactions occur in the system, each will have its own extent of reaction and chemical-potential constraint like Eqn. (20.12). In turn, the number of macroscopic degrees of freedom will be reduced exactly by the number of independent reactions.

20.3 Reactions involving ideal gases

In earlier chapters we derived expressions for the chemical potential of simple multi-component systems, which allows us to further develop Eqn. (20.12) in specific cases. For ideal gas mixtures, we can use $\mu_i = \mu_i^\circ(T) + k_B T \ln P_i$, where P_i is the partial pressure of component i, so as to obtain

$$\sum_i v_i(\mu_i^\circ + k_B T \ln P_i) = 0 \tag{20.13}$$

We omit the dependence of the standard chemical potentials μ° on temperature for simplicity. Isolating all of these on one side of the equation gives

$$-\frac{1}{k_B T} \sum_i v_i \mu_i^\circ = \sum_i v_i \ln P_i = \ln\left(\prod_i P_i^{v_i}\right) \tag{20.14}$$

Upon exponentiation, we then find

$$\prod_i P_i^{v_i} = \exp\left[-\frac{\sum_i v_i \mu_i^\circ}{k_B T}\right] \tag{20.15}$$

Notice that the RHS is independent of pressure and composition because μ_i° is a function only of temperature. This allows us to introduce a general equilibrium constant and relation of a familiar form,

$$\prod_i P_i^{v_i} = K_P(T), \quad \text{where} \quad K_P(T) \equiv \exp\left[-\frac{\sum_i v_i \mu_i^\circ}{k_B T}\right] \tag{20.16}$$

Equation (20.16) shows that the equilibrium state for a mixture of ideal gases is given by the point at which the product of the component partial pressures, raised to their stoichiometric coefficients, equals a temperature-dependent equilibrium constant. If this had been the case for a nonideal gas, the mixture fugacities would have replaced the partial pressures. Note that the equilibrium constant is notated as K_P to indicate that it relates to the gas partial pressures. In addition, we can define a *standard change of free energy of reaction* according to

$$\Delta g^\circ(T) = \sum_i v_i \mu_i^\circ(T) \tag{20.17}$$

such that $K_P = \exp(-\Delta g^\circ/k_B T)$. An interpretation of Δg° is that it gives the part of the free energy change per instance of the reaction due to intramolecular interactions within the molecular species.

We can convert the equilibrium relation to one involving compositions because a partial pressure is defined by $P_i \equiv y_i P$, where y_i is the mole fraction of component i. For an ideal gas mixture, we can express the partial pressure as $P_i = \rho_i k_B T$, where ρ_i denotes the concentration of i in molecules per volume. We can then rewrite (20.16) by defining a new equilibrium constant K_C,

$$\prod_i \rho_i^{v_i} = K_C(T), \quad \text{where} \quad K_C(T) \equiv \exp\left[-\frac{\sum_i v_i \mu_i^\circ}{k_B T}\right](k_B T)^{-\sum_i v_i} \tag{20.18}$$

Again, K_C depends only on the temperature; its subscript C stems from its relation to the compositions of the components rather than partial pressures. Alternatively, we could have expressed the composition in terms of molarity and made the substitution $P_i = [i]RT$, where $R = k_B \mathcal{N}_A$. This would have simply changed the final $k_B T$ term on the RHS of (20.18) to RT.

For the example reaction introduced at the beginning of this chapter, these considerations show that the equilibrium conditions reduce to

$$\frac{P_C}{P_A P_B^2} = \frac{\rho_C}{\rho_A \rho_B^2}(k_B T)^{-2} = \frac{[C]}{[A][B]^2}(RT)^{-2} = K_P(T)$$

$$\text{where } K_P(T) = \exp\left[-\frac{\mu_C^\circ(T) - \mu_A^\circ(T) - 2\mu_B^\circ(T)}{k_B T}\right] \tag{20.19}$$

20.4   Reactions involving ideal solutions

Chapter 12 introduced a general expression for the chemical potential of a component in an ideal solution, $\mu_i = \mu_i^* + k_B T \ln x_i$. In this case Eqn. (20.12) becomes

$$\sum_i \nu_i(\mu_i^* + k_B T \ln x_i) = 0 \tag{20.20}$$

Again we omit the dependence of the pure-state chemical potentials μ_i^* on T and P for clarity, but it should be kept in mind. Rearranging and exponentiating Eqn. (20.20) as we did for the ideal gas case gives

$$\prod_i x_i^{\nu_i} = \exp\left[-\frac{\sum_i \nu_i \mu_i^*}{k_B T}\right] \tag{20.21}$$

The RHS of this equation is composition-independent since the quantities μ_i^* depend only on T and P. However, as discussed in Chapter 12, the exact significance of μ_i^* can differ depending on whether a component is dilute or nearly pure. In either case, we can express the reaction equilibrium as

$$\prod_i x_i^{\nu_i} = K_x(T, P), \quad \text{where} \quad K_x(T, P) \equiv \exp\left[-\frac{\sum_i \nu_i \mu_i^*}{k_B T}\right] \tag{20.22}$$

We find that the mole fractions of each component form a multiplicative term that equals an equilibrium constant, as in Eqn. (20.16). This is also the rationale for the subscript x in K_x. If we were to consider nonideal solutions, the LHS would involve activity coefficients in addition to mole fractions. As with the case for gases, we can define another *standard change in free energy of reaction* using the pure-phase chemical potentials,

$$\Delta g^*(T, P) = \sum_i \nu_i \mu_i^*(T, P) \tag{20.23}$$

If all of the participating components in a reaction are dilute, we can make further approximations by considering the relationship of the mole fractions to the concentrations,

$$x_i = \frac{\text{molecules } i}{\text{molecules total}}$$

$$\approx \frac{\text{molecules } i}{\text{molecules solvent}} = \frac{\text{molecules } i}{V \rho_{\text{solvent}}} = \frac{\rho_i}{\rho_{\text{solvent}}} \tag{20.24}$$

where ρ_i indicates the concentration of i in molecules per volume, or, equivalently, the number density. In any case, the approximation of Eqn. (20.24) then allows us to rewrite the equilibrium condition using a new constant K_C,

$$\prod_i \rho_i^{\nu_i} = K_C(T, P), \quad \text{where} \quad K_C(T, P) \equiv \exp\left[-\frac{\sum_i \nu_i \mu_i^*}{k_B T}\right] \rho_{\text{solvent}}^{\sum_i \nu_i} \tag{20.25}$$

For dilute reactants and products, the solvent density does not change appreciably with the concentrations of these components, and hence we can say quite generally that K_C is also concentration-independent. As in the ideal-gas case, we could equally define the equilibrium relation using molarities rather than molecular number densities for the concentrations $[i]$, in which case the quantity ρ_{solvent} would also need to have units of moles per volume in Eqn. (20.25).

For our simple reaction example, Eqns. (20.22) and (20.25) for the ideal solution case bear out the relations

$$\frac{x_C}{x_A x_B^2} = \frac{\rho_C}{\rho_A \rho_B^2} \rho_{\text{solvent}}^2 = K_x(T,P)$$

$$\text{where } K_x(T,P) = \exp\left[-\frac{\mu_C^*(T,P) - \mu_A^*(T,P) - 2\mu_B^*(T,P)}{k_B T}\right] \tag{20.26}$$

Reaction equilibrium in ideal gases and dilute ideal solutions can be expressed in a form involving a product of concentrations raised to stoichiometric coefficients, as

$$\prod_i \rho_i^{\nu_i} = K_{\text{eq}}$$

This result stems directly from the fact that these systems have non-interacting components and thus their behavior is dominated by a balance of the mixing entropy and the free energies of the pure components. For other systems, the equilibrium relations must be modified to replace the concentrations with either fugacities for gases or activities for solutions.

20.5   Temperature and pressure dependence of K_{eq}

Consider the temperature derivative of the logarithm of the ideal gas equilibrium constant,

$$\frac{d\ln K_P}{dT} = \frac{d}{dT}\left(-\frac{\sum_i \nu_i \mu_i^\circ}{k_B T}\right) \tag{20.27}$$

This is a complete derivative because K_P depends only on temperature. We can now extend the general thermodynamic identity $\partial(\mu/T)/\partial T = -h/T^2$ by simply adding superscripts "$\circ$" to μ and h. This gives

$$\frac{d\ln K_P}{dT} = \frac{1}{k_B T^2}\sum_i \nu_i h_i^\circ \tag{20.28}$$

We can define a *standard change of enthalpy of reaction* such that

$$\frac{d\ln K_P}{dT} = \frac{\Delta h^\circ}{k_B T^2}, \quad \text{where} \quad \Delta h^\circ \equiv \sum_i v_i h_i^\circ \tag{20.29}$$

Notice that Δh° depends on the pure-phase ideal gas enthalpies since, for an ideal gas, $h = h^\circ$. However, this is not likely to be the case for a nonideal gas described with a fugacity. In any case, Eqn. (20.29) is more commonly written as

$$\frac{d\ln K_P}{d(1/T)} = -\frac{\Delta h^\circ}{k_B}, \quad \text{where} \quad \Delta h^\circ \equiv \sum_i v_i h_i^\circ \tag{20.30}$$

This equation gives the so-called *van 't Hoff relation* and is often used to construct a *van 't Hoff plot*. Namely, Eqn. (20.30) shows that a graph of the logarithm of the equilibrium constant versus the inverse temperature has a slope related to the reaction enthalpy. In this sense, measurements of the T dependence of equilibrium concentrations, and hence of equilibrium constants, give a way to determine enthalpies of reaction. When these kinds of plots are linear, they indicate that Δh° is independent of temperature. In such cases, one can integrate (20.30) to obtain a simpler dependence of the equilibrium constant on temperature, assuming that Δh° is constant. The definite integral between T_1 and T_2 shows that

$$\frac{K_P(T_1)}{K_P(T_2)} = \exp\left[-\frac{\Delta h^\circ}{k_B}\left(\frac{1}{T_1} - \frac{1}{T_2}\right)\right] \tag{20.31}$$

or, perhaps more simply,

$$K_P \propto \exp\left(-\frac{\Delta h^\circ}{k_B T}\right) \tag{20.32}$$

If we had considered an ideal solution rather than an ideal gas mixture, we would have obtained a similar van 't Hoff relationship,

$$\left(\frac{\partial \ln K_x}{\partial(1/T)}\right)_P = -\frac{\Delta h^*}{k_B}, \quad \text{where} \quad \Delta h^* \equiv \sum_i v_i h_i^* \tag{20.33}$$

Note that the h_i^* give the per-particle enthalpies for the pure components in the case that the ideal solution model holds over the entire composition range. For dilute reacting species, the h_i^* are the partial molar enthalpies for infinitely dilute components i in the solvent in accordance with the dilute convention for μ_i^*.

The derivative in Eqn. (20.33) is a partial one because the ideal solution equilibrium constant is also affected by the pressure. We can uncover the relationship by taking the P derivative of $\ln K_x$,

$$\left(\frac{\partial \ln K_x}{\partial P}\right)_T = \frac{\partial}{\partial P}\left(-\frac{\sum_i v_i \mu_i^*}{k_B T}\right)_T \tag{20.34}$$

For a pure, single-component system, we have the fundamental relation $(\partial \mu / \partial P)_T = v$, where v is the per-particle volume. Application to all of the μ^* terms above gives

$$\left(\frac{\partial \ln K_x}{\partial P} \right)_T = -\frac{\Delta v^*}{k_B T}, \quad \text{where} \quad \Delta v^* \equiv \sum_i v_i v_i^* \tag{20.35}$$

Here Δv^* involves the per-particle volumes of the pure phase components for one instance of the reaction, or, if the species are dilute, the partial molar volumes of infinitely dilute components in the solvent.

Both for ideal gases and for ideal solutions, the **van 't Hoff relation** shows that the dependence of $\ln K_{eq}$ on $1/T$ gives a measure of the standard enthalpy of reaction. For ideal solutions, the dependence of $\ln K_{eq}$ on P gives a measure of the standard volume change of the reaction. For ideal gases, the equilibrium constant has no dependence on pressure.

20.6   Reaction equilibrium at the microscopic level

We now take a molecular perspective of reaction equilibrium. For simplicity we will examine the case at constant volume rather than pressure so that we can examine equilibrium in the canonical ensemble. At constant T and V, the Helmholtz free energy is minimized at equilibrium. An essentially identical derivation to that in Eqn. (20.11) shows that the macroscopic condition for reaction equilibrium in this case leads to the same general chemical-potential condition as in Eqn. (20.12), $\sum_i v_i \mu_i = 0$.

We make a connection to microscopic properties by finding expressions for the μ_i in terms of partition functions. We specifically examine the case of *independent molecules* that have no intermolecular interactions. This does not mean, however, that each molecule cannot possess a molecular structure and internal degrees of freedom and interactions. Thus before proceeding, let us more generally define a *molecular* partition function in the canonical ensemble that sums over the states of a single independent molecule,

$$q = \sum_{\substack{\text{molecular} \\ \text{states } i}} e^{-\beta \epsilon_i} \tag{20.36}$$

In Chapter 16, we showed that this summation generally involves enumeration of all combinations of the molecule's microscopic degrees of freedom. While in Chapter 18 we showed the exact form of this equation for a monatomic particle, here we must emphasize that q can be defined for an arbitrary independent molecule. It can account for many internal degrees of freedom and intramolecular potential energies in addition to the molecule's position in space: bond stretching, angle bending, rotations around torsional angles, etc. Not all of these are well described classically, so the molecular partition function generally requires one to treat some degrees of freedom quantum-mechanically and others classically. One important fact is that the translational degrees of freedom – namely, the variables x, y, and z for the center of mass of the molecule – usually can be taken in the classical limit. In the absence of any kind of field, the integral over these leads to a factor of V in Eqn. (20.36) as we saw in Chapter 18. The details of these models are beyond the scope of this text, but the essential message is the following.

It is possible to define a molecular partition function q for a non-interacting, inde-pendent, non-monatomic molecule that accounts for **translational, rotational**, and **intramolecular** degrees of freedom. Typically, the translational component is well described classically and contributes a factor of V to q, as it does for the monatomic case. However, many of the other degrees of freedom must be treated quantum-mechanically in order to achieve accurate models; they contribute additional factors to q that are molecule-specific and have a temperature dependence.

Armed with the concept of a molecular partition function, we can find expressions for the chemical potentials of components in ideal gas reactions. For the example system above, the canonical partition function for a set of independent molecules is

$$Q(T, V, N_A, N_B, N_C) = \frac{q_A^{N_A} q_B^{N_B} q_C^{N_C}}{N_A! N_B! N_C!}$$

$$\approx \left(\frac{q_A e}{N_A}\right)^{N_A} \left(\frac{q_B e}{N_B}\right)^{N_B} \left(\frac{q_C e}{N_C}\right)^{N_C} \tag{20.37}$$

Here the functions $q_i(T, V)$ give the single-molecule partition functions for each species, and the factorial terms in the denominator account for indistinguishability. The second line applies Stirling's approximation. Equation (20.37) is easily generalized to arbitrary multicomponent systems,

$$Q(T, V, \{N\}) = \prod_i \frac{q_i^{N_i}}{N_i!} \approx \prod_i \left(\frac{q_i e}{N_i}\right)^{N_i} \tag{20.38}$$

We are now in a position to determine the chemical potentials,

$$\mu_i = \left(\frac{\partial A}{\partial N_i}\right)_{T, V, N_{j \neq i}}$$

$$= -k_B T \frac{\partial \ln Q}{\partial N_i}$$

$$= -k_B T \ln \left(\frac{q_i}{N_i}\right) \tag{20.39}$$

We now substitute Eqn. (20.39) into the reaction-equilibrium condition of (20.12),

$$-k_B T \sum_i \nu_i \ln \left(\frac{q_i}{N_i} \right) = 0 \tag{20.40}$$

To simplify, we exponentiate this expression and rearrange terms slightly,

$$\prod_i N_i^{\nu_i} = \prod_i q_i^{\nu_i} \tag{20.41}$$

Finally, we divide by $\prod_i V^{\nu_i}$ on both sides,

$$\prod_i \left(\frac{N_i}{V} \right)^{\nu_i} = \prod_i \left(\frac{q_i}{V} \right)^{\nu_i} \tag{20.42}$$

The purpose of this last manipulation is to obtain an expression on the LHS involving species concentrations. As a result, we find that we can rewrite the entire expression in terms of an equilibrium constant,

$$\prod_i \rho_i^{\nu_i} = K_{eq}, \quad \text{where} \quad K_{eq} \equiv \prod_i \left(\frac{q_i}{V} \right)^{\nu_i} \tag{20.43}$$

Importantly, this derivation shows that we can write the equilibrium constant for independent molecules as a product of single-molecule partition functions for the species. It actually provides a way to determine K_{eq} on the basis of molecular interactions. For the example described by Eqn. (20.1), we would have

$$\frac{\rho_C}{\rho_A \rho_B^2} = K_{eq}, \quad \text{where} \quad K_{eq} = \frac{q_C}{q_A q_B^2} V^2 \tag{20.44}$$

Note that, because we expect translational degrees of freedom to lead to a factor of V in each component's molecular partition function, q_i/V should be intensive and thus so should the equilibrium constant according to its definition in Eqn. (20.43). Of course, we expect this on macroscopic grounds. Moreover, if any of the individual species in the reaction scheme are monatomic ideal gases, we can immediately write $q_i = V/\Lambda_i(T)^3$ for them, where $\Lambda_i(T)$ is the mass- and temperature-dependent thermal de Broglie wavelength.

20.7   Fluctuations

The considerations above used the macroscopic condition for reaction equilibrium, in terms of chemical potentials, to express the equilibrium constant in terms of partition functions. The use of the macroscopic condition, which depended on a minimization of the Helmholtz free energy, neglects fluctuations. At a microscopic level, we know that the numbers of A, B, and C molecules in our example can fluctuate around an average equilibrium value because random, discrete instances of the forward and reverse reactions occur with time. As we have seen many times before, however, the magnitude of these fluctuations is typically too small to notice on a macroscopic scale.

Let us perform a full analysis of reactions from the molecular perspective, including fluctuations. To make things easy, we consider a simpler reaction case study. Our approach closely mimics the example discussed by Hill,

$$A \overset{K_{eq}}{\leftrightarrow} B \tag{20.45}$$

The conversion of A to B would likely involve some kind of isomerization: for example, a *cis*-to-*trans* conversion around a stiff bond, or a change in the molecular connectivity such as n-butane to isobutane. Regardless, the system has a fixed total number of molecules $N = N_A + N_B$ and is thus at constant T, V, N conditions. After reaching equilibrium through the reaction above, N will remain unchanged, but N_A and N_B will each experience fluctuations around an equilibrium value. Specifically, the populations of different states will be determined by Boltzmann factors. We must sum up microstates that differ not only in the configurations of the molecules, but also in the numbers of A and B molecules. At the same time N is constant such that $N_B = N - N_A$ at any given time. Therefore, the full canonical partition function for this system is

$$Q(T, V, N) = \sum_{N_A=0}^{N} \sum_{\substack{\text{all } n \\ \text{at } N_A, N_B, V}} e^{-\beta E_n}$$

$$= \sum_{N_A=0}^{N} Q(T, V, N_A, N_B) \tag{20.46}$$

The second line simply shows that the inner sum over all microstates gives the canonical partition function for a system of N_A and $N_B = N - N_A$ molecules. Provided that the molecules are independent, as in ideal gases, Eqn. (20.46) can be further simplified, giving

$$Q(T, V, N) = \sum_{N_A=0}^{N} \frac{q_A^{N_A} q_B^{N_B}}{N_A! N_B!}$$

$$= \frac{1}{N!} \sum_{N_A=0}^{N} \frac{N!}{N_A!(N - N_A)!} q_A^{N_A} q_B^{N - N_A} \tag{20.47}$$

The sum simply gives the binomial expansion so that we can finally write

$$Q(T, V, N) = \frac{(q_A + q_B)^N}{N!} \tag{20.48}$$

At this point, we are interested in the probability with which the system visits different values of N_A. We can construct this by examining the form of Eqn. (20.46),

$$\wp(N_A) = \frac{Q(T, V, N_A, N_B)}{Q(T, V, N)} = \frac{1}{Q(T, V, N)} \frac{q_A^{N_A} q_B^{N_B}}{N_A! N_B!} \tag{20.49}$$

Clearly, $\sum_{N_A} \wp(N_A) = 1$ from the way $Q(T, V, N)$ is constructed. Indeed, $Q(T, V, N_A, N_B)$ sums the probability weight for microstates of a given value of N_A, while $Q(T, V, N)$ is the normalization factor. The average number of A molecules is therefore

$$\langle N_A \rangle = \sum_{N_A=0}^{N} \wp(N_A) N_A$$

$$= \frac{1}{Q(T, V, N)} \sum_{N_A=0}^{N} \frac{q_A^{N_A} q_B^{N_B}}{N_A! N_B!} N_A \tag{20.50}$$

Notice that the summation looks quite similar to $Q(T, V, N_A, N_B)$ except for an additional factor of N_A. This could be extracted using a derivative with respect to q_A,

$$\frac{q_A^{N_A} q_B^{N_B}}{N_A! N_B!} N_A = q_A \frac{\partial}{\partial q_A} \left(\frac{q_A^{N_A} q_B^{N_B}}{N_A! N_B!} \right)_{T, V, N, q_B} \tag{20.51}$$

With this mathematical trick, we can rewrite the average in terms of a partition function derivative,

$$\langle N_A \rangle = q_A \left(\frac{\partial \ln Q(T, V, N)}{\partial q_A} \right)_{T, V, N, q_B} \tag{20.52}$$

On substituting Eqn. (20.48) for the full partition function, we find

$$\langle N_A \rangle = N \left(\frac{q_A}{q_A + q_B} \right) \tag{20.53}$$

We could have repeated the entire derivation above, swapping indices A for B to show that

$$\langle N_B \rangle = N \left(\frac{q_B}{q_A + q_B} \right) \tag{20.54}$$

On taking the ratio of (20.54) with (20.53), we find that the macroscopic, maximum-term equilibrium case is not surprising,

$$\frac{\rho_B}{\rho_A} = \frac{\langle N_B \rangle / V}{\langle N_A \rangle / V} = \frac{q_B / V}{q_A / V} \tag{20.55}$$

which is in agreement with Eqn. (20.43).

On the other hand, Eqn. (20.49) provides an expression for the full distribution in particle numbers and we can examine fluctuations in detail. Consider the variance of N_A. It is left as an end-of-chapter problem to show that the following relationship holds:

$$\sigma_{N_A}^2 = q_A \left(\frac{\partial \langle N_A \rangle}{\partial q_A} \right)_{T, V, N, q_B}$$

$$= \frac{\langle N_A \rangle \langle N_B \rangle}{N} \tag{20.56}$$

It must also be true that $\sigma_{N_A}^2 = \sigma_{N_B}^2$ because the sum $N_A + N_B$ is constant. Equation (20.56) shows that the magnitude of the fluctuations in N_A scales as $N^{1/2}$. Compared with the average value, which scales as N as per Eqn. (20.53), the fluctuations therefore become negligible in the macroscopic limit.

Chemical reactions experience microscopic fluctuations in their extents of reaction, just like any fluctuating thermodynamic quantities in an ensemble. These are undetectable in macroscopic systems, scaling in a relative manner as the familiar $N^{-1/2}$ trend, but they can become important for small-scale systems.

Problems

Conceptual and thought problems

20.1. Is any reaction in a mixture ever truly irreversible? Consider that a mixture will have a contribution stemming from an ideal entropy of mixing, even if there are nonideal contributions as well. How much must the Gibbs free energy change in order that $x_i \to 0$ for any of the reactants?

20.2. A friend suggests that the third law must play a role in reaction equilibrium because of the way Eqn. (20.7) involves chemical potentials. If each chemical potential is shifted with respect to a reference entropy, as $\mu'_A = \mu_A + Ts_{ref}$, then the final equation involves a nonzero number of s_{ref} terms that affect the equilibrium equation. Prove that your friend is wrong.

20.3. Consider Gibbs' phase rule for a system involving C components, n_{phase} phases, and $n_{reaction}$ reversible reactions.
  (a) Prove that a general form for the rule is
  $$n_{DOF} = C - n_{phase} + 2 - n_{reaction}$$
  (b) A system of water could be considered to have a single species H_2O, or two atomic species H and O. Should one use $C = 1$ or $C = 2$ in Gibbs' phase rule? Explain how both can be reasonable answers, provided a suitable analysis.
  (c) The terms C and $n_{reaction}$ are intimately related. How do both depend on what one defines as a "species"?
  (d) What is the maximum number of independent reactions for a single-phase system of C components? Explain your result on physical grounds.

20.4. A large number of chemical reactions in a complex gaseous mixture are proposed. However, some may be redundant. For example, the reactions

$$CH_4 + (3/2)O_2 \to CO + 2H_2O, \quad CO + (1/2)O_2 \to CO_2,$$
$$CH_4 + 2O_2 \to CO_2 + 2H_2O$$

are not all independent because the last one can be expressed as the sum of the first two. How can you test whether or not you have redundant reactions in your proposed list? Hint: consider the form of Eqn. (20.8) and how one tests mathematically whether a set of linear equations is independent.

20.5. The Gibbs free energy for the example reaction presented earlier in the chapter can also be written in integrated form as $G = N_A \mu_A + N_B \mu_B + N_C \mu_C$. Show that minimization of this quantity with respect to N_C also leads to Eqn. (20.7). Keep in mind that the chemical potentials in general depend on the compositions and thus also implicitly on N_C, so your derivative must take this into account.

20.6. Consider the chemical reaction $A \leftrightarrow 2B$. What determines the direction in which the reaction will proceed at constant T, P conditions? How about at constant T, V conditions? For each case, find an inequality and prove it mathematically. Hint: consider which quantities can only increase or decrease as equilibrium is approached.

20.7. Write out a general expression for the equilibrium relation $K_P = \ldots$ in the case of nonideal gases. How does this expression simplify if the Lewis–Randall approximation is used?

20.8. Write out a general expression for the equilibrium relation $K_x = \ldots$ in the case of nonideal solutions.

20.9. Equation (20.16) indicates that the equilibrium constant of a reaction need not be dimensionless if $\sum_i \nu_i \neq 0$. How is this possible when it is defined as the exponential of a seemingly dimensionless term? Explain.

20.10. Show that $h = h^\circ$ for an ideal gas, where h is the pure-phase per-particle enthalpy and h° the standard part related to μ°. Then show that the equality does not hold for the general case of a nonideal gas described by a fugacity.

$$-T^2 \left(\frac{\partial(\mu^\circ/T)}{\partial T} \right)_P = h^\circ \overset{?}{=} h = -T^2 \left(\frac{\partial(\mu/T)}{\partial T} \right)_P$$

20.11. Equations (20.16) and (20.32) look remarkably similar considering that $\Delta g^\circ = \sum_i \nu_i \mu_i^\circ$. On this basis one might be tempted to suggest that $\Delta g^\circ = \Delta h^\circ$ and hence $\Delta s^\circ = 0$. Explain what is wrong with this argument and how Δs° can be nonzero.

20.12. In a system with one or more reactions, the numbers of different types of molecules can fluctuate. Why then is the relevant thermodynamic potential the Gibbs free energy? Why is it not some Legendre transform of G with respect to the particle numbers?

20.13. Consider a three-component system of particles of types A, B, and C. The system is held at constant T and P, and is coupled to an infinite reservoir of A and B particles that freely undergo the reaction $3A \leftrightarrow B$. Write down expressions for the following quantities:
  (a) the partition function for this ensemble in terms of the microcanonical one, $\Omega(E, V, N_A, N_B, N_C)$;
  (b) the thermodynamic potential to which (a) relates; and
  (c) the microstate probability distribution, $\wp_m$.
  Be careful to take note of Gibbs' phase rule.

20.14. A particular reaction takes the form A $\leftrightarrow$ B + C, where, for a wide range of conditions, A and B form immiscible, pure solid phases and C is a gas. For example, A may be a hydrate such that C is water vapor. How does the fraction of solid material that is A vary with the partial pressure of C? Sketch a graph showing the variation schematically.

Fundamentals problems

20.15. Derive a general relation involving component chemical potentials that indicates in which direction a reaction will proceed, if it is not yet at equilibrium.

20.16. Give an argument in support of Le Chatelier's effect of temperature on reaction equilibrium: increases in temperature shift an exothermic reaction towards the reactants, where the opposite is true for endothermic ones.

20.17. Show that the temperature dependence of K_C for ideal gases can be related to a standard change in energy of reactions,

$$\Delta e = \sum_i \nu_i e_i$$

where e_i gives the per-particle pure-phase internal energy of species i.

20.18. Explicitly derive the fluctuation formula in Eqn. (20.56). Moreover, demonstrate that $\sigma_{N_A}^2 = \sigma_{N_B}^2$ by a simple argument: show that, if two variables $x + y =$ constant, then $\sigma_x^2 = \sigma_y^2$.

20.19. Consider two separate volumes containing identical compositions of an ideal mixture of A and B. A catalyst is added to each that enables the reaction A $\leftrightarrow$ B to come to equilibrium. Subsequently, the catalyst is removed from container 1 such that the reaction can no longer take place. The catalyst is left in container 2, however. Afterwards, the temperature of both containers is increased by a small amount ΔT, isobarically. It is found that the heat added to container 2 is always greater than that for container 1. Show that the difference in added heat is given by

$$Q_2 - Q_1 = \frac{(\Delta h^*)^2 \, \Delta T}{k_B T^2} \frac{K_{eq}}{(1 + K_{eq})^2} N_0$$

where Δh^* is the enthalpy of reaction, T is the initial temperature, K_{eq} is the equilibrium constant, and N_0 is the total number of moles of A and B.

20.20. A porous, solid catalyst facilitates the reaction 2A $\leftrightarrow$ B, where A and B exist in a bulk gas phase around the catalyst. The reaction mechanism involves the adsorption of the species onto a catalytic site at the interface of the solid surface and the gas phase. The solid interface can be considered a two-dimensional surface with volume V and density of sites ρ_S such that the total number of sites is $N_S = \rho_S V$. No reactions take place in the gas phase; reactions occur only in the adsorbed phase at the surface. The temperature throughout is T and the gas-phase pressure is maintained at constant P.

(a) Consider first the adsorbed phase. Let N_A and N_B denote the numbers of adsorbed species. How many microstates are there due to degeneracies in placing these molecules on N_S sites?

(b) Assume that the species are present at dilute concentrations such that the adsorbed molecules can be considered independent of one another. Write an expression for the adsorbed-phase free energy A as a function of N_S and the fractional coverages $x_A = N_A/N_S$ and $x_B = N_B/N_S$. Your answer should make use of Stirling's approximation and include the single-molecule adsorbed-phase partition functions q_A and q_B. Note that you will have to account for your result in part (a).

(c) Find an expression for the chemical potentials of the adsorbed phase species, $\mu_A(T, x_A, x_B)$ and $\mu_B(T, x_A, x_B)$. Notice that $x_A + x_B \neq 1$ in this case.

(d) Show that the reaction equilibrium in the adsorbed phase takes the form

$$\frac{x_B(1 - x_A - x_B)}{x_A^2} = K_{eq}$$

and find an expression for K_{eq}.

20.21. Consider an ideal binary solution containing two components, A and B. Initially, there are N_0 moles of A in the container, with no B. When a catalyst is added, the reversible isomerization reaction A $\leftrightarrow$ B can take place. The following information is known about the system. For the reaction, the equilibrium constant at T_0 is $K_{eq,0}$ and the enthalpy of reaction Δh^* is independent of temperature. The pure-component properties are given in Table 20.1.

(a) The catalyst is added to the solution, which is held at temperature T_0. Find the equilibrium extent of reaction ξ, which denotes the number of B molecules that are present such that ξ can vary between 0 and N_0.

(b) The solution (with catalyst) is then heated very slowly at constant pressure well below its boiling temperature. The net heat flow is measured and an apparent heat capacity is computed from $c_{P,\mathrm{app}} = \delta Q/dT$. Note that this heat capacity accounts for changes in enthalpy due to both sensible heat and reaction heats. Write an expression for $c_{P,\mathrm{app}}$ in terms of $K_{eq}(T)$ and the variables defined above. Hint: recall that for an ideal solution the enthalpy of mixing is zero, so one can write $H = N_A h_A + N_B h_B$, where h_A and h_B are the pure-component molar enthalpies.

Table 20.1 Pure-component properties for Problem 20.21

Property	Pure A	Pure B
Boiling temperature	$T_{b,A}$	$T_{b,B}$
Vaporization enthalpy	$\Delta h_{\mathrm{vap},A}$	$\Delta h_{\mathrm{vap},B}$
Molar liquid-state heat capacity (T-independent)	$c_{P,A}$	$c_{P,B}$

(c) The solution continues to be heated. At some temperature T_b, the mixture will begin to boil and an infinitesimally small amount of a vapor phase will be formed. Assume that the vapor phase is ideal and that no reaction takes place in it. Write three equations that can be solved simultaneously to predict T_b and the corresponding liquid and vapor compositions (x_A and y_A) at this point.

20.22. Repeat part (a) of the previous problem for the case that the solution had not been ideal but had followed the behavior of a binary regular solution, for which $G_{ex} = -wNx_A(1 - x_A)$, where w is a positive constant. Would your answer be modified? If so, provide a new equation that could be solved to determine the value of ζ.

Applied problems

20.23. A semi-permeable, immovable membrane separates two compartments of equal volumes V. On the LHS there is an ideal mixture of A and B. On the RHS there is pure A. The membrane is impermeable to B but permeable to A. Also on the LHS there is a catalyst that enables the reaction $A \leftrightarrow 2B$, whose equilibrium constant K_{eq}° is known at $T = T_0$ and $P = P_0$. Both reaction and osmotic equilibrium are established in this system.

(a) If A and B are liquids with equal molar volumes v, find an expression for the equilibrium constant as a function of pressure, $K_{eq}(P)$, using its known value at P_0. Assume ideal-solution behavior and constant temperature at T_0.

(b) Repeat (a) for the case that A and B are ideal gases.

(c) Consider the case that A and B are ideal gases and the system is held at T_0. Assume N_0 molecules of A are added to the system originally, with no B present. At equilibrium, how many molecules of A and B are to be found on each side of the container?

20.24. Two containers exposed to the atmosphere both initially hold pure liquid species A at temperature $T_0 = 300$ K. A catalyst that permits the decomposition reaction $A \leftrightarrow 2B$, where B is also a liquid, is then added identically to each. The reaction is endothermic and Δh_{rxn} is, to a very good approximation, independent of temperature. The first container is maintained at a constant temperature, and at equilibrium it is found to have a mole fraction $x_A = 0.05$. The second container, on the other hand, is insulated and is found to have $x_A = 0.17$ and a lower final temperature. Note that the heat capacity of A is constant at $c_{P,A} = 350$ J/mol K, and that the solutions behave ideally.

(a) Estimate the final temperature of the second container and the enthalpy of reaction.

(b) You become concerned A might freeze in the second container during this process. If pure A has a heat of fusion of roughly 5 kJ/mol and a melting temperature of $T_m = 270$ K, should you be worried?

20.25. Consider the liquid–vapor equilibrium of a particular substance X. It is found that, upon addition of a catalyst to the liquid, X can exist in two forms, either as a monomer A or as a dimer B. The reaction 2A $\leftrightarrow$ B is reversible, very rapid, and occurs only in the liquid. The equilibrium constant $K = x_B/x_A^2$ is known to have a value $K_1 = 6$ at T_1. You might want to assume ideal behaviors.

(a) At a given temperature T_1 separate catalyst-free experiments are used to measure the vapor pressures of pure A (P_A^{vap}) and pure B (P_B^{vap}). The latter is found to be negligible at T_1. Subsequently, catalyst is added and the vapor pressure of X is measured, $P_1 = P_X^{\text{vap}}$. Find an expression for it in terms of the other properties defined here.

(b) For the system with catalyst, the temperature is increased by a small amount ΔT to $T_2 = T_1 + \Delta T$, and a change in vapor pressure is measured, $\Delta P = P_2 - P_1$. The enthalpy of vaporization of pure A is known, Δh_A^{vap}. Find an expression you could use to estimate the enthalpy of reaction, Δh_{rxn}. Hint: find the T derivative of the expression you determined in part (a).

(c) For what kind of reaction – endothermic or exothermic – could the vapor pressure actually decrease with temperature? Explain.

20.26. Consider the decomposition reaction involving ideal gases A and B:

$$A \overset{K_{\text{eq}}}{\leftrightarrow} 2B$$

A rigid, well-insulated container containing only pure A is prepared at temperature $T_1 = 300$ K and pressure $P_1 = 1$ bar. A catalyst is added and the container is immediately sealed. Eventually the gas is found to be 0.55 A by mole fraction and with $T_2 = 350$ K. It is known that the molar heat capacity of A is $c_{V,A} \approx 3k_B$ and roughly twice that of B.

(a) Estimate the fraction of A that has reacted, f, and the final pressure, P_2.

(b) Estimate the heat of reaction on an A basis, $\Delta h^* = 2h_B^* - h_A^*$, at temperature T_2 and in units of J/mol K.

(c) Estimate the final mole fraction of A if the container had instead been maintained at constant T_1 and P_1.

20.27. Binding of one molecule to another can be described in terms of reactions even if no chemical bonds are formed. It may be possible for a given receptor molecule to bind multiple ligand molecules, as is common for proteins and their ligands inside cells. In cases where the binding of multiple ligands to a single receptor happens in an all-or-none fashion, the binding is termed *cooperative*. The Hill model poses a general reaction scheme for this kind of behavior:

$$R + nL \leftrightarrow RL_n$$

where R denotes receptor and L ligand. This reaction mechanism states that the binding can happen only if n different ligands all bind at once, and n specifies the degree of cooperativity. For noncooperative transitions, $n = 1$. For cooperative ones, $n > 1$.

(a) Assume a solution is prepared with total concentration of receptors $[R]_0$. To this solution ligand is then added in excess such that its free concentration is $[L]$. Find an expression for the fraction of receptors bound to ligand, $f = [RL_n]/[R]_0$, as a function of the ligand concentration and the dissociation equilibrium constant $K_D \equiv [R][L]^n/[RL_n]$.

(b) If f were measured as a function of $[L]$, propose a way to determine the degree of cooperativity n using a linear regression analysis.

(c) Sketch f versus $[L]$ for noncooperative binding ($n = 1$), cooperative binding ($n > 1$), and negative or anti-cooperative binding ($0 < n < 1$). What are the slopes of these three curves at the origin, $[L] = 0$, in terms of K_D and n?

20.28. Surfactants are molecules that have both hydrophilic and hydrophobic groups. In aqueous solution, they can self-assemble into micelles in order to reduce the exposure of their hydrophobic groups to water. This process can be described using the reaction scheme:

$$nS \overset{K_{eq}}{\leftrightarrow} M$$

where S denotes a free surfactant molecule and M a micelle made of n surfactant molecules. A solution has a constant total molar concentration of surfactants $[S]_0 = [S] + n[M]$. Define the fraction of free surfactants as x and let the concentration $[S]_0$ at which $x = 1/2$ be denoted by c^*, the so-called critical micelle concentration (CMC). Because n is typically large, it is found that almost all surfactants exist as free monomers in solution when $[S]_0$ is below the CMC; above it, almost all participate in micelles.

(a) Find an equation that can be solved to give x as a function of the dimensionless ratio $[S]_0/c^*$ and n. Be sure to provide an expression for c^* in terms of n and K_{eq}.

(b) Consider two different temperatures T_1 and T_2. Show that the ratio of the CMCs at these two temperatures is approximately

$$\ln\left(\frac{c_2^*}{c_1^*}\right) \approx \frac{\Delta h}{k_B}\left(\frac{1}{T_2} - \frac{1}{T_1}\right)$$

where Δh is the standard enthalpy change *per monomer* for forming a micelle.

(c) Is it favorable for a free monomer with water-exposed hydrophobic groups to transfer into a micelle? Does this imply that the CMC increases or decreases with temperature?

FURTHER READING

H. Callen, *Thermodynamics and an Introduction to Thermostatistics*, 3rd edn. New York: Wiley (1985).

K. Denbigh, *The Principles of Chemical Equilibrium*, 4th edn. New York: Cambridge University Press (1981).

K. Dill and S. Bromberg, *Molecular Driving Forces: Statistical Thermodynamics in Biology, Chemistry, Physics, and Nanoscience*, 2nd edn. New York: Garland Science (2010).

T. L. Hill, *An Introduction to Statistical Thermodynamics*. Reading, MA: Addison-Wesley (1960); New York: Dover (1986).

D. A. McQuarrie, *Statistical Mechanics*. Sausalito, CA: University Science Books (2000).

A. Z. Panagiotopoulos, *Essential Thermodynamics*. Princeton, NJ: Drios Press (2011).

21 Reaction coordinates and rates

21.1 Kinetics from statistical thermodynamics

We introduced thermodynamics as the science of equilibrium and hence the science of the systems in the limit of infinite time, or when time is an irrelevant variable. In this chapter, we will now take a slight departure from that perspective and consider how one might determine *rates* of molecular processes. In order to extract such quantities, we will have to make certain assumptions about the equilibrium behavior and molecular trajectories that take us slightly beyond the mere province of thermodynamics and into the realm of kinetics. We will find, however, that thermodynamic quantities like the free energy are strong determinants of kinetic rate coefficients.

The archetypical picture for the role thermodynamics plays in the rates of chemical reactions is shown in Fig. 21.1. We envision some coordinate that takes us along a reaction pathway, from a reactant to a product state. Along this coordinate we track a free energy, which connects to the underlying interactions among the species involved. There is a barrier in free energy the system must overcome, $\Delta F^{\ddagger}$, that determines how fast the reaction can proceed. The barrier is often called the *activation free energy*. The state of the system when it is at the height of this barrier

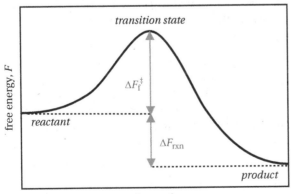

Figure 21.1. A generic reaction diagram shows how the free energy of a system changes as one moves along a reaction pathway. A free energy barrier $\Delta F_f^{\ddagger} > 0$ exists at the transition state, while an overall change in free energy occurs for the entire reaction $\Delta F_{rxn} < 0$. For the reverse reaction (product to reactant state), the free energy barrier is $\Delta F_r^{\ddagger} = \Delta F_f^{\ddagger} - \Delta F_{rxn}$, where "r" and "f" indicate the barriers with respect to the reverse and forward cases, respectively.

is called the *transition state*, or the *rate-limiting state*, and it is usually denoted with a superscript ‡ symbol. After reacting, if the process is favorable, there will be a net lowering in free energy at the product state.

In this chapter we will understand the origins of such diagrams in terms of microscopic properties. We will find that the emergence of this kind of picture really stems from many simplifications about molecular interactions and dynamics, but quite often it is able to give a quantitative description of the kinetic rates of simple reaction systems. This kind of analysis is called *classical transition-state theory*.

21.2   Macroscopic considerations for reaction rates

Let us first review the usual introductory approach to chemical kinetics. For so-called elementary chemical reactions, we often express a reaction rate as the product of a concentration-independent prefactor and the concentrations of reactants raised to their respective stoichiometric coefficients. Consider the simple reaction A + 2B ↔ C. The rates of change of the component concentrations might be written as

$$\frac{d\rho_A}{dt} = -k_f \rho_A \rho_B^2 + k_r \rho_C$$

$$\frac{d\rho_B}{dt} = -2k_f \rho_A \rho_B^2 + 2k_r \rho_C \tag{21.1}$$

$$\frac{d\rho_C}{dt} = k_f \rho_A \rho_B^2 - k_r \rho_C$$

Here, k_f and k_r are the *forward and reverse rate coefficients*, and the concentrations are expressed in terms of number densities. Of course, with a simple change of units, we could equally well express the rates in terms of molarities, $[i] = \rho_i / \mathcal{N}_A$. Each component changes concentration due to forward and reverse reactions. At equilibrium, setting any of the time derivatives in (21.2) equal to zero gives

$$\frac{\rho_C}{\rho_A \rho_B^2} = \frac{k_f}{k_r} \tag{21.2}$$

Chapter 20 showed that one condition for reaction equilibrium in ideal gases and solutions is that the combination of variables on the LHS of Eqn. (21.2) must be equal to the equilibrium constant. Therefore, we find that the forward and reverse rate coefficients must satisfy

$$\frac{k_f}{k_r} = K_C \tag{21.3}$$

where K_C is the equilibrium constant on a concentration basis. Thus the rate constants, and even the forms of the forward and reverse rate expressions themselves in (21.1), are constrained by equilibrium considerations. That is, in the limit of long times the rate laws must lead to final concentration values that result in a minimization of the overall Gibbs free energy.

A natural question is why the rate laws should have the form suggested by the example in Eqn. (21.1), involving concentrations raised to stoichiometric coefficients. For an elementary reaction involving only a single step, these forms can be anticipated on a qualitative basis. Namely, one might think of concentration as related to the probability that a molecule is at a particular location in space. For the example above, therefore, the probability that a forward reaction event will occur at some location is proportional to the probability that an A molecule and two B molecules are there simultaneously. Hence, the total probability scales as $[A][B]^2$, where the coefficient k_f then accounts for the details of the actual mechanism once all three molecules (one A and two Bs) are co-located. A similar argument could be made for the reverse rate expression.

Of course, these are merely rough arguments. While this recipe for writing the rate laws is supported by experimental data for many simple reactions, it is not a rigorous principle. There are reactions that do not follow the stoichiometric coefficients in their rate laws because they have more complicated microscopic mechanisms that cannot be described by such simple probability considerations. However, whatever the expression for the rate equation, we must find that the concentrations in the equilibrium limit – when the forward and reverse rates balance – satisfy the equilibrium prediction. If the system can be treated as an ideal gas or ideal solution, then of course the equilibrium prediction adopts one of the simple forms described in Chapter 20 involving ratios of concentrations raised to stoichiometric coefficients.

Rate coefficients can be temperature- and pressure-dependent, and, because their ratio gives an equilibrium constant, the dependences are constrained to follow specific forms. In Chapter 20 we found that the van 't Hoff relation gives the temperature dependence of the equilibrium constant,

$$\left(\frac{\partial \ln K_{eq}}{\partial T}\right)_P = \frac{\Delta h}{k_B T^2} \tag{21.4}$$

where Δh is a kind of standard enthalpy of reaction, depending on the nature of K_{eq}. It has been found that similar expressions are able to describe the temperature dependences of k_f and k_r. The so-called Arrhenius relationship postulates that

$$\left(\frac{\partial \ln k_f}{\partial T}\right)_P = \frac{E_f}{k_B T^2}, \qquad \left(\frac{\partial \ln k_r}{\partial T}\right)_P = \frac{E_r}{k_B T^2} \tag{21.5}$$

where E_f and E_r are called the *forward and reverse activation energies*. On subtracting the second of these expressions from the first, we obtain

$$\left(\frac{\partial \ln(k_f/k_r)}{\partial T}\right)_P = \frac{E_f - E_r}{k_B T^2} \tag{21.6}$$

Because $K_{eq} = k_f/k_r$, comparison of Eqn. (21.6) with (21.4) shows that the activation energies must satisfy

$$\Delta h = E_f - E_r \tag{21.7}$$

For temperature-independent activation energies, one can integrate Eqn. (21.5) to find an explicit form for the rate coefficients,

$$k_f = A \exp\left(-\frac{E_f}{k_B T}\right), \qquad k_r = B \exp\left(-\frac{E_r}{k_B T}\right) \tag{21.8}$$

where the ratio A/B is also constrained by the equilibrium constant. These expressions embody the typical *Arrhenius form* of the coefficients' dependence on temperature. It is important to note that Eqn. (21.8), while widely used, must be taken as an empirical relationship. Shortly we will find that a more general form, called the *Eyring equation*, can be found from molecular principles:

$$k_f = \frac{k_B T}{h} \exp\left(-\frac{\Delta F_f^{\ddagger}}{k_B T}\right), \qquad k_r = \frac{k_B T}{h} \exp\left(-\frac{\Delta F_r^{\ddagger}}{k_B T}\right) \tag{21.9}$$

where $\Delta F_f^{\ddagger}$ gives the activation free energy on going from the reactant to the product state, whereas $\Delta F_r^{\ddagger}$ gives that for the reverse process. (We have not yet specified what type of free energy, since this will depend on the particular constant conditions.) In Fig. 21.1, for example, $\Delta F_f^{\ddagger}$ is shown explicitly. That diagram also easily shows that the overall change in free energy for the process would be

$$\Delta F_{rxn} = \Delta F_f^{\ddagger} - \Delta F_r^{\ddagger} \tag{21.10}$$

This is clearly consistent with the expressions in (21.9). We can see that by dividing one of the two by the other and taking the logarithm,

$$\ln\left(\frac{k_f}{k_r}\right) = \frac{\Delta F_r^{\ddagger} - \Delta F_f^{\ddagger}}{k_B T} \tag{21.11}$$

and using Eqn. (21.10), we obtain

$$\ln K_{eq} = -\frac{\Delta F_{rxn}}{k_B T} \tag{21.12}$$

In other words, the Eyring form for the rate coefficients naturally shows that the equilibrium constant is related to a standard free energy of reaction, which we know must be the case from the considerations in Chapter 20. We now proceed to show the molecular origins of this rate law.

21.3   Microscopic origins of rate coefficients

To perform a microscopic analysis of reaction rates, we need to introduce the concept of a *potential energy landscape* (PEL). The PEL is simply the potential energy as a function of the configurational degrees of freedom, envisioned as a surface in a highly dimensional space. If we have N spherically symmetric particles, then there are $3N$ positional degrees of freedom. We envision the PEL as a plot of the function $U(\mathbf{r}^N)$ in a

$(3N + 1)$-dimensional world, where there is one axis for each positional degree of freedom and one for the potential energy.

> A **potential energy landscape** is the representation of the potential energy function U (r^N) in a high-dimensional space. For a system of N structureless particles, the landscape is a hypersurface in a $(3N + 1)$-dimensional space.

What is the rationale for thinking in terms of such a high-dimensional projection? The landscape perspective implies that we are interested in the *topography* of the PEL: what kinds of minima, maxima, and saddle points it has, and how they are connected to each other. Any point on the landscape specifies a particular configuration of all atoms and its corresponding potential energy. As a result, the dynamic evolution of a system corresponds to a pathway on this bumpy, multidimensional surface. It is this topography and connectivity of features that determines the dynamics of a system since, as we will see, the rate-limiting steps typically involve hops over saddles in the landscape.

To keep things at a manageable level with a number of dimensions that we can partially visualize, we consider a very simple reaction. The treatment that follows is based on the text of Hill, to which the reader is referred for more details. The reaction is

$$A + BC \leftrightarrow AB + C \qquad (21.13)$$

For simplicity, we will assume that the species A, B, and C are atoms and that the reaction involves the breaking and formation of chemical bonds. This is not necessary to develop the subsequent theory, but it will simplify the presentation. For example, the species could be molecular and the reaction would then entail more complex bonding patterns. Alternatively, the complexes BC and AB need not involve covalent bonds at all, but could be stabilized by physical interactions, such as van der Waals forces or hydrogen bonding. In any case, we should note that the empirical rules for elementary reaction rates discussed in the previous section would suggest a forward rate law of the form

$$\frac{\text{molecules reacted}}{\text{time} \cdot \text{volume}} = k_f \rho_A \rho_{BC} \qquad (21.14)$$

where, as indicated explicitly, the rate is in units of molecules of product formed per volume and per time.

In this simple reaction there are three species and thus nine configurational coordinates contributing to the PEL. However, six of these are usually redundant because of translational and rotational symmetries; we could specify the potential energy of the system simply by knowing the three pairwise separation distances r_{AB}, r_{BC}, and r_{AC}. In this case, therefore, the PEL can be envisioned in a four-dimensional space. The problem becomes even simpler if we assume that the reaction proceeds via the mechanism shown in Fig. 21.2: A approaches BC in collinear form and the three atoms form a linear transition state, or *activated complex*, before dissociation of C occurs. We will denote the activated complex by the notation ABC, although $(ABC)^{\ddagger}$ could be appropriate as well.

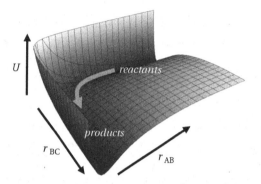

Figure 21.2. A simplified mechanism assumes that the reaction in Eqn. (21.13) occurs in a collinear fashion. In this case, only two degrees of freedom are needed to describe the system at any point in time, for instance r_{AB} and r_{BC}.

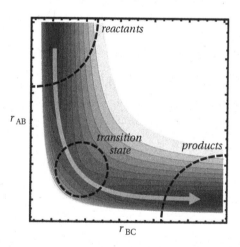

Figure 21.3. The potential energy landscape (PEL) of the simple reaction of Fig. 21.2 can be projected in three-dimensional space. In the reactant state, r_{BC} is small and r_{AB} is large, but the opposite is true for the products. During the reaction, the system moves along a pathway in the PEL (gray arrow) requiring a temporary increase in potential energy and concerted motions in r_{AB} and r_{BC}.

Figure 21.4. The three-dimensional PEL of Fig. 21.3 can be represented as a contour plot, where darker coloring indicates states of lower potential energy. The particular reaction mechanism shown by the gray arrow takes the lowest-energy pathway and passes through the transition state.

Figures 21.3 and 21.4 show that the PEL of the collinear reaction mechanism is simple enough to be visualized. In principle, these are *projections* of the true PEL along the coordinates r_{AB} and r_{BC} given the constraint that $r_{AC} = r_{AB} + r_{BC}$. Such projections from higher- to lower-dimensional subspaces along relevant degrees of freedom are common in analyses of PELs. In any case, notice that the landscape for this system has two deep basins. One lies at small values of r_{AB} and large values of r_{BC} and hence is the reactant state. The other basin lies at the product state, with opposite conditions for the distances. In order to proceed from one state to another, the system must traverse this surface along a path that connects the two basins. There is a single *saddle point* separating the two minima and this occurs in the *transition-state region*.

Now that we have seen a specific example, we can ask a more general question: from where does a PEL like that shown in Figs. 21.3 and 21.4 come? Recall that PELs give the projection of the system potential energy as a function of the configurational coordinates. For chemical reactions, we must almost always use a quantum description in order to compute these potential energies because bond breaking and bond formation are not well described by simple classical expressions. This typically entails solving Schrödinger's equation for the electrons in the system for each nuclear configuration (specified in the example by r_{AB} and r_{BC}) to obtain the ground-state electronic energy. Then, the results for all states (all combinations of r_{AB} and r_{BC}) are used to construct an energy hypersurface. This can be a challenging task, even for the kind of simple reaction of Eqn. (21.13), and the computation of PELs using numerical methods remains an active research field.

As mentioned earlier, reactions do not always involve the breaking and formation of covalent bonds. For events involving the association or dissociation of molecules driven by *physical* interactions, such as the complexation of two DNA strands into a helical structure, approximate classical expressions for the potential energy are often sufficient to produce an accurate PEL and no quantum electronic structure calculations are needed. Still, such cases can also be extremely challenging due to the numbers of degrees of freedom that need to be considered and the high dimensionality of the PEL, particularly if the reaction occurs in a solvent like water.

We will now make a number of approximations that describe the system's trajectory over the PEL. These ideas constitute classical transition-state theory, a field originated by Eyring, Evans, Polanyi, and Wigner in the 1930s.

Classical transition state theory makes the following assumptions about the trajectory a system takes in the PEL during a reactive process.

(1) In going from reactants to products, a system takes the lowest-energy trajectory in its PEL. This involves it passing through the saddle point on the landscape.

(2) The system at the transition state – the so-called activated complex – can be considered a distinct species in the system. It corresponds to the region of the PEL immediately surrounding the saddle point.

(3) The activated complex is in equilibrium with the reactant state.

(4) Once the system has reached the transition state, regardless of how it arrived there, it is 50% likely to convert to a product and 50% likely to convert back to a reactant.

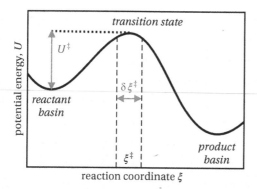

Figure 21.5. The projection of the PEL along the reaction coordinate ξ shows a reactant basin, a product basin, and a transition-state saddle point at $\xi^{\ddagger}$. The transition state has energy $U^{\ddagger}$ with respect to the reactants and its width along the reaction coordinate is designated by $\delta\xi$.

Many of these assumptions may seem less severe for systems that are thermalized, that is, maintained at constant temperature by coupling to a heat bath. For example, the Boltzmann factor $\exp(-\beta U)$ suggests that configurations of lower potential energy have exponentially higher probability. In turn, the assumption that the system follows the lowest-energy trajectory might be viewed as it following the most likely trajectory. Moreover, the assumption that the transition state is in equilibrium with the reactant state simply means that both are thermalized by the same bath, such that the relative populations of each are given by partition functions and Boltzmann factors.

Figure 21.5 shows that we can project the PEL to an even lower-dimensional space than that of Figs. 21.3 and 21.4 if we consider the reaction progress along the lowest-energy trajectory. We use the variable ξ to indicate a *reaction coordinate* that quantifies the position on this trajectory. The exact numerical range for ξ is immaterial at this point, except for two aspects. First, we say that it adopts a specific value $\xi^{\ddagger}$ at the saddle point. Second, the transition state involves the region in the PEL in the immediate vicinity of the saddle, as shown in Fig. 21.4, which we quantify by a small width $\delta\xi^{\ddagger}$ along the reaction coordinate centered around $\xi^{\ddagger}$. This is illustrated in Fig. 21.5.

The most important assumption stated above is the equilibrium condition between reactants and the transition state. We will further assume that the different components are independent and non-interacting; for example, they are ideal gases or solutions. According to Chapter 20, therefore, the concentrations of these species at equilibrium can be related to simple partition functions,

$$\frac{\rho_{ABC}}{\rho_A \, \rho_{BC}} = \frac{q_{ABC}/V}{(q_A/V)(q_{BC}/V)} \tag{21.15}$$

where ρ gives the number density in molecules per volume and the q terms are the single-molecule partition functions of each component. Some simple rearrangement gives an expression for the concentration of the activated complex, which ultimately we will use to compute a rate. Indeed one might anticipate that the rate of reaction would be proportional to the concentration of complex present,

$$\rho_{ABC} = \rho_A \rho_{BC} \frac{q_{ABC}/V}{(q_A/V)(q_{BC}/V)} \qquad (21.16)$$

It is interesting to note that Eqn. (21.16) already bears some resemblance to the empirical rule expected in (21.14) in that it contains a product of reactant concentrations. The coefficient then involves the molecular partition functions. To calculate the q terms, one sums microstates that are restricted to each component's corresponding region in the PEL: the basins for the products and reactants, and near the saddle point for the transition state. Often the microstate sums need to be determined from quantum-mechanical considerations, certainly if chemical bonds are involved in the reaction. Depending on the complexity of the reaction, the q terms could also include other degrees of freedom due to nuclear states as well.

However, there is a subtlety with these partition functions. Each represents a sum of states in the PEL, but how does one determine the region of configurations over which this sum should be performed? Figure 21.4, for example, shows dashed lines that delineate states, and the question becomes that of whether or not the exact locations of such lines matter. For the reactants and products, it turns out that we do not need to pick an exact boundary because these states sit in basins in the landscape. As one moves away from their minima, states make exponentially lower and ultimately vanishing contributions to the partition function according to the Boltzmann factor. Therefore, as long as the boundary is wide enough, it is easy to find values for partition functions like q_A and q_{BC}.

On the other hand, the partition function q_{ABC} is more complicated. The activated complex represents an *unstable* state since it is at a saddle point in the PEL. Small perturbations away from it along the reaction coordinate ξ will lower the energy and lead to configurations of increased probability. However, this is only the case for movement along the reaction coordinate. Perturbations away from the saddle point in directions that are perpendicular or *orthogonal* to ξ increase the energy, as in the case of the reactants. This means that the boundary we choose for q_{ABC} along ξ must be treated with some care. Examples 21.1 and 21.2 illustrate these complexities.

Example 21.1 *Consider a simple minimum in a toy energy landscape of three dimensions, as shown in Fig. 21.6. Let the potential near the minimum be harmonic, with $U(x, y) = a(x^2 + y^2)$, where a is a constant characterizing the basin's curvature. Show that the configurational partition function is insensitive to the exact boundaries so long as they are several $k_B T$ above the minimum in energy.*

Let us construct the configurational part of the partition function by assuming that each of x and y varies between $-l$ and l:

$$q = \int_{-l}^{l} \int_{-l}^{l} e^{-a\beta(x^2+y^2)} \, dx \, dy$$

Table 21.1 Convergence of the partition function in Example 21.1

	$n = 1$	$n = 2$	$n = 3$	$n = 4$
q/q_{all}	0.71014	0.99067	0.99996	0.99999...

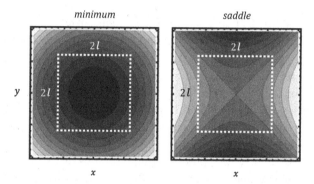

Figure 21.6. Two "toy" energy landscapes in a contour-plot representation, with darker colors giving lower potential energies U. The left landscape is a harmonic minimum, while the right one is a harmonic saddle. These PELs are examined in Examples 21.1 and 21.2, in which partition functions are computed for the regions $x \in [-l, l]$ and $y \in [-l, l]$, which are illustrated by the dotted lines.

The two integrals are separable and each evaluates to the error function,

$$q = \left[\sqrt{\frac{\pi}{\alpha\beta}} \, \mathrm{erf}(l\sqrt{\alpha\beta}) \right]^2$$

If the integral were performed over all of space, $l \to \infty$, we would find

$$q_{all} = \frac{\pi}{\alpha\beta}$$

For finite l, the relative value of the partition function is

$$\frac{q}{q_{all}} = \mathrm{erf}(l\sqrt{\alpha\beta})^2$$

When $l = n/\sqrt{\alpha\beta}$, the minimum energy along the boundary of the integral is nk_BT. Therefore, we might write

$$\frac{q}{q_{all}} = \mathrm{erf}(n)^2$$

This expression gives the relative value of the partition function if we integrate over the PEL by the same range in each x, y direction to include all energies less than nk_BT. Let us take some values for n to see how quickly the partition function converges to its "full" value Table 21.1 shows that even small values of n quickly converge the partition function relative to its "full" value over all of space. We see that, if we include regions in the PEL that are only $2k_BT$ in energy above the minimum of a harmonic basin, the partition function has converged to 99% of its full value. This means that extending the boundaries further, in this case by increasing l, does not significantly change the result for q.

Table 21.2 Convergence of partition functions in Example 21.2

	$n = 1$	$n = 2$	$n = 3$	$n = 4$
$q_x/\sqrt{q_{\text{all}}}$	0.84270	0.99532	0.99998	0.99999...
$q_y/\sqrt{q_{\text{all}}}$	1.6504	18.565	1,630.0	1.2970×10^6
q/q_{all}	1.3908	18.478	1,630.0	1.2970×10^6

Example 21.2 *Repeat the previous example for the case of a harmonic saddle, with*
$u = \alpha(x^2 - y^2)$.

In this case the partition function is still separable, but the x and y integrals evaluate differently,

$$
\begin{aligned}
q &= \left(\int_{-l}^{l} e^{-\alpha\beta x^2} \, dx \right) \left(\int_{-l}^{l} e^{\alpha\beta y^2} \, dy \right) \\
&= \left[\sqrt{\frac{\pi}{\alpha\beta}} \, \text{erf}(l\sqrt{\alpha\beta}) \right] \left[\sqrt{\frac{\pi}{\alpha\beta}} \, \text{erfi}(l\sqrt{\alpha\beta}) \right] \\
&= q_x q_y
\end{aligned}
$$

where erfi is the imaginary error function. In the last line, we defined two contributions to the partition function, one each from the x and y degrees of freedom. Let us perform a similar analysis to what we did before, letting $q_{\text{all}} = \pi/(\alpha\beta)$ and describing the integration domain in $k_B T$ units using $l = n/\sqrt{\alpha\beta}$. Table 21.2 gives some numerical results. We see that the q_x contribution to the total partition function converges rapidly with the boundary around the saddle, as was the case for Example 21.1. In contrast, the q_y contribution grows in an unbounded fashion. This is due to the fact that energies increase in the direction of x but decrease along y; in the latter case, the Boltzmann factors become exponentially larger as one moves away from the saddle point.

The implication of Examples 21.1 and 21.2 is that a partition function computed for a state corresponding to a basin in the PEL is insensitive to the exact boundaries of that state, so long as one includes energies a few $k_B T$ above the minimum. On the other hand, partition functions computed for saddles are very sensitive to the boundary, but only along the direction for which the potential energy decreases. As in Example 21.2, if we can isolate this direction, then the other contributions to the partition function converge for lenient enough boundaries on the remaining degrees of freedom.

This is exactly what we will attempt to do for the transition-state complex ABC: we will isolate the direction of negative curvature in the PEL around the saddle point from the others that have positive curvature. In fact, we know exactly this direction in that it is given by the reaction coordinate ξ at the transition state $\xi^{\ddagger}$. For the purposes of illustration, we will assume that the molecular partition functions can be treated classically and that all three species have the same mass and hence identical thermal de Broglie wavelengths $\Lambda(T)$. For the activated complex, we then can write

$$q_{ABC} = \frac{1}{\Lambda(T)^9} \int e^{-\beta U} \, d\mathbf{r}^3 \tag{21.17}$$

where $\mathbf{r}^3$ contains the coordinates of the three atoms and the integral is restricted to only those conformations lying close to the saddle point on the PEL. To compute this partition function, we perform a change of coordinates so that one of the integrals is exactly along the direction of ξ and the remaining eight are conjugate to it, denoted by directions $s_1, s_2, \ldots, s_8$ as conceptually illustrated in Fig 21.7. This enables us to write the integral as

$$q_{ABC} = \frac{1}{\Lambda(T)^9} \int e^{-\beta U(\mathbf{r})} \, d\xi \, ds_1 \, ds_2 \ldots ds_8 \tag{21.18}$$

It is worth noting that, in general, the change of variables could introduce a Jacobian determinant, although we will not consider that possibility here. Since the integral in Eqn. (21.18) is performed over the transition state conformations, the ξ coordinate varies over the very small differential slice $\delta\xi^{\ddagger}$ surrounding $\xi^{\ddagger}$. We can do that part of the integral explicitly and factor the partition function as

$$
\begin{aligned}
q_{ABC} &= \left[\frac{1}{\Lambda(T)^8} \int e^{-\beta U} \, ds_1 \, ds_2 \ldots ds_8 \right] \frac{\delta\xi^{\ddagger}}{\Lambda(T)} \\
&= \left[\frac{1}{\Lambda(T)^8} \int e^{-\beta(U-U^{\ddagger})} \, ds_1 \, ds_2 \ldots ds_8 \right] \frac{\delta\xi^{\ddagger}}{\Lambda(T)} e^{-\beta U^{\ddagger}} \\
&= q^{\ddagger} \frac{\delta\xi^{\ddagger}}{\Lambda(T)} e^{-\beta U^{\ddagger}}
\end{aligned}
\tag{21.19}
$$

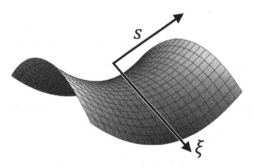

Figure 21.7. A three-dimensional schematic representation of the coordinate charge used to describe the transition state. Instead of the usual particle coordinates, directions in the energy landscape are described by the reaction coordinate ξ and conjugate directions s.

In the second line, we simply factored out the potential energy at the saddle point, $U^{\ddagger}$, as a constant. In the last line, we performed the integration over the conjugate degrees of freedom s_i and let this be equal to the partial transition state partition function $q^{\ddagger}$. Since the energy increases along these directions, we can converge $q^{\ddagger}$ for any reasonable boundaries around the transition state, just as we did with q_x in Example 21.2.

The rate of reaction is given by the amount of product produced per volume and per time. To assess the time component, we need to determine how long the system spends in the transition state. First, we find the speed with which the system moves along the reaction coordinate. The average value of the absolute velocity along any one molecular coordinate is given by the Maxwell-Boltzmann distribution; here the velocity of interest is that corresponding to the degree of freedom ξ,

$$\langle v_{\xi} \rangle = \int |v_{\xi}| \wp(v_{\xi}) dv_{\xi}$$

$$= \left(\frac{2k_B T}{\pi m} \right)^{1/2} \tag{21.20}$$

The amount of time it takes a molecule, on average, to leave the transition state region is then given by its width $\delta \xi^{\ddagger}$ divided by the speed,

$$\delta t = \delta \xi^{\ddagger}/v_{\xi}$$

$$= \delta \xi^{\ddagger} \left(\frac{2k_B T}{\pi m} \right)^{-1/2} \tag{21.21}$$

Finally, the net rate of molecules moving to the product state is given by the number of activated complexes times half of the average rate at which they cross it. The half factor accounts for the 50% probability that the activated complex will actually proceed to the product state. Combining Eqns. (21.16), (21.19), and (21.21) gives

$$\frac{\text{molecules reacted}}{\text{time} \cdot \text{volume}} = \rho_{ABC} \times \frac{1}{\delta t} \times \frac{1}{2}$$

$$= \rho_A \rho_{BC} \frac{q_{ABC}/V}{(q_A/V)(q_{BC}/V)} \times \frac{v_{\xi}}{d\xi^{\ddagger}} \times \frac{1}{2}$$

$$= \rho_A \rho_{BC} \frac{q^{\ddagger}/V}{(q_A/V)(q_{BC}/V)\Lambda(T)} \frac{\delta \xi^{\ddagger}}{e^{-\beta U^{\ddagger}}} \times \left(\frac{2k_B T}{\pi m} \right)^{1/2} \frac{1}{\delta \xi^{\ddagger}} \times \frac{1}{2} \tag{21.22}$$

After expanding the de Broglie wavelength and simplifying, Eqn. (21.22) becomes

$$\frac{\text{molecules reacted}}{\text{time} \cdot \text{volume}} = \rho_A \rho_{BC} \frac{q^{\ddagger}/V}{(q_A/V)(q_{BC}/V)} \left(\frac{k_B T}{h} \right) e^{-\beta U^{\ddagger}} \tag{21.23}$$

This implies that we can express the rate law in a form similar to Eqn. (21.14),

$$\frac{\text{molecules reacted}}{\text{time} \cdot \text{volume}} = k_f \rho_A \rho_{BC}$$

$$\text{where } k_f \equiv \frac{q^{\ddagger}/V}{(q_A/V)(q_{BC}/V)} \left(\frac{k_B T}{h} \right) e^{-\beta U^{\ddagger}} \tag{21.24}$$

Equation (21.24) is particularly powerful because it gives an expression for the rate constant in terms of molecular properties, namely, through the topography of the PEL. Note that we can consider the quantity $U^{\ddagger}$ as the potential energy of the transition state above that of the reactants since we can shift the energy surface so that the energy at the reactant minimum corresponds to $U = 0$. Therefore, this expression shows that the rate constant depends on the elevation in potential energy of the saddle point. It also, however, depends on the molecular partition functions for all of the species involved, which correspond to *integrals* over regions in the PEL. These partition functions depend not only on the energy of the reactant minimum and saddle point, but also on the curvature of the energy surface and the temperature. All of these contributions mean that a general expression for the rate constant is given by

$$k_{\mathrm{f}} = \left(\frac{k_{\mathrm{B}}T}{h}\right) e^{-\beta \Delta F_{\mathrm{f}}^{\ddagger}}$$

$$\text{where } \Delta F_{\mathrm{f}}^{\ddagger} \equiv -k_{\mathrm{B}}T \ln\left[\frac{q^{\ddagger}/V}{(q_{\mathrm{A}}/V)(q_{\mathrm{BC}}/V)}\right] + U^{\ddagger} \tag{21.25}$$

where $\Delta F_{\mathrm{f}}^{\ddagger}$ is a *free energy* corresponding to the forward reaction, as shown by the transition state barrier of Fig. 21.1. This shows that the rate constant depends on the relative free energy of the transition state complex, which includes both entropic and energetic contributions. The former are contained in the temperature dependence of $\Delta F_{\mathrm{f}}^{\ddagger}$, that is, in the first term of Eqn. (21.25).

In some cases, the assumption that half of the molecules at the transition state make it to the product state turns out to be poor, since one often finds fewer molecules passing over the saddle than would be expected from a purely random point of view. For such instances, a correction factor called a *transmission coefficient*, κ, is usually introduced. Then the rate coefficient is defined as

$$\frac{\text{molecules reacted}}{\text{time} \cdot \text{volume}} = k_{\mathrm{f}} \rho_{\mathrm{A}} \rho_{\mathrm{BC}}$$

$$\text{where } k_{\mathrm{f}} \cong \kappa \frac{q^{\ddagger}/V}{(q_{\mathrm{A}}/V)(q_{\mathrm{BC}}/V)} \left(\frac{k_{\mathrm{B}}T}{h}\right) e^{-\beta U^{\ddagger}} \tag{21.26}$$

21.4   General considerations for rates of rare-event molecular processes

The above considerations were applied to a specific, simple chemical reaction in order to illustrate the basic concepts of PELs and classical transition-state theory. However, the approach taken to find the rate – involving movement over a saddle point in some kind of low-dimensional landscape – can be applied to many rate problems, including those not involving chemical bond breaking or formation. For example, transition-state theory can be applied to the rate at which molecules hop from one adsorption site on a surface to another, the rate of propagation of crystal defects, or the rate of non-covalent binding.

The general philosophy of transition state theory is to determine a reaction coordinate ξ for the process of interest that takes the system along the most likely kinetic trajectory. The coordinate must be well defined so that it has a specific value for any given microscopic configuration, or, equivalently, each location on the PEL. It is usually defined so that it varies between 0 and 1 as the system moves from reactants to products. In the simplest of reactions, this coordinate typically follows the trajectory of lowest potential energy on the PEL. In general, however, determining the appropriate reaction coordinate for a molecular process is a challenging task and is the subject of much modern research.

If a kinetically relevant form of the reaction coordinate and pathway can be identified, it becomes possible to construct a free energy as a function of the internal degree of freedom ξ. Essentially, one divides the total partition function (and hence the free energy) into contributions from parts of the landscape that correspond to different values of ξ. Conceptually, for a single-component system, we write

$$Q(T, V, N, \xi) = \sum_{\text{all } n} e^{-\beta E_n} \delta_{\xi_n, \xi} \tag{21.27}$$

Here, the sum extends over all microstates of the system, namely all locations in the PEL. The delta function, however, selects out and includes only those points on the PEL where the reaction coordinate adopts a specified value. With (21.27) we can define a free energy as a function of the reaction coordinate at constant T, V, N conditions,

$$A(T, V, N, \xi) = - k_B T \ln Q(T, V, N, \xi) \tag{21.28}$$

This free energy corresponds to a partition sum over all degrees of freedom orthogonal to ξ. The relationship of this free energy to the true one is the following. We begin by summing the "reaction-coordinate partition function" in (21.27) over all possible values of ξ to recover the usual canonical one,

$$\sum_{\xi} Q(T, V, N, \xi) = \sum_{\xi} \sum_{\text{all } n} e^{-\beta E_n} \delta_{\xi_n, \xi}$$

$$= \sum_{\text{all } n} e^{-\beta E_n} \sum_{\xi} \delta_{\xi_n, \xi}$$

$$= \sum_{\text{all } n} e^{-\beta E_n}$$

$$= Q(T, V, N) \tag{21.29}$$

The free energy along the internal degree of freedom ξ in Eqn. (21.28) is therefore related to the total system free energy by the logarithm of (21.29),

$$A(T, V, N) = -k_B T \ln \left[\sum_{\xi} e^{-\beta A(T, V, N, \xi)} \right] \tag{21.30}$$

Many PELs are continuous or at least approximated as such, so these expressions really entail integrals rather than sums and Dirac delta functions rather than Kronecker ones. Conceptually, the properties of the coordinate partition function are no different. A general form might be

$$Q(T, V, N, \xi) = \frac{1}{\Lambda(T)^{3N} N!} \int \delta[\hat{\xi}(\mathbf{r}^N) - \xi] e^{-\beta U(\mathbf{r}^N)} \, d\mathbf{r}^N$$

(21.31)

$$Q(T, V, N) = \int Q(T, V, N, \xi) d\xi$$

where the function $\hat{\xi}(\mathbf{r}^N)$ returns the value of the reaction coordinate for any particular configuration $\mathbf{r}^N$.

The probability that the system is at a particular value of the reaction coordinate is given by a ratio of the partition functions,

$$\wp(\xi) = \frac{Q(T, V, N, \xi)}{Q(T, V, N)}$$

$$= \frac{e^{-\beta A(,\xi)}}{Q}$$

(21.32)

In the second line, we suppressed the dependence on all variables except for the reaction coordinate. Typically, we find that the probability $\wp(\xi)$ has a minimum at some value $\xi^{\ddagger}$ corresponding to a maximum in $A(\xi)$. Of course, this is the transition state. It is a bottleneck in the microstate probability in going from $\xi = 0$ to $\xi = 1$.

The major assumption of the classical theory is that the transition state is in equilibrium with the reactant state. We can invoke a similar but more general argument here in terms of probabilities,

$$\frac{\wp(\xi = \xi^{\ddagger})}{\wp(\xi = 0)} = e^{-\beta[A(\xi^{\ddagger}) - \beta A(0)]} = e^{-\beta \Delta A^{\ddagger}}$$

(21.33)

where $\Delta A^{\ddagger}$ gives the free energy increase in moving from reactants to the transition state. Equation (21.33) simply implies that both the reactant states and the transition states obey equilibrium probabilities as determined by Boltzmann factors.

The second assumption is that the bottleneck probability at the transition state is the key factor determining the rate at which the system can pass from $\xi = 0$ to $\xi = 1$. We assume that the rate of the molecular process is given by the probability that the system is at the transition state times the rate at which it leaves it to go to the products. The latter quantity sometimes depends linearly on the temperature, since T determines the average velocities of all the particles and hence the average speed with which the system traverses the PEL. Therefore, we might propose the general expression

$$\text{rate} = f(T) \exp[-\beta \Delta A^{\ddagger}]$$

(21.34)

To maintain generality, we introduce a prefactor $f(T)$ that is a simple function of temperature; it may be linear as in (21.25). It is also important to note that the concentrations of species involves are included implicitly in the free energy difference; consider, for example, that the Helmholtz free energy of an ideal gas is logarithmic in the molar density.

Equation (21.34) is rather striking because it emphasizes the tremendous importance that free energies along the reaction pathway play in determining the rates of molecular processes. The prefactor $f(T)$ is usually much less important than the quantity $\Delta A^{\ddagger}$ because the latter is exponentiated and thus has a more significant effect on the magnitude of the rate.

> Free energies often play a strong role in determining the kinetics of molecular processes because they determine the bottleneck probabilities of intermediate states, provided that the system is at equilibrium in those intermediate states.

Certainly these statements are predicated on the applicability of the assumptions of the theory, and the classical transition state picture does break down in many cases. One instance occurs when the transition state cannot be assumed to exist in thermal equilibrium with the reactants, and the probability that the system will cross it is strongly correlated with the specific trajectory the system is on. Moreover, there are molecular events that proceed not by crossing a single barrier with high energy, but by crossing a very large number of small barriers. Similarly, some processes do not take a single trajectory from reactant to product, but can follow one of a highly degenerate number of trajectories. Though not discussed here, there are many other kinetic theories that build on thermodynamic considerations, and indeed work on their development and generalization remains an active field.

Problems

Conceptual and thought problems

21.1. True or false? The entropies of each of the species do not contribute to the reaction rate coefficient. Explain.

21.2. Explain the error in logic of the following statements that describe classical transition-state kinetics.
   (a) The determinant of the rate of a reaction is the transition state's elevation in potential energy in the PEL.
   (b) The main temperature dependence of the rate of reaction stems from the average velocity of molecules according to the Maxwell–Boltzmann form.

21.3. Why does the curvature of the PEL along the reaction coordinate at the transition state not contribute to the reaction rate?

21.4. If the potential energy barrier at the transition state approaches zero, do you think classical transition state theory will give a reasonable description of the rate constant? Explain. Hint: when there is no longer a probability bottleneck, what kinetic process will determine the relevant time scales?

Fundamentals problems

21.5. Find a way to formally determine the direction of the reaction coordinate at the saddle point in an energy landscape. Let $\mathbf{r}_{\ddagger}^N$ give the coordinates of all of the atoms at the saddle and $U(\mathbf{r}_{\ddagger}^N)$ its potential energy. Hint: consider the concept of the Hessian matrix of second derivatives introduced in Chapter 14.

21.6. Explicitly derive $\wp(\xi) = Q(\xi)/Q$ from Eqn. (21.32). Hint: begin by summing microstate probabilities.

Applied problems

21.7. A dilute gas of molecules of species X is placed inside a cubic container of volume $V = L^3$. The overall bulk number density of X is ρ_b and the system is maintained at constant temperature T. On one of the six interior container interfaces, at $z = 0$, there is a catalytic surface that enables an isomerization reaction of X to form Y,

$$X \xrightarrow{k} Y$$

where k is the forward reaction rate coefficient and has units of inverse time. The reaction is irreversible, its rate is first order (in the number density of X at the surface), and no Y is present initially. The reaction takes place only within a region a distance λ away from the catalytic interface. In this region, X molecules also experience an attractive interaction with the interface of magnitude $-\epsilon$. Beyond λ, molecule–surface interactions are zero. The other walls do not interact with the species.

(a) Find an expression for the time $t_{1/2}$ that it takes for half of the X molecules to have reacted, using the constants above. In terms of varying attractive strength ϵ, what are the minimum and maximum possible values for this time? You may assume that the reaction occurs slowly enough that thermodynamic equilibrium throughout the container is maintained at all times for the density distribution of species X.

(b) Assume that the temperature dependence of the rate coefficient k is well described by an expression of the form $k = k_0 \exp(-E_a/k_B T)$, where k_0 and E_a are constants. Show that the time $t_{1/2}$ then has a minimum as a function of temperature if $\epsilon > E_a$.

21.8. A cubic volume of $V = L^3$ initially contains N_0 molecules of an ideal gas X. A strong electric field of magnitude $\mathcal{E}$ exists in the x direction of the container. Each X molecule has a charge q such that it experiences a potential energy of $q\mathcal{E}x$. In addition, X molecules react essentially irreversibly at constant temperature, according to the synthesis

$$2X \xrightarrow{k_f} Y$$

where k_f is the forward rate constant such that the rate of loss of X per volume per time is $k_f \rho_X^2$, where ρ_X is the local number density of X.

(a) Find an expression for the time $t_{1/2}$ it takes for half of the X to have reacted. Assume that the gas reaches equilibrium with respect to the field much faster than the reaction proceeds.

(b) If the rate constant has the form $k_f = k_0 \exp(-E_a/k_B T)$, where k_0 and E_a are both positive constants, then show that there can be a temperature at which $t_{1/2}$ is a minimum. Find an expression that could be solved to give this temperature.

21.9. An enzyme is a protein catalyst that operates in aqueous solutions. To a very rough approximation, an enzyme can be considered to exist in two possible states: an active form that is predominant at low temperatures and an inactive form at higher temperatures. One might consider the conversion between these two states to be a reaction of the form $A \leftrightarrow I$. Let the temperature at which an enzyme is equally likely to be in either state be denoted T_0. For many proteins, the difference in molar enthalpies between the two states ($\Delta h = h_A - h_I$) can be considered constant. In the following, you can also assume that the enzyme is dilute in solution.

(a) Find an expression for the fraction of enzyme that is active, f, as a function of T. What is the sign of Δh?

(b) A solution is prepared with a total molar concentration of enzyme c_0 and an initial dilute amount of a substrate $[S]_0$. The rate of decomposition of S in moles per concentration per time is approximately $-k[A][S]$, where $[A]$ indicates the concentration of active enzyme. The enzyme is not decomposed. Find an expression for the time it takes for half of the substrate to be decomposed, $t_{1/2}$, at a given temperature T. Present your answer in terms of $[A]$, k, c_0, and the enzyme-specific constants T_0 and Δ_h.

(c) The temperature dependence of the rate coefficient k might be described by the functional form $k = k_0 \exp(-E_a/k_B T)$, where the activation energy E_a is a positive number and k_0 is a constant. There can be an optimal temperature at which $t_{1/2}$ is at a minimum, i.e., where the net reaction is the fastest. Given that such a temperature exists, find an expression for it.

FURTHER READING

H. Callen, *Thermodynamics and an Introduction to Thermostatistics*, 3rd edn. New York: Wiley (1985).

K. Denbigh, *The Principles of Chemical Equilibrium*, 4th edn. New York: Cambridge University Press (1981).

P. Hänggi, P. Talkner, and M. Borkovec, "Reaction-rate theory: fifty years after Kramers," *Reviews of Modern Physics* **62**, 251 (1990).

T. L. Hill, *An Introduction to Statistical Thermodynamics*. Reading, MA: Addison-Wesley (1960); New York: Dover (1986).

D. A. McQuarrie, *Statistical Mechanics*. Sausalito, CA: University Science Books (2000).

22 Molecular simulation methods

Throughout this text we have presented a molecular foundation for the principles of thermodynamics that has considered many highly simplified molecular models of a variety of systems, including idealized gases, solutions, polymers, and crystals. It might not be immediately obvious how we extend these ideas to more realistic models that, for example, might entail structured molecules with complex potential energy functions including bonded, electrostatic, and van der Waals energies all at once. Even the seemingly simple task of developing the thermodynamics of a monatomic fluid at high densities, such as liquid argon, can be challenging owing to the difficulty of treating the detailed pairwise interactions in the configurational partition function.

Two routes enable one to move beyond the general statistical-mechanical considerations of Chapters 16–19 to solve molecular models of nontrivial complexity. The first is the large collection of mathematical approximations and conceptual approaches that comprises the framework of statistical-mechanical theory. These techniques often give closed-form but approximate analytical expressions for the properties of a system that are valid in certain limits (e.g., the high-density one), or, alternatively, sets of equations that can be solved using standard numerical tools. The particular approaches are usually system-specific because they hinge on simplifications motivated by the physics of the interactions at hand. We will not discuss this body of work in any detail, but refer the reader to the excellent introductory texts by Hill and McQuarrie.

The second approach to extracting the properties of molecular models is through detailed computational means, a field broadly termed *molecular simulations*. Many distinct types of methods, distinguished by the resolution of the model, fall into this category. At the highest resolution, *ab initio* or *quantum chemical* techniques predict the properties of molecules from first principles by solving Schrödinger's equation for the electronic structure. *Classical molecular simulations* instead represent atoms as point sites and rely on semi-empirical force fields of the kind introduced in Chapter 3. At an even lower resolution, *coarse-grained* models represent entire molecular units using *pseudoatoms* that lump multiple atoms into single sites. For example, a coarse-grained polymer may represent monomers as spherical beads connected by harmonic springs. In such coarse models, the challenge is to determine the effective interactions between pseudoatoms in a systematic manner. In any case, the scope of a molecular simulation is limited by its resolution: ab initio techniques, for instance, can reach simulation sizes (numbers of atoms) and time scales that are far less than those achievable with classical techniques. Therefore, a trade-off between accuracy and scale is always inherent to these approaches.

Molecular simulations have become increasingly important to and common in scientific and engineering research. Their prevalence is due to several factors. First, computational platforms have become progressively more powerful and inexpensive. For example, it is now relatively cheap and easy to build a modest cluster of computers that exceeds in power the supercomputers of just a decade past. Second, there have been and continue to be significant advances in algorithms that take advantage of such clusters using parallel-computing techniques or that accelerate simulations using novel theoretical ideas. Finally, there has been a community-wide effort to develop shared force fields, codes, and computational packages that enable the novice user to rapidly begin performing molecular simulations of many different kinds. Many of these codes are open-source and freely available.

In this chapter, we present a very brief introduction to basic concepts in molecular simulations that is designed to give a flavor of the field. It would be impossible to be comprehensive in a single text, let alone a single chapter, in this ever-growing area. For that reason, we will focus on two of the most fundamental and widespread techniques: molecular dynamics and Monte Carlo simulations. We will also consider these solely as applied to classical molecular models because they provide the most natural case study. However, some excellent molecular simulation texts provide a more detailed exposition and are referenced at the end of this chapter.

22.1 Basic elements of classical simulation models

The most important component of any classical molecular simulation is the molecular model that describes the connectivity of atoms and their interactions. The latter is the potential energy function U, which is often called the *force field* because its derivatives give the forces on each atom. Liquid argon, for example, is well described by a Lennard-Jones pair potential that accounts for van der Waals and excluded-volume interactions,

$$U(\mathbf{r}^N) = \sum_{i<j} 4\epsilon \left[\left(\frac{r_{ij}}{\sigma} \right)^{-12} - \left(\frac{r_{ij}}{\sigma} \right)^{-6} \right] \tag{22.1}$$

Here, the summation is performed over all unique pairs of atoms i and j, and $r_{ij} = |\mathbf{r}_i - \mathbf{r}_j|$ gives their scalar separation distance. The parameters ϵ and σ give the energy and length scales of the pair interaction; two atoms contribute a minimum possible energy of amount $-\epsilon$ when their separation distance is $2^{1/6}\sigma$. In general, the form of (22.1) can be cast as

$$U(\mathbf{r}^N) = \sum_{i<j} u(r_{ij}) \tag{22.2}$$

where $u(r)$ is a *pair potential* that could be the Lennard-Jones expression or one of many alternative forms. Its general behavior should always have the following limits motivated on physical grounds: $u(r) \rightarrow +\infty$ as $r \rightarrow 0$ and $u(r) \rightarrow 0$ as $r \rightarrow \infty$. The first ensures that no two atoms overlap while the second guarantees that atoms separated by long distances no longer interact or "see" each other. The behavior in between can be attractive or repulsive, and can involve positive or negative energies.

More complex molecules can introduce additional interactions: electrostatic terms due to partial charges on different atoms, and energies due to bond stretching, bending, and rotation. As shown in Chapter 3, a minimal force field might be

$$
\begin{aligned}
U(\mathbf{r}^N) = &\sum_{\text{bonds } i} a_i(d_i - d_{i,0})^2 \\
&+ \sum_{\text{angles } j} b_j(\theta_j - \theta_{j,0})^2 \\
&+ \sum_{\text{torsions } k} \left[\sum_n c_{k,n}\left[1 + \cos(\omega_k n + \gamma_k)\right] \right] \\
&+ \sum_{\text{pairs } ij} \frac{q_i q_j}{4\pi\epsilon_0 r_{ij}} + 4\epsilon_{ij}\left[\left(\frac{r_{ij}}{\sigma_{ij}}\right)^{-12} - \left(\frac{r_{ij}}{\sigma_{ij}}\right)^{-6} \right]
\end{aligned} \tag{22.3}
$$

Equation (22.3) is often called a flexible-bond, fixed-partial-charge potential because the bonds lack rigidity and each atom i has a constant partial charge q_i. The potential energy is a function of the atomic positions through the bond distances d_i, angles θ_j, torsions ω_k, and pairwise distances r_{ij}, all of which are functions of $\mathbf{r}^N$. Of the four sums here, by far the most computationally expensive is the pairwise atomic sum in the last line since the number of terms scales as N^2 rather than N as in the others.

Equation (22.3) contains a large number of parameters: a_i, $d_{i,0}$, b_j, $\theta_{j,0}$, $c_{k,n}$, ω_k, γ_k, q_i, ϵ_{ij}, and σ_{ij}. The subscripts i and ij indicate that there can be different values of parameters for different types of bonds, angles, torsions, partial charges, and repulsive/dispersive interactions depending on the kinds of atoms involved and their chemical environment (i.e., a carbonyl oxygen behaves differently than a hydroxyl one). This means that a typical force field even for molecules of modest complexity can contain a huge set of adjustable parameters. Values are typically determined from a combination of electronic-structure calculations on small molecules and experimental data. The inclusion of experimental data tends to improve accuracy because it fits properties to bulk phases rather than the very small systems that ab initio methods can treat. As a result, these force fields are *semi-empirical*. One additional goal and constraint in fitting these parameters is *transferability*, that is, the ability to reuse the same sets of parameters for different molecules and systems. Transferability is often informed by chemistry: there might be one set of parameters for sp^2-hybridized carbons, one for sp^3-hybridized carbons, one for aromatic carbons, etc. Force-field development is an extensive task pursued by a very large research community. Various force-field efforts exist and have specific names such as AMBER, CHARMM, and OPLS. Moreover, multiple versions of force fields can exist as they are refined over time.

The force field is so called because its negative gradient with respect to the configurational coordinates gives the forces on each atom,

$$f_{i,x} = -\frac{\partial U(\mathbf{r}^N)}{\partial x_i}, \quad f_{i,y} = -\frac{\partial U(\mathbf{r}^N)}{\partial y_i}, \quad f_{i,z} = -\frac{\partial U(\mathbf{r}^N)}{\partial z_i} \tag{22.4}$$

In vector notation, we might also write Eqn. (22.4) as

$$\mathbf{f}_i = -\frac{\partial U(\mathbf{r}^N)}{\partial \mathbf{r}_i} = -\nabla_{\mathbf{r}_i} U(\mathbf{r}^N) \tag{22.5}$$

Alternatively, we can consider the $3N$-dimensional vector of forces on all N atoms at once,

$$\mathbf{f}^N = -\frac{\partial U(\mathbf{r}^N)}{\partial \mathbf{r}^N} = -\nabla U(\mathbf{r}^N) \tag{22.6}$$

If the force field consists only of simple pairwise interactions as in Eqn. (22.2), it can be simplified considerably. Consider the force in the x direction on atom i. There are $N-1$ terms in the pairwise summation of (22.2) that include the coordinate x_i, and thus the energy derivative with respect to it is

$$f_{i,x} = -\frac{\partial U}{\partial x_i}$$

$$= -\frac{\partial}{\partial x_i} \sum_{j \neq i} u(r_{ij})$$

$$= -\sum_{j \neq i} \frac{\partial r_{ij}}{\partial x_i} \frac{du(r_{ij})}{dr} \tag{22.7}$$

Note that

$$r_{ij}^2 = (x_i - x_j)^2 + (y_i - y_j)^2 + (z_i - z_j)^2 \tag{22.8}$$

such that

$$2r_{ij} \frac{\partial r_{ij}}{\partial x_i} = 2(x_i - x_j) \tag{22.9}$$

or, alternatively,

$$\frac{\partial r_{ij}}{\partial x_i} = \frac{x_i - x_j}{r_{ij}} \tag{22.10}$$

With Eqn. (22.10), we can simplify the force expression in (22.7) to

$$f_{x,i} = -\sum_{j \neq i} \frac{x_i - x_j}{r_{ij}} \frac{du(r_{ij})}{dr} \tag{22.11}$$

By doing the same for the y and z coordinates, we can finally express the force compactly in vector notation as

$$\mathbf{f}_i = \sum_{j \neq i} \frac{\mathbf{r}_{ij}}{r_{ij}} \frac{du(r_{ij})}{dr} \tag{22.12}$$

Here, $\mathbf{r}_{ij} \equiv \mathbf{r}_j - \mathbf{r}_i$ gives the vector from atom i to atom j. For example, if our pair potential is the so-called *soft-sphere* interaction,

$$u(r) = \epsilon \left(\frac{r}{\sigma}\right)^{-n} \tag{22.13}$$

where n is a positive integer, then the force takes the form

$$\mathbf{f}_i = \sum_{j \neq i} \frac{\mathbf{r}_{ij}}{r_{ij}} [-n\epsilon\sigma^n r_{ij}^{-n-1}]$$

$$= -n \sum_{j \neq i} \frac{\mathbf{r}_{ij}}{r_{ij}^2} \left[\epsilon \left(\frac{r_{ij}}{\sigma}\right)^{-n}\right] \tag{22.14}$$

If we switch indices, the force acting on atom i due to atom j is the negative of the force acting on atom j due to atom i,

$$\mathbf{f}_{ij} = -\mathbf{f}_{ji} \tag{22.15}$$

This, of course, is a manifestation of Newton's third law. In general, once a form for the potential energy function has been determined, the expression for the forces on each atom follows from differentiation. Clearly, with a more detailed force field like that in Eqn. (22.3), the forces will involve multiple contributions and summations. However, conceptually the approach is no different than the simple example above.

It is important to keep in mind that the flexible-bond, fixed-partial-charge approach embodied by the generic force field in Eqn. (22.3) is not a unique classical model of molecular systems. Frequently, models may involve fixed bond lengths and angles, or entire molecules that are rigid. Such approaches reduce computational expense by eliminating the time needed to resolve motion in fast-vibrating degrees of freedom. In addition, some models allow charges or dipoles on each atom to change with configuration, that is, each atom becomes *polarizable* and can respond to its electrostatic environment. This often permits better transferability and agreement with quantum-mechanical calculations, but it costs in terms of simulation speed. For the same reasons, there are many specialized force-field forms that have been developed to capture the behavior of specific interactions and systems, such as hydrogen bonding, water, metals and semiconductors (delocalized electrons), and ionic solids.

While *all-atom* or *fully atomistic* simulation models of the kind described above can provide quantitatively accurate predictions, in many simulation studies we are interested in more basic physical behaviors and thus consider simpler *coarse-grained* systems involving pseudoatoms that represent combined groups of multiple atoms. A classic example is that of a bead–spring Lennard-Jones polymer, with force field

$$U = \sum_{\substack{i<j, \\ ij \text{ not bonded}}} 4\epsilon \left[\left(\frac{r_{ij}}{\sigma}\right)^{-12} - \left(\frac{r_{ij}}{\sigma}\right)^{-6}\right] + \sum_{\substack{i<j, \\ ij \text{ bonded}}} \frac{k}{2}(r_{ij} - r_0)^2 \tag{22.16}$$

Here each site i represents one monomer; consecutive monomers in the same polymer are bonded and experience a harmonic potential, while nonbonded monomers interact through the Lennard-Jones term. A second prominent example is that of *implicit solvation models*. In many cases, a solute like a protein exists in a vast sea of solvent molecules like water. Rather than represent each water molecule explicitly, implicit approaches use a coarse background, continuum solvent that modifies the effective solute interatomic interactions.

Compared with all-atom ones, coarse-grained models eliminate degrees of freedom that are deemed irrelevant to the physics of interest, such as individual atomic motions within a polymer monomer or detailed configurations of solvent molecules. In turn, they enable more thorough and larger-scale simulation studies: the models can be far less expensive in terms of computational requirements. However, it must be cautioned that there is no rigorous recipe for determining which degrees of freedom and how much detail a coarse model should contain. Typically one considers what the dominant interactions relevant to a given behavior will be, and, through trial and error, builds a simple model in light of their length and energy scales.

Coarse-grained models are widely used to construct fundamental pictures of the thermodynamic properties of broad classes of systems, and in turn, to show the emergence of universal behaviors in such classes. They embody a philosophy that, by identifying the minimal model and features necessary to reproduce a particular kind of thermodynamic behavior, one has actually gained a basic physical understanding of it. Coarse-grained models are frequently used to uncover scaling laws (e.g., dependence of properties on system size, chain length, molecular size or energy scales), microscopic mechanisms (e.g., of diffusion, binding, or conformational changes), relative magnitudes of different driving forces (e.g., electrostatic, dispersive, excluded-volume, hydrogen-bonding, and hydrophobic interactions), microscopic structure (e.g., conformational fluctuations, molecular packing, structure in bulk liquids, and degree of geometric ordering), and trends in various properties' dependences on state conditions (e.g., temperature, pressure, density, and composition).

A central feature of any classical simulation is the molecular model employed and, in particular, the **force field** used to represent interatomic interactions. The latter is typically parameterized to small-scale quantum chemical calculations and/or experimental data. Quantitative predictions of **all-atom models** can be improved by introducing more detail, for example, in terms of the complexity of the force field. However, a trade-off is computational expense, since very detailed models become difficult to simulate with large numbers of atoms and for long time scales. On the other hand, **coarse-grained models** eliminate "irrelevant" microscopic degrees of freedom and permit much faster simulations. Such systems typically inform qualitative rather than quantitative behaviors, such as scaling laws, trends, and molecular mechanisms.

22.2   Molecular-dynamics simulation methods

The basic principle of a molecular-dynamics (MD) simulation is straightforward: reproduce the deterministic evolution of an isolated system at constant E, V, N conditions as
described by Newton's equations of motion,

$$\frac{d\mathbf{p}_i}{dt} = -\frac{\partial U\,(\mathbf{r}^N)}{\partial \mathbf{r}_i} \quad \text{for all } i \tag{22.17}$$

Alternatively,

$$\frac{d\mathbf{p}^N}{dt} = -\nabla U\,(\mathbf{r}^N) \tag{22.18}$$

Equations (22.17) and (22.18) describe a set of $3N$ second-order, nonlinear, coupled
partial differential equations for the coordinates of all of the atoms. As shown in
Chapter 3, these equations conserve total energy, that is, $dH/dt = 0$, where H is the
classical Hamiltonian. They also conserve net linear and angular momentum.

A simple MD simulation reproduces the microcanonical ensemble at long times in
ergodic systems that are able to reach equilibrium, exploring all microstates that are
consistent with the constant total energy. Technically, MD simulations are of finite length
and therefore never exhaustively explore microstates, but they can sample a subset of
microstates that are statistically identical or that are *representative*. Importantly, they
determine a history of states in the form of a *trajectory* of atomic momenta and positions
as a function of time, $\mathbf{p}^N(t)$ and $\mathbf{r}^N(t)$. From this detailed information, one can determine
the average of any property in the microcanonical ensemble by performing a time
average.

In reality, we solve Eqn. (22.18) and the trajectory in a discrete-time approximation by
taking small steps in time repeatedly. At each time point, we compute the forces on each
atom and predict the configuration at the next time step; the former task is the most
expensive since it typically requires an analysis of $o(N^2)$ pair interactions. In turn, this
approach gives atomic positions at specific time intervals, such as $\mathbf{r}^N(0)$, $\mathbf{r}^N(\delta t)$, $\mathbf{r}^N(2\,\delta t)$,
$\mathbf{r}^N(3\,\delta t)$, ... and so on. Here, δt is called the *time step*. As δt becomes smaller, the
numerical solution of Newton's equations of motion becomes increasingly accurate.
However, this comes at the expense of a shorter simulation time and overall trajectory
duration for the same computational effort (i.e., number of time steps taken).

The exact time-stepping procedure can be developed on the basis of expansions of the
continuous function $\mathbf{r}^N(t)$. For the sake of simplicity, let us abbreviate the position vector
$\mathbf{r}^N$ as r, and do similarly for the velocity and force vectors. A Taylor expansion for an
increment in time δt from the current time t gives

$$r\,(t+\delta t) = r(t) + \frac{dr(t)}{dt}\,\delta t + \frac{d^2 r(t)}{dt^2}\frac{\delta t^2}{2} + \frac{d^3 r(t)}{dt^3}\frac{\delta t^3}{6} + \vartheta(\delta t^4)$$

$$= r(t) + v(t)\delta t + \frac{f(t)}{m}\frac{\delta t^2}{2} + \frac{d^3 r(t)}{dt^3}\frac{\delta t^3}{6} + \vartheta(\delta t^4) \tag{22.19}$$

In the second line, we used Newton's law to replace the acceleration with the force. Similarly, by considering a step backwards in time, we have

$$r(t - \delta t) = r(t) - v(t)\delta t + \frac{f(t)}{m}\frac{\delta t^2}{2} - \frac{d^3 r(t)}{dt^3}\frac{\delta t^3}{6} + \vartheta(\delta t^4) \qquad (22.20)$$

By adding Eqns. (22.19) and (22.20) and moving the $r(t - \delta t)$ term to the RHS, we obtain an equation that predicts the atomic positions at the next time step,

$$r(t + \delta t) = 2r(t) - r(t - \delta t) + \frac{f(t)}{m}\delta t^2 + \vartheta(\delta t^4) \qquad (22.21)$$

Equation (22.21) forms the basis of the so-called *Verlet algorithm* for molecular dynamics. To propagate a system forward in time by an amount δt, it uses the positions at the previous two time steps as well as the forces at the current time step. To obtain the forces, we use the force-field derivative and the current positions at time t, through Eqn. (22.6). We see from the analysis that the accuracy of this approach is of order δt^4. The Verlet time-stepping equation does not use the velocities, but we can approximate them using the expression

$$v(t) = \frac{r(t + \delta t) - r(t - \delta t)}{2\delta t} + \vartheta(\delta t^2) \qquad (22.22)$$

A possible disadvantage of the Verlet algorithm is that it requires one to start a simulation knowing two sets of positions, at $r(t)$ and $r(t - \delta t)$. An alternative is the so-called *velocity Verlet algorithm*, which uses the velocity directly in two separate update steps,

$$r(t + \delta t) = r(t) + v(t)\delta t + \frac{f(t)}{2m}\delta t^2$$
$$v(t + \delta t) = v(t) + \frac{f(t + \delta t) + f(t)}{2m}\delta t \qquad (22.23)$$

The time-step update in Eqn. (22.23) shows that the velocity Verlet algorithm requires only one set each of positions and velocities at each time step, and thus a simulation is more conveniently started using initial values for these two vectors. For this reason, the velocity Verlet algorithm is one of the most frequently used approaches in molecular simulations. In practice, each time-step update proceeds in the following manner.

1. Given $r(t)$ and $v(t)$ at one time point t, first compute the forces $f(t)$ on each atom using the force field.
2. Do an update of the positions:

$$r \leftarrow r + v\,\delta t + \frac{f}{2m}\delta t^2$$

3. Do a partial update of the velocity array on the basis of the current forces:

$$v \leftarrow v + \frac{f}{2m}\delta t$$

4. Compute the new forces $f(t + \delta t)$ using the new positions:

$$r\,(t + \delta t)$$

5. Finish the update of the velocity array:

$$v \leftarrow v + \frac{f}{2m} \, \delta t$$

The Verlet and velocity Verlet algorithms are called *integrators* because they solve the equations of motion by performing a discrete-time integration. Many other integrators exist. In particular, there are higher-accuracy schemes in which the Taylor expansions are taken out to the δt^4 term or further, or in which the new forces after a time step are used to back-correct for the finite extrapolation in time. Generally speaking, such methods are not frequently used because they impose additional computational overheads; the same accuracy might be achieved with a simpler algorithm by shrinking the size of the time step.

This leads to an important question: how can the accuracy of an integrator be diagnosed? A convenient metric is the degree of conservation of energy since, rigorously, Newton's equations lead to a constant total energy $E = K + U$. Numerical solutions to the equations of motion are approximate and can show small fluctuations in E that shrink as the time step grows smaller. Typically, it is desirable to conserve energy to one part in 10^4–10^5 for a stable simulation; that is, one would like to see at least four or five significant figures in E that do not fluctuate over the course of the simulation. Two kinds of energy conservation are often considered. In the short term, fluctuations in total energy from step to step can be significant. The algorithms of the Verlet family are actually not as good at short-term energy conservation as many higher-accuracy schemes, but this is not terribly problematic. On the other hand, long-term energy fluctuations are more serious. Over large numbers of time steps, one finds that the total energy can *drift* away from its initial value. The Verlet-like algorithms are actually quite good at avoiding energy drift, while some other schemes are not. Drift is problematic because it changes the macroscopic E, V, N conditions for the system when simulated at long times.

For numerical stability and accuracy in conserving the energy, one typically needs to pick a time step that is at least an order of magnitude smaller than the fastest time scale in the system. If a system has flexible bonds, then the time step must be of the order 1 fs $= 10^{-15}$ s to resolve these motions. If bonds are made rigid but angles can bend, the time step can be increased by a factor of 2–3. Systems with completely rigid molecules or without bonds, such as liquid argon, often can use a time step of the order 5–10 fs to resolve the purely translational motion. Too large a time step can cause an MD simulation to become unstable at some point, with the total energy rapidly increasing with time. This behavior is colloquially termed "blowing up," and it is caused by devastating atomic collisions that occur when a large time step erroneously propagates the positions of two atoms to be nearly overlapping; the strong repulsive interactions then create a massive force that propels these atoms apart.

Practically speaking, the time step limits the length attainable in an MD trajectory. At each step, the slowest computational operation is the loop over all pairwise interactions used to compute the potential energy and forces, which ostensibly scales as N^2 in computational expense. A total MD trajectory length of n time steps is $t_{\text{total}} = n \, \delta t$. The computational time is $t_{CPU} = n\, \delta t_{CPU}$. For liquid argon modeled by the Lennard-Jones potential, for example, a thousand time steps for $N = 2{,}048$ may require on the order of a few minutes for a single processing core.

A final practicality is how one initializes a simulation. For the initial positions, there are several common approaches. In bulk fluid phases, we can place all molecules on a crystalline lattice, the simplest being the cubic lattice. If the simulation conditions correspond to the liquid phase, we then must first simulate for long enough until the system melts. A seeming alternative would be to place molecules randomly in the simulation volume; while perhaps intuitive, this approach is generally avoided because there is a high probability that two or more atoms will be placed too close together and create a *core overlap*. Atoms within molecules can be arranged in idealized or approximate geometries that satisfy known bond lengths and angles. On the other hand, one might take an initial configuration from experiment (typically X-ray diffraction or NMR). This is common practice when simulating large, structured macromolecules such as proteins. In any case, we usually energy-minimize the initial configuration to relax the system, remove any core overlaps, and fix up bond lengths and angles.

For the initial velocities, we typically draw these randomly from a Maxwell–Boltzmann distribution at the desired temperature,

$$\wp(v_{x,i}) = (2\pi\sigma_v^2)^{-1/2} \exp\left(-\frac{v_{x,i}^2}{2\sigma_v^2}\right), \quad \text{with } \sigma_v^2 = \frac{k_B T}{m_i} \tag{22.24}$$

Equation (22.24) shows that the distribution is Gaussian with zero mean and variance σ_v^2. Random velocities must be drawn for each of the $3N$ components, and are both positive and negative. However, this procedure can lead to a nonzero net momentum of the system, which would produce a systematic translational drift. Thus, after assignment of random velocities, we typically shift all of them by whatever constant amount is needed to remove the net momentum. For systems without boundaries or fields, we typically also remove the net angular momentum, which prevents the system from rotating.

A molecular-dynamics simulation attempts to reproduce the time evolution of a classical system described by Newton's equations of motion. It corresponds to the microcanonical ensemble at constant E, V, N conditions. These simulations use **integrators** to take repeated small but finite time steps, evolving the system along its dynamic trajectory. The time steps must be chosen carefully in order to ensure accuracy and stability.

22.3   Computing properties

Often we start a simulation with initial velocities and positions that are not representative of the state condition of interest, e.g., as specified by the desired temperature and density. As such, we must *equilibrate* our system by first running the simulation for an

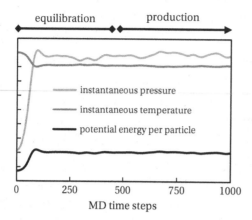

equilibration   production

instantaneous pressure

instantaneous temperature

potential energy per particle

0   250   500   750   1000

MD time steps

Figure 22.1. Molecular simulations generally consist of equilibration and production stages. The former is a period of time required for the system to lose memory of its initial configuration and velocities, and to approach an equilibrium state. Shown here are several metrics (units irrelevant) for $N = 864$ Lennard-Jones argon atoms initialized on a face-centered cubic lattice. The crystal rapidly melts to a liquid phase in only ∼500 time steps, with an increase in potential energy, and thus the equilibration period is relatively short. For more complex systems, the equilibration time can be orders of magnitude longer.

amount of time that lets it evolve to configurations representative of the target state conditions. Once we are sure we have equilibrated, we then perform a *production* period of simulation time that we use to study the system and/or compute properties at the target state conditions.

How do we know whether we have equilibrated our system properly? One approach is to monitor the time dependences of simple properties like the potential energy or pressure. An example for liquid argon at a relatively high temperature is shown in Fig. 22.1. After equilibration, such quantities will still fluctuate, and should fluctuate if we are correctly reproducing the properties of the statistical-mechanical ensemble of interest (here, the microcanonical ensemble). We must therefore be careful to separate long-time drifts in these properties from inherent ensemble fluctuations. Many systems can require very long equilibration times, needing 10^6 time steps or more. Indeed, some systems cannot be simulated at all because the number of time steps required for equilibration can be out of the range of practicality. Generally speaking, the longest *relaxation time* for molecular motion in a system determines the necessary number of steps. If one is simulating bulk fluids, the order of the relaxation time can be found if an approximate diffusion constant D is known. Then, $\tau_{\text{relax}} \sim \sigma^2/D$ gives an approximate time for one molecule to diffuse a distance equal to one molecular diameter, σ.

During the production phase of a simulation, we can compute many kinds of properties or *observables* in the form of time averages. For example, the average kinetic and potential energies are simply

$$\langle K \rangle = \frac{1}{n}\sum_i K_i, \qquad \langle U \rangle = \frac{1}{n}\sum_i U_i \qquad (22.25)$$

where we sum n samples of the instantaneous kinetic and potential energies at different time points in the simulation. The samples need not be at every time step because of correlations in properties at short times; rather, samples should be spaced apart by an amount of the order of a relaxation time. The behavior of any simple average like that in Eqn. (22.25) is such that the error in the estimate goes as $n_{\text{eff}}^{-1/2}$, where n_{eff} gives the number of *statistically independent* samples.

To make a connection with temperature, we need to use statistical mechanics. We saw in Chapter 18 that the average kinetic energy has a relation to the temperature in the canonical ensemble. A more general result stemming from the so-called equipartition theorem, not discussed in detail here, gives the following result that is appropriate to the simulated ensemble,

$$\langle K \rangle = n_{\text{DOF}} \frac{k_B T}{2} \tag{22.26}$$

Here, n_{DOF} is the number of degrees of freedom in the system. For a system of N atoms that conserves net momentum, $n_{\text{DOF}} = 3N - 3$. For large systems, the subtraction of the three center-of-mass degrees of freedom has little effect since it is small relative to $3N$. If rigid bonds are present in the system, we also lose one degree of freedom per rigid bond. Equation (22.26) thus permits a *kinetic estimate of the temperature,*

$$T = \frac{2\langle K \rangle}{k_B n_{\text{DOF}}} \tag{22.27}$$

Operationally, we can define an *instantaneous kinetic temperature,*

$$T_{\text{inst}} = \frac{2K}{k_B n_{\text{DOF}}}, \quad \text{with} \quad T = \langle T_{\text{inst}} \rangle \tag{22.28}$$

Because K fluctuates during a simulation, T_{inst} also fluctuates. Note that T_{inst} is an *estimator* of the temperature in that we must take an average to compute the latter. It should be noted that Eqn. (22.27) is not a unique way to determine the temperature, but it is the most convenient.

Instantaneous temperature estimators allow us to perform a simulation at a specified T. For a constant E, V, N simulation, we would like to adjust the total energy such that the average temperature is equal to the one of interest. The most common approach is to rescale the velocities at periodic time intervals based on the deviation of the instantaneous temperature from a set-point temperature. This is perhaps the simplest form of a *thermostat*, of which there are many. We can rescale the velocities by a factor λ such that T_{inst} is equal to the target temperature,

$$\mathbf{v}^N \leftarrow \lambda \mathbf{v}^N \tag{22.29}$$

where

$$\lambda = \sqrt{\frac{k_B T n_{\text{DOF}}}{\sum_i m_i |\mathbf{v}_i|^2}} \tag{22.30}$$

The rescaling is not done at every time step but only periodically, e.g., every 10^2–10^4 time steps. One problem with such rescaling is that it affects the dynamics of the simulation and is an artificial interruption to Newton's equations of motion. In particular, it prevents correct computation of transport properties like the self-diffusivity and viscosity. If dynamic properties are of interest, one typically performs rescaling only during the equilibration portion of the simulation, then switching to the true microcanonical dynamics during the production period.

To compute the pressure, one can use the virial result from Chapter 18,

$$P = \frac{1}{3V}\left\langle 3Nk_BT + \sum_i \mathbf{f}_i \cdot \mathbf{r}_i \right\rangle$$

$$= \frac{1}{3V}\left\langle 2K + \sum_i \mathbf{f}_i \cdot \mathbf{r}_i \right\rangle \tag{22.31}$$

While this expression stems from the canonical ensemble (constant T, V, and N), it is often applied to constant E, V, N molecular dynamics simulations; the difference between the two is negligible for large enough systems. A more important caveat is that Eqn. (22.31) cannot be used directly for systems of pairwise-interacting molecules subject to periodic boundary conditions (described below). Instead, it can be reformulated as

$$P = \frac{1}{3V}\left\langle 2K - \sum_{i<j} \mathbf{f}_{ij} \cdot \mathbf{r}_{ij} \right\rangle \tag{22.32}$$

The sum of pairwise interactions in the average is usually computed in the same pairwise loop as the forces and energies.

Many other equilibrium properties can be determined by time-averaging. To attain a high degree of statistical precision, it is important that the simulation be performed long enough to well exceed the relaxation and correlation times in the system, so that the values used in averaging stem from *statistically uncorrelated* or *independent* time points. It is also common practice to perform multiple *trials* or *runs* of the same simulation with different initial states (i.e., random velocities), as a way to produce results for averaging that have greater statistical independence.

Because they attempt to reproduce the true microscopic equations of motion, MD simulations permit the determination of dynamic or transport properties in addition to equilibrium ones. The former cannot be developed from simple averages, but require an examination of the time dependences of certain observables. For example, the self-diffusivity D of a molecule in a simulation of a bulk fluid can be determined from the behavior of its *mean-squared displacement* with time,

$$D = \frac{1}{6}\lim_{t\to\infty}\frac{d}{dt}\left\langle |\mathbf{r}(t) - \mathbf{r}_0|^2 \right\rangle \tag{22.33}$$

Here, $\mathbf{r}_0$ gives the position of the molecule at the start of the production period and $\mathbf{r}(t)$ is the position at some later point in time. The self-diffusivity can be determined from the slope of the average mean-squared displacement, $\langle |\mathbf{r}(t) - \mathbf{r}_0|^2 \rangle$, with time at long times.

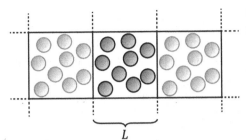

Figure 22.2. Periodic boundary conditions realize an infinite set of copies of the simulation box. Shown here are image repetitions of a cubic central periodic box with side length L.

Many thermodynamic properties can be measured directly from an MD simulation in the form of time or **trajectory averages**. Transport and kinetic properties can also be extracted from the average time dependences of different metrics. In general, properties should be measured only during the production period of the simulation, after the system has equilibrated.

22.4 Simulations of bulk phases

In simulations of bulk phases, the presence of explicit interfaces or "walls" in a simulation box would have a profound effect on the resulting properties, since these interactions would be significant compared with the necessarily small size of the system. Instead, for bulk phases we typically implement *periodic boundary conditions* that realize an infinite number of copies of the simulation box without interfaces, as shown in Fig. 22.2. A particle that exits one side of the simulation boundary has a duplicate that re-enters on the opposite face. In other words, particles have copies of themselves inside every periodic repetition of the simulation box, called *image particles*.

For any given particle, its position in the *central box* can be found using the nearest-integer operation,

$$\mathbf{r}^0 = \mathbf{r} - L\,\text{nint}(\mathbf{r}/L) \tag{22.34}$$

where L is the length of the imaginary cubic volume in which the particles reside. It may also be a vector for non-regular but cuboidal simulation boxes. This equation implies a separate operation for each of the x, y, z coordinates. After its application, each coordinate will be in the range $-L/2$ to $L/2$.

Interactions between two particles require the vector and scalar distances between them. We examine the *minimum image distance*, the smallest distance between any two images of the particles. It can be found from

$$\mathbf{r}^0_{ij} = \mathbf{r}_{ij} - L\,\text{nint}(\mathbf{r}_{ij}/L) \tag{22.35}$$

as

$$r_{ij}^0 = |\mathbf{r}_{ij}^0|$$

where $\mathbf{r}_{ij} = \mathbf{r}_j - \mathbf{r}_i$ as before. Note that we must first find the vector minimum image distance in order to compute its scalar norm. That is, we must first minimum-image each of the x, y, z distances separately.

While a cubic box is the most frequently used geometry for periodic boundary conditions, other simulation cell geometries are possible, including cuboid, parallelepiped, hexagonal prism, truncated octahedron, and rhombic dodecahedron. All of these will regularly tile space and thus can serve to replicate the infinite number of periodic images. The choice of one of these alternate geometries can be motivated by the desire to apply a particular symmetry or to save on the number of simulation atoms required. It should be noted that the expression for the minimum image distance is distinct for each one.

Closely related to the simulation box size is the manner in which nonbonded pairwise interactions are *truncated*. Many pairwise potentials decay rapidly with separation distance, and we save simulation time by ignoring interactions between pairs of particles that are separated by large distances. We typically implement a *distance cutoff* r_c at which the energy is approximately only one or two percent of the minimum energy. Then, we use a modified, *cut pair potential* that takes the form

$$u_{\text{cut}}(r) = \begin{cases} u(r) & r < r_c \\ 0 & r \geq r_c \end{cases} \tag{22.36}$$

With periodic boundary conditions, such a cutoff allows us to avoid accounting for multiple periodic images of the same particle in the pairwise interaction loop. This is possible if $r_c < L/2$ in cubic simulation geometries. For distances larger than $L/2$, a particle i begins to interact with multiple images of a neighboring particle j, rather than only the minimum-image-distance one. Therefore, simulations are generally designed so that all cutoff distances obey $r_c < L/2$. Typically, the cutoff is chosen first and the minimum size of the simulation box is then computed. Given a desired bulk density, this will then also determine a minimum number of particles.

The simple truncation of the potential in Eqn. (22.36) leads to a discontinuity in the pair potential at r_c at which the forces are undefined. A better approach is to shift the potential for $r < r_c$ so that the energy continuously approaches a value of zero at r_c,

$$u_{\text{cut}}(r) = \begin{cases} u(r) - u(r_c) & r < r_c \\ 0 & r \geq r_c \end{cases} \tag{22.37}$$

This *cut-and-shift* approach to potential truncation is perhaps the most popular treatment, and results in better energy conservation in molecular dynamics than does simple cutting alone. More sophisticated treatments are possible, including cutoffs that smoothly interpolate the potential to zero at r_c such that the forces remain continuous.

Interestingly, some *long-ranged* pair interactions do not decay fast enough with distance to be truncated in this simple manner. The most important case is that of Coulomb interactions. A central atom i with partial charge q_i interacts with other particles j with an interaction that scales as $q_i q_j / r_{ij}$. However, the number of particles a distance r from *i* scales faster, as r^2. The net result is that contributions from particles farther from the central one seemingly make a more significant contribution to the pair energy than those nearby, ultimately summing to an infinite energy. It turns out that electrostatic energies for Coulombic interactions, while long-ranged, do in fact converge for systems that are net charge-neutral. However, these interactions require special treatment and simulations must actually sum the interactions over an infinite number of periodic replicas of the simulation cell. Most approaches are based on a technique called the *Ewald summation*.

> Simulations of bulk phases use **periodic boundary conditions** to avoid the presence of explicit interfaces. Typically, pairwise interaction potentials are **truncated** at large distances to save computational time spent on otherwise weakly interacting atoms. The truncation distance must be considered carefully with respect to the box size in order to avoid interactions involving multiple images of atom pairs.

22.5 Monte Carlo simulation methods

While molecular dynamics generates a *deterministic* trajectory of atomic positions as a function of time, another major method for generating atomic trajectories is the Monte Carlo (MC) approach. MC simulations are *stochastic* in nature – the time progression of the atomic positions proceeds randomly and is not predictable given a set of initial conditions. In an MC simulation, the dynamic principles by which we evolve the atomic positions incorporate random moves or perturbations of our own design. As such, the dynamics are not representative of the true microscopic dynamics and instead depend on the kinds of random moves that we perform. However, MC methods rigorously generate correct ensemble properties because they are designed by construction to do so.

MC and MD approaches are complementary. If we are interested in kinetic properties or microscopic mechanisms, or if molecules are too complicated for us to develop good *random move sets*, MD is the natural choice. On the other hand, MC methods can offer several attractive features. Foremost, they naturally and easily treat different thermodynamic ensembles. For example, it is quite simple to perform rigorously a constant-temperature simulation using MC, while MD requires special thermostat techniques. MC methods also offer great flexibility in choosing the random moves by which the system evolves. This can sometimes greatly speed equilibration in complex systems, e.g., in dense polymeric systems. From a practical point of view, MC methods are not subject to inaccuracies due to discrete-time approximations of the equations of motion, and require only potential energy evaluations in order to generate the atomic trajectories. These approaches, therefore, do not involve expensive force calculations, and can handle continuous and discrete intermolecular potentials in an equivalent way.

Before diving into the details of the MC approach, let us see how it might be applied to a simple case study: simulation of a monatomic liquid at constant *T*, *V*, *N* conditions. This might be, for example, the Lennard-Jones system with a force field given by

Eqn. (22.1). Like MD, the simulation progresses through repeated iterations of steps to generate a trajectory. In this case, however, a basic MC step involves the following.

1. Randomly pick one of the N particles.
2. Perturb each of its x, y, z coordinates separately by three random values taken from the uniform distribution on the interval $[-\delta r_{max}, \delta r_{max}]$. Here, δr_{max} is the *maximum displacement*.
3. Compute the change in potential energy due to the particle move, $\Delta U = U_2 - U_1$.
4. Use a rule called the *Metropolis acceptance criterion* to decide whether or not to keep the move or instead revert back to the original configuration before step 2:
   - If $\Delta U < 0$, accept the move.
   - If $\Delta U > 0$, compute $P^{acc} = \exp(-\Delta U / k_B T)$. Draw a random number r on the interval $[0.0, 1.0)$ and accept the move if and only if $P^{acc} > r$.
5. If the move is accepted, keep the new configuration as the next state in the trajectory. If the move is rejected, discard the new configuration and use the original configuration again as the next state in the trajectory.

Ultimately, aggregated over many MC steps, such a simulation produces configurations that obey the canonical distribution at temperature T. In other words, configurations in the trajectory appear with probability

$$\wp(\mathbf{r}^N) \propto e^{-U(\mathbf{r}^N)/k_B T} \tag{22.38}$$

Any configurational property of interest, such as the average potential energy or pressure, can then be computed as a "time" average over the sequence of configurations produced during the Monte Carlo steps. Such an average is a *canonical* average corresponding to the imposed T, V, N conditions. Like MD simulations, MC runs require separate equilibration and production periods. Initial structures can be chosen in a variety of ways, but it is important to allow the system to lose memory of them before computing averages during the production period.

In each step, the perturbation applied to the coordinates of an atom is called a *single-particle displacement*, and it is one of many possible kinds of Monte Carlo moves. Figure 22.3 gives a schematic representation of a single-particle displacement move. Notice that the maximum displacement δr_{max} is a free parameter that we can tune to adjust the efficiency of the moves. If it is too large, particles will be displaced far from their original positions and will likely have core overlaps with other particles, resulting in large values of ΔU and hence frequent rejections. If it is too small, ΔU will be small and the move likely accepted, but the evolution of the configuration with MC steps will be very slow. To compromise, maximum displacements are generally adjusted such that the average *acceptance rate* of proposed moves is 30%–50%.

In the fourth part of an MC move as outlined above, we compute the acceptance probability and draw a random number to determine whether or not we accept the proposed move. In addition to the random move proposals, this is another stochastic element in the simulation. We can draw random numbers in the computer using pseudorandom-number generators. These numbers are not truly random – they follow a specific mathematical sequence that is ultimately deterministic – but they have

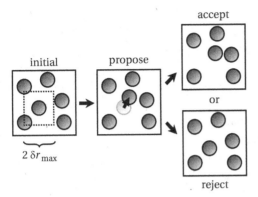

Figure 22.3. In a Monte Carlo simulation, random moves perturb the configuration, such as the single-particle displacement shown here. Proposed moves are accepted or rejected depending on the change in energy and a Boltzmann-like criterion. If a rejection occurs, the state of the trajectory at MC step $i + 1$ will be the same as that at step i.

statistical properties that are reminiscent of actual random variables. In the computer, a *random-number generator* is used to accomplish this task. The basic random-number generator will produce a random real number from the uniform distribution in the range [0.0, 1.0).

MC simulations generate configurations according to the canonical probabilities $\wp$ $(\mathbf{r}^N)$. This is called *importance sampling* because configurations with larger Boltzmann factors appear more frequently. The theory of *Markov chains* underlies this approach. A Markov chain describes a stochastic process in which the state of a system (here, the instantaneous atomic configuration) changes randomly with time but does so without long-term memory. At each step in time, the system can move randomly to another state (another configuration). An important property of Markov processes is that the future state of a system after one step depends only on the current state, not on any previous ones. It is easy to see that MC simulations are Markov processes that generate Markov chains.

The *acceptance criterion* in an MC move is constructed so that the long-time trajectory correctly generates the correct distribution of states. Let $\wp_m(i)$ denote the probability that the system at step i will have configuration m. Then let π_{mn} be the *transition probability* that, given the system is currently in state m, it will move to state n. Although a classical system actually has access to an infinite number of molecular configurations m and n, here we adopt a finite-state notation for convenience. If a system has evolved for some number of MC steps i, we can only characterize the probability that it will be in any given state, $\wp_m(i)$, since the simulation is not deterministic. Ultimately we want this probability distribution to converge to the canonical one,

$$\lim_{i \to \infty} \wp_m(i) = \frac{e^{-\beta U_m}}{Z} \tag{22.39}$$

Because the simulation is a Markov process, it is straightforward to show how the probabilities $\wp_m$ change as the system moves one step forward from i to $i + 1$. For a given state m, we consider both the decrease in probability associated with the system transitioning from m to other states n and the increase in probability associated with the system transitioning from other states n into m,

$$\wp_m(i+1) = \wp_m(i) - \sum_n \pi_{mn}\wp_m(i) + \sum_n \pi_{nm}\wp_n(i) \tag{22.40}$$

In the long-time limit at equilibrium, the probabilities become constant with time, $\wp_m(i+1) = \wp_m(i)$, and we therefore must have

$$\sum_n \pi_{mn}\wp_m = \sum_n \pi_{nm}\wp_n \quad \text{for all } m \tag{22.41}$$

Here, we have omitted the step dependence to signify that these are equilibrium, time-independent probabilities. Equation (22.41) is termed the *balance equation*. One way to satisfy it is through the stricter *detailed-balance* condition, which is given by

$$\pi_{mn}\wp_m = \pi_{nm}\wp_n \quad \text{for all } m, n \tag{22.42}$$

Detailed balance applies a constraint to the transition and state probabilities for every pair of states. All Markov processes that obey the detailed-balance condition automatically obey general balance; however, the reverse is not true.

Importantly, detailed balance provides a way to determine an appropriate MC acceptance criterion for a given kind of move. We can write the transition probability as the product of two quantities,

$$\pi_{mn} = \alpha_{mn} P^{acc}_{mn} \tag{22.43}$$

Here α_{mn} is the *move-proposal probability*. It gives the probability that we will propose a random move from state m to n, and depends entirely on the kind of Monte Carlo move that we perform. Once we pick a kind of MC move, such as a single particle displacement, this probability has a specific form. On the other hand, P^{acc}_{mn} is the *acceptance probability* to be determined. It gives the probability that we should accept a proposed move, and is the critical function to determine in MC simulations. If we impose detailed balance as in Eqn. (22.42), the acceptance probability should obey

$$\frac{P^{acc}_{mn}}{P^{acc}_{nm}} = \frac{\alpha_{nm}\wp_n}{\alpha_{mn}\wp_m} \quad \rightarrow \quad \frac{P^{acc}_{12}}{P^{acc}_{21}} = \frac{\alpha_{21}\wp_2}{\alpha_{12}\wp_1} \tag{22.44}$$

Equation (22.44) gives a starting point for correctly performing an MC simulation. To keep the notation simple, on the RHS we simply chose a specific situation in which the system is at state $m = 1$ initially, and we proposed a move to state $n = 2$. The subscripts 1 and 2 actually correspond to two atomic configurations, $\mathbf{r}_1^N$ and $\mathbf{r}_2^N$, respectively.

So-called *symmetric moves* have proposal probabilities that are equal in the forward and reverse directions, $\alpha_{12} = \alpha_{21}$. Such is the case for single-particle displacements. For this class of moves, the acceptance criterion must satisfy

$$\frac{P_{12}^{\text{acc}}}{P_{21}^{\text{acc}}} = \frac{\wp_2}{\wp_1}$$

$$= \frac{e^{-\beta U_2}/Z(T,V,N)}{e^{-\beta U_1}/Z(T,V,N)}$$

$$= e^{-\beta(U_2 - U_1)} \tag{22.45}$$

The second and third lines specialize to the canonical ensemble and its configurational distribution function, as described in Chapter 18. If we wanted to simulate in another ensemble, we would use a different expression for the distribution $\wp_m$. In any case, we can choose any acceptance criterion P_{12}^{acc} that satisfies the ratio in Eqn. (22.45). The most common is the *Metropolis criterion*, which in this case takes the form

$$P_{12}^{\text{acc}} = \min[1, e^{-\beta(U_2 - U_1)}] \tag{22.46}$$

The Metropolis criterion is not a unique solution for the acceptance criterion, but it is the most frequently employed because it always accepts moves to lower energies, which helps speed equilibration. In general, for any move type and ensemble, the Metropolis criterion is given by

$$P_{12}^{\text{acc}} = \min\left[1, \frac{\alpha_{21}\wp_2}{\alpha_{12}\wp_1}\right] \tag{22.47}$$

One of the great strengths of MC methods is their flexibility to incorporate many different kinds of Monte Carlo moves. For polyatomic molecules, one can perform moves that make random perturbations to the molecular orientations. For polymeric systems, there are many different moves that make concerted changes to backbone conformations, such as snaking polymers through the dense fluid. Indeed, a huge range of move types continues to be developed; many are system-specific. Importantly, moves need not be reminiscent of any physical molecular process, a freedom that can sometimes allow MC simulations to achieve much more rapid equilibration than MD ones given clever enough moves. That is, advanced MC move sets can enhance the rate at which a simulation *samples* its relevant configuration space. Frequently, multiple kinds of MC moves are used in the same simulation run. The selection of the move type at each MC step must be done randomly in order not to break the Markov property of the simulation.

MC methods are also a natural starting point for simulating other ensembles, beyond the microcanonical and canonical ones. For example, in an isothermal–isobaric MC simulation, a system is placed at constant T, P, N conditions and all of these variables are specified as inputs. In addition to moves that change the particle coordinates and hence potential energy, such simulations also involve moves that randomly perturb the simulation volume. A typical approach is to pick a random change in volume ΔV in the range $[-\delta V_{\text{max}}, \delta V_{\text{max}}]$ and propose a new volume $V + \Delta V$, scaling all particle coordinates uniformly. The Metropolis condition of Eqn. (22.47), along with the isothermal–isobaric configurational probability distribution, then shows that the appropriate acceptance criterion for a volume move is

$$P_{12}^{\text{acc}} = \min\left[1, \left(\frac{V_2}{V_1}\right)^N e^{-\beta\Delta U - \beta P\Delta V}\right] \tag{22.48}$$

In contrast, a grand–canonical MC simulation specifies a system at constant T, V, μ conditions, and involves fluctuations both in the energy and in the number of particles. To accomplish the latter, one uses moves that involve random additions and deletions of particles. The corresponding acceptance criteria again derive from Eqn. (22.47) and are found to be

$$P_{12}^{\text{acc}} = \min\left[1, \frac{V}{N+1} e^{-\beta\Delta U + \beta\mu'}\right] \tag{22.49}$$

for a random particle insertion and

$$P_{12}^{\text{acc}} = \min\left[1, \frac{N}{V} e^{-\beta\Delta U - \beta\mu'}\right] \tag{22.50}$$

for a random particle deletion, where $\mu' \equiv \mu - k_{\text{B}}T \ln \Lambda^3(T)$ is a relative chemical potential. There are, of course, many possibilities for simulating different ensembles using MC methods, even nonphysical ones. The general approach is to specify the configurational probability distribution $\wp_m$ and a set of moves (and corresponding proposal probabilities α_{mn}), and then determine the corresponding acceptance criteria using the detailed-balance condition.

Monte Carlo simulations impose artificial, random dynamics in the form of repeatedly performed **moves** to generate configurations according to a specified ensemble probability. During the production period, ensemble averages can then be approximated by trajectory averages over the history of Monte Carlo steps. Many different **move sets** and ensembles are possible; in each case, an appropriate acceptance criterion can be determined by use of Markov theory and the **detailed-balance** condition.

Problems

Conceptual and thought problems

22.1. Consider a pair potential $u(r)$. What determines whether two atoms separated by a distance r experience an attractive or repulsive interaction? Must it be negative, $u(r) < 0$, in order to be attractive? Explain with examples.

22.2. Show that the error in a simple average $\overline{X} = (\sum_i X_i)/n$ goes as $n^{-1/2}$ if each sample X_i for $i = 1, 2, \ldots, n$ is independent. Hint: consider the probability distribution $\wp(X)$. The true average for an infinite number of samples is given by $\langle X \rangle = \int \wp(X)X\,dX$. Find the expected variance,

$$\langle(\overline{X} - \langle X \rangle)^2\rangle = \int \cdots \int \wp(X_1, X_2, \ldots, X_n)(\overline{X} - \langle X \rangle)^2 \, dX_1 \, dX_2 \ldots dX_n$$

Note that, for independent samples,

$$\wp(X_1, X_2, \ldots, X_n) = \wp(X_1)\wp(X_2) \times \cdots \times \wp(X_n)$$

22.3. Derive Eqn. (22.30) for the velocity rescaling factor λ.

22.4. Show that the heat capacity can be estimated from an MD simulation as

$$C_V = \frac{n_{\text{DOF}}k_B}{2} + \frac{\sigma_U^2}{k_B T^2}$$

22.5. Prove that the Metropolis criterion, Eqn. (22.46), satisfies the detailed-balance condition, Eqn. (22.45).

22.6. Explain why single-particle displacement MC moves are symmetric $(\alpha_{12} = \alpha_{21})$ if $\delta r_{\max}$ is constant.

Fundamentals problems

22.7. Find and simplify the expression for the pairwise force exerted on atom i by atom j, $\mathbf{f}_{ij}$, for the Lennard-Jones system.

22.8. An approximate form for the effective interactions between two ions in an implicit solvent is the screened Coulomb potential,

$$u(r_{ij}) = \frac{q_i q_j}{4\pi\epsilon_0 \epsilon r_{ij}} e^{-r_{ij}/\lambda_D}$$

where λ_D is the Debye length, which is dependent on the salt concentration of the solution, and ϵ is the dielectric constant of the solvent. Find an expression for $\mathbf{f}_{ij}$.

22.9. An angle potential for atoms i, j, and k has the form $u(\mathbf{r}_i, \mathbf{r}_j, \mathbf{r}_k) = b(\theta - \theta_0)^2$, where b and θ_0 are parameters and θ is the angle subtended by atoms i, j, and k with j as the central atom. Find expressions for the forces $\mathbf{f}_i$, $\mathbf{f}_j$, and $\mathbf{f}_k$ on each atom due to this potential. Hint: Newton's third law demands that $\mathbf{f}_i + \mathbf{f}_j + \mathbf{f}_k = 0$.

22.10. Show that the velocities predicted by the Verlet algorithm as in Eqn. (22.22) are accurate to $o(\delta t^2)$.

22.11. For pairwise-interacting systems that adhere to Eqn. (22.2), show that the sum $\sum_i \mathbf{f}_i \cdot \mathbf{r}_i$ in Eqn. (22.31) for the pressure is equal to

$$-\sum_{i<j} \mathbf{f}_{ij} \cdot \mathbf{r}_{ij} = -\sum_{i<j} \frac{du(r_{ij})}{dr} r_{ij}$$

Then find an expression that could be used to calculate the pressure in a simulation of the Lennard-Jones system.

22.12. Show that Eqn. (22.33) can be used to estimate the self-diffusion coefficient D. Start with the diffusion equation for the probability that a molecule will be located at some point in space at a given time, $\wp(\mathbf{r}; t)$,

$$\frac{\partial \wp(\mathbf{r}; t)}{\partial t} = -D\nabla^2 \wp(\mathbf{r}; t)$$

The molecule is known to reside at $\mathbf{r}_0$ at $t = 0$ such that

$$\wp(\mathbf{r}; t = 0) = \delta(\mathbf{r} - \mathbf{r}_0) = \delta(x - x_0)\delta(y - y_0)\delta(z - z_0)$$

where δ is the Dirac delta function. Show that the solution to the diffusion equation is a Gaussian in the distance $|\mathbf{r} - \mathbf{r}_0|$, with a time-dependent variance. Then, find the average $\langle |\mathbf{r} - \mathbf{r}_0|^2 \rangle$.

22.13. For what cutoff distance will the Lennard-Jones potential reach 1% of its minimum value? Express your answer in terms of the Lennard-Jones parameter σ.

Applied problems

22.14. Using your favorite programming language, write code to perform an MD simulation of the Lennard-Jones liquid described by Eqn. (22.1). Let $\sigma = 1$ and $\epsilon = 1$ such that distances are measured in units of σ and energies in that of ϵ. Similarly, let the mass of an atom be $m = 1$. Use cubic periodic boundary conditions and choose a potential cutoff of $r_c = 2.5$. Let the number of atoms be 216.

(a) A good starting point is to write a subroutine that returns the potential energy and force array for a given atomic configuration. In pseudo-code, this might look like

```
set the potential energy variable and elements of the force array
   equal to zero
loop over atom i = 1 to N - 1
   loop over atom j = i + 1 to N
            calculate the pair energy of atoms i and j
            add the energy to the potential energy variable
            calculate the force on atom i due to atom j
            add the force to the array elements for atom i
            subtract the force from the array elements for atom j
```

(b) Write a second routine that returns a new set of position and velocities, given current arrays for each, using the velocity Verlet integrator for a single time step δt.

(c) Write a third routine that initializes the coordinates of the atoms on a cubic lattice.

(d) Choose $\rho = N/V = 0.8$ such that $L = (N/\rho)^{1/3}$. Initialize and perform an MD simulation for 2,000 time steps with $\delta t = 0.001$. On a graph show the potential energy as a function of time for $t = 0$ to $t = 2$. When has the system equilibrated? Determine a reasonable equilibration time.

(e) Write a fourth routine that rescales the velocities to a target temperature T. Note that $k_B = 1$ in these units such that temperature is measured in units of of ϵ/k_B. Modify your simulation to rescale velocities to a target temperature $T = 1.0$ every 100 time steps during the equilibration period. Then, turn off velocity rescaling for the production period and monitor the total energy, which should be constant with time. Compute the fractional fluctuation in total energy $\sigma_E/\langle E \rangle$ as a function of the size of the time step $\delta t = 0.0001$, 0.0002, 0.0004, 0.0008, 0.0016, 0.0032, 0.0064, and 0.0128. Place your results on a second graph that is a log–log plot of $\sigma_E/\langle E \rangle$ versus Δt. Is there any discernible trend?

(f) With $\delta t = 0.001$, create a third graph that gives the average mean-squared displacement of the atoms as a function of time during the production period. Find an approximate value for the self-diffusion constant using the slope at long times using Eqn. (22.33).

(g) Compute the self-diffusion coefficient for $T = 1.0$, 1.5, 2.0, 2.5, and 3.0. Make a final plot of $\ln (D/T)$ as a function of $1/T$. If the diffusion constant follows an Arhennius relationship, to first order the slope of this line should be linear, $D \sim T \exp(-E_a/k_B T)$, where E_a is an activation energy. To what extent does this behavior hold?

22.15. Repeat the previous MD simulation exercise for the Lennard-Jones (LJ) polymer described by Eqn. (22.16). Define the following variables: M gives the number of monomers (LJ atoms) per polymer, N_{poly} the number of polymers in the simulation, and $N = MN_{poly}$ the total number of atoms in the system. Let $k = 3,000$ and $r_0 = 1$ in the bonded potential, and $N = 240$ throughout.

(a) Modify your energy and force calculation to account for the fact that some atom pairs are bonded and interact through a harmonic potential rather than the Lennard-Jones one. One way to construct your code is the following:

```
loop over atom i = 1 to N - 1
     loop over atom j = i + 1 to N
          if j = i + 1 and modulo (j - 1, M) > 0 then
               treat the pair as a bonded interaction;
               update the energy and forces using the harmonic
               potential
          else
               treat the pair as a nonbonded interaction;
               update the energy and forces using the Lennard-Jones
               potential
```

The code modulo $(j - 1, M)$ will return 0 every time the i and $j = i + 1$ particles are in different molecules: when the modulo is zero, $j - 1$ is a multiple of M and therefore it is the first atom of one polymer molecule while i is the last atom of the previous one.

(b) Again choose $\rho = N/V = 0.8$ and $\delta t = 0.001$. Use velocity rescaling in the equilibration period to thermostat the system at $T = 1.0$. Perform MD simulations for $M = 2, 4, 6, 8, 12$, and 16, where $N_{poly} = N/M = 240/M$. On a single graph show the potential energy as a function of time. How does the equilibration time depend on M?

(c) Perform equilibration and long production runs (at least 100,000 steps) for the cases $M = 2, 4, 6, 8, 12$, and 16. Create a second graph that gives the mean-squared displacement of monomers as a function of time for $t_{tot} = 100$. Estimate the diffusion coefficient from the slopes and plot these on a third graph as a function of chain length. Does the diffusion coefficient appear to obey any obvious scaling law, $D \sim M^\nu$, where ν is an exponent? You may want to benchmark your results with those in the literature [Reis et al., *Fluid Phase Equilibria* **221**, 25 (2004)].

(d) Compute the self-diffusion coefficient for $M = 4, 8$, and 16, and for $T = 1.0$, 1.5, 2.0, 2.5, and 3.0. Make a final plot with $\ln(D/T)$ as a function of $1/T$ for each value of M. To what extent does the diffusion constant follow an Arhennius relationship, $D \sim T \exp(-E_a/k_B T)$?

22.16. Write code to perform a Monte Carlo simulation of the Lennard-Jones liquid described by Eqn. (22.1). Let $\sigma = 1$, $\epsilon = 1$, and $m = 1$. Note that, in these units, $k_B = 1$. Use cubic periodic boundary conditions and choose a potential cutoff of $r_c = 2.5$. Let the number of atoms be 216.

(a) Start by writing two subroutines: one to calculate the total potential energy and one to calculate the potential energy of a specific atom i with all other atoms. The latter will be useful because you do not need to compute the entire energy every time a single atom is moved. Rather, you can compute the energy of the moved atom alone. You will have to do this twice in order to get the change in potential energy ΔU: once before the atom is moved and once after.

(b) Write a subroutine that performs a single-atom displacement by random amounts in each of the x, y, z directions. The routine should both propose such a move (pick a random particle and make random displacements) and accept or reject it according to the Metropolis criterion. That is, the subroutine should advance time by a single MC step. A possible pseudocode template is

```
pick a random atom i between 1 and N
save the old position of atom i
save the old potential energy of the system
compute the old energy of i with all other particles,Uiold
displace atom i by random amounts in each dimension from -drmax to +drmax
```

```
compute the new energy of i with all other particles,Uinew
calculate the change in energy, DeltaU = Uinew - Uiold
decide to accept/reject on the basis of DeltaU / T
if the move is accepted:
    update the new energy with U = U + DeltaU
else:
    return i to its original position
```

(c) Starting with atoms placed on a cubic lattice, perform a Monte Carlo simulation at $T = 2.0$ and $\rho = 0.90$. For what value of δr_{max} does the simulation achieve $\sim 50\%$ acceptance at long times? Using this value, what is a reasonable equilibration time, as assessed by the progression of the potential energy U with the number of MC "sweeps" – the number of steps divided by the number of particles?

(d) Modify your code to periodically compute the pressure using the virial route in Eqn. (22.32). Using multiple simulations at four different densities, $\rho = 0.75, 0.80, 0.85,$ and 0.90, find the average pressure during the production period. Plot the equation of state $P(\rho, T)$ at $T = 2.0$.

22.17. Repeat the previous MC simulation exercise for the Lennard-Jones polymer described by Eqn. (22.16). Let $k = 3,000$, $r_0 = 1$, and $N = 240$ throughout.

(a) Perform MC simulations at $T = 2.0$, $\rho = 0.90$ for the cases $M = 1, 2,$ and 4. How do the equilibration time and average per-particle potential energy $\langle U \rangle / N$ vary with M?

(b) Generally, MC simulations with single-particle displacements are less efficient than MD for molecules with many stiff bonds. By efficient, we mean the rate at which the simulation explores configuration space, which we might measure by a mean-squared displacement. To make a fair comparison, we need to examine how far the system moves as a function of equivalent computational effort: either one MC sweep (N MC moves) or one MD step. From the production periods for $M = 1, 2,$ and 4, compute the mean-squared displacement as a function of the number of MC sweeps. Compare this with the mean-squared displacement as a function of the number of MD steps for the same conditions. Make a graph with these six curves and compare the rates at which the two methods explore configuration space.

FURTHER READING

M. P. Allen and D. J. Tildesley, *Computer Simulation of Liquids*. New York: Oxford University Press (1989).

D. Frenkel and B. Smit, *Understanding Molecular Simulation: From Algorithms to Applications*, 2nd edn. San Diego, CA: Academic (2002).

A. R. Leach, *Molecular Modelling: Principles and Applications*, 2nd edn. New York: Prentice-Hall (2001).

Index

Printed in the United States
By Bookmasters